Practical Machinery Management for Process Plants
VOLUME 2 ▪ THIRD EDITION

Machinery Failure Analysis and Troubleshooting

Practical Machinery Management for Process Plants:

Volume 1: Improving Machinery Reliability, 2nd edition
Volume 2: Machinery Failure Analysis and Troubleshooting, 3rd edition
Volume 3: Machinery Component Maintenance and Repair, 2nd edition
Volume 4: Major Process Equipment Maintenance and Repair, 2nd edition

Other Machinery Engineering Texts from the Same Authors:

Introduction to Machinery Reliability Assessment, 2nd edition
Reciprocating Compressors: Operation and Maintenance

Gulf Publishing Company
Houston, Texas

THIRD EDITION

Practical Machinery Management for Process Plants

Machinery Failure Analysis and Troubleshooting

VOLUME 2

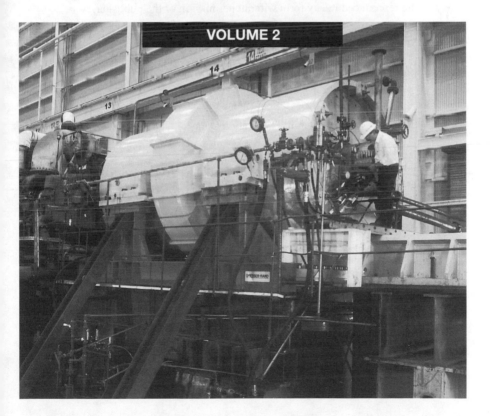

Heinz P. Bloch · Fred K. Geitner

Dedication

To Gerritt Bloch, Courtney Bavender,
and Peter and Derek Geitner
—As encouragement to go the full distance,
and then the extra mile. You will not encounter
any traffic jams on your way.
(Freely quoted from Ecclesiastes 9:10)

Practical Machinery Management for Process Plants
Volume 2, 3rd Edition

Machinery Failure Analysis
and Troubleshooting

Gulf Publishing Company
Book Division
P.O. Box 2608 □ Houston, Texas 77252-2608

10   9   8   7   6   5   4   3   2   1

Library of Congress Cataloging-in-Publication Data

Bloch, Heinz P., 1933–
    Machinery failure analysis and troubleshooting / Heinz P. Bloch, Fred K. Geitner. —3rd ed.
        p.   cm. —(Practical machinery management for process plants; v. 2)
    Includes index.
    ISBN 0-88415-662-1
    1. Machinery—Maintenance and repair.   2. Plant maintenance.   I. Geitner, Fred K.   II. Title.   III. Series: Bloch, Heinz P., 1933–   Practical machinery management for process plants ; v. 2.
TS191.B56   1994
658.2′dc20
                                                                93-4776
                                                                CIP
Printed on Acid-Free Paper (∞)

Contents

Troubleshooting as an Extension of Failure Analysis. Causes of
Machinery Failures. Root Causes of Machinery Failure. References.

Metallurgical Failure Analysis Methodology. Failure Analysis of
Bolted Joints. Shaft Failures. Stress Raisers in Shafts. Analysis of
Surface-Change Failures. Analyzing Wear Failures. Preventive
Action Planning Avoids Corrosion Failure. References.

Bearings in Distress. Rolling-Element Bearing Failures and Their
Causes. Patterns of Load Paths and Their Meaning in Bearing
Damage. Troubleshooting Bearings. Journal and Tilt-Pad Thrust
Bearings. Gear Failure Analysis. Preliminary Considerations.
Analytical Evaluation of Gear Theoretical Capability. Metallurgi-
cal Evaluation. General Mechanical Design. Lubrication. Defects
Induced by Other Train Components. Wear. Scoring. Surface
Fatigue. Failures from the Manufacturing Process. Breakage.
Lubricated Flexible-Coupling Failure Analysis. Gear-Coupling
Failure Analysis. Gear-Coupling Failure Mechanisms. Determin-
ing the Cause of Mechanical Seal Distress. Troubleshooting and
Seal-Failure Analysis. Summary of Mechanical Seal Failure
Analysis. Lubricant Considerations. Lubrication Failure Analysis.
Why Lube Oil Should Be Purified. Six Lube-Oil Analyses Are
Required. Periodic Sampling and Conditioning Routines Imple-
mented. Calculated Benefit-to-Cost Ratio. Wear-Particle Analy-
sis. Grease Failure Analysis. Magnetism in Turbomachinery.
References.

Acknowledgments

An experienced machinery engineer usually has a few file cabinets filled with technical reports, course notes, failure reports, and a host of other machinery-related data. But these files are rarely complete enough to illustrate all bearing failure modes, all manners of gear distress, etc. Likewise, we may have taken problem-solving courses, but cannot lay claim to recalling all the mechanics of problem-solving approaches without going back to the formal literature.

Recognizing these limitations, we went to some very knowledgeable companies and individuals and requested permission to use some of their source materials for portions of this book. We gratefully acknowledge the help and cooperation we received from:

American Society of Lubrication Engineers, Park Ridge, Illinois (ASLE Paper 83-AM-1B-2, Bloch/Plant-Wide Turbine Lube Oil Reconditioning and Analysis).

American Society of Mechanical Engineers, New York, New York (Proceedings of 38th ASME Petroleum Mechanical Engineering Workshop, Bloch/Setting Up a Pump Failure Reduction Program).

American Society for Metals, Metals Park, Ohio (Analysis of Shaft Failures, etc.).

American Gear Manufacturers Association, Arlington, Virginia (Gear Failures, etc.).

William G. Ashbaugh, Houston, TX (Corrosion Failures)

Beta Machinery Analysis, Ltd. Calgary, Canada (Problem Analysis on Reciprocating Machinery).

Durametallic Company, Kalamazoo, Michigan (Mechanical Seal Distress).

Dean L. Gano, Apollo Associated Services, Richland, Washington (Cause Analysis by Pursuing the Cause and Effect Relationship).

Glacier Metal Company, Ltd., Alperton/Middlesex, England (Journal and Tilt-Pad Bearing Failure Analysis).

T.J. Hansen Company, Dallas, Texas (Generalized Problem-Solving Approaches).

Robert M. Jones and SKF Condition Monitoring, Newnan, GA (Vibration Monitoring and Pattern Identification).

Dr. Wayne Nelson, General Electric Company, Schenectady, New York and American Society for Quality Control (Statistical Methods of Failure Analysis and Hazard Plotting).

Paul Nippes, Magnetic Products and Services, Inc., Holmdel, New Jersey (Magnetism in Turbomachinery).

Rome Air Development Center, Rome, New York (Sneak Analysis Methods).

SKF Industries, Inc., King of Prussia, Pennsylvania (Bearing Distress—Recognition and Problem Solving).

Neville W. Sachs, PE, Sachs, Salvaterra & Associates, Inc., Syracuse, N.Y. (Failure Analysis of Mechanical Components)

John S. Sohre, Sohre Turbomachinery, Inc., Ware, Massachussets (Magnetism in Turbomachinery).

Brian Turner, Fort Meyers, Florida (Distinguishing Between Bad Repairs and Bad Designs).

Preface

The prevention of potential damage to machinery is necessary for safe, reliable operation of process plants. Failure prevention can be achieved by sound specification, selection, review, and design audit routines. When failures do occur, accurate definition of the root cause is an absolute prerequisite to the prevention of future failure events.

This book concerns itself with proven approaches to failure definition. It presents a liberal cross section of documented failure events and analyzes the procedures employed to define the sequence of events that led to component or systems failure. Because it is simply impossible to deal with every conceivable type of failure, this book is structured to teach failure identification and analysis methods that can be applied to virtually all problem situations that might arise. A uniform methodology of failure analysis and troubleshooting is necessary because experience shows that all too often process machinery problems are never defined sufficiently; they are merely "solved" to "get back on stream." Production pressures often override the need to analyze a situation thoroughly, and the problem and its underlying cause come back and haunt us later.

Equipment downtime and component failure risk can be reduced only if potential problems are anticipated and avoided. Often, this is not possible if we apply only traditional methods of analysis. It is thus appropriate to employ other means of precluding or reducing consequential damage to plant, equipment, and personnel. This objective includes, among others, application of redundant components or systems and application of sneak circuit analysis techniques for electrical/electronic systems.

The organizational environment and management style found in process plants often permits a "routine" level of machinery failures and breakdowns. This book shows how to arrive at a uniform method of assessing what level of failure experience should be considered acceptable and achievable. In addition, it shows how the organizational environment can be better prepared to address the task of thorough machinery failure analysis and troubleshooting, with resulting maintenance incident reduction. Finally, by way of successful examples, this book demonstrates how the progress and results of failure analysis and troubleshooting efforts can be documented and thus monitored.

H.P. Bloch
Montgomery, Texas

F. K. Geitner
Brights Grove, Ontario

Chapter 1

The Failure
Analysis and
Troubleshooting System

Troubleshooting as an Extension of Failure Analysis

For years, the term "failure analysis" has had a specific meaning in connection with fracture mechanics and corrosion failure analysis activities carried out by static process equipment inspection groups. Figure 1-1 shows a basic outline of materials failure analysis steps.[1] The methods applied in our context of process machinery failure analysis are basically the same; however, they are not limited to metallurgic investigations. Here, failure analysis is the determination of failure modes of machinery components and their most probable causes. Figure 1-2 illustrates the general significance of machinery component failure mode analysis as it relates to quality, reliability, and safety efforts in the product development of a major turbine manufacturer.[2]

Very often, machinery failures reveal a reaction chain of cause and effect. The end of the chain is usually a performance deficiency commonly referred to as the symptom, trouble, or simply "the problem." Troubleshooting works backward to define the elements of the reaction chain and then proceeds to link the most probable failure cause based on failure (appearance) analysis with a root cause of an existing or potential problem. For all practical purposes, failure analysis and troubleshooting activities will quite often mesh with one another without any clear-cut transition.

However, as we will see later, there are numerous cases where troubleshooting alone will have to suffice to get to the root cause of the problem. These are the cases that present themselves as performance deficiencies with no apparent failure modes. Intermittent malfunctions and faults are typical examples and will tax even the most experienced troubleshooter. In these cases, troubleshooting will be successful only if the investigator knows the system he is dealing with. Unless he is thoroughly familiar with component interaction, operating or failure modes, and functional characteristics, his efforts may be unsuccessful.

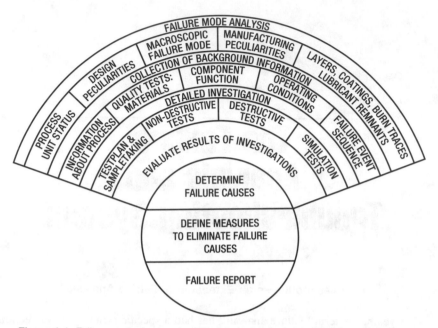

Figure 1-1. Failure analysis steps—materials technology (modified from Ref. 1).

There are certain objectives of machinery failure analysis and troubleshooting:

1. Prevention of future failure events.
2. Assurance of safety, reliability, and maintainability of machinery as it passes through its life cycles of:
   a. Process design and specification.
   b. Original equipment design, manufacture, and testing.
   c. Shipping and storage.
   d. Installation and commissioning.
   e. Operation and maintenance.
   f. Replacement.

From this it becomes very obvious that failure analysis and troubleshooting are highly co-operative processes. Because many different parties will be involved and their objectives will sometimes differ, a systematic and uniform description and understanding of process machinery failure events is important.

Causes of Machinery Failures

In its simplest form, failure can be defined as any change in a machinery part or component which causes it to be unable to perform its intended function

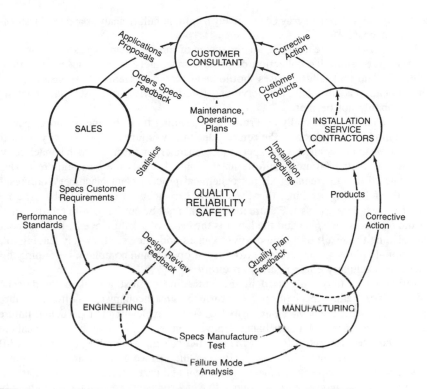

Figure 1-2. Failure analysis and the "wheel of quality."[2]

satisfactorily. Familiar stages preceding final failure are "incipient failure," "incipient damage," "distress," "deterioration," and "damage," all of which eventually make the part or component unreliable or unsafe for continued use.

Meaningful classifications of failure causes are:

1. Faulty design.
2. Material defects.
3. Processing and manufacturing deficiencies.
4. Assembly or installation defects.
5. Off-design or unintended service conditions.
6. Maintenance deficiencies (neglect, procedures).
7. Improper operation.

All statistics and references dealing with machinery failures, their sources and causes, generally use these classifications. And, as will be shown in Chapter 4, remembering these seven classifications may be extremely helpful in failure analysis and troubleshooting of equipment.

For practical failure analysis, an expansion of this list seems necessary. Table 1-1 shows a representative collection of process machinery failure causes. The table makes it clear that failure causes should be allocated to areas of responsibilities. If this allocation is not made, the previously listed objectives of most failure analyses will probably not be met.

Failure causes are usually determined by relating them to one or more specific failure modes. This becomes the central idea of any failure analysis activity. Failure mode (FM) in our context is the appearance, manner, or form in which a machinery component or unit failure manifests itself. Table 1-2 lists the basic failure modes encountered in 99 percent of all petrochemical process plant machinery failures.

In the following sections, this list will be expanded so that it can be used for day-to-day failure analysis. Failure mode should not be confused with failure cause, as the former is the *effect* and the latter is the *cause* of a failure event. Failure mode can also be the result of a long chain of causes and effects, ultimately leading to a functional failure, i.e a symptom, trouble, or operational complaint pertaining to a piece of machinery equipment as an entity.

Other terms frequently used in the preceding context are "kind of defect," "defect," or "failure mechanism." The term "failure mechanism" is often described as the metallurgical, chemical, and tribological process leading to a particular failure mode. For instance, failure mechanisms have been developed to describe the chain of cause and effect for fretting wear (FM) in roller bearing assemblies, cavitation (FM) in pump impellers, and initial pitting (FM) on the surface of a gear tooth, to name a few. The basic agents of machinery component and part failure mechanisms are *always* force, time, temperature, and a reactive environment. Each of these can be subdivided as indicated in Table 1-3.

For our purpose, failure mechanisms thus defined will have to stay part of the failure mode definition: They will tell how and why a failure mode might have occurred in chemical or metallurgical terms, but in so doing, the root cause of the failure will remain undefined.

Root Causes of Machinery Failure

The preceding pages have shown us that there will always be a number of causes and effects in any given failure event. We need to arrive at a practical point—if not all the way to the beginning—of the cause and effect chain where removal or modification of contributing factors will solve the problem.

A good example would be scuffing (FM) as one of the major failure modes of gears. It is a severe form of adhesive wear (FM) with its own well-defined failure mechanism. Adhesive wear cannot occur if a sufficiently thick oil film separates the gear tooth surfaces. This last sentence—even though there is a long chain of cause and effect hidden in the adhesive wear failure mechanism—will give us the clue as to the root cause. What then is the root cause? We know that scuffing usually occurs quite suddenly, in contrast to the time-dependent failure mode of pitting. Thus, we cannot look for the root cause in the design of the lube oil system or in the lube oil

Table 1-1
Causes of Failures

**Design and Specification
  Responsibility**

Application
- [ ] Undercapacity
- [ ] Overcapacity
- [ ] Incorrect physical conditions
      (temperature pressure, etc.)
- [ ] Incorrect physical prop.
      (mol. wt., etc.)
- [ ] _____
- [ ] _____

Specifications
- [ ] Inadequate lubrication system
- [ ] Insufficient control instrumentation
- [ ] Improper coupling
- [ ] Improper bearing
- [ ] Improper seal
- [ ] Insufficient shutdown devices
- [ ] _____
- [ ] _____

Material of Construction
- [ ] Corrosion and/or erosion
- [ ] Rapid wear
- [ ] Fatigue
- [ ] Strength exceeded
- [ ] Galling
- [ ] Wrong hardening method
- [ ] _____
- [ ] _____

Design
- [ ] Unsatisfactory piping support
- [ ] Improper piping flexibility
- [ ] Undersized piping
- [ ] Inadequate foundation
- [ ] Unsatisfactory soil data
- [ ] Liquid ingestion
- [ ] Inadequate liquid drain
- [ ] Design error
- [ ] _____
- [ ] _____

Vendor Responsibility

Material of Construction
- [ ] Flaw or defect
- [ ] Improper material
- [ ] Improper treatment
- [ ] _____

Design
- [ ] Improper specification
- [ ] Wrong selection
- [ ] Design error
- [ ] Inadequate or wrong lubrication
- [ ] Inadequate liquid drain
- [ ] Critical speed
- [ ] Inadequate strength
- [ ] Inadequate controls and protective devices
- [ ] _____

Fabrication
- [ ] Welding error
- [ ] Improper heat treatment
- [ ] Improper hardness
- [ ] Wrong surface finish
- [ ] Imbalance
- [ ] Lub. passages not open
- [ ] _____

Assembly
- [ ] Improper fit
- [ ] Improper tolerances
- [ ] Parts omitted
- [ ] Parts in wrong
- [ ] Parts/bolts not tight
- [ ] Poor alignment
- [ ] Imbalance
- [ ] Inadequate bearing contact
- [ ] Inadequate testing
- [ ] _____

Shipping and Storage Responsibility

Preparation for Shipment
- [ ] Oil system not clean
- [ ] Inadequate drainage
- [ ] Protective coating not applied
- [ ] Wrong coating used
- [ ] Equipment not cleaned
- [ ] _____

Protection
- [ ] Insufficient protection
- [ ] Corrosion by salt
- [ ] Corrosion by rain or humidity
- [ ] Poor packaging
- [ ] Dessicant omitted
- [ ] Contamination with dirt, etc.
- [ ] _____

Physical Damage
- [ ] Loading damage
- [ ] Transport damage
- [ ] Insufficient support
- [ ] Unloading damage
- [ ] _____

Installation Responsibility

Foundations
- [ ] Settling
- [ ] Improper or insufficient
      grouting
- [ ] Cracking or separating
- [ ] _____

Piping
- [ ] Misalignment
- [ ] Inadequate cleaning
- [ ] Inadequate support
- [ ] _____

Assembly
- [ ] Misalignment
- [ ] Assembly damage (crafts)

- [ ] Defective material
- [ ] Inadequate bolting
- [ ] Connected wrong
- [ ] Foreign material left in
- [ ] General poor workmanship
- [ ] _____

**Operations and Maintenance
  Responsibility**

Shock
- [ ] Thermal
- [ ] Mechanical
- [ ] Improper startup
- [ ] _____

Operating
- [ ] Slugs of liquid
- [ ] Process surging
- [ ] Control error
- [ ] Controls deactivated
      or not put in service
- [ ] Operating error
- [ ] _____

Auxiliaries
- [ ] Utility failure
- [ ] Insufficient instrumentation
- [ ] Electronic control failure
- [ ] Pneumatic control failure
- [ ] _____

Lubrication
- [ ] Dirt in oil
- [ ] Insufficient oil
- [ ] Wrong lubricant
- [ ] Water in oil
- [ ] Oil pump failure
- [ ] Low oil pressure
- [ ] Plugged lines
- [ ] Improper filtration
- [ ] Contaminated oil
- [ ] _____

Craftsmanship
- [ ] Improper tolerances
- [ ] Welding error
- [ ] Improper surface finish
- [ ] Improper fit
- [ ] General poor workmanship
- [ ] _____

Assembly
- [ ] Mechanical damage
- [ ] Parts in wrong
- [ ] Parts omitted
- [ ] Misalignment
- [ ] Improper bolting
- [ ] Imbalance
- [ ] Piping stress
- [ ] Foreign material left in
- [ ] Wrong material of construction
- [ ] _____

(table continued on next page)

Table 1-1 (cont.)

Preventive Maintenance		
☐ Postponed	☐ Seal	☐ Blades
☐ Schedule too long	☐ Coupling	☐ Blade root
☐ _____	☐ Shaft	☐ Blade shroud
	☐ Pinion gear	☐ Labyrinth
	☐ Bull gear	☐ Thrust bearing
	☐ Turning gear	☐ Pivoted pad bearing
	☐ Casing	☐ Roller/ball bearing
Distress, Damages, or Failed Components	☐ Rotor	☐ Cross-head piston
	☐ Impeller	☐ Cylinder
☐ Vibration	☐ Shroud	☐ Crankshaft
☐ Short circuit	☐ Piston	☐ _____
☐ Open circuit	☐ Diaphragm	☐ _____
☐ Sleeve bearing	☐ Wheel	☐ _____

Comments:_____
_____
_____
_____
_____
_____

itself—that is, if scuffing was not observed before on that particular gear set. Sudden and intermittent loss of lubrication could be the cause. Is it the root cause? No, we still have to find it because we are looking for the element that, if removed or modified, will prevent recurrence or continuation of scuffing. Is it because this particular plant is periodically testing their standby lube oil pumps, causing sudden and momentary loss of lube oil pressure? Eventually, we will arrive at a point where a change in design, operation, or maintenance practices will stop the gear tooth scuffing.

Removal of the root cause of machinery failures should take place in design and operations-maintenance. Quite often the latter, in its traditional form, is given too much emphasis in failure analysis and failure prevention. In our opinion, long-term reductions in failure trends will only be accomplished by specification and design modifications. We will see again in Chapter 7 that only design changes will achieve the required results. How then does this work? After ascertaining the failure mode, we determine whether or not the failed machinery component could be made more resistant to the failure event. This is done by checking design parameters such as the ones shown in Table 1-4 for possible modification. Once a positive answer has been obtained, the root cause has also been determined and we can specify whatever is required to impart less vulnerability to the material, component, assembly, or system. As we formulate our action plan, we will test whether the mechanic's axiom holds true:

> *When in doubt*
> *Make it stout*
> *Out of something*
> *You know about.*

Table 1-2
Machinery Failure Mode Classification

Deformation—i.e. plastic, elastic, etc.
Fracture—i.e. cracks, fatigue fracture, pitting, etc.
Surface changes—i.e. hairline cracks, cavitation, wear, etc.
Material changes—i.e. contamination, corrosion, wear, etc.
Displacement—i.e. loosening, seizure, excessive clearance, etc.
Leakage
Contamination

Table 1-3
Agents of Machinery Component and Part Failure Mechanisms

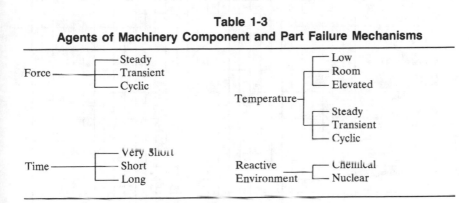

Table 1-4
Process Machinery Design Properties

Material-of-Construction Level

1. Material properties. i.e. ductility, creep resistance, heat resistance, etc.
2. Properties derived from processing, i.e. cast, rolled, forged, etc.
3. Properties resulting from heat treatment, i.e. not heat treated, hardened, stress relieved, etc.
4. Surface properties, i.e. machined, ground, lapped, etc.
5. Properties derived from corrosion and wear protection measures, i.e. overlayed, enameled, painted, etc.
6. Properties resulting from connecting method, i.e. welded, shrunk, rolled-in, etc.

Part and Component Level

7. Properties derived from shape and form, i.e. cylindrical, spherical, perforated, etc.

Part, Component, and Assembly Level

8. Suitability for service, i.e. prone to plugging, wear, vibration, etc.
9. Properties resulting from assembly type, i.e. riveted, pinned, bolted, etc.
10. Assembly quality, i.e. countersunk, flush, tight, locked, etc.

Table 1-5
Machinery Failure Modes—Process Plant

Left-margin group labels: *Table 1-4—Material Properties* (Tensile Strength … Thermal Conductivity); *Table 1-4—Design Properties* (Properties from Manufacturing Process … Properties from Type and Quality of Assembly); *Table 1-3* (Static Force … Chemical Environment).

Design Parameter Failure Resistance "Immediate" Cause	Yielding	Bending—Warping	Buckling	Brinelling	Creep—Bowing	Expansion/shrinking	Melting	Ductile Fracture	Brittle Fracture	Fatigue Fracture	Cracking	Bond Failure—Babbitt	Aging (bending/creep)	Rusting	Uniform Corrosion	Stress Corrosion	Pitting—Surface Fatigue	Spalling—Surface Fatigue	Cavitation Erosion	Erosion—Corrosion	Abrasive Wear	Fretting	Scuffing	Scoring	Galling	Rubbing/seizing	Jamming/blocking/sticking	Shifting	Exc. Clearance	Loosening	Leakage—Contamination
Tensile Strength								•		•	•		•			•															
Yield Strength	•																														
Compressive Yield Strength			•														•	•				•									
Shear Yield Strength	•							•																							
Fatigue Properties										•							•	•				•									
Ductility %RA								•	•	•	•																				
Charpy Energy								•		•																					
Modulus of Elasticity	•	•																													
Creep Rate					•								•																		
K$_{ic}$									•	•																					
Erosion Resistance																			•												
EMF Series															•					•	•										
Hardness				•						•							•	•	•	•	•	•	•	•	•	•					•
Coefficient of Thermal Expansion						•					•	•													•	•	•	•	•	•	•
Melting Point							•																			•					
Thermal Conductivity							•					•														•					
Properties from Manufacturing Process								•	•	•	•	•				•	•	•			•	•	•	•		•	•	•			
Properties from Corrosion Protection														•	•	•															
Properties from Heat Treatment		•	•					•	•	•	•						•	•				•	•	•	•						
Properties from Wear Protection																					•		•	•	•	•	•				
Properties from Shape and Form																						•	•			•	•	•			
Properties from Type and Quality of Assembly																									•	•	•	•	•	•	•
Static Force	•	•	•	•				•		•	•					•												•	•	•	
Dynamic Force		•	•	•				•	•	•	•					•	•	•	•	•	•	•	•	•	•	•	•	•	•	•	
Coefficient of Sliding Friction						•																•	•	•	•	•	•				
Time	•	•	•	•	•			•	•	•	•	•	•	•	•	•	•	•	•	•	•	•							•	•	•
Temperature						•	•			•	•	•			•	•									•		•	•	•	•	•
Chemical Environment										•		•	•	•	•	•					•			•							•

We will keep in mind our inability to influence machinery failures by simply making the part stronger in every conceivable situation. A flexibly designed component may, in some cases, survive certain severe operating conditions better than the rigid part.

Table 1-5 concludes this section by summarizing machinery failure modes as they relate to their immediate causes or design parameter deficiencies.

References

1. VDI Guidelines No. 3822., *Der Maschinenschaden,* Vol. 54, No. 4, 1981, p. 131.

2. Ludwig, G. A., "Tests Performed by the Builder on New Products to Prevent Failure", *Loss Prevention of Rotating Machinery,* The American Society of Mechanical Engineers, New York, N.Y. 10017, 1972, p. 3.

Chapter 2

Metallurgical Failure Analysis

Failure analysis of metallic components has been the preoccupation of the metallurgical community for years.[1] Petrochemical plants usually have an excellent staff of "static equipment" inspectors, whose services prove invaluable during machinery component failure analysis. The strengths of the metallurgical inspectors lie in solving service failures with the following primary failure modes and their causes:

1. Deformation and distortion
2. Fracture and separation
   a. ductile fractures
   b. brittle fractures
   c. fatigue fractures
   d. environmentally affected fractures
3. Surface and material changes
   a. corrosion
      ● uniform corrosion
      ● pitting corrosion
      ● intergranular corrosion
4. Stress-corrosion cracking
5. Hydrogen damage
6. Corrosion fatigue
7. Elevated temperature failures
   a. creep
   b. stress rupture

The detailed analysis of these machinery component failures lies in metallurgical inspection, a highly specialized field. For an in-depth discussion of these analyses, refer to the references listed at the end of this chapter.

However, in more than 90 percent of industrial cases, a trained person can use the basic techniques of failure analysis to diagnose the mechanical causes behind a failure, without having to enlist outside sources and expensive analytical tools like electron microscopes.* Then, knowing how a failure happened, the investigator can pursue the human roots of why it happened.

There are times, however, when 90 percent accuracy is not good enough. When personal injury or a large loss is possible, a professional should guide the analysis.

To interpret a failure accurately, the analyst has to gather all pertinent facts and then decide what caused the failure. Also, to be consistent, the analyst should develop and follow a logic path that ensures a critical feature will not be overlooked.

- Decide what to do. How detailed an analysis is necessary? Before starting, try to decide how important the analysis is. If the failure is relatively insignificant, in cost and inconvenience, it deserves a cursory analysis; the more detailed steps can be ignored. But this strategy increases the chance of error. Some failures deserve a 20 minute analysis with an 80 percent probability of being correct, but critical failures require true root cause failure analysis (RCFA), in which no questions are left unanswered. RCFA may require hundreds of man-hours, but it guarantees an accurate answer.
- Find out what happened The most important step in solving a plant failure is to seek answers soon after it happened and talk to the people involved. Ask for their opinions, because they know the everyday occurrences at their worksite and their machinery better than anyone. Ask for their opinions, because they know the everyday occurrences at their worksite and their machinery better than anyone. Ask questions and try to get first-person comments. Do not leave until you have a good understanding of exactly what happened and the sequence of events leading up to it.
- Make a preliminary investigation. At the site, examine the broken parts, looking for clues. Do not clean them yet because cleaning could wash away vital information. Document the conditions accurately and take photographs from a variety of angles of both the failed parts and the surroundings.
- Gather background data. What are the original design and the current operating conditions? While still at the site, determine the operating conditions: time, temperatures, amperage, voltage, load, humidity, pressure, lubricants, materials, operating procedures, shifts, corrosives, vibration, etc. Compare the difference between actual operating conditions and design conditions. Look at everything that could have an effect on machine operation.
- Determine what failed. After you leave the site and the immediate crush of the failure, look at the initial evidence and decide what failed first—the primary failure—and what secondary failures resulted from it. Sometimes these decisions are very difficult because of the size of analysis that is necessary.
- Find out what changed. Compare current operating conditions with those in the past. Has surrounding equipment been altered or revised? Some failure examples have their mechanical roots in changes that took place years before the parts actually failed.

---

*Excerpted, by permission, from source material furnished by Neville Sachs, Sachs, Salvaterra & Associates, Inc., Syracuse, New York.

- Examine and analyze the primary failure. Clean the component and look at it under low-power magnification, 5x to 50x. What does the failure surface look like? From the failure surface, determine the forces that were acting on the part. Were conditions consistent with the design? With actual operation? Are there other cracks or suspicious signs in the area of the failure? Important surfaces should be photographed and preserved for reference.
- Characterize the failed piece and the support material. Perform hardness tests, dye penetrant and ultrasonic examinations, lubricant analysis, alloy analysis, etc. Examine the failed part and the components around it to understand what they are. Check to see if the results agree with design conditions.
- Conduct detailed chemical and metallurgical analyses. Sophisticated chemical and metallurgical techniques may reveal clues to material weaknesses or minute quantities of chemicals that may cause unusual fractures.
- Determine the failure type and the forces that caused it. Review all the steps listed. Leaving any questions unasked or unanswered reduces the accuracy of the analysis.
- Determine the root causes. Always ask, "Why did the failure happen in the first place?" This question usually leads to human factors and management systems. Typical root causes like "The shaft failed because of an engineering error" or "The valve failed because we decided not to PM it" or "The shaft failed because it was not aligned properly" expose areas where huge advances can be realized. However, these problems have to be dealt with differently; people will have to recognize personal errors and to change the way they think and act.

Types of Failures

Different analysts use different systems, but the most practical way for plant people to categorize failures is by overload, fatigue, corrosion-influenced fatigue, corrosion, and wear.

- Overload. Applying a single load causes the part to deform or fracture as the load is applied.
- Fatigue. Fluctuating loads over a relatively long time cause this type of failure and usually leave obvious clues.
- Corrosion-influenced fatigue. Corrosion substantially reduces the fatigue strength of most metals and eventually causes failure at relatively light loads.
- Corrosion. The failure is the result of the electrical or biological action of the corrosion, causing a loss of material.
- Wear. A variety of mechanisms result in loss of material by mechanical removal.

Corrosion and wear are complicated subjects and may deserve the input of experts.

Metallurgical Failure Analysis Methodology

Even though the machinery failure analyst will lack the expertise to perform a detailed metallurgical analysis of failed components, he nevertheless has to stay in charge of all phases of the analysis. His job is to define the root cause of the failure incident and to come up with a corrective or preventive action. A checklist of what should be accomplished during a metallurgical failure analysis is shown in Table 2-1.

It is absolutely necessary to plan the failure analysis before tackling the investigation. A large amount of time and effort may be wasted if insufficient time is spent carefully considering the background of the failure and studying the general features before the actual investigation.[2]

In the course of the various steps listed in Table 2-1, preliminary conclusions will often be formulated. If the probable fundamental cause of the metallurgical failure has become evident early in the examination, the rest of the investigation should focus on confirming the probable cause and eliminating other possibilities. Other investigations will follow the logical sequence shown in Figure 2-1, and the results of each stage will determine the following steps. As new facts change first impressions, different failure hypotheses will surface and be retained or rejected as dictated. Where suitable laboratory facilities are available, the metallurgical failure analyst should compile the results of mechanical tests, chemical analyses, fractography, and microscopy before preliminary conclusions are formulated.

There is always the temptation to curtail work essential to an investigation. Sometimes it is indeed possible to form an opinion about a failure cause from a single aspect of the analysis procedure, such as the visual examination of a fracture surface or the inspection of a single metallographic specimen. However, before final

Table 2-1
Main Stages of a Metallurgical Failure Analysis (Modified from Ref. 1)

1. Collection of background data and selection of samples.
2. Preliminary examination of failed part (visual examination and record keeping).
3. Nondestructive testing.
4. Mechanical testing (including hardness and toughness testing).
5. Selection, identification, preservation and/or cleaning of all specimens.
6. Macroscopic examination and analysis (fracture surfaces, secondary cracks and other surface phenomena).
7. Microscopic examination and analysis.
8. Selection and preparation of metallographic sections.
9. Examination and analysis of metallographic sections.
10. Determination of failure mechanism.
11. Chemical analyses (bulk, local, surface corrosion products, deposits or coatings and microprobe analysis).
12. Analysis of fracture mechanics.
13. Testing under simulated service conditions (special tests).
14. Analysis of all evidence leading to formulation of conclusions.
15. Writing of report including recommendations.

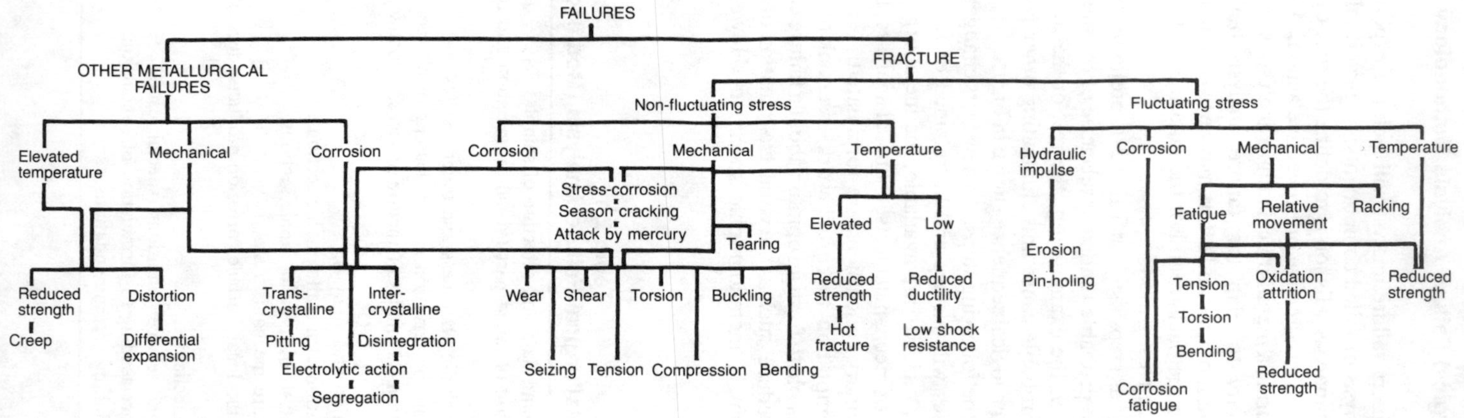

Figure 2-1. Classifications of failure causes in metals.

conclusions are reached, supplementary data confirming the original opinion should be looked for. Total dependence on the conclusions that can be drawn from a single specimen, such as from a metallographic section, may be readily challenged unless a history of similar failures can be drawn upon.[3]

Table 2-2 is a checklist that has been used as an aid in analyzing the evidence derived from metallurgical examinations and tests and in postulating conclusions.

As in other types of failure analyses, the end product of a metallurgical failure investigation should be the written failure analysis report. One experienced investigator has proposed that the report be divided into the main sections shown in Table 2-3. A detailed discussion of failure reports is given in Chapter 9.

Table 2-2
Metallurgical Failure Examination Checklist[3]

1. Has failure sequence been established?
2. If the failure involved cracking or fracture, have the initiation sites been determined?
3. Did cracks initiate at the surface or below the surface?
4. Was cracking associated with a stress concentrator?
5. How long was the crack present?
6. What was the intensity of the load?
7. What was the type of loading: static, cyclic, or intermittent?
8. How were the stresses oriented?
9. What was the failure mechanism?
10. What was the approximate service temperature at the time of failure?
11. Did temperature contribute to failure?
12. Did wear contribute to failure?
13. Did corrosion contribute to failure? What type of corrosion?
14. Was the proper material used? Is a better material required?
15. Was the cross section adequate for class of service?
16. Was the quality of the material acceptable in accordance with specification?
17. Were the mechanical properties of the material acceptable in accordance with specification?
18. Was the component that failed properly heat treated?
19. Was the component that failed properly fabricated?
20. Was the component properly assembled or installed?

Table 2-3
Main Sections of a Metallurgical Failure Report (Modified from Ref. 2)

1. Description of the failed component.
2. Service conditions at time of failure.
3. Prior service history.
4. Manufacturing and processing history of component.
5. Mechanical and metallurgical study of failure.
6. Metallurgical evaluation of quality.
7. Summary of failure-causing mechanism(s).
8. Recommendations for prevention of similar failures.

Practical Hints

In Chapter 1, Figure 1-1, we presented the major steps of a successful failure analysis. Together with the points made in Table 2-1 and 2-2 they can be a very practical approach to metallurgical failure analysis.

Failure mode inventory. The most useful first step, the visual inspection, cannot be emphasized enough. This, in conjunction with the part history, can frequently provide useful "clues" to the failure cause. Like a detective, the failure analyst must view the scene. A visual inspection should include observations of colors, corrosion products, presence of foreign materials, surface conditions (such as pits and other marks), dimensions, and fracture characteristics in order to attempt to answer the questions in Table 2-2. It goes without saying that careful notes and photographs or sketches should be made during this phase of failure mode inventory.

If a metallurgist is not available to help with metals analysis, it would be well to recognize the difference between brittle failures and ductile failures. Very simply, in the case of a brittle failure, the broken pieces behave like china. They are visually smooth and sharp and they fit back together. With a ductile failure, the pieces are more like taffy. They are distorted and, even if they fit back together, they are no longer the right shape. Figure 2-2 illustrates the two different failure modes. Brittle failure in a part that should be ductile, such as a compressor frame or a crankshaft, is a sign of fatigue. Is the particular part notched? Was the right alloy used for a weld? Was it heat treated? Was it misaligned? How did the failure progress? Is the fracture surface discolored or corroded?

Ductile failure in a part that should be hard, like control rods or gears, is a sign of either the use of the wrong material or faulty heat treatment. A good rule of thumb concerning the ductile/brittle relationship is shown in Table 2-4. Here the factors influencing either brittle or ductile failure are evaluated. For example, Table 2-4 shows that brittle failures tend to occur at lower temperatures and ductile failures at higher temperatures.

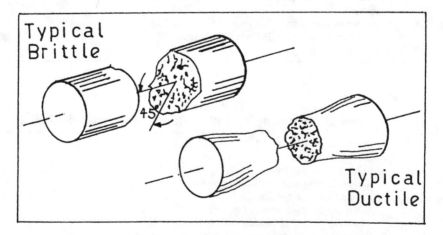

Figure 2-2. Distinguishing between brittle failure and ductile failure.

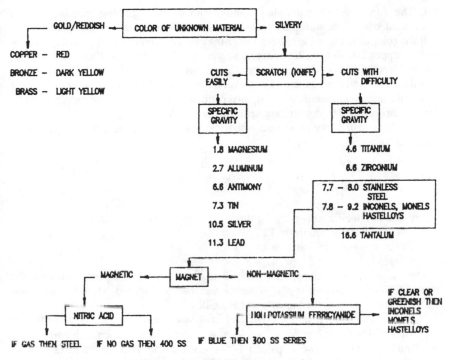

Figure 2-3. Identification of metals.

Table 2-4
The Ductile-Brittle Relationship

Factors	Ductile	Brittle
Temperature	Higher	Lower
Rate of loading	Lower	Higher
Geometry	No stress concentration	Stress concentration
Size	Smaller or thinner	Larger or thicker
Type of loading	Torsion	Tension or compression
Pressure (hydrostatic)	Higher	Lower
Strength of metal	Lower	Higher

Source: D. J. Wulpi

Qualitative tests. The next step might be the detailed investigation and diagnosis involving qualitative and perhaps also quantitative tests. If, for instance, you are faced with a failure caused by unexpected corrosion, you would suspect the use of an unspecified or unsuitable material. Usually a mass spectrograph or similar instrument for positive metal component identification will yield the desired answer. In absence of such instruments the analyst would have to resort to quick tests as indicated in Figure

2-3. This figure describes the alloy family, with its distinguishing characteristics of color, hardness as determined by scraping with a knife, magnetism, and spot tests.

If the color of the metal is reddish rather than silvery, you are most probably dealing with a copper containing alloy. Several sources (4, 5, 6) describe chemical spot tests for the identification of various copper alloys as well as of other alloys.

If the color is silvery, the use of a pocket knife allows the identification of easily cut alloys like aluminum, antimony, lead, silver, tin, and magnesium. The metals could be further distinguished by determining specific gravity. Weighing a specimen and dividing by its volume will allow us to obtain specific gravity. A different method would be using the formula:

$$SG = \frac{W_N}{W_N - W_W}$$

where SG = specific gravity in kg/dm^3
W_N = normal weight in kg
W_W = weight in water in kg

The use of a magnet allows us to differentiate between ferromagnetic alloys like the steels, 400-series stainless steels, or nickel and the nonmagnetic 300-series austenitic stainless steels, such as the Inconels and Hastelloys.

If the metal is silvery, nonmagnetic and hard, another fast identification method is the spot test. The procedure prescribed for it is to clean the specimen with emery cloth and then to place one or two drops of 1:1 hydrochloric acid, HCl, on the surface. After a reaction time, apply a watery solution of 10% potassium ferricyanide onto the HCl. A blue color indicates the presence of iron-base alloys. Yellow or green indicates a nickel-base alloy. There are commercial spot-test kits available that allow the identification of the stainless steels of the 300 and 400 series, of Monel, nickel, steel, and many other alloys.

Failure Analysis of Bolted Joints

At some time in the course of his career, the machinery failure analyst will have to deal with failures of threaded fasteners or bolted joints. This is also the time when he will find that the basic subject of "nuts and bolts" suddenly becomes complicated beyond his wildest dreams.

Anyone in the business of machinery failure analysis should be up-to-date on the design and behavior of bolted joints, for they are frequently the weakest links in engineered structures. Here is where machinery leaks, wears, slips, ruptures, loosens up, or simply fails.

Many factors contribute to failures of bolted joints. A look at available statistics reveals that problems encountered with threaded fasteners vary greatly. Consider the following: During the period 1964–1970 the research center of a large European machinery insurance company, ATZ*, analyzed 132 cases where failures of threaded fasteners had

---

*The Allianz Center for Technology, Ismaning/Munich (Germany).

caused damage to machinery.[7] Distribution of failure causes and failure modes are shown in Table 2-5.

Table 2-5
Failure Causes and Modes of Threaded Fasteners (Modified from Ref. 7)

Cause of Failure	Failure Distribution %
Product problems	50.0
Operational problems	40.0
Assembly problems	10.0
Failure Mode	
Fatigue failures	40.0
Creep failures	20.0
Sudden failures	
• brittle	10.0
• plastic	20.0
• corrosion	10.0

Now consider a study of joint failures during live missions on the U.S. Aerospace Skylab program, which produced the statistics shown in Table 2-6.[8]

From this we can see that in order to solve our problems, we have to list and document machinery failures in our own plants to obtain the necessary insight into prevailing failure causing factors specific to our environment.

Why Do Bolted Joints Fail?

It is beyond the scope of this text to give an exhaustive answer to this question. However, an overview will be provided to enable readers to ask the necessary questions when faced with a bolted-joint failure.

Table 2-6
Summary of the Causes of Bolted Joint Failure on the Skylab Program
(All Fasteners Had Been Torqued. Modified from Ref. 8)

Failure Causes	Failure Distribution, %
Product problems	
inadequate design	24.0
parts damaged in handling	23.0
faulty parts	10.0
Assembly problems	
improper assembly	29.0
incorrect preload	14.0

Joints fail in many ways, but in all cases failure has occurred because joint members behave this way:

1. They slip in relationship to each other (displacement).
2. They simply separate (displacement).
3. Bolts and/or joint members break (fracture).

These basic failure modes are preceded in turn by the failure modes listed in Table 2-7. Table 2-7 will convey an idea of probable causes or factors that will, depending on circumstances, contribute to bolted-joint failures.

Table 2-7
Failure Modes of Bolted Joints

	Primary Causes (Factors)	Fracture Under Static Load	Fatigue Failure	Vibration Loosening	Joint Leakage
Design and Manufacturing	Direction of bolt axis relative to vibration axis			•	
	Damping in Joint			•	
	Relaxation Effects		•	•	
	Radius of Thread Roots		•		
	Bolt/Joint Stiffness Ratio		•		•
	Thread Run-Out		•		
	Fillet Size and Shape		•		
	Nut Dilation	•			
	Poor Fits	•			
	Galling	•			
	Finish of Parts		•		
	Improper Heat Treatment	•			
	Tool Marks		•		
Assembly Practices	Condition of Joint Surfaces				•
	Condition of Gaskets				•
	Bolt-Up Procedure				•
	Thread Lubrication	•	•	•	
	Type of Tool Used	•			
	Improper Preload			•	•
Operating Conditions	Magnitude of Load Excursions	•	•		
	Temperature Cycling				•
	Corrosion	•	•	•	•

Fastener problems on machinery in the petrochemical industry will arise if the following are not considered:

1. Proper joint component selection suitable for the application.
2. Proper joint detail design parameters.
3. Importance of installation and maintenance procedures.

Some significant examples from our experience are:

Use of low-grade cap screws. If we use a cap screw with a yield strength too low for the forces being applied, it will stretch, causing "necking-out" (Figure 2-4). When the load is relaxed, the increased length will result in a loose nut which is free to vibrate off the bolt.[9] If, during preventive maintenance, the loose nut is discovered and tightened, reapplication of the load will cause the bolt to stretch at a lower load because there is less metal in the necked-out section. In many cases, the cap screw will fail completely while a mechanic is retightening the nut. Since he assumes that failure occurred because he pulled his wrench too hard, he will replace the bolt and nut with a new one of the same grade, and a vicious circle has begun.

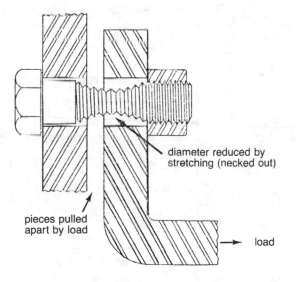

diameter reduced by stretching (necked out)

pieces pulled apart by load

load

Figure 2-4. Necking-out of a cap screw.[6]

Use of mismatched joint components. All components in a bolted-joint assembly must be matched to each other to achieve the desired holding power and service life.

Proper joint design. The kind and direction of forces to be transmitted—static or cyclic—is extremely important in the design of threaded fasteners.[10] Frequently, however, there is little known about the actual forces and loads that will be encountered in service. Consequently, the designer has to start with common assumptions regarding possible forces and moments, such as those shown in Figure 2-5.

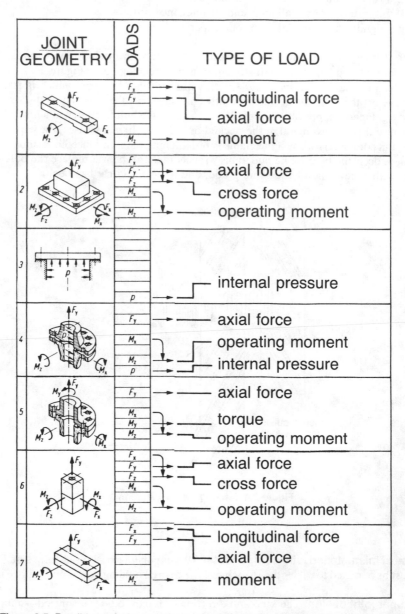

Figure 2-5. Possible operating loads encountered by bolted joints (modified from Ref. 10).

Long-term cyclic loads as encountered in rotating/reciprocating process machinery can only be transmitted by high-tensile-strength fastener components. In order to obtain high-strength fasteners, heat treatment after fabrication is necessary. Heat treatment, however, makes steel susceptible to fatigue failure when used under variable (vibration) load conditions. The higher the grade of heat treatment, the greater the danger of fatigue if the fastener is not properly preloaded.

A properly designed and preloaded bolted joint is extremely reliable without additional locking devices. This is especially true for high-strength steels, provided there is sufficient bolt resilience and a minimum of joint interfaces. Design measures to increase effective bolt lengths or their resilience are shown in Figure 2-6. These measures not only have the advantage of achieving more favorable bolt load distributions but also provide greater insurance against loosening.

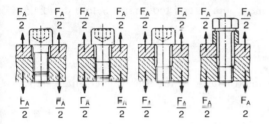

Figure 2-6. Bolted joint designs with increasing fatigue strength and resistance to loosening (modified from Ref. 10).

Failure to apply proper preload. Applying proper preload to a bolt or nut assembly is the crucial phase of many bolted joints in process machinery. J. H. Bickford[11] refers to the difficulties of bolt preload and torque control as he lists the problems associated with using a torque wrench to assemble a joint:

$\underline{F}$riction
$\underline{O}$perator
$\underline{G}$eometry $\qquad$ FOGTAR*
$\underline{T}$ool $\underline{A}$ccuracy
$\underline{R}$elaxation

Most of us have wrestled with these problems, and if we do not know everything about "FOGTAR" we should get acquainted with J. H. Bickford's delightful book on the behavior of bolted joints.

Carefully consider reusing fastener components in critical applications. Some critical applications are:

1. Piston-rod locknuts—reciprocating compressors
2. Crosshead-pin locknuts—reciprocating compressors
3. Impeller locknuts—centrifugal pumps
4. Thrust-disc locknuts—centrifugal compressors
5. Bolts and nuts—high-performance couplings

*A Tibetan word for trouble (Ref. 11, p. 77).

Typical fastener components not to be reused in any application are:

1. Prevailing torque locknuts (nylon insert)
2. Prevailing torque bolts (interference fit threads)
3. Anaerobic adhesive secured fasteners
4. Distorted thread nuts
5. Beam-type self-locking nuts
6. Castellated nut and cotter key
7. Castellated nut and spring pin
8. Flat washers
9. Tab washers
10. Vibration-resistant washers
11. Lock wire

Finally, inspect bolt and nuts for nicks, burrs, and tool marks before deciding to reuse!

Failure Analysis Steps

Failure analysis of bolted joints should consist of the following essential steps:

1. *Definition of failure mechanism.*
   a. The bolt failed under static load. Did it occur while tightening? The fracture surface will usually be at an angle other than 90° to the bolt axis. This is because the strength of the bolt has been exceeded by a combination of tension and torsional stress. A failure in pure tension will usually be at a right angle to the bolt axis.
   b. The bolt failed in fatigue under variable and cyclic loads. High cycle fatigue will usually be indicated by "beach marks" on the fracture surface (see Figure 2-7). This might not be conclusive, as the absence of these marks will not rule out a fatigue-related failure mechanism.
   c. Static or fatigue failure from corrosion.
   d. The joint failed to perform its design function because clamping forces fell below design requirements. Possible failure modes are partial or total separation (displacement), joint slippage (displacement), fretting of the joint surfaces (corrosion), and vibration loosening (displacement) of the nut. Consequential failure mode in all these cases is "leakage."

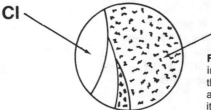

Figure 2-7. Fracture surface of a bolt that failed in fatigue. The surface is smooth and shiny in those regions that failed during crack initiation and growth (CI); it is rough in those areas where it failed rapidly (RF).

2. *Design review.*
   a. The analyst will now estimate or calculate the operating loads and possible preloads on joint components. If failure was static, he can refer to suitable references such as the ones listed at the end of this chapter.
   b. If the failure has been caused by cyclic loads, follow up work will be much more difficult: The analyst will have to determine the endurance limit of the parts involved in the failure. This may require experiments, as published data are rare.
3. *Special-variables check.* Consider and check the factors that could contribute to the fastener failure, as shown in Table 2-7.

Shaft Failures

Causes of Shaft Failures*

Shafts in petrochemical plant machinery operate under a broad range of conditions, including corrosive environments, and under temperatures that vary from extremely low, as in cold ethylene vapor and liquid service, to extremely high, as in gas turbines.

Shafts are subjected to one or more of the following loads: tension, compression, bending, or torsion. Additionally, shafts are often exposed to high vibratory stresses.

With the exception of wear as consequential damage of a bearing failure, the most common cause of shaft failures is metal fatigue. Fatigue failures start at the most vulnerable point in a dynamically stressed area—typically a stress raiser, which may be either metallurgical or mechanical in nature, and sometimes both.

Occasionally, ordinary brittle fractures are encountered, particularly in low-temperature environments. Some brittle fractures have resulted from impact or a rapidly applied overload. Surface treatments can cause hydrogen to be dissolved in high-strength steels and may cause shafts to become embrittled even at room temperature.

Ductile fracture of shafts usually is caused by accidental overload and is relatively rare under normal operating conditions. Creep, a form of distortion at elevated temperatures, can lead to stress rupture. It can also cause shafts with close tolerances to fail because of excessive changes in critical dimensions.

Fracture Origins in Shafts

Shaft fractures originate at stress-concentration points either inherent in the design or introduced during fabrication. Design features that concentrate stress include ends of keyways, edges of press-fitted members, fillets at shoulders, and edges of oil holes. Stress concentrators produced during fabrication include grinding damage,

---

* Adapted by permission from course material published by the American Society for Metals, Metals Park, Ohio 44073. For additional information on metallurgical service failures, refer to the 15-lesson course *Principles of Failure Analysis,* available from the Metals Engineering Institute of the American Society for Metals, Metals Park, Ohio 44073.

machining marks or nicks, and quench cracks resulting from heat-treating operations.

Frequently, stress concentrators are introduced during shaft forging; these include surface discontinuities such as laps, seams, pits, and cold shuts, and subsurface imperfections such as bursts. Subsurface stress concentrators can be introduced during solidification of ingots from which forged shafts are made. Generally, these stress concentrators are internal discontinuities such as pipe, segregation, porosity, shrinkage, and nonmetallic inclusions.

Fractures also result from bearing misalignment, either introduced at assembly or caused by deflection of supporting members in operation; from mismatch of mating parts; and from careless handling in which the shaft is nicked, gouged, or scratched.

To a lesser degree, shafts can fracture from misapplication of material. Such fractures result from using materials having high ductile-to-brittle transition temperatures; low resistance to hydrogen embrittlement, temper embrittlement, or caustic embrittlement; or chemical compositions or mechanical properties other than those specified.

Stress Systems in Shafts

The stress systems acting on a shaft must be understood before the cause of a fracture in that shaft can be determined. Also, both ductile and brittle behavior under static loading or single overload and the characteristic fracture surfaces

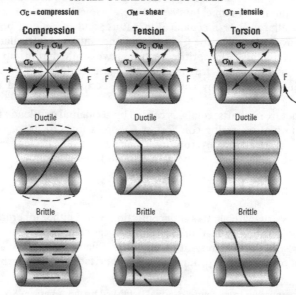

Figure 2-8. Both material properties and type of overload failure affect the appearance of the fracture surface.

produced by these types of behavior must be clearly understood for proper analysis of shaft fractures.

Figure 2-8 gives simplified, two-dimensional diagrams showing the orientations of the normal-stress and shear-stress systems at any internal point in a shaft loaded in pure tension, torsion, and compression. Also, the single-overload-fracture behavior of both ductile and brittle materials is illustrated with the diagram of each type of load.*

A free-body stress system may be considered as a square of infinitely small dimensions. Tensile and compressive stresses act perpendicularly to each other and to the sides of the square to stretch and squeeze the sides, respectively. The shear, or sliding stresses act on the diagonals of the square, 45° to the normal stresses. The third-dimension radial stresses are ignored in this description.

The effects of the shear and normal stresses on ductile and brittle materials under the three types of loads illustrated in Figure 2-8 and those under bending load are discussed below.

Tension. Under tension loading, the tensile stresses (σ_T) are longitudinal, whereas the compressive-stress components (σ_C) are transverse to the shaft axis. The maximum-shear-stress components (σ_M) are at 45° to the shaft axis.

In ductile material, shear stresses developed by tensile loading cause considerable deformation prior to fracture, which originates near the center of the shaft and propagates toward the surface, ending with a conical shear lip usually about 45° to the shaft axis.

In a brittle material, a fracture from a single tensile overload is roughly perpendicular to the direction of tensile stress, but involves little permanent deformation. The fracture surface usually is rough and crystalline in appearance.

The elastic stress distribution in pure tension loading, in the absence of a stress concentration, is uniform across the section. Thus, fracture can originate at any point within the highly stressed volume.

Torsion. The stress system rotates 45° counterclockwise when a shaft is loaded in torsion, as also shown in Figure 2-8. Both the tensile and compressive stresses are 45° to the shaft axis and remain mutually perpendicular. One shear-stress component is parallel with the shaft axis; the other is perpendicular to the shaft axis.

In a ductile material loaded to failure in torsion, shear stresses cause considerable deformation prior to fracture. This deformation, however, usually is not obvious because the shape of the shaft has not been changed. If a shaft loaded in torsion is assumed to consist of a number of infinitely thin disks that slip slightly with respect to each other under torsional stress, visualization of deformation is simplified. Torsional single-overload fracture of a ductile material usually occurs on the transverse plane, perpendicularly to the axis of the shaft. In pure torsion, the final-fracture region is at the center of the shaft; the presence of slight bending will cause it to be off center.

A brittle material in pure torsion will fracture perpendicularly to the tensile-stress component, which is 45° to the shaft axis. The resulting fracture surfaces usually have the shape of a spiral.

---

*Figures 2-8 through 2-19 and accompanying narrative courtesy of Mr. Neville Sachs, Sachs/Salvaterra & Associates, Syracuse, New York.

The elastic-stress distribution in pure torsion is maximum at the surface and zero at the center of the shaft. Thus, in pure torsion, fracture normally originates at the surface, which is the region of highest stress.

Compression. When a shaft is loaded in axial compression (see Figure 2-8), the stress system rotates so that the compressive stress (σ_C) is axial and the tensile stress (σ_T) is transverse. The shear stresses (τ_M) are 45° to the shaft axis, as they are during axial tension loading.

In a ductile material overloaded in compression, shear stresses cause considerable deformation but usually do not result in fracture. The shaft is shortened and bulges under the influence of shear stress. If a brittle material loaded in pure compression does not buckle, it will fracture perpendicularly to the maximum tensile-stress component. Because the tensile stress is transverse, the direction of brittle fracture is parallel to the shaft axis.

The elastic-stress distribution in pure compression loading, in the absence of a stress concentration, is uniform across the section. If fracture occurs, it likely will be in the longitudinal direction because compression loading increases the shaft diameter and stretches the metal at the circumference.

Overload failures happen immediately as the load is being applied. However, the two common forms of overload failures, ductile and brittle, have very different appearances, as was illustrated in Figure 2-8.

The most important point to understand when conducting failure analysis on a fractured part is that the crack always grows perpendicular to the plane of maximum stress. However, both the nature of the material and the type of failure affect the appearance of the failure face. A compressive overload on a ductile material, for example, a low carbon steel nail, causes the nail to bend. But if that same type of overload were applied to a more brittle material, like drill steel or some types of cast iron, it would shatter. Figures 2-8, 2-9, and 2-10 show three ways in which ductile and brittle materials react differently to the same forces because they create different internal stresses.

In the failure of a 5½ in. diameter agitator shaft, the keyway looked like a barber pole. The shaft was made from AISI 1020, a low-strength, very ductile carbon steel. It had twisted through six complete revolutions before the final failure. Ductile material of this type frequently allows a great deal of deformation, but with brittle materials there is essentially no deformation. Brittle fracture pieces frequently look as if they could be glued back together.

There are often "chevron marks" on the face of a brittle fracture that show the progression of the failure across the piece. These chevrons or "arrows" always point to where the crack started, Figure 2-10.

Figure 2-9. The application of identical forces caused each shaft to fail. Both were severely overtorqued, but they have very different appearances. The top shaft is ductile and has twisted off. The bottom shaft shows brittle fracture.

Failure
origin

Figure 2-10. The chevrons or arrows on the face of a brittle fracture always point to the origin of a crack.

Fatigue is the primary failure mode for more than 90 percent of mechanical failures. The term originated during the 1800s when it was thought that metal parts failed because, like our muscles, they grew tired after long use. Actually, fatigue failures are caused by repeated stress cycles, that is, by fluctuating stresses. Four point are important to understanding fatigue:

- Without stress fluctuations fatigue cannot happen
- Fatigue happens at stress levels well below the tensile strength of the material
- Where corrosion is present, the fatigue strength of metals continuously decreases
- The crack takes measurable time to progress across the fracture face.

Interpretation of the failure face can disclose the forces that caused the crack, the amount of time elapsed from initiation to final failure, the relative size and type of the load, and the severity of the stress concentrations. The features of a typical fatigue failure face and their significance are shown in Figure 2-11.

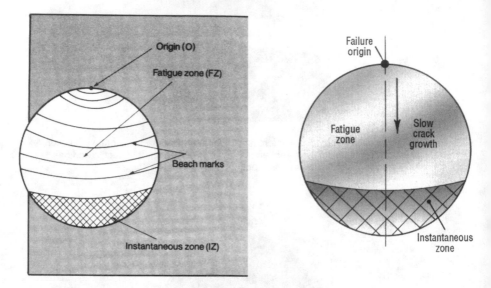

Figure 2-11. Interpretation of the failure face can disclose the forces that caused the crack, the amount of time elapsed from initiation to final failure, the relative size and type of the load, and the severity of the stress concentrations.

In a fatigue failure the fracture face always shows separate slow and fast failure zones. Figures 2-12a and 2-12b illustrate reverse and rotational bending failures, respectively, while Figure 2-12c shows the face of an actual bolt failure. The crack slowly progressed across the shaft face from the point of origin until it reached the boundary of the fast failure (or instantaneous) zone. At this point, crack growth accelerated tremendously and traveled the rest of the way at extremely high speeds.

The rate at which the crack grows across the face of the part varies with the load on the part. It may take only a few cycles, but in most industrial applications it takes millions of stress applications before the part finally breaks. On a 3600 rpm motor the interval may be only a day, but on a large mixer or press shaft it may be months or even years.

When the amplitude of the stress fluctuations changes, it frequently causes a phenomenon called beachmarks. A typical example is shown in Figure 2-13. These beachmarks show how the fatigue loads varied during the life of the failure. Frequently, significant load changes show up as beachmarks that can be read as though they were the rings on a tree.

Electron microscopy can be used to view the fatigue zone in many materials and estimate the number of cycles the crack took to cross the fatigue zone. However, in a

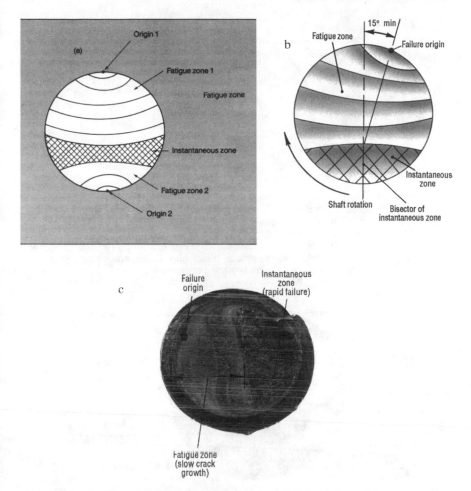

Figure 2-12. The failure slowly propagated across the fatigue zone, and then very rapidly crossed the instantaneous failure zone: (a) reverse bending, (b) rotational bending, (c) actual shaft.

more practical vein, a visual inspection of the face also can be used as a guide. The older the crack, the smoother the fracture surface. This rule is complicated by the type of material because fine-grained materials, like heat-treated steels, tend to have smoother cracks, but similar materials can be compared.

Figure 2-14 shows a component that failed from fatigue. The relative size of the fatigue and instantaneous zones tells how heavily loaded the part was. If the small area held the final load, the part was not heavily loaded. If conditions were reversed—small fatigue zone and large instantaneous zone—it would show that much more strength was needed to carry the load and the part was heavily loaded.

The fatigue failure shown in Figure 2-14 resulted from one-way bending, the kind of stress a floor beam or a leaf spring may be subjected to. Because the stress was

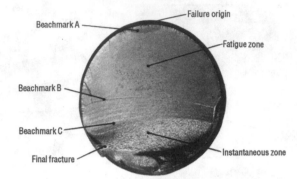

Figure 2-13. The crack started at the failure origin, grew for a short time, and then stopped at beachmark A for a long time. Across the fatigue zone, the crack grew slowly and uniformly. At beachmark B, it stopped growing for a while because the stress level was reduced. During the next period of growth, the machine was alternately run at very high and moderate loads. When the loads decreased, at beachmark C, the crack stopped growing for a while. The final fracture shows a heavily loaded component.

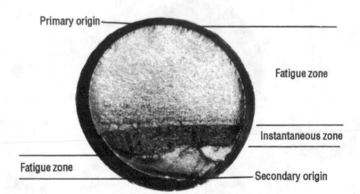

Figure 2-14. The relative size of the fatigue and instantaneous zones tell how heavily loaded the part was. It is easy to see the huge fatigue zone and the tiny instantaneous zone.

most severe on one side of the part, the cracks started at one point and grew uniformly across it. Other types of stresses cause different failure appearances. For instance, the component in Figure 2-15, actually a geartooth, shows the effect of two-way bending, because it was loaded in both directions. The unequal size of the fatigue zones shows that the stress in one direction was greater than the stress in the other.

Fasteners, bearings, and shafts are the most common victims of fatigue. Fastener failures are usually caused by fluctuating tension loads and look very similar to the bending failures shown in Figure 2-15. Bearings usually develop fatigue cracks parallel to the rolling surfaces, and shafts almost always fail from reversed (rotational) bending.

If rotational bending occurs and each part of the shaft is first exposed to tension and then to compression, such as a motor shaft subjected to side loads (like a belt drive), the crack could start anywhere on the surface. Because of the rotation, as it

Figure 2-15. The less heavily loaded side has two failure origins, and the more heavily loaded side show that cracks started at several points and worked across the face. The unequal size of the fatigue zones shows that the stress in one direction was greater than the stress in the other.

progressed across the face it would grow more on one side than the other. As a result, the bisector of the instantaneous zone would point off to one side of the origin, as shown earlier in Figure 2-12b.

However, if the shaft were more heavily loaded or if stress concentrations were present, cracks would start from a number of points around the shaft, Figure 2-16.

Stress concentrations increase the stress in one area so it is much higher than the average stress in the part. One example of stress concentration is the transition area in a bolt from the straight shank to the threaded section, or in a shaft at a shoulder. The relatively high stress concentration in this area is the reason most bolts fail at the first thread off the shank, and most shafts fail at a step such as a shoulder.

Figure 2-17 shows a crack that starts at a keyway, a stress concentration where the stress is about four times that in the rest of the shaft. The beachmarks show how the crack progressed across the shaft, and the comparative sizes of the fatigue and instantaneous zones show the relative size of the load.

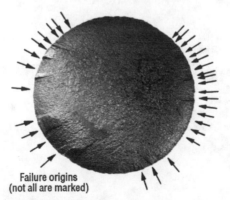

Figure 2-16. Heavily loaded shaft was subjected to rotational loading. It also had a severe stress concentration all the way around that caused the many failure origins.

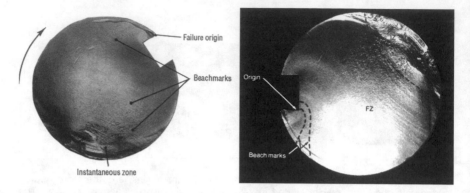

Figure 2-17. (a) The crack origin was caused by stress concentration from the keyway. Eccentric growth pattern shows the shaft was rotating in the direction of the arrow. (b) Failure of this reducer input shaft was caused by a loose, misaligned coupling and aggravated by stress concentration at the keyway corner. Beachmarks start from the keyway corner and work outward.

Unusual or irregular cracks (Figure 2-18) are almost always indicative of more elusive problems and may require expert assistance to solve. Figure 2-19 depicts the effect of loading on fracture appearance. These sketches of bending fatigue failures illustrate how the type of loading, the load (or stress) magnitude, and stress concentrations affect the fracture surface.

Corrosion has a tremendous effect on the fatigue strength of metals. Most shafts are not subject to general corrosion or chemical attack. Corrosion, which takes place as general surface pitting, may uniformly remove metal from the surface or may uniformly cover the surface with scale or other corrosion products.

Figure 2-18. Jagged, detached, irregular cracks tend to indicate unusual metallurgical problems.

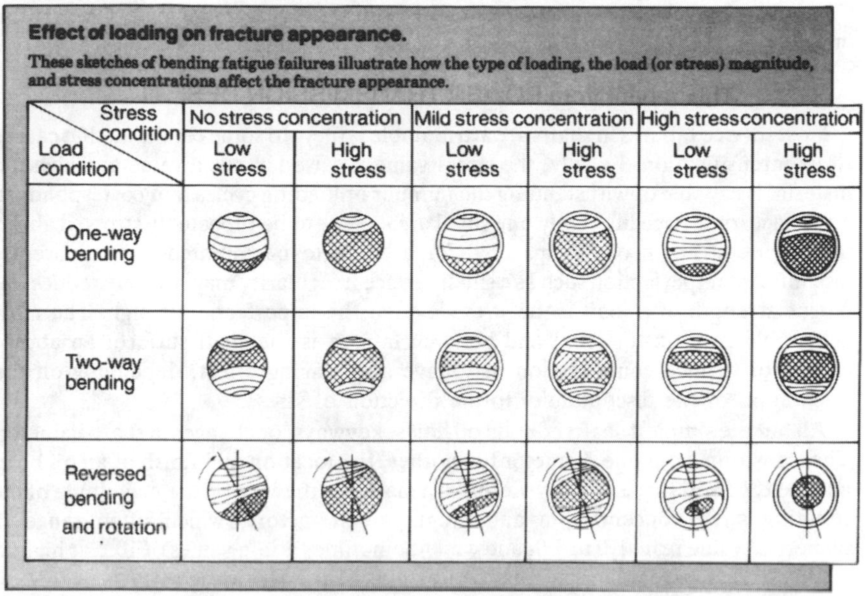

Effect of loading on fracture appearance.

These sketches of bending fatigue failures illustrate how the type of loading, the load (or stress) magnitude, and stress concentrations affect the fracture appearance.

Stress condition / Load condition	No stress concentration		Mild stress concentration		High stress concentration	
	Low stress	High stress	Low stress	High stress	Low stress	High stress
One-way bending						
Two-way bending						
Reversed bending and rotation						

Figure 2-19. Effect of loading on fracture appearance.

Corrosion pits have a relatively minor effect on the load-carrying capacity of a shaft, but they do act as points of stress concentration at which fatigue cracks can originate.

A corrosive environment will greatly accelerate metal fatigue; even exposure of a metal to air results in a shorter fatigue life than that obtained under vacuum. Steel shafts exposed to salt water may fail prematurely by fatigue even if they are thoroughly cleaned periodically. Aerated salt solutions usually attack metal surfaces at the weakest points, such as scratches, cut edges, and points of high strain. To minimize corrosion fatigue, it is necessary to select a material resistant to corrosion in the service environment, or to provide the shaft with a protective coating.

Most large shafts and piston rods are not subject to corrosion attack. Centrifugal process compressors frequently handle gases that contain moisture and small amounts of a corrosive gas or liquid. If corrosion attack occurs, a scale is often formed that may be left intact and increased by more corrosion, eroded off by entrained liquids (or solids), or slung off from the rotating shaft.

Stress-corrosion cracking occurs as a result of corrosion and stress at the tip of a growing crack. Stress-corrosion cracking often is accompanied or preceded by surface pitting, but general corrosion often is absent, just as rapid, overall corrosion does not accompany stress-corrosion cracking.

Corrosion fatigue results when corrosion and an alternating stress, neither of which is severe enough to cause failure by itself, occur simultaneously and thus can cause failure. Once such a condition exists, shaft life will probably be days or weeks rather than years. Corrosion-fatigue cracking usually is transgranular; branching of the main cracks occurs, although usually not as much as in stress-corrosion cracking.

Corrosion generally is present in the cracks, both at the tips and in regions nearer the origins.

Stress Raisers in Shafts

Most service failures in shafts are attributable largely to some condition that causes stress intensification. Locally, the stress value is raised above a value at which the material is capable of withstanding the number of loading cycles that corresponds to a satisfactory service life. Only one small area needs to be repeatedly stressed above the fatigue strength of the material for a crack to be initiated. An apparently insignificant imperfection such as a small surface irregularity may severely reduce the fatigue strength of a shaft if the stress level at the imperfection is high. The most vulnerable zone in torsional and bending fatigue is the shaft surface; an abrupt change of surface configuration may have a damaging effect, depending on the orientation of the discontinuity to the direction of stress.

All but the simplest shafts contain oil holes, keyways, or changes in shaft diameter. The transition from one diameter to another, the location and finish of an oil hole, and the type and shape of a keyway exert a marked influence on the magnitude of the resulting stress-concentration and fatigue-notch factors, which often range in numerical value from 1.0 to 5.0, and which sometimes attain values of 10.0 or higher.

Most stress raisers can be classified as one of the following:

1. Nonuniformities in the shape of the shaft, such as steps at changes in diameter, broad integral collars, holes, abrupt corners, keyways, grooves, threads, splines, and press-fitted or shrink-fitted attachments.
2. Surface discontinuities arising from fabrication practices or service damage, such as seams, nicks, notches, machining marks, identification marks, forging laps and seams, pitting, and corrosion.
3. Internal discontinuities such as porosity, shrinkage, gross nonmetallic inclusions, cracks, and voids.

Most shaft failures are initiated at primary stress raisers (see number 1), but secondary stress raisers (number 2 or 3) may contribute to a failure. For instance, a change in shaft diameter can result in stress intensification at the transition zone. If there is a surface irregularity or other discontinuity in this zone, the stress is sharply increased around the discontinuity.

Changes in Shaft Diameter

A change in shaft diameter concentrates the stresses at the change in diameter and in the small-diameter portion.

The effects on stress concentration of an abrupt change and three gradual changes in diameter are shown schematically in Figure 2-20. The sharp corner at the intersection of the shoulder and shaft in Figure 2-20, a, concentrates the stresses at the corner as they pass from the large to the small diameter. The large-radius fillet shown in Figure 2-20, d, permits the stresses to flow with a minimum of restriction.

However, the fillet must be tangent with the smaller-diameter section; otherwise, a sharp intersection will result, overcoming the beneficial effect of the large-radius fillet.

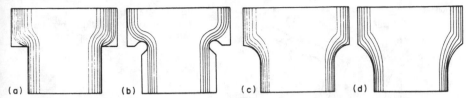

Figure 2-20. Effect of size of fillet radius on stress concentration at a change in shaft diameter.

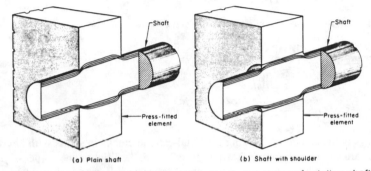

(a) Plain shaft   **(b) Shaft with shoulder**

Figure 2-21. Schematic diagram of stress distribution in two types of rotating shafts with press-fitted elements under a bending load.

Press and Shrink Fitting

Components such as gears, pulleys, and impellers often are assembled onto shafts by press or shrink fitting, which can result in stress raisers under bending stress. Typical stress flow lines in a plain shaft at a press-fitted member are shown in Figure 2-21, a. Enlarging the shaft at the press-fitted component and using a large-radius fillet would produce a stress distribution such as that shown schematically in Figure 2-21, b. A small-radius fillet at the shoulder would result in a stress pattern similar to that shown in Figure 2-21, a.

Longitudinal Grooves

Longitudinal grooves, such as keyways and splines, often cause failure in shafts subjected to torsional stress. Generally, these failures result from fatigue fracture where a small crack originated in a sharp corner because of stress concentration. The crack gradually enlarges as cycles of service stress are repeated until the remaining section breaks. A sharp corner in a keyway can cause the local stress to be as much as ten times the average nominal stress.

Failures of this kind can be avoided by using a half-round keyway, which permits the use of a round key, or by using a generous fillet radius in the keyway. Good results are obtained by the use of fillets having radii equal to approximately one-half the depth of the keyway. A half-round keyway produces a local stress of only twice the average stress, thus providing greater load-carrying ability than that permitted by a square keyway. Many shafts with square keyways do not fracture in service because stresses are low or because fillets with generous radii are used.

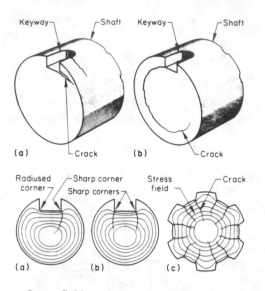

Figure 2-22. Stress fields and corresponding torsional-fatigue cracks in shafts with keyways or splines.

Stress fields and corresponding torsional-fatigue cracks in shafts with keyways or splines are shown in Figure 2-22. In Figure 2-22 a, the fillet in one corner of the keyway was radiused and the fillet in the other corner was sharp, which resulted in a single crack. Note that this crack progresses approximately normal to the original stress field. In Figure 2-22, b, both fillets in the keyway corners were sharp, which resulted in two cracks that did not follow the original stress field, a condition arising from the cross effect of cracks on the stress field.

A splined shaft subjected to alternating torsion can crack along the bottom edge of the splines, as shown in Figure 2-22, c. This is another instance of a highly localized stress field strongly influencing crack development.

Failures Due to Manufacturing Processes

Surface discontinuities produced during manufacture of a shaft and during assembly of the shaft into a machine can become points of stress concentration and thus contribute to shaft failure. Operations or conditions that produce this type of stress raiser include:

1. Manufacturing operations that introduce stress raisers such as tool marks and scratches.
2. Manufacturing operations that introduce high tensile stresses in the surface, such as improper grinding, repair welding, electromachining, and arc burns.
3. Processes that introduce metal weakening, such as forging flow lines that are not parallel with the surface; plating, causing hydrogen embrittlement; or heat treating, causing decarburization.

Fatigue strength may be increased by imparting high compressive residual stresses to the surface of the shaft. This can be accomplished by surface rolling or burnishing, shot peening, tumbling, coining, induction hardening, and sometimes case hardening.

Influence of Metallurgical Factors

Internal discontinuities such as porosity, large inclusions, laminations, forging bursts, flakes, and centerline pipe will act as stress concentrators under certain conditions and may originate fatigue fracture.

In order to understand the effect of discontinuities, it is necessary to realize that fracture can originate at any location, surface or interior, where the stress first exceeds material strength. The stress gradient must be considered in torsion and bending because the stress is maximum at the surface but zero at the center, or neutral axis. In tension, however, the stress is essentially uniform across the section.

If discontinuities such as those just noted occur in a region highly stressed in tension by bending or torsional loading, fatigue cracking may be initiated. However, if the discontinuities are in a low-stress region, such as near a neutral axis, they will be less harmful. Similarly, a shaft stressed by repeated high tensile loading must be free from serious imperfections, for there is no neutral axis. Here, any imperfection can be a stress concentrator and can be the origin of fatigue cracking if the stress is high with respect to the strength.

Nonmetallic inclusions oriented parallel to the principal stress do not usually exert as great an effect upon fatigue resistance as those 90° to the principal stress.

Surface Discontinuities

In mill operations, a variety of surface imperfections often result from hot plastic working of material when lapping, folding, or turbulent flow is experienced. The resultant surface discontinuities are called laps, seams, and cold shuts. Similar discontinuities also are produced in cold working operations, such as fillet and thread rolling. Other surface imperfections develop from foreign material embedded under high pressures during the working process. For example, oxides, slivers, or chips of the base material are occasionally rolled or forged into the surface.

Most of the discontinuities are present in the metal prior to final processing and are open to the surface. Standard nondestructive testing procedures such as liquid-penetrant and magnetic-particle inspections will readily reveal most surface discontinuities. If not detected, discontinuities may serve as sites for corrosion or crack initiation.

Because fatigue-crack initiation is the controlling factor in the life of most small shafts, freedom from surface imperfections becomes progressively more important in more severe applications. Similarly, internal imperfections, especially those near the surface, will grow under cyclic loading and result in cracking when critical size is attained. Service life can be significantly shortened when such imperfections cause premature crack initiation in shafts designed under conventional fatigue-life considerations. Surface or subsurface imperfections can cause brittle fracture of a shaft after a very short service life when the shaft is operating below the ductile-to-brittle transition temperature. When the operating temperature is above the transition temperature, or when the imperfection is small relative to the critical flaw size, especially when the cyclic-loading stress range is not large, service life may not be affected.

Examination of Failed Shafts

In general, the procedures for examining failed shafts are similar to those discussed in connection with Table 2-2. However, in summary of the foregoing, some unique characteristics of shafts require special attention:

Potential stress raisers or points of stress concentration, such as splines, keyways, cross holes, and changes in shaft material, mechanical properties, heat treatment, test locations, nondestructive examination used, and other processing requirements, also should be noted.

Special processing or finishing treatments, such as shot peening, fillet rolling, burnishing, plating, metal spraying, and painting, can have an influence on performance, and the analyst should be aware of such treatments.

Mechanical conditions. The manner in which a shaft is supported or assembled in its working mechanism and the relationship between the failed part and its associated members can be valuable information. The number and location of bearings or supports and how their alignment may be affected by deflections or distortions that can occur from mechanical loads, shock, vibrations, or thermal gradients should be considered.

The method of connecting the driving or driven member to the shaft, such as press fitting, welding, or use of a threaded connection, a set screw, or a keyway, can influence failure. Also influential is whether power is transmitted to or taken from the shaft by gears, splines, belts, chains, or torque converters.

Operating history. Checking the operating and maintenance records of an assembly should reveal when the parts were installed, overhauled, and inspected. These records also should show whether maintenance operations were conducted in accordance with the manufacturer's recommendations.

We are concluding our discussion of shaft failure analysis with an example of a typical shaft failure case and analysis.

The Case of the Boiler Fan Turbine
Shaft Fracture (Figures 2-23 and 2-24)

After examining a failed turbine rotor shaft in a Canadian refinery, the investigating metallurgical laboratory prepared the following report for the owners:

> We have completed our examination of the F-703 turbine shaft fracture faces submitted from your refinery. The fracture faces were considerably damaged as a result of contact and relative motion subsequent to fracture, but nevertheless three distinct zones were apparent:
>
> 1. An outer ring approximately 5mm (3/16 in.) wide of relatively undamaged surface normal to the axis of the shaft. Close inspection showed that this zone had suffered some abrasion by relative motion.
> 2. A mid-radius ring extending about 20mm (3/4 in.) into the center, which displayed severe abrasion caused by contact and relative motion of the two halves of the shaft. The original fracture features

were virtually obscured by the abrasion. The fracture profile in this region displayed a shallow "cup and cone" appearance.

3. A central region about 25mm (1 in.) in diameter in which abrasion caused by the contact and relative motion of the two halves of the shaft had resulted in continuous gouging or ploughing of the surface to produce a relatively smooth and shiny surface.

The surface of the shaft was heavily scored and there was some slight surface oxidation which may have occurred as a result of frictional heating. There were no obvious corrosion products on either the shaft or the fracture surfaces.

Metallographic examination showed the microstructure of the shaft to be of tempered martensite with some proeutectoid ferrite. The fracture profile was covered with what appeared to be wear transformation products formed by abrasion. Figure 2-23 shows these products at the surface and also entrained in the sublayers by severe abrasion. The figure also shows the general microstructure. Damage to the surface was too severe to tell if the original fracture was intergranular or transgranular. No corrosion at the fracture surface was apparent. Transformation products, similar to those observed on the fracture surface, were observed in patches on the scored surface of the shaft. The appearance of these products was attributed to scoring and abrasion of the shaft surface during service.

The mean hardness was 266 HV/10, corresponding to 25 RC. There was no significant variation in hardness between the edge and the center of the shaft section.

Figure 2-23. Turbine shaft fracture profile and general microstructure. Etchant: 2% nital (200x).

An attempt was made to examine the fracture surface by scanning electron microscopy. Most of the surface was damaged by abrasion in some way. In the few areas in which the effects of abrasions after fracture were not apparent, the fracture surface appeared smooth and relatively featureless.

Although the shaft fracture surfaces were so badly abraded as to preclude any conclusive statement on the nature of the failure, certain characteristics do suggest that the shaft failed by fatigue. The outer zone around the periphery of the fracture face suggests that fatigue cracking was initiated at multiple points under a fairly severe stress concentration, with crack propagation occurring as the result of a bending moment applied to the rotating shaft. It appears that stress concentration arose as a result of scoring of the shaft surface during service. Further comments are not possible because of the damage sustained by the fracture surfaces.

If you have any questions concerning this failure, I shall be pleased to answer them.

Yours truly,

James Bond,
Senior Metallurgical Engineer

Making use of this report, the owner's engineer focused his attention on root cause of failure and remedial action. He commented as follows:

The report is suggesting fatigue cracking of the shaft material as a result of a bending moment applied to the shaft while rotating. It is in agreement with our earlier assumptions as to the most probable cause of the failure.

We think that a failure like this can be avoided by careful alignment of the three sleeve bearings supporting the turbine/pinion rotor (Figure 2-24). We are now using a mandrel technique where we assemble a dummy shaft in the bearings to obtain a blueing pattern as an indication for internal alignment. This approach is recommended. The mandrel should have oversize journal diameters to take up actual bearing clearance, and the bearing caps should be torqued down at final assembly. This would allow an indication of proper bearing alignment not only in the vertical down direction but in all other directions.

Analysis of Surface-Change Failures

Failures through material surface changes are generally caused by wear, fatigue, and corrosion. Each of these failure forms is affected by many conditions, including environment, type of loading, relative speeds of mating parts, lubrication, temperature, hardness, and surface finish.

Wear, or the undesired removal of material from rubbing surfaces, causes many surface failures. Wear can be classified as either "abrasive" or "adhesive." If corrosion is present, a form of adhesive wear occurs that is termed "corrosive wear."

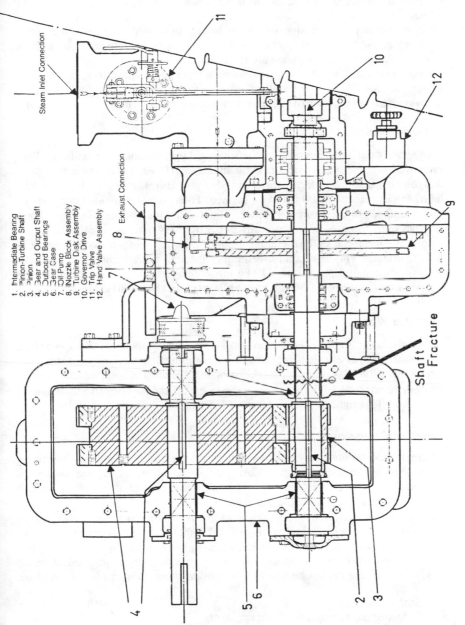

1. Intermediate Bearing
2. Pinion-Turbine Shaft
3. Pinion
4. Gear and Output Shaft
5. Outboard Bearings
6. Gear Case
7. Oil Pump
8. Nozzle Block Assembly
9. Turbine Disk Assembly
10. Governor Drive
11. Trip Valve
12. Hand Valve Assembly

Steam Inlet Connection

Exhaust Connection

Shaft Fracture

Figure 2-24. Horizontal section of a built-in type geared turbine.

Abrasive Wear

Abrasive wear occurs when a certain material scratches or gouges a softer surface. It has been estimated[13] that abrasion is responsible for 50% of all wear-related failures.

A typical example of abrasive wear is the damage of crankshaft journals in reciprocating compressors. Hard dirt particles will break through the lubricant film and cut or scratch the journal's comparatively softer surface.

Wear, or the undesired removal of material from rubbing surfaces, causes many surface failures. Wear can be classified as either "abrasive" or "adhesive." If corrosion is present, a form of adhesive wear occurs that is termed "corrosive wear."

Adhesive Wear

Adhesive-wear failure modes are scoring, galling, seizing, and scuffing. They result when microscopic projections at the sliding interface between two mating parts weld together under high local pressures and temperatures. After welding together, sliding forces tear the metal from one surface. The result is a minute cavity on one surface and a projection on the other, which will cause further damage. Thus, adhesive wear initiates microscopically but progresses macroscopically.

Adhesive wear can be best eliminated by preventing metal-to-metal contact of sliding surfaces. This is accomplished either through a lubricant film or suitable coatings or through deposits such a PTFE infusion layers, for instance. In the section on gear failure analysis in Chapter 3, adhesive wear will be discussed further.

Corrosive Wear

Corrosive wear is usually encountered on lubricated components. For example, a major source of gas-engine component wear in connection with frequent stops and starts is corrosion by water in the oil. Lube-oil management of rotating equipment is mainly directed toward eliminating the water in the oil and the corrosive wear associated with it. This topic is further developed in Chapter 3.

Effects of Corrosion

Experienced failure analysts know that corrosion often complicates machinery component failure investigations. Corrosion is the chemical or electrochemical reaction at the surface of the material which results in a change of the original surface finish or in local damage to the material.

In the case of corrosion of a metal in water, there is a reduction process taking place side by side with oxidation. Figure 2-25 shows typical types of corrosion. Porous or needle-shaped forms are referred to as "pitting corrosion."

If a corrosion-resistant material is technically or economically justified, it should be selected. If one cannot be justified, corrosion protection can be achieved by:

1. Surface coatings.
2. Inducing the formation of surface layers.
3. Corrosion preventive additives in the fluid.
4. Anodic or cathodic polarization.

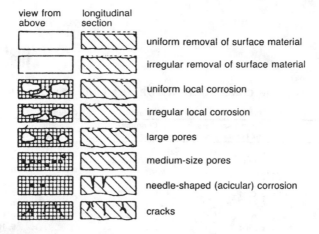

Figure 2-25. Typical forms of corrosion.

Every material exhibits a given resistance to chemical or electrochemical attack. Earlier, we talked about increasing resistance to failure as a failure-fighting strategy. Naturally, in the context of corrosion, this strategy is most effective. Consequently, various materials have been evaluated for construction, and the troubleshooter may consult one of several tables published, such as those for pumps, shown in Figure 2-26.

Fretting Corrosion

Fretting wear, or fretting corrosion, is a failure mode of considerable practical significance. Fretting corrosion is also called "friction oxidation," "bleeding," "red mud," and "fit corrosion." It is defined as accelerated surface damage occurring at the interface of contacting materials subjected to small oscillatory displacement. Fretting corrosion is found in all kinds of press fits, spline connections, bearings, and riveted and bolted joints, among other places.

One important effect of fretting wear is its contribution to fatigue failures. Examinations of surface fractures have shown that fatigue cracks originate in or at the edge of a fretted area. An example of the effect of fretting on fatigue of an aged Al-Cu-Mg alloy is shown in Figure 2-27.

Detailed explanations of fretting wear have been given by Waterhouse.[15] Factors influencing fretting wear are given in the following list[16]:

1. *Slip:* No fretting can occur unless relative motion is sufficient to produce slip between the surfaces.
2. *Frequency:* Fretting wear rates increase at lower frequencies and become almost-constant as frequency increases.
3. *Normal Load:* Fretting wear generally increases with applied load, provided the amplitude of slip is maintained constant.
4. *Duration:* Fretting wear increases almost linearly with the number of cycles.
5. *Temperature:* In general, fretting wear tends to increase with decreasing temperature.

materials

MATERIALS OF CONSTRUCTION FOR PUMPING VARIOUS LIQUIDS

Column 1 Liquid	Column 2 Condition of Liquid	Column 3 Chemical Symbol	Column 4 Specific Gravity	Column 5 Material Selection
Acetaldehyde	....	C_2H_4O	0.78	C
Acetate Solvents	....	....	....	A, B, C, 8, 9, 10, 11
Acetone	....	C_3H_6O	0.79	B, C
Acetic Anhydride	....	$C_4H_6O_3$	1.08	8, 9, 10, 11, 12
Acid, Acetic	Conc. Cold	$C_2H_4O_2$	1.05	8, 9, 10, 11, 12
Acid, Acetic	Dil. Cold	....	....	A, 8, 9, 10, 11, 12
Acid, Acetic	Conc. Boiling	....	....	9, 10, 11, 12
Acid, Acetic	Dil. Boiling	....	....	9, 10, 11, 12
Acid, Arsenic, Ortho-	....	$H_3AsO_4, \frac{1}{2}H_2O$	2.0-2.5	8, 9, 10, 11, 12
Acid, Benzoic		$C_7H_6O_2$	1.27	8, 9, 10, 11
Acid, Boric	Aqueous Sol.	H_3BO_3	....	A, 8, 9, 10, 11, 12
Acid, Butyric	Conc.	$C_4H_8O_2$	0.96	8, 9, 10, 11
Acid, Carbolic	Conc. (M.P. 106 F)	C_6H_6O	1.07	C, 8, 9, 10, 11
Acid, Carbolic	(See Phenol)	....	....	B, 8, 9, 10, 11
Acid, Carbonic	Aqueous Sol.	CO_2+H_2O	....	A
Acid, Chromic	Aqueous Sol.	$Cr_2O_3 + H_2O$	....	8, 9, 10, 11, 12
Acid, Citric	Aqueous Sol.	$C_6H_8O_7 + H_2O$	....	A, 8, 9, 10, 11, 12
Acids, Fatty (Oleic, Pal-mitic, Ste-				A, 8, 9, 10, 11
Acid, Formi			1.22	9, 10, 11
Acid, Fruit			..	A, 8, 9, 10, 11, 14
Acid, Hydro			1.19 (38%)	11, 12
Acid, Hydro			..	10, 11, 12, 14, 15
Acid, Hydro			...	11, 12
Acid, Hydro			0.70	C, 8, 9, 10, 11
Acid, Hydro				3, 14
Acid, Hydro				A, 14
Acid, Hydro			1.30	A, 14
Acid, Lactic			1.25	A, 8, 9, 10, 11, 12
Acid, Mine			..	A, 8, 9, 10, 11
Acid, Mixed			..	C, 3, 8, 9, 10, 11, 12
Acid, Muria			..	....
Acid, Naph			..	C, 5, 8, 9, 10, 11
Acid, Nitric			1.50	6, 7, 10, 12
Acid, Nitric				5, 6, 7, 8, 9, 10, 12
Acid, Oxali			1.65	8, 9, 10, 11, 12
Acid, Oxali				10, 11, 12
Acid, Ortho			1.87	9, 10, 11
Acid, Picric			1.76	8, 9, 10, 11, 12
Acid, Pyrog			1.45	8, 9, 10, 11
Acid, Pyroli			..	A, 8, 9, 10, 11
Acid, Sulfu			1.69-1.84	C, 10, 11, 12
Acid, Sulfu			..	11, 12
Acid, Sulfu			..	10, 11, 12
Acid, Sulfu			..	10, 11, 12
Acid, Sulfu			..	A, 10, 11, 12, 14
Acid, Sulfu			1.92-1.94	3, 10, 11
Acid, Sulfu				A, 8, 9, 10, 11
Acid, Tanni			..	A, 8, 9, 10, 11, 14

Summary of Material Selections and National Society Standards Designations

Material Selection	Corresponding National Society Standards Designation ASTM	ACI	AISI	Remarks
1.....	A48, Classes 20, 25, 30, 35, 40 & 50	......	......	Gray Iron—Six Grades
1(a)..	A536 & A395	......	......	Ductile Cast Iron—Six Grades
2.....	B143, 1A, 1B & 2A; B144, 3A; B145, 4A	......	......	Tin Bronze & Leaded Tin Bronze—seven alloys (includes 2 alloys not covered by ASTM Specifications, as explained above under Selection #2)
3.....	A216-WCB	......	1026	Carbon Steel
4.....	A217-C5	......	501	5% Chromium Steel
5.....	A296-CA15	CA15	410	12% Chromium Steel
6....	A296-CB30	CB30	......	20% Chromium Steel
7....	A296-CC50	CC50	446	28% Chromium Steel
8....	A296-CF-8	CF-8	304	19-9 Austenitic Steel
9....	A296-CF-8M	CF-8M	316	19-10 Molybdenum Austenitic Steel
10.....	A296-CN-7M	CN-7M	......	20-29 Chromium Nickel Austenitic Steel with Copper & Molybdenum
11....				A series of nickel-base alloys
12.....	A518	......	......	Corrosion Resistant High-silicon cast iron
13.....	A436	......	......	Austenitic cast iron—2 types
13(a)	A439	......	......	Ductile Austenitic Cast Iron
14.....				Nickel-Copper alloy
15.....				Nickel

*ASTM—denotes American Society for Testing Materials
ACI—denotes Alloy Casting Institute
AISI—denotes American Iron and Steel Institute

Figure 2-26. Example for listing of materials of construction for pumps.[14]

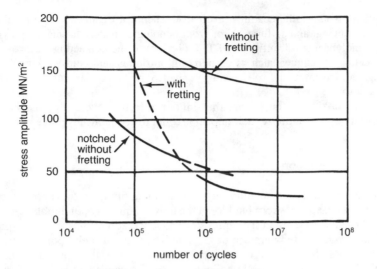

Figure 2-27. Effect of fretting on the fatigue of an aged Al-Cu-Mg alloy.[16]

6. *Atmosphere.* Fretting wear that occurs in air or oxygen atmospheres is far more severe than that produced in an inert atmosphere.
7. *Humidity:* Increased relative humidity decreases fretting wear for most metals. This is because the moisture in the air provides a lubricating film which promotes the removal of debris from contact areas.
8. *Surface Finish:* Generally, fretting wear is more serious when the surfaces are smooth, as a smooth finish will have smaller and fewer "lubricant pockets" on its surface.
9. *Lubricants:* Fretting wear is reduced by lubricants which restrict the access of oxygen. Lubricants remove debris from the fretted zone and change the coefficient of friction.
10. *Hardness:* It is generally accepted that an increase in materials hardness decreases fretting wear.
11. *Coefficient of Friction:* Fretting wear generally decreases with a decreasing coefficient of friction because of a reduced slip amplitude.

Consequently, the prevention of fretting corrosion and wear lies in the elimination of relative displacement. One way to achieve this is to decrease fit clearances. Troubleshooters encountering fretting corrosion would do well to see, for instance, that improved shaft tolerances in repair work are adopted. The tighter fits obtained by such measures will increase contact areas between shafts and bores and consequently increase frictional force.

Increasing the coefficient of friction is another preventive measure. This can be achieved by coating the contact surfaces with materials with suitable frictional properties. These coatings could be metallic or nonmetallic. Metallic coatings that have been successfully employed are cadmium, silver, gold, tin, lead, copper, and chromium. Nonmetallic coatings result from chemical treatments such as phosphatizing, anodizing, and sulfurizing, or from bonding of materials such as polymer of MoS2 (molybdenum-disulfide) or PTFE (Teflon) to the contacting surfaces. Diffusion-coating techniques such as carburizing, nitriding, and cold working (i.e. cold rolling, shot-peening, and roll peening) enhance the fatigue strength of the contacting members.

Finally, looking at the factors that influence fretting wear, the analyst should always attempt to minimize their impact in order to economically prevent this failure mode.

Cavitation Corrosion

Cavitation, or the formation of cavities in a fluid, is the appearance and subsequent collapse of vapor bubbles in the flow of liquid. Vapor bubbles are formed when the static pressure in the liquid sinks so low that it attains the vapor pressure associated with the temperature of the liquid at that particular point. If the static pressure rises after this above the vapor pressure along the flow path, the vapor bubble collapses quite suddenly, followed by sudden condensation in the form of an implosion. If the implosion occurs not in the body of flowing liquid but at the wall of a component containing the flowing liquid, cavitation will result in material erosion.

Recent research in the field of cavitation has indicated that the vapor bubble inverts at first, once the implosion begins. After that, a fluid microjet is formed, directed toward the interior of the bubble, which pierces the opposite wall of the bubble. Slow-motion pictures[17] of the phenomenon indicate that where bubbles are close to a wall, liquid microjets are always directed against the wall, striking it at high speed. This causes material disintegration, which is in turn intensified by chemical action: The microjet entrains the dissolved oxygen in the liquid, which is then liberated in the vapor and forced at high pressure between the grain boundaries of the material at the wall surface. This process increases the corrosion of the wall material.

The first sign of cavitation corrosion will be pitting. In its progressed form, it will have a honey-combed, spongy appearance and structure. The amount of material removed by cavitation could be determined by:

1. Geometric measurements.
2. Weight loss.
3. Measurement of amount of built-up repair materials, i.e. metal deposited by welding or quantity of molecular metal used.
4. Mean-time-to-repair.

Figures 2-28 and 2-29 show a bronze, mixed-flow pump impeller from a cooling tower pump in a petrochemical plant after five years of operation under cavitation conditions.

Usually the troubleshooter will attempt to curtail the effects of cavitation corrosion by design or operational changes. Quite often, it will be impossible to shift the collapse of the vapor bubbles away from the wall toward the center of the flow path, just as it will be uneconomical, for instance, to change existing unfavorable

Figure 2-28. Cast-bronze mixed-flow pump impeller failed from cavitation corrosion.

Figure 2-29. Close-up of pump impeller from Figure 2-28.

Table 2-8
Cavitation-Induced Loss-of-Weight Indexes
of Pump Materials[11]

Grey cast iron	1.0
Cast steel	0.8
Bronze	0.5
Cast chrome steel	0.2
Multicomponent bronze	0.1
Chrome nickel steel	0.05

submersion conditions in a vertical cooling tower pump installation. A change of materials of construction will be appropriate under these circumstances.

Materials resistant to cavitation are those with high fatigue strength and ductility, together with a high corrosion resistance. Comparative cavitation resistance indexes are listed in Table 2-8.

Analyzing Wear Failures*

An accurate analysis of a wear failure depends largely upon three sources of evidence: the worn surfaces, the operating environment, and the wear debris.

Surface damage can range from polished or burnished conditions to removal of a relatively large volume of material. Examining the worn surface can provide much information, including the amount of material removed, the type of damage or failure mode, and the existence and character of surface films. Further, it will tell whether certain constituents are being attacked preferentially, the direction of relative motion between a worn surface and abrading particles, and whether abrading particles have become embedded in the surface.

Environmental Conditions

The environment in which the failure takes place greatly affects the mechanism and rate of metal removal, and detailed knowledge of these conditions should always be sought. For instance, a coke crusher experiences erratic wear, with increased wear occurring at certain abnormal operating conditions. It would stand to reason to investigate what circumstances lead to these operating conditions and then, by eliminating them, stop the accelerated wear of the coke crusher.

Wear environments may be corrosive; may have been altered during service, such as by breakdown of a lubricant; may provide inadequate lubrication; or may differ from the assumed environment on which the original material selection was made.

---

*Adapted by permission from course material published by the American Society of Metals, Metals Park, Ohio 44073. For additional information on metallurgical service failures, refer to the 15-lesson course *Principles of Failure Analysis,* available from the Metals Engineering Institute of the American Society for Metals, Metals Park, Ohio 44073.

Wear debris, whether found between the worn surfaces, embedded in a surface, suspended in the lubricant, or beside the worn part, can provide clues to the wear mechanism. A wear particle consisting of a metallic center with an oxide covering probably was detached from the worn surface by abrasive or adhesive wear and subsequently oxidized by exposure to the environment. On the other hand, a small wear particle consisting solely of oxide may result from corrosion on the worn surface which was subsequently removed mechanically.

Analysis Procedure

Though conditions may vary greatly, the general steps in wear-failure analysis are described in Table 2-9.

Operating Conditions

Probably the first step in wear-failure analysis is to identify the type of wear or, if more than one type can be recognized, to evaluate the relative importance of each type as quantitatively as possible. This identification of the type or types of wear requires a detailed description of the operating conditions based on close observation and adequate experience. A casual and superficial description of operating conditions will be of little value.

Descriptions of operating conditions often are incomplete, thus imposing a serious handicap on the failure analyst, especially if he is working in a laboratory remote from the plant site. For instance, assume that an analyst must study the problem of a

Table 2-9
Wear-Failure Analysis

1. Identify the actual materials in the worn part, the environment, the abrasive, the wear debris, and the lubricant.
2. Identify the mechanism or combination of mechanisms of wear: adhesive, abrasive, corrosive, surface fatigue, or erosive.
3. Define the surface configuration of both the worn surface and the original surface.
4. Define the relative motions in the system, including direction and velocity.
5. Define the force or pressure between mating surfaces or between the worn surface and the wear environment, both macroscopically and microscopically.
6. Define the wear rate.
7. Define the coefficient of friction.
8. Define the effectiveness and type of lubricant: oil, grease, surface film, naturally occurring oxide layer, adsorbed film, or other.
9. Establish whether the observed wear is normal or abnormal for the particular application.
10. Devise a solution, if required.

badly seized engine cylinder. Obviously this is an instance of adhesive metal-to-metal wear or lubricated wear, because it is assumed that a suitable engine oil was used. Furthermore, assume that during an oil change the system had been flushed with a solvent to rinse out the old oil, and that inadvertently it had been left filled with the solvent instead of new oil. Also assume that a slow leak, resulting in loss of the solvent, was not detected during the operating period immediately preceding seizure. The analyst probably would receive the failed parts (cylinder block and pistons) after they had been removed from the engine, cleaned, and packed. If evidence of the substitute "lubricant" could not be established clearly, determination of the cause of failure would be extremely difficult or perhaps impossible.

Similarly, incomplete descriptions of operating conditions can be misleading in the analysis of abrasive wear. For example, in describing the source of abrasion that produces wear of a bottoms pump impeller, generalized references to the abrasive in question, such as coke particles, might be too vague. Unless the abrasives are studied both qualitatively and quantitatively, a valid assessment of wear, whether normal or abnormal, is impossible.

Solving Wear Problems

As stated earlier, wear failures differ from many other types of failures, such as fatigue, because wear takes place over a period of time. Seldom does the part suddenly cease to function properly. In most instances, wear problems are solved by two approaches: The service conditions are altered to provide a less destructive environment, or materials more wear resistant for the specific operating conditions are selected. Because the latter method is easier and less expensive, it is chosen more frequently.

Laboratory Examination of Worn Parts

Analysis of wear failures depends greatly on knowledge of the service conditions under which the wear occurred. However, many determining factors are involved that require careful examination, both macroscopically and microscopically.

Wear failures generally result from relatively long-time exposure, yet certain information obtained at the time the failure is discovered is useful in establishing the cause. For example, analyzing samples of the environment—often the lubricant—can reveal the nature and amount of wear debris or abrasive in the system.

Procedures. Examination of a worn part generally begins with visual observation and measurement of dimensions. Observations of the amount and character of surface damage often must be made on a microscopic scale. An optical comparator, a toolmaker's microscope, a recording profilometer, or other fine-scale measuring equipment may be required to adequately assess the amount of damage that has occurred.

Weighing a worn component or assembly and comparing its weight with that of an unused part can help define the amount of material lost. This material is lost in abrasive wear or transferred to an opposing surface in adhesive wear. Weight-loss estimates also help define relative wear rates for two opposing surfaces that may be made of different materials, or that have been worn by different mechanisms.

Screening the abrasives or wear debris to determine the particle sizes and weight percentage of particles of each size is often helpful. Both a determination of particle size and a chemical analysis of the various screenings can be useful when one component in an abrasive mixture primarily causes wear, or when wear debris and an abrasive coexist in the wear environment. The combination of screening with microscopy often can reveal details such as progressive alteration of the size and shape of abrasive particles with time, as might occur in a fluid coker ball mill.

Physical measurements can define the amount and location of wear damage, but they seldom provide enough information to establish either the mechanism or the cause of the damage.

Microscopy can be used to study features of the worn surface, including the configuration, distribution, and direction of scratches or gouges and indications of the preferential removal of specific constituents of the microstructure. Abrasive particles or wear debris are viewed under the microscope to study their shape and the configuration of their edges (sharp or rounded), and to determine if they have fractured during the wear process.

Metallography can determine whether or not the initial microstructure of the worn part was to specifications. It also will reveal the existence of localized phase transformation, cold-worked surface layers, and embedded abrasive particles.

Taper sectioning. For valid analyses, techniques such as taper sectioning are needed to allow metallographic observations or microhardness measurements of very thin surface layers. Almost always, special materials that support the edge of a specimen in a metallographic mount (for example, nickel plating on the specimen or powdered glass in the mounting material) must be used, and the mounted specimen must be polished with care.

Etchants, in addition to preparing a specimen for the examination of its microstructure, also reveal characteristics of the worn surface. Two features that can be revealed by etching a worn surface are phase transformations caused by localized adhesion to an opposing surface and the results of overheating caused by excessive friction, such as the "white layer" (untempered martensite) that sometimes develops on steel or cast iron under conditions of heavy sliding contact.

Macroscopic and microscopic hardness testing indicate the resistance of a material to abrasive wear. Because harder materials are likely to cut or scratch softer materials, comparative hardness of two sliding surfaces may be important. Microhardness measurements on martensitic steels may indicate for example, that frictional heat has overtempered the steel, and when used in conjunction with a tempering curve (a plot of hardness versus tempering temperature) they will give a rough estimate of surface temperature. Hardness measurements also can indicate whether a worn part was heat treated correctly.

Chemical analysis. One or more of the various techniques of chemical analysis—wet analysis, spectroscopy, calorimetry, x-ray fluorescence, atomic absorption, or electron-beam microprobe analysis—usually is needed to properly analyze wear failures. The actual compositions of the worn material, the wear debris, the abrasive, and the surface film must be known in order to devise solutions to most wear problems. An analysis of the lubricant, if a worn lubricated component is involved, should always be a matter of course.

Recording Surface Damage

Sometimes the experienced troubleshooter would like to record and preserve the appearance of surface failures. Several methods have been applied. One is the use of dental impression materials* that are cast in liquid form onto failed or damaged gear-tooth surfaces, for instance. These materials will yield highly accurate prints of surface damage.[18]

Another method, which has been successfully applied by one of the authors, is described in Ref. 12. This method is simple and can be used to record a variety of surface failures such as pitting, scoring, and general wear. With it, replicas of entire cylindrical surfaces or gear tooth flanks can be recorded permanently on a single sheet of paper. The procedure is as follows:

1. Clean the surface and leave a light oil film.
2. Brush the surface to be reproduced with graphite powder. Brush off the excess with a soft brush or tissue.
3. Apply pressure-sensitive transparent tape (matte Scotch tape), adhesive down, on the surface of interest. Rub the tape so that it adheres to the surface.
4. Strip off the tape and apply it to a sheet of white paper or a glass slide so that the surface appearance is permanently recorded.

Table 2-10 lists machinery components and their potential surface failure modes. Tables 2-11 and 2-12 will help the failure analyst identify major wear failure modes by registering their appearance and nature of debris. This is a first approach leading eventually to improved material selection as indicated in Table 2-13.

---

*Silicone rubbers, i.e., Reprosil or addition-cured polyvinylsiloxanes.

Table 2-10
Process Machinery Components and Wear Failure Modes

	Scuffing	Fretting wear	Galling	Corrosion	Erosion	Sticking
Trip valve stems on turbines (all)	•	•	•			•
Governor valve stems on turbines		•				•
Governor linkage pivots on turbines		•				•
Pump wear rings	•		•			
Pump impellers					•	•
Sleeves or shafts under mechanical seals		•				•
Seal faces, high-viscosity fluid						•
Seal faces, metal on metal	•		•			
Antifriction bearings		•		•		
Centrifugal compressor impellers				•		
Centrifugal compressor labyrinths				•	•	
Centrifugal compressor shaft sleeves				•	•	
Centrifugal compressor breakdown bushings	•					
Centrifugal compressor shafts under bearings and seals		•				
Centrifugal compressor guide vanes and diaphragms		•		•		•
Pump casings					•	•
Pump sleeves	•					
Non-lube compressor valve plates, seals, springs	•					
Compressor piston rods	•					
Compressor piston rings	•				•	
Engine pistons	•					
Engine piston rings	•					
Engine camshafts	•					
Engine valve stems	•		•			•
Coupling-hub shaft fits		•				
Gear couplings	•					
Gears	•					
Scrapers for exchangers	•					•
Gear pumps	•					
Cams or followers for turbine valve gear	•					
Reciprocating plungers	•					
Reciprocating pistons	•					
Reciprocating valves	•					
Reciprocating steam pistons	•					
Reciprocating valve gear	•					
Progressive cavity pump impellers	•					
Screw pump rotors and idlers	•					
Sliding piping supports at machinery	•	•				

Table 2-11
Surface Appearance and Failure Mode in Wear Analysis
(Modified from Ref. 16)

	Description of Worn Surface	Staining	Pitting	Spalling	Cavitation	Erosion	Abrasive Wear	Fretting	Polishing	Scratching	Scuffing	Gouging	Scoring	Grooving	Galling	Exfoliation	Melting
Micro-Smooth	Progressive loss and reformation of surface films by fine abrasion and/or tractive stresses, mutually imposed by adhesive or viscous interaction.							•	•		•		•	•			
	Very fine abrasion, with loss of substrate in addition to loss of surface film.					•			•								
	Melting																•
Micro-Rough	Due to tractive stresses resulting from adhesion.									•					•		
	Micro-pitting by fatigue.		•		•												
Macro-Smooth	Abrasion by medium-coarse particles.						•			•							
	By abrasive held on or between solid backing.						•		•	•		•					
Macro-Rough	Abrasion by coarse particles, including carbide and other hard inclusions in the sliding materials.											•		•			
	Abrasion by fine particles in turbulent fluid, producing scallops, waves, etc.					•											
	Severe adhesion.										•				•		
	Local fatigue failure resulting in pits or depressions by repeated rolling contact stress, and high friction sliding or impact by hard particles as in erosion.			•	•											•	
Shiny	Advanced stages of micro-roughening, where little unaffected surface remains between pits.		•					•									
	Very thin—or no—surface film of oxide, hydroxide, sulfide, chloride, or other.							•	•					•		•	
Dull	Thick films resulting from aggressive environments, including high temperatures due to corrosion.	•						•									

Table 2-12
Debris and Material-Loss Mechanisms
(Modified from Ref. 19)

Wear Debris	Material-Loss Mechanisms*
Long, often curly, chips or strings	Abrasion—involves particles (of some acute angular shapes but mostly obtuse) that cause wear debris, some of which forms ahead of the abrasive particle (*cutting* mechanism) but most of which is material that has been plowed aside repeatedly by passing particles and that breaks off by *low-cycle fatigue*.
	Adhesion—a strong bond that develops between two surfaces (either between coatings and/or substrate materials), which, with relative motion, produces tractive stress that may be sufficient to deform materials to fracture. The mode of fracture will depend on the property of the material, involving various amounts of energy loss or ductility to fracture, i.e.:
Solid particles, often with cleavage surfaces	low energy and ductility *brittle fracture*
	or
Severely deformed solids, sometimes with oxide clumps mixed in	high energy and ductility *ductile fracture*
Solid particles, often with cleavage surfaces with ripple pattern	Fatigue—due to cyclic strains, usually at stress levels below the yield of the material, also called *high-cycle fatigue*.

* In italics.

Table 2-13
Material-Loss Mechanisms and Wear Resistance (Modified from Ref. 19)

Material-Loss Mechanisms	Appropriate Material Characteristics to Resist Wear	Precautions to be Observed when Selecting a Material*
Corrosion	Reduce corrosiveness of surrounding region; increase corrosion resistance of material by alloy addition or by selecting a soft, homogeneous material.	Soft materials tend to promote galling and seizure.
Cutting	Use material of high hardness, with very hard particles or inclusions such as carbides, nitrides, etc. and/or overlaid or coated with materials that are hard or that contain very hard particles.	All methods of increasing cutting resistance cause brittleness and lower fatigue resistance.
Ductile Fracture	Achieve high strength by any method other than by cold working or by heat treatments that produce internal cracks or large and poorly bonded intermetallic compounds.	
Brittle Fracture	Minimize tensile residual stress for cold temperature; ensure low-temperature brittle transition; temper all martensite; use deoxidized metal; avoid carbides; effect good bond between fillers and matrix to deflect cracks.	
Low-cycle Fatigue	Use homogeneous and high-strength materials that do not strain-soften; avoid overaged materials or two-phase systems with poor adhesion between filler and matrix.	Soft materials will not fail through brittleness and will not resist cutting.
High-cycle Fatigue	For steel and titanium, use stresses less than half the tensile strength; for other materials to be load-cycled < 10^8 times, allow stresses less than $1/4$ the tensile strength; avoid decarburization of surfaces; avoid tensile stresses or form-compressive residual stresses by carburizing or nitriding.	Calculation of stress should include the influence of tractive stress.

* Materials of high hardness or strength usually have decreased corrosion resistance, and all materials with multiple and carefully specified properties and structure are expensive.

Preventive Action Planning Avoids Corrosion Failure*

A review of many years of work on plant failures indicates a pattern of time/frequency occurrences that fits the classical bathtub curve (Figure 2-30). This means that in the history of a process unit there will be a large number of failures during the construction and start-up periods, which then diminishes to a smaller number of occurrences during the middle years. After a number of years when the plant becomes older the failure occurrences start to rise again.

Why do we have this pattern? What is going on that is affecting the failure occurrences? By looking at some of the underlying causes of the early and late failures, we can find ways to reduce the frequency of problems on the two ends of the curve.

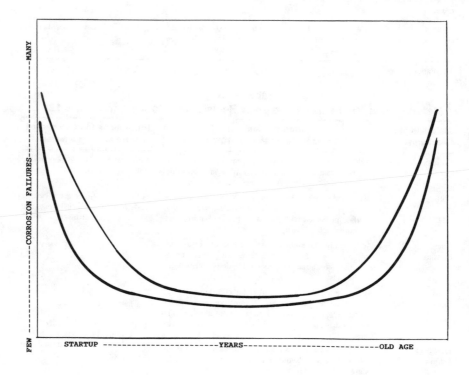

Figure 2-30. Typical process industry failure history.

---

*Contributed by William G. Ashbaugh, Ashbaugh Corrosion Consulting Services, The Woodlands, Texas.

This early period in the history of a process unit is most fraught with failures. We find that these problems can go back as far as the research and development (R&D) phase of the project, followed by engineering, construction, and start-up. Each phase of the project can produce a root cause that will result in a failure. We will examine some case histories of failures and from them see what actions could be taken to prevent such failures.

Corrosion Events Related to Research and Development

A batch process had been used to produce several specialty chemicals. R&D developed 12 new recipes for a new broadened line of products. These were turned over to production, which ran them successfully in the all-stainless-steel process equipment. After the second series of batches, tube leaks developed in a circulation heat exchanger. Failure analysis indicated chloride stress corrosion cracking from the process side (Figure 2-31). None of the 12 products was a chloride bearing chemical. However, it turned out that one of the recipes called for a pH adjustment using HCL.

This had not caused any problems at the R&D facility because the pilot size unit was all-glass-lined steel or high nickel alloys. Further inspection of the production unit showed stress corrosion cracking (SCC) of the batch reactor and associated piping, which was caused by the one recipe that used just a little HCl.

This start-up corrosion failure was caused by an unsuspected corrosive agent added to a previously innocuous chemical mixture. Neither R&D nor Operations was

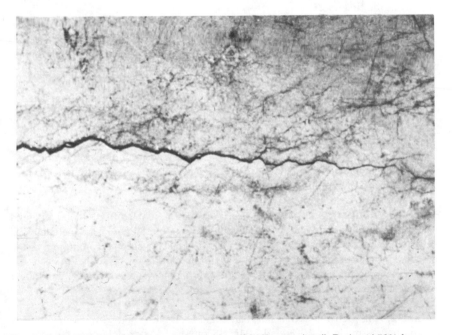

Figure 2-31. Chloride stress corrosion cracking of 316L vessel wall. Reduced 50% from actual size.

corrosion knowledgeable, and they did not have the new recipes checked by a corrosion specialist.

Preventive Action Plan. Inform a corrosion specialist of any project that will introduce chemical changes to an existing process. If the organization has a procedures manual, be sure that the corrosion engineer is involved at the correct times in the program procedures between R&D and production.

Events Related to Engineering

Overlooking details during the project engineering phase will often lead to early failures.

Case 1. As part of a new process, a reciprocating pump was specified to feed product to the reactor. The PI & D (piping instrumentation and design) diagram showed a pressure gauge and a drain valve on a side branch from the pump discharge. All the piping materials were 316 stainless steel and the small diameter piping was of screwed construction while the larger pipe was flanged.

Within several days of the start-up of the process, the pipe nozzle connecting the gauge and drain assembly broke off. The failure occurred on Saturday, and so maintenance personnel simply replaced the broken pipe nipple with a new one. On Tuesday the new pipe nipple broke off in the same manner as the first one.

A quick, on-the-spot examination indicated the two stainless pipe nipples failed due to fatigue cracking in the threads (Figure 2-32). Further discussion with the operators revealed that this pump caused the line to shake and jump with each stroke. This caused the heavy pipe assembly with associated valves and gauge to vibrate. All the stress was focused on the pipe nipple threads at the discharge line. Fatigue cracking of the low strength stainless components developed quickly.

Although such small piping arrangements are not uncommon around plants, the vigorous shaking of this piping was unusual. More consideration should have been given to clamping of the line and to minimizing overhang of branch connections. This design detail had never been put into the criteria for either piping or instrument engineering.

Preventive Action Plan. A detail design for small pipe takeoffs was prepared and placed in both the piping and instrumentation design guidelines. Also included was the recommendation to back weld the threads of the screwed pipe coming out of the line. As a result of this incident, a notice went out calling for personnel in each unit to inspect piping for vibration and threaded, overhanging connections. If found, they were to be corrected according to the new guidelines.

Case 2. A new chemical transport barge had been fabricated with several compartments made of 10% clad 304L SS on steel. After delivery of its first cargo of glacial acetic acid, the compartments were inspected. An observant

Figure 2-32. The threaded 304 stainless piping assembly broke off at the main line connection due to fatigue cracking caused by pipe vibrations.

inspector noticed a dark line at the top wall to roof fillet weld (Figure 2-33). Upon closer examination, he found he could insert a knife blade deep into the crevice.

Before attempting repair, a cross section was taken of the solid 304 SS deck and the tee joint made by the double clad steel. A portion of the cross section, shown in Figure 2-34, shows how corrosion of the steel base metal has occurred where the very thin cladding was undercut during welding. The fillet weld on the other side of the tee joint showed no undercutting or corrosion.

Considerable discussion followed concerning the advisability of using thin clad steel with this type of fabrication. It was concluded that with proper welding procedures, this design was an economical one for cargos not corrosive to the stainless steel.

Preventive Action Plan. Before starting repairs, a welding procedure was prepared and tested with the clad material. With a workable procedure, the welders were then trained and tested to make sure they knew how to make a good weld. They were qualified steel and stainless welders but needed to practice on this particular welding procedure. Welding of the thin cladding took more than normal care and concentration. A rigorous, daily inspection of the welding work was done and the job successfully completed.

The weld design, procedure, and welder training were written up and issued as special engineering fabrication criteria for thinly clad metals.

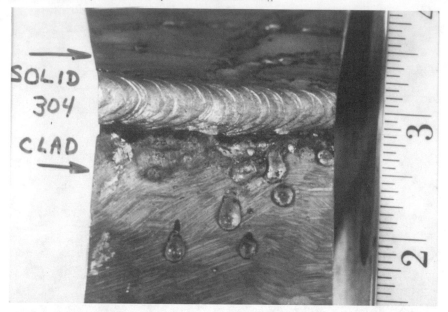

Figure 2-33. The fillet weld between the 304L roof/deck plate and clad 304L wall partition plate. The dark line at the base of the weld on the clad side is undercutting.

Figure 2-34. This is a side view of the sample shown in Figure 2-33. The steel below the undercut area is badly corroded.

Case 3. Within months after start-up, leaks occurred between the tubes and tube sheet of a heat exchanger. Corrosive chemicals on the tube side required 316SS; steam was on the shell side. Due to the process pressure, the tube sheet was several inches thick. As an economy move, a 316 clad steel tube sheet was used with the 316 tubes rolled into the steel and the 316 clad face. The result is shown in Figure 2-35, where the corrosive material has leaked past the stainless clad and corroded away the steel base metal.

Tubes that are expanded into tube sheets of hot exchangers will eventually lose their tightness and leak. Expecting the expanded tube end to hold a seal against the thin clad was a mistake. The tubes should have been seal welded to the stainless cladding.

Preventive Action Plan. A guide was developed for use of clad tube sheets. In any service that is corrosive to the base metal, the tubes must be seal-welded. Additional guidelines were developed for the seal-welding procedures and for testing the welds.

Corrosion Events Related to Construction

Case 1. Some of the most pervasive problems involving construction and fabrication activities are materials mix-ups. When a variety of alloys is used in a chemical process, industry experience shows that 2 or 3 percent of those materials will not be as specified. Fortunately, this is often not a problem

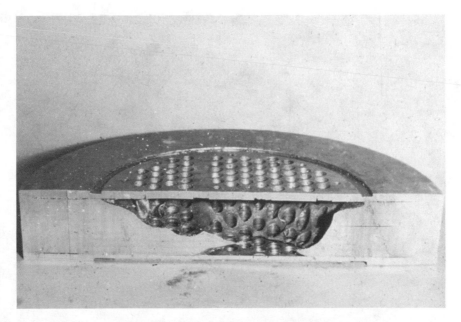

Figure 2-35. This 316 double clad steel tube sheet corroded when the 316 tubes rolled into the top clad plate leaked corrosive process chemicals to the underlying steel.

because prudent selection led to an upgrade in properties or corrosion resistance and the error thus goes unnoticed. However, where a weaker or less corrosion-resistant material is mixed in, early failure often results. Figures 2-36 and 2-37 show two examples in which substitution of 304SS resulted in corrosion failure. In the case of the pipe elbow, which was plainly stamped "type 304," the fabricator did not pay attention to the materials supplied to the welders. One might ask where the welder was, since he is normally expected to check his materials against the job sheet. In this case, the mix-up was not isolated because there were several incorrect fittings used.

The corroded thermowell is a more difficult case in that the parts are clearly stamped 316SS. Unfortunately, they are not 316 but are 304. When installed, one failed quickly in hot acetic acid service. In this instance, the machine shop producing the thermowells did not pay attention to their bar stock identification. The machinist pulled a bar out of the rack and assumed it to be 316SS.

One might blame these incidents on the fabricators because alloy materials leaving the producing mill are correctly identified and have both heat numbers and lab analysis for traceability. Problems occur when dealers or shops cut up the materials or separate them from their mill identity. Thoughtless individuals then identify them with numbers that may or may not be correct.

Another all too common mix-up in stainless steels is the inordinate use of regular carbon grades where either "L" or stabilized grades are specified.

Figure 2-36. A 304L SS elbow mistakenly welded into a 316L piping unit. The elbow was actually stenciled 304L.

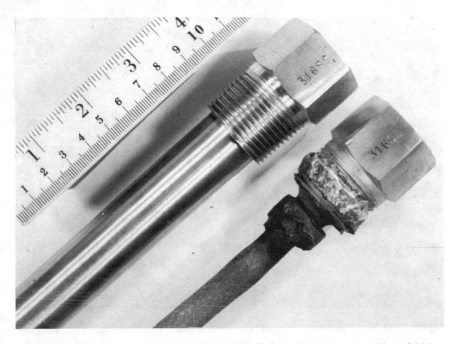

Figure 2-37. Before and after views of thermowells not of 316, as stamped, but of 304.

Figure 2-38 shows an extreme case of weld-heat-affected-zone corrosion where the "L" grade alloy was not used.

These problems are different from the widely publicized "counterfeit bolts" situation in which some manufacturers of bolts deliberately made substitutions or took shortcuts, or deliberately mislabeled their products.

Preventive Action Plan. Materials mix-ups are a complicated matter and cannot be covered in just a few sentences. Each project has its own materials quality control needs. In some cases, there are relatively few areas where materials quality control issues are critical while in others, quality control can be a major concern for process reliability and safety.

The necessary levels and types of quality control should be determined during the design phase and will have to be incorporated into all parts of the specifications, purchasing, fabrication, and construction activities. Mix-ups will occur unless deliberate and specific efforts are made to prevent them.

Case 2. Some of the most frustrating failures are those caused by seemingly mundane conditions. One example is the repeated pitting problems caused by microbiologically induced corrosion (MIC) occurring in hydrotest water. The failure shown in Figure 2-39 involves a 304L storage tank that was fabricated and hydrotested. The associated piping was not completed for another two months. The tank leaked badly when it was subsequently put into service.

Figure 2-38. Type 316 plate was used here instead of the specified 316L. This is the classic HAZ corrosion of sensitized stainless steel.

Figure 2-39. Examples of some of the 1,000 pits found in the bottom of a new 304L tank. Hydrotest water left in the tank resulted in this MIC attack.

Upon cleaning and internal inspection of the tank, over 1,000 pits were found in the bottom. The top and sides of the tank were unaffected.

It was established that untreated river water used for hydrotesting was drained, but the tank was not emptied or dried out. A thin layer of water and silt remained at the bottom and the MIC went to work.

Preventive Action Plan. MIC attack by stagnant hydrotest water is easily avoided by emptying and drying out equipment after hydrotesting. Clear and specific hydrotesting instructions should be prepared and included with engineering and construction contracts. Furthermore, some communication must also inform the tank builder about the criteria. It may also be necessary to communicate these requirements to plant maintenance and inspection departments who may be involved in the hydrotesting.

Case 3. A 25% sodium hydroxide piping system was fabricated of stress relieved A106 carbon steel pipe. The system developed several leaks within months of start-up.

Figure 2-40 shows a section of the failed pipe that has stress cracked at locations of welded-on tee bars. The pipe was steam traced and insulated and the tee bars were used to hold the insulated pipe away from the support structure.

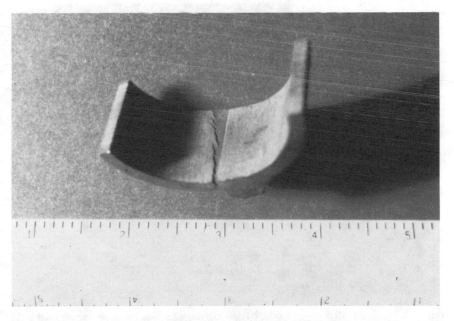

Figure 2-40. Stress corrosion cracking of this steel pipe in 25% NaOH service was caused by the non-stress-relieved weld joining the external support bar.

The tee bars were welded to the stress-relieved pipe by the insulation contractor. Nobody had told the contractor that his external welds needed to be stress relieved. The non-stress-relieved weld areas failed by stress corrosion cracking.

Preventive Action Plan. The stress relieving, of course, must be applied to any and all welding on vessels or piping that require stress relieving. This special fabrication instruction that is surely found in the vessel and piping standards of competent companies also should be put into the insulation and perhaps even the instrument standards. The instrument/electrical department, and perhaps other technology groups, must be informed about special needs and circumstances so they can pass them on to their subcontractors. All too often, subcontractors are left to do "their thing" outside the realm of detailed engineering specifications, and costly mistakes can result.

Case 4. One fabrication related failure was created by a situation that turned out to be difficult to prevent. A leak in a 316L stainless line was investigated and found to be the result of external stress corrosion cracking (ESCC). The line had been painted, according to specifications, to protect it from an atmosphere conducive to ESCC. The failed section shown in Figure 2-41 reveals the obvious problem.

Figure 2-41. This 316L process line failed due to external stress corrosion cracking under a fabricator sticker that was then painted over. Chlorides from the sticker or the atmosphere attacked the unprotected pipe.

Something on the pipe under the paint caused or allowed the ESCC to happen. Further investigation of other pipe sections showed them all to have a *Welders Union* sticker on the pipe; this had simply been painted over.

Other investigators have reported that certain sticker adhesives contain water hydrolyzable chlorides that could cause cracking whenever piping becomes hot and wet. In addition, process temperatures can cause stickers to wrinkle and come loose, thus forming crevices and traps for water and chlorides present in most atmospheres.

Preventive Action Plan. An interesting discussion was held to find a preventive solution to this situation. It would be impractical to ban stickers or markings on stainless steel products. Also, to require them to be chloride-free would be difficult to police when subcontractors are involved.

A good solution would be to clean stainless steel before painting in an effort to remove all foreign materials. The choice of cleaning method would depend upon the types of materials to be removed. Many engineers have opted to require light abrasive blasting of stainless steel to achieve both cleaning and better paint adhesion.

Case 5. Some failures are more serious than others. An example is this case involving the failure of a steam heated exchanger. Figure 2-42 shows that the longitudinal seam weld has failed catastrophically. While it was only 300

Figure 2-42. This steel exchanger shell ruptured in 300 psi steam service. The failure was along a vertical seam weld.

psig steam and no one was injured, the rupture was potentially hazardous, and the loss of the exchanger shut down the unit for several days.

A look at the failed weld quickly reveals the problem—a very poor weld with no root pass and poor fusion to the base metal (Figure 2-43). A review of the fabrication drawing indicated there was no non-destructive examination (NDE) of the weld.

Preventive Action Plan. The action needed to prevent such incidents is rather obvious—tight inspection of the vendors' work and qualifications. Large projects may include many pressure vessels and the cost of inspection may cause the inspection effort to be concentrated on the more critical and larger items.

If the owner's inspector is not able to inspect a vessel, the selection of a quality fabricator and clear specification requiring documentation of code compliance of workers as well as the vessel itself are required. Pressure vessels must be subjected to quality controls and appropriate documentation.

Events Related to Plant Start-up

While a certain number of problems are to be expected upon start-up of a process unit, corrosion-related problems should not be among them. The following two case histories illustrate how events can take place outside of the engineering and construction domain and cause early failures.

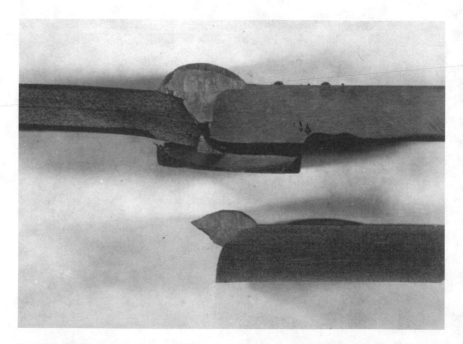

Figure 2-43. A cross section of the steam shell seam weld shows the gross lack of fusion and penetration that resulted in the failure.

Case 1. Liquid naphtha was fed through a 304SS vaporizer coil made of two-inch schedule 40 pipe. After only a few weeks, the stainless coil began to leak. One of the leaking return bends was cut in half and is shown in Figure 2-44.

The cracking was examined metallographically and found to be transgranular. A high chloride concentration was also found. This, of course, is chloride stress corrosion cracking, but there were no chlorides or water in the naphtha analysis.

The original purchase contract for naphtha was with a nearby refinery and specified delivery by pipeline. By the time the plant was built, a more favorable contract was developed for offshore naphtha to be delivered by barge. This changed the material selection concept. If it had been known that the feed stock was being delivered by barge, a high nickel alloy material would have been selected instead of the 304SS. Experience has taught that all too often barge operators, when washing down tanks, do not drain them totally dry. A little sea water in the bottom apparently does not make any difference to most users. But with the hot 304SS pipe evaporating the naphtha and then the water, even small droplets of sea water can cause rapid failure!

Preventive Action Plan. A situation in which a raw material contract is changed but the same specifications seemingly are left in force often creates a problem. We can only suggest that all groups involved in the project be educated to look for the potential for corrosion situations. A deeper level of communication should be put in

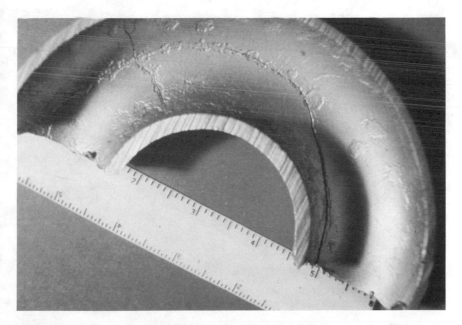

Figure 2-44. This 304L return bend in a naphtha vaporizer failed when sea water contaminated the naphtha. Failure is chloride SCC.

place within the project team, which should include a corrosion specialist, to discuss all facets of the project. Why would a corrosion engineer be interested in a new raw materials contract? Well, why not? Doesn't the case speak for itself?

Case 2. A particular process start-up was fraught with many mechanical problems, due in part to the difficulties inherent in using the local labor force in an emerging technology country. Once the new plant was running smoothly, it did not take long for a corrosion problem to develop.

A large tube and shell exchanger with steam on the shell side was used to heat the evaporator liquid in a metal salt crystallizer. The temperature and salt concentration were critical to the development of good salt crystals. These parameters were monitored and controlled carefully by the newly trained local operators. A third factor, less critical but necessary, was pH control with the acidity kept just on the acid side. The instructions were "never go alkaline," that is, never go above 7.

Leaks developed in the exchanger tubes within 60 days of start-up. A tube section is shown in Figure 2-45. In this case the as-welded one-and-one-half-inch tubes of stainless steel developed pitting in the weld metal. The parent metal, while lightly etched, showed no thinning.

A check of the operators' records of the pH control levels showed that they had not been diligent in taking and recording the pH as instructed. Large

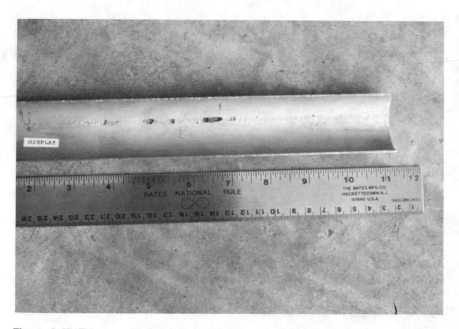

Figure 2-45. This as-welded stainless tube corroded when low pH conditions occurred in the process.

swings in the pH from 7 to 3 indicated their acid additions were too few and too large. The result was that for many hours the hot exchanger tubes were exposed to a pH of about 5. The tube as-welded grain structure was the first to be affected, with pitting starting in the cast structure.

The tube bundle was re-tubed with seamless 316L tubes and tight controls placed on the pH of the solution.

Preventive Action Plan. The operators were good operators but had no previous experience in a chemical plant. Their chemistry training was general and specific for the process. The operating instructions for running the crystallizer were specific for the high pH, but no low side pH was set since a low pH excursion would not upset the crystallization process. The people who prepared the operating instructions and trained the new operators knew how to add the acid to control the pH; however, they overlooked this simple step in training and assumed the operators learned it during the hands-on start-up.

Additional training sessions and more precise operating instructions were implemented. Supervisors were also instructed to keep a close watch on the logs and records to spot operating parameters about to go out of control.

Events Relating to Mature Plants

Many process units remain viable producers for more than 25 years. However, they usually have undergone a variety of changes over the years and also may simply be suffering from the ravages of time.

For these general reasons, corrosion-related problems often begin to show up at these later stages of life. Can these be prevented? In most cases, yes, although sometimes it's too late when the failure occurs.

Case 1. A long established process unit had a corrosion-erosion problem with a control valve in concentrated sulfuric acid. This high alloy stainless valve had been in service for many years and was due for replacement. It finally failed on a Friday night. On Saturday it was removed to be replaced by a spare. There was, however, no spare of this particular style and size. The unit maintenance chief looked at the old valve and remembered he had at one time worked in another unit that used the same valve in sulfuric acid. However, in that experience, the acid was more dilute and the valve was made of zirconium. He also remembered that there had always been a spare available. He borrowed the spare zirconium control valve body with the promise to return it as soon as the new stainless valve was delivered.

By Tuesday it was too late; the zirconium valve had all but dissolved. One half of the split body is shown in Figure 2-46. The concentrated sulfuric acid had corroded the valve to failure in two days. In 60–65% sulfuric acid, zirconium has almost nil corrosion rate. The corrosion break point is about 70%. In concentrated sulfuric acid, the corrosion rate is over 200 mils per year.

These corrosion properties were known to the people in the process area using zirconium components. The person that transferred to another area forgot the details and only remembered how well zirconium performed in sulfuric acid.

Figure 2-46. This zirconium control valve body was badly attacked by 92% H_2SO_4 in only two days.

Preventive Action Plan. This problem in some ways exemplifies things that can happen when plant personnel get moved around. The transfer of personnel between departments is inevitable and broadens the experience of the people involved. However, spur-of-the-moment materials substitutions should not be made without some cross-check or confirmation as to acceptability.

A plant procedure was implemented to preclude the unauthorized usage of zirconium, and a plant technical bulletin was issued explaining both the new procedure and the corrosion limitations of zirconium.

    Case 2. After many years of uneventful service, a 600 psig stainless line ruptured. The failure was found to be chloride stress corrosion cracking along the bottom on the inside of the 304L pipe. This, by the way, is an example that proves that catastrophic failures of stress cracked stainless steel do occur.

       The chlorides were traced back to leaking tubes in a seawater-cooled compressor aftercooler. These 90-10 copper-nickel tubes had never given any trouble before—why were they now a problem? It was noticed that the failure was inlet end pitting and that the zinc anodes in the water box were missing. Without the cathodic protection that had always been there, the tube ends failed.

       Unit procedures called for the zinc anodes to be replaced every two years at the unit turnaround. At the last shutdown, the storehouse had run out of zinc anodes and, under pressure of time, the decision was made to button up the water box without such protection. The contract maintenance people did not know the importance of zinc anodes, and no one had checked their work.

Preventive Action Plan. Work that had always been done by the plant maintenance crew was now being done by contract maintenance. The contractor did not know why this little task was important and, thus, did not make the right decision when a problem came up. Furthermore, there was no check with the plant contact to see if they should wait for a special delivery order. In addition, the plant inspector missed this exchanger and assumed it had been taken care of.

When organizations and people change, it is most important that clear task instructions are issued. No deviations should be permitted without the approval of the plant inspector or an assigned representative.

New guidelines for communication between contractor and plant people were established. Contractor supervision was instructed on how to use the guidelines and why guidelines were necessary. An inspection checklist of contractors' work was devised; it had to be signed off by both inspector and contractor.

Case 3. An extreme case of old age corrosion with serious consequences occurred in a plastics producing plant on the Texas Gulf Coast. The plant manager received a process technology bulletin that made note of stress cracking under insulation of a stainless vessel in a sister plant. A cursory dye penetrant check of a top head weld on one of the vessels showed several "indications."

A corrosion engineer specialist was called in to comment on the indications. They were confirmed to have the earmarks of external stress cracks. An inspection plan was developed for the vessel and inspections carried out the next day. At every inspection site, stress cracks were found. At one location, the crack blew bubbles when the dye was applied—the cracking had already penetrated the one-inch-thick 304L wall! (Figures 2-47 and 2-48).

Further insulation removal showed the stress cracking to be universal and that no part of the 60-foot-tall by 10-foot-diameter column could be salvaged.

Inspection of three more identical vessels and associated stainless piping in the same structure showed them to be in the same totally stress cracked condition. The associated piping, much of which operated at cooler temperatures, had limited cracking. This unit was then just 25 years old and producing in a very profitable market. It was located about one mile from a salt water estuary, but perhaps more importantly, it was just downwind of a large vinyl chloride monomer (VCM) plant. Over the years, small or large emissions of chlorine, ethylene dichloride, or vinyl chloride enriched the chloride atmosphere around the unit and penetrated the cellular glass insulation.

The polymer plant personnel had never inspected their stainless equipment under its insulation and were totally unaware of the corrosion cracking destroying the plant equipment.

Prevention of Failure. In this case there was not much left to prevent. However, as the new columns were installed and new piping fabricated, they were given a protective coating to prevent future chloride SCC. An annual inspection of the paint with

Figure 2-47. Catastrophic failure of this 304L pipe was by chloride stress cracking caused by sea water leaking into the gas stream.

Figure 2-48. A cross section of the 304L one-inch-thick vessel wall cracked through by external chloride stress corrosion cracking under insulation.

touch-ups as necessary was written into the maintenance procedures. Engineering criteria were prepared calling for painting under thermal insulation at the construction stage.

In the past decade, many inspectors have looked under insulation and to their dismay found corrosion of steel or stress cracking of stainless steel. These incidents have caused a sharp rise at the extended time end of the bathtub curve.

Summary

Prevention of failure is neither a simple task nor is it easily accomplished. Failure prevention must be the goal of everyone in the organization. Constant efforts to make it happen require the application of both new technology and lessons learned from failure analysis.

The examples we have used could all have been prevented by better ***communications, training, and documentation.*** These activities are necessary to help keep people from making mistakes. Failures are not the fault of the material but rather caused by decisions, actions, or omissions involving people.

It is thus important to communicate, train, and document on an ongoing basis. The input information must be constantly updated, and that is where failure analysis comes in. When failures happen and are analyzed, the root causes should be identified and steps taken to change the conditions that allowed an event to happen.

These ideas are not new and have been widely practiced in many fields. In order to affect the failure rate in process plants, information on corrosion and metallurgy must be fully incorporated into an organization. Those working in the fields of corrosion and materials engineering must be *proactive* so that *prevention* of failure will be achieved every step of the way in the production and manufacturing process.[20]

References

1. *Metals Handbook,* Vol. 10, "Failure Analysis and Prevention," published by the American Society for Metals, Metals Park, Ohio 44073, 8th Edition, 1975, pp. 10–26.
2. Hutchings, F. R., "The Laboratory Examinations of Service Failures," Technical Report, Vol. III, British Engine, Boiler, and Electrical Insurance Co., Ltd., London, U.K., May 1957.
3. Van der Voort, G. F., "Conducting the Failure Examination," *Met. Eng. Quart.,* May 1975.
4. *Identification of Metals and Alloys,* International Nickel Company, New York, N.Y.
5. "Symposium on Rapid Methods for Identification of Metals," ASTM Publication 98, American Society for Testing and Materials, Philadelphia, PA.
6. Wilson, M. L., *Nondestructive Rapid Identification of Metals and Alloys by Spot Test,"* Langley Research Center, Hampton, VA.
7. *Handbook of Loss Prevention,* Editors: Allianz Versicherungs-AG, translated from the German by P. Cahn-Speyer, 1978, Berlin-Heidelburg-New York: Springer-Verlag, p. 321.
8. Investigation of Threaded Fastener Structural Integrity, Final report, Contract No. NAS9-15140, DRL-T-1190, CLIN3, DRDMA 129TA, Southwest Research Institute, Project No. 15-4665, Oct. 1977.
9. Bowman Products Division, "Fastener Facts," Cleveland, Ohio, 1974.
10. *Dubbel, Taschenbuch für den Machinenbau,* Editors: W. Beitz & K. H. Küttner, Berlin-Heidelberg-New York: Springer Verlag, 14th Edition, 1981, pp. 382–384.
11. Bickford, H. J., *An Introduction to the Design and Behavior of Bolted Joints,* New York and Basel: Marcel Dekker, Inc., 1981, p. 79.
12. Wulpi, D. J., Understanding How Components Fail, American Society for Metals, Metals Park, Ohio 44073, 1985.
13. Eyre, T. S., "Wear Characteristics of Metals," *Tribology International,* October 1976, pp. 203–212.
14. Hydraulic Institute Standards, 13th Edition, Hydraulic Institute, Cleveland, Ohio, 1975, "Materials of Construction of Pumping Various Liquids," pp. 285–296.
15. Waterhouse, R. B., *Fretting Corrosion,* Pergamon Press, 1972.
16. Braunovic, M., "Tribology: Fretting Wear,"*Engineering Digest,* Toronto, October 1981, pp. 48–54.
17. Klein, Schanzlin, and Becker, A. G., *Centrifugal Pump Lexicon,* Frankenthal, Germany, 1975.
18. Hennings, Cr., "Ermittlung des Schädigungszustandes an Zahnflanken," *Schmierungstechnik* 10 (1979) 10, pp. 307–310.
19. Ludema, K. E., "How to Select Material Properties for Wear Resistance,"*Chemical Engineering,* July 27, 1981, pp. 89–92.
20. Ashbaugh, William G., "Preventive Action Planning Avoids Corrosion Failures in the HPI," Proceedings of the Fourth International Conference on Process Plant Reliability, Houston, Texas, 1995.

Chapter 3

Machinery Component Failure Analysis

Bearings in Distress

In the preceding pages, we looked at a variety of wear-related failures. We can readily conclude that a large portion of lubricated machinery components will experience wear damage through lubrication failure. Bearings are no exception. Lubrication-related bearing problems, according to our experience, are most frequently caused by *lack* of lubrication or lubricant *contamination*. Two examples will illustrate this.

As the first example, analysis of a pump bearing failure showed that a rolling-element bearing failed from lack of oil. However, there was a copious supply of *fresh* oil in the housing. It is not too difficult to conclude what really happened. The positive thinker will not despair but will take heart in knowing that the operating crew still owns up to their duty of keeping oil in the bearing. Perhaps somewhat late in *this* case—but better late that never as far as the rest of the many pumps in their care is concerned.

The second example: A certain engineer once described the person with the water hose as the rolling-element-bearing salesman's best friend in the paper mill.[1] No doubt, the same can sometimes be said in the chemical process industry. Nevertheless, given a proper lubrication environment and proper design service conditions, plain journal bearings will last indefinitely. However, rolling-element bearings, even under optimum conditions, will fail in wear-out mode.[2] In other words, the ultimate end of all rolling-element bearings would be fatigue if it were possible to run them long enough. In any group of rolling-element bearings, if all were run under identical conditions until all failed by fatigue, there would be a considerable spread between the longest and the shortest "lives." Figure 3-1 illustrates this fact in the form of a life dispersion curve. In short and simple terms, given proper design and application, rolling-element bearings will sooner or later fail in wear-out mode, plain bearings will last almost indefinitely, but all will fail early from abuse.

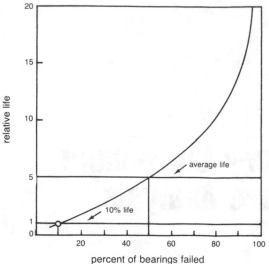

Figure 3-1. Life dispersion curve for typical group of bearings run under identical conditions.[2]

Table 3-1
Failure Causes of Rolling and Plain Bearings[3]

Failure Cause	Occurrence, %	
	Rolling Bearings	Plain Bearings
Vendor problems	30.1	23.4
Workmanship	14.4	10.7
Errors in design/applications	13.8	9.1
Wrong material of construction	1.9	3.6
User-induced problems	65.9	69.6
Operational errors, maintenance deficiencies, failure of monitoring equipment	37.4	39.1
Wear	28.5	30.5
External problems	4.0	7.0
Contaminated lubricants; intermittent failure of oil supply system	4.0	7.0

Typical bearing-failure causes have been compiled by Allianz Insurance of Munich, Germany.[3] Table 3-1 contains the results of investigations of the causes of 1400 rolling-bearing failures and 530 plain-bearing failures. It can be seen that for these components, about 30% of the functional failures are due to wear- or lubrication-related processes.

Based on our experience, we find these statistics somewhat controversial as far as the ratio of wear-related failures to other user-induced problems is concerned. Also, only a very sophisticated analysis strategy could come up with a similar separation of

Table 3-2
Antifriction Bearing Failure Modes and Their Causes

Category	Failure Causes	Spalling/flaking	Cracks/heat cracks	Smearing → seizing	Cage fracture	Cage deformation	Indentations	Fragment denting	Brinelling/false brinelling	Ball Path—widened	Ball Path—skew	Ball Path—uneven load zones	Fluting	Cage wear	Abrasive wear/wear	Overheating → burning → scuffing	Corrosion/etching	Fit Corrosion/fretting	Rust staining
		Fracture/Separation				**Deformation**								**Wear**			**Corrosion**		
Assembly	Excessive, uneven heat application	•																	
Assembly	Hammer blows		•				•	•											
Assembly	Improper tooling					•	•	•											
Assembly	Loose/tight fits		•									•			•			•	
Assembly	Distorted bearing housing	•							•	•									
Assembly	Rotor unbalance											•							
Assembly	Misalignment			•	•						•	•		•					
Operating Conditions	Vibration — Moving / Stationary	•							•	•					•			•	
Operating Conditions	Current Passage														•		•		
Operating Conditions	Life attainment/fatigue	•																	
Operating Conditions	Overload	•	•							•						•			
Operating Conditions	Design error			•	•											•			
Seal	Contamination						•	•						•	•		•		
Seal	Moisture ingress																	•	•
Lubrication	Lack of lubricant			•	•									•	•	•			•
Lubrication	Excess of lubricant			•														•	
Lubrication	Unsuitable lubricant			•							•			•	•	•			•

user-induced problems and external problems. Nevertheless, similar data should be compiled for your own plant, because knowledge of failure-cause distributions at the component and parts levels will eventually result in better specifications.

Tables 3-2 and 3-3 will help relate both rolling-element and plain-bearing defects or failures modes to their most probable causes.

Rolling-Element Bearing Failures and Their Causes*

Accurate and complete knowledge of the causes responsible for the breakdown of a machine is as necessary to the engineer as similar knowledge concerning a breakdown in health is to the physician. The physician cannot effect a lasting cure

---

*Copyright © SKF Industries, King of Prussia, PA. Reprinted by permission.

Table 3-3
Plain (Journal) Bearing Failure Modes and Their Causes

Group	Failure Causes	Spalling	Cracking	Seizing	Deformation	Embedments	Uneven load patterns	Scoring → galling	Overheating → scuffing	Wear	Abrasive wear/scratching/grooving	Erosion	Cavitation	Corrosion
Assembly and Manufacture	Insufficient clearance			•						•				
	Misaligned journal bearing				•		•							
	Rough surface finish on journal							•	•					
	Pores and Cavities in bearing metal		•											
	Insufficient metal bond	•												
Operating Conditions and Design	General operating conditions											•	•	
	Overload/fatigue	•	•										•	
	Overload/vibration	•	•	•					•	•	•		•	
	Current passage												•	•
	Unsuitable bearing material			•						•				
Lubrication	Contamination of lubricant			•		•		•		•	•			•
	Insufficient or lack of lubricant			•				•	•	•	•			
	Oil viscosity too low									•				
	Oil viscosity too high									•				
	Improper lubricant selection							•						
	Lubricant deterioration									•				•

unless he knows what lies at the root of the trouble, and the future usefulness of a machine often depends on correct understanding of the causes of failure. Since the bearings of a machine are among its most vital components, the ability to draw the correct inferences from bearing failures is of utmost importance.

In designing the bearing mounting the first step is to decide which type and size of bearing shall be used. The choice is generally based on a certain desired life for the bearing. The next step is to design the application with allowance for the prevailing service conditions. Unfortunately, too many of the ball and roller bearings installed never attain their calculated life expectancy because of something done or left undone in handling, installation, and maintenance.

The calculated life expectancy of any bearing is based on the assumptions that good lubrication in proper quantity will always be available to that bearing, that the bearing will be mounted without damage, that dimensions of parts related to the bearing will be correct, and that there are no defects inherent in the bearing.

However, even when properly applied and maintained the bearing will still be subjected to one cause of failure: fatigue of the bearing material. Fatigue is the result of shear stresses cyclically applied immediately below the load-carrying surfaces and is observed as spalling away of surface metal, as seen in Figures 3-2 through 3-4. However, material fatigue is not the only cause of spalling. There are causes of premature spalling. So, although the observer can identify spalling, he must be able to discern between spalling produced at the normal end of a bearing's useful life and that triggered by causes found in the three major classifications of premature spalling: lubrication, mechanical damage, and material defects. Most bearing failures can be attributed to one or more of the following causes:

1. Defective bearing seats on shafts and in housings.
2. Misalignment.
3. Faulty mounting practice.
4. Incorrect shaft and housing fit.
5. Inadequate lubrication.

Figure 3-2. Incipient fatigue spalling.

Figure 3-3. More-advanced spalling.

Figure 3-4. Greatly advanced spalling.

6. Ineffective sealing.
7. Vibration while the bearing is not rotating.
8. Passage of electric current through the bearing.

The actual beginning of spalling is invisible because the origin is usually below the surface. The first visible sign is a small crack, and this too is usually indiscernible. The crack cannot be seen nor its effects heard while the machine operates. Figures 3-2 through 3-4 illustrate the progression of spalling. The spot on the inner ring in Figure 3-2 will gradually spread to the condition seen in the ring of Figure 3-4, where spalling extends around the ring. Figure 3-2 illustrates incipient fatigue spalling. Figure 3-3 shows more advanced spalling, with still further degrees of deterioration shown in Figure 3-4. By the time spalling reaches proportions shown in Figure 3-3, the condition should announce itself by noise. If the surrounding noise level is too great, a bearing's condition can be learned by listening to the bearing through its housing by means of a metal rod. The time between incipient and advanced spalling varies with speed and load, but it is not a sudden condition that will cause destructive failure within a matter of hours. Total destruction of the bearing and consequent damage to machine parts is usually avoided because of the noise the bearing will produce and the erratic performance of the shaft carried by the bearing.

Patterns of Load Paths
and Their Meaning in Bearing Damage

There are many ways bearings can be damaged before and during mounting and in service. The pattern or load zone produced on the internal surfaces of the bearing by the action of the applied load and by the rolling elements is a clue to the cause of failure. To benefit from a study of load zones, one must be able to differentiate between normal and abnormal patterns. Figure 3-5 illustrates how an applied load of constant direction is distributed among the rolling elements of a bearing. The large arrow indicates the applied load, and the series of small arrows show the share of this load that is supported by each ball or roller in the bearing. The rotating ring will have a continuous 360° zone, while the stationary ring will show a pattern of less than 180°. Figure 3-6 illustrates the load zone found inside a ball bearing when the inner ring rotates and the load has a constant direction. Figure 3-7 illustrates the load zone resulting if the outer zone rotates relative to a load of constant direction, or when the inner ring rotates and the load also rotates in phase with the shaft. Figure 3-8 illustrates the pattern found in a deep-groove ball bearing carrying an axial load, and Figure 3-9 shows the pattern from excessive axial load. Uniformly applied axial load and overload are the two conditions where the load paths are the full 360° of both rings. Combined thrust and radial load will produce a pattern somewhere between the two, as shown in Figure 3-10. With combined load, the loaded area of the inner ring is slightly off-center, and the length in the outer is greater than that produced by

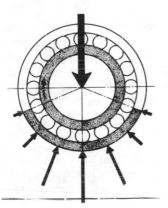

Figure 3-5. Load distribution within a bearing.

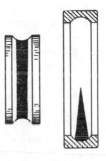

Figure 3-6. Normal-load zone, inner ring rotating relative to load.

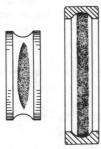

Figure 3-7. Normal-load zone, outer ring rotating relative to load, or load rotating in phase with inner ring.

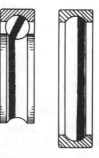

Figure 3-8. Normal-load zone, axial load.

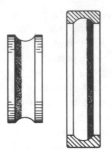 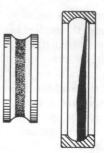

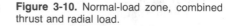

Figure 3-9. Load zone when thrust loads are excessive.

Figure 3-10. Normal-load zone, combined thrust and radial load.

radial load, but not necessarily 360°. In a two-row bearing, a combined load will produce zones of unequal length in the two rows of rolling elements. If the thrust is of sufficient magnitude, one row of rolling elements can be completely unloaded.

When an interference fit is required, it must be sufficient to prevent the inner ring from slipping on the shaft but not so great as to remove the internal clearance of the bearing. There are standards defining just what this fit should be for any application and bearing type, and a discussion of fitting practice appears later in this chapter, under "Damage Due to Improper Fits." If the fit is too tight, the bearing can be internally preloaded by squeezing the rolling elements between the two rings. In this case, the load zones observed in the bearing indicate that this is not a normal-life failure, as Figure 3-11 shows. Both rings are loaded through 360°, but the pattern will usually be wider in the stationary ring, where the applied load is superimposed on the internal preload.

Distorted or out-of-round housing bores can radially pinch an outer ring. Figure 3-12 illustrates the load zone found in a bearing where the housing bore was initially out-of-round or became out-of-round from the housing being bolted to a concave or convex surface. In this case, the outer ring will show two or more load zones, depending on the type of distortion. This is actually a form of internal preload. Figure 3-13 is a photograph of a bearing that had been mounted in an out-of-round housing that pinched the stationary outer ring. This is a mirror view and shows both sides of the outer ring raceway.

Certain types of rolling bearings can tolerate only very limited amounts of misalignment. A deep-groove ball bearing, when misaligned, will produce load zones not parallel to the ball groove on one or both rings, depending on which ring is misaligned. Figure 3-14 illustrates the load zone when the outer ring is misaligned relative to the shaft. When the inner ring is misaligned relative to the housing, the pattern is as shown in Figure 3-15. Cylindrical roller bearings, tapered roller bearings, and angular-contact ball bearings are also sensitive to misalignments, but it is more difficult to detect this condition from the load zones. Against the foregoing background of failure patterns the following failure descriptions should be meaningful.

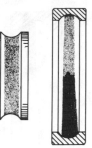

Figure 3-11. Load zone from internally pre-loaded bearing supporting radial load.

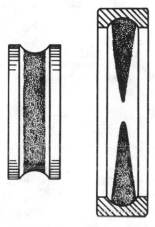

Figure 3-12. Load zones produced by an out-of-round housing pinching the bearing outer ring.

Figure 3-13. Mirror view of outer ring distorted by housing.

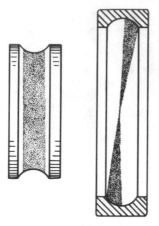

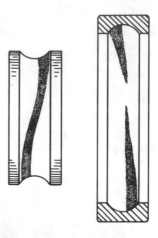

Figure 3-14. Load zone produced when outer ring is misaligned to shaft.

Figure 3-15. Load zones when inner ring is misaligned relative to housing.

Failure Due to Defective Bearing Seats on Shafts and in Housings

The calculated life expectancy of a rolling bearing presupposes that its comparatively thin rings will be fitted on shafts or in housings that are as geometrically true as modern machine-shop technique can produce. There are, unfortunately, factors that produce shaft seats and housing bores that are oversize or undersize, tapered or oval. Figure 3-16 shows another mirror view of a spherical roller bearing outer ring that has been seated in an out-of-round housing. Notice how the widest portions of the roller paths are directly opposite each other. The same condition can be produced by seating the bearing in a housing with a correctly made bore but where the housing is distorted when it is secured to the machine frame. An example would be a pillow block bolted to a pedestal that is not plane.

Figure 3-17 shows the condition resulting when a bearing outer ring is not fully supported. The impression made on the bearing OD by a turning chip left in the housing when the bearing was installed is seen in the left-hand view. This outer ring was subsequently supported by the chip alone, with the result that the entire load was borne by a small portion of the roller path. The heavy specific load imposed on that part of the ring immediately over the turning chip produced the premature spalling seen in the right-hand illustration. On either side of the spalled area is a condition called fragment denting, which occurs when fragments from the flaked surface are trapped between the rollers and the raceway.

When the contact between a bearing and its seat is not intimate, relative movement results. Small movements between the bearing and its seat produce a condition called fretting corrosion. Figure 3-18 illustrates a bearing outer ring that has been subjected to fretting corrosion. Figure 3-19 illustrates an advanced state of this condition. Fretting started the crack, which in turn triggered the spalling.

Figure 3-16. Mirror view of spherical roller-bearing outer ring pinched by housing.

Figure 3-17. Fatigue from chip in housing bore.

Fretting corrosion can also be found in applications where machining of the seats is accurate, but where on account of service conditions the seats deform under load. Railroad journal boxes are an example of this condition. Experience shows that this type of fretting corrosion on the outer ring does not as a rule detrimentally affect the life of the bearing. Shaft seats or journals as well as housing bores can yield and produce fretting corrosion. Figure 3-20 illustrates damage by movement on a shaft. The fretting corrosion covers a large portion of the surface on both the inner ring and the journal. The axial crack through the inner ring started from surface damage caused by the fretting.

Figure 3-18. Wear due to fretting corrosion.

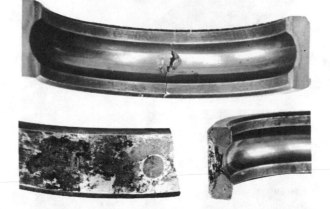

Figure 3-19. Advanced wear and cracking due to fretting corrosion.

Figure 3-20. Fretting caused by yield in the shaft journal.

Figure 3-21. Crack caused by faulty housing fit.

Bearing damage is also caused by bearing seats that are concave, convex, or tapered. On such a seat, a bearing ring cannot make contact throughout its width. The ring therefore deflects under loads, and fatigue cracks commonly appear axially along the raceway. Cracks caused by faulty contact between a ring and its housing are seen in Figure 3-21.

Misalignment

Misalignment is a common source of premature spalling. Misalignment occurs when an inner ring is seated against a shaft shoulder that is not square with the journal, or when a housing shoulder is out-of-square with the housing bore. Misalignment arises when two housings are not on the same center line. A bearing ring can be misaligned even though it is mounted on a tight fit if it is not pressed against its shoulder and thus left cocked on its seat. Bearing outer rings in slip-fitted housings risk being cocked across their opposite corners.

Some of the foregoing misalignment faults are not cured by using self-aligning bearings. When the inner ring of a self-aligning bearing is not square with its shaft seat, the inner ring wobbles as it rotates. This results in smearing and early fatigue. If an outer ring is cocked in its housing across corners, a normally floating outer ring can become axially held as well as radially pinched in its housing. The effect of a pinched outer ring was shown in Figures 3-13 and 3-16.

Ball thrust bearings suffer early fatigue when mounted on supports not perpendicular to the shaft axis, because one short load zone of the stationary ring carries all of the load. When the rotating ring of the ball thrust bearing is mounted on an out-of-square shaft shoulder, the ring wobbles as it rotates. The wobbling ring loads only a small portion of the stationary ring and causes early fatigue.

Figure 3-22. Smearing in ball thrust bearing.

Figure 3-22 illustrates the smearing within a ball thrust bearing when either one of two conditions occurs: first, if the two rings are not parallel to each other during operation; and second, if the load is insufficient at the operating speed to hold the bearing in its designed operational position. Under the first condition, the smearing seen in Figure 3-22 occurs when the balls pass from the loaded into the unloaded zone. Under the latter condition, even if the rings are parallel to each other, centrifugal force causes each ball to spin instead of roll upon contact with the raceway, resulting in smearing. Smearing from misalignment will be localized in one zone of the stationary ring, whereas smearing from gyral forces will be general around both rings.

Where two housings supporting the same shaft do not have a common centerline, only self-aligning ball or roller bearings will be able to function without inducing bending moments. Cylindrical and tapered roller bearings, although crowned, can accommodate only very small misalignments. If misalignment is appreciable, edgeloading results, and this is a source of premature fatigue. Edgeloading from misalignment was responsible for spalling in the bearing ring seen in Figure 3-23. Advanced spalling due to the same cause can be seen on the inner ring and roller of the tapered roller bearing in Figure 3-24.

Faulty Mounting Practice

The origin of premature fatigue, or of any failure, lies too many times in abuse and neglect before and during mounting. Prominent among the causes of early fatigue is the presence of foreign matter in the bearing and in its housing during operation. We have already seen the effect of trapping a chip between the OD of the bearing and the bore of the housing, as shown in Figue 3-17. Figure 3-25 shows the inner ring of a bearing where foreign matter has been trapped between the raceway and the rollers, causing brinelled depressions. This condition is called fragment denting. Each of these small dents is the potential start of premature fatigue. Foreign matter of small particle size results in wear, and when the original internal geometry is changed, the calculated life expectancy cannot be achieved. This is illustrated later in this chapter under "Ineffective Sealing" (Figures 3-48, 3-49, and 3-50).

Figure 3-23. Fatigue caused by edge-loading.

Figure 3-24. Advanced spalling caused by edge-loading.

Figure 3-25. Fragment denting.

Figure 3-26. Fatigue caused by impact damage during handling or mounting.

Impact damage during handling or mounting results in brinelled depressions that become the start of premature fatigue. An example of this is shown in Figure 3-26, where the spacing of flaked areas corresponds to the distance between the balls. The bearing has obviously suffered impact, and if it is installed, the fault should be apparent by the noise or vibration during operation.

Cylindrical roller bearings are easily damaged in mounting, especially when the rotating part with the inner ring mounted on it is assembled into a stationary part having its outer ring and roller set assembled. Figure 3-27 shows the inner ring of a cylindrical roller bearing that has been damaged because the rollers had to slide forceably across the inner ring during assembly. Here again the spacing of the damage marks on the inner ring is the same as the distance between rollers. The smeared streak in Figure 3-27 is shown enlarged eight times in Figure 3-28.

If a bearing is subjected to loads greater than those calculated for life expectancy, premature fatigue results. Unanticipated or parasitic loads can arise from faulty mounting practice. One example of a parasitic load is found in mounting the front wheel of an automobile and not backing off the locknut after applying the specified torque to seat the bearing. Another example is a bearing that should be free in its housing, but because of pinching or cocking cannot move with thermal expansions, inducing a parasitic thrust load on the bearing. Figure 3-29 shows the effect of a

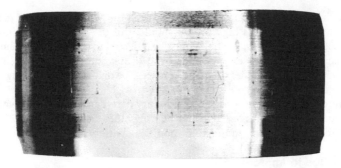

Figure 3-27. Smearing caused by excessive force in mounting.

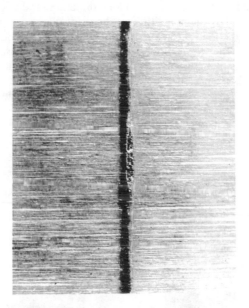

Figure 3-28. Smearing, enlarged eight times from Figure 3-27.

Figure 3-29. Spalling from excessive thrust.

parasitic thrust load. The damaged area is not in the center of the ball groove, as it should be normally, but is high on the shoulder of the groove. The ring shown in Figure 3-30 is of a self-aligning ball bearing subjected to an abnormally heavy thrust load. Usually, in such cases, evidence of axial restraint will appear either as the imprint of a housing shoulder on the outer ring face or as areas of fretting on the OD of the bearing.

Interference between rotating and stationary parts can result in destructive cracks in the rotating bearing ring. The roller bearing inner ring in Figure 3-31 shows the effect of contact with an end cover while the bearing ring rotated.

Damage Due to Improper Fits

To decide whether a bearing ring, either inner or outer, should be mounted with an interference or a slip fit on its shaft or in its housing, determine whether the ring rotates or is stationary with reference to load direction. The degree of tightness or looseness is governed by the magnitude of the load and the speed. If a bearing ring rotates relative to the load direction, an interference fit is required. If the ring is stationary with reference to the load, it is fitted with some clearance and is called a slip fit. The degree of fit is governed by the concept that heavier loads require greater interference. The presence of shock or continuous vibration calls for a heavier interference fit of the ring that rotates relative to the load. Lightly loaded rings, or even those rings with considerable load but operating at extremely slow speeds, that rotate relative to the load, may use a lighter fit or, in some cases, a slip fit.

Figure 3-30. Spalling from parasitic thrust.

Figure 3-31. Cracks caused by contact with the end cover while the bearing ring rotated.

Figure 3-32. Scoring of inner ring caused by "creep."

Consider two examples. In an automobile front wheel, the direction of the load is constant, i.e. the pavement is always exerting an upward force on the wheel. In this case the outer rings or cups are rotating and are press-fitted into the wheel hub, while the inner rings or cones are stationary and are slip-fitted on the spindle. On the other hand, the bearings of a conventional gear drive have their outer rings stationary relative to the load and so are slip fitted, but the inner rings rotate relative to the load and are mounted with an interference fit. There are some cases where it is necessary to mount both inner and outer rings of a bearing with interference fits because of a combination of stationary and rotating loads or loads of indeterminate characteristics. Such cases require bearings that can allow axial expansion at the rollers rather than at a slip-fitted ring. Such a mounting would consist of a cylindrical roller bearing at one end of the shaft and a self-contained bearing (deep-groove ball or spherical roller bearing) at the other end.

Some examples of the effect of departure from good fitting practice follow. Figure 3-32 shows the bore surface of an inner ring that has been damaged by relative movement between it and its shaft while rotating under a constant direction load. This relative movement, called creep, can result in the scoring shown in Figure 3-32.

If lubricant can penetrate the loose fit, the bore as well as the shaft seat will appear brilliantly polished (see Figure 3-33). When a normally tight-fitted inner ring does creep, the damage is not confined to the bore surface but can have its effect on the faces of the ring. Contact with shaft shoulders or spacers can result in either wear or severe rubbing cracks (Figure 3-31), depending on the lubrication condition. Figure 3-32, also shows how a shaft shoulder wore into the face of a bearing inner ring when relative movement occurred. Wear between a press-fitted ring and its seat is an accumulative damage. The initial wear accelerates the creep, which in turn produces more wear. The ring loses adequate support and develops cracks, and the products of wear become foreign matter to fragment dent and internally wear the bearing.

Excessive fits also result in bearing damage by internally preloading the bearing, as seen in Figure 3-11, or by inducing dangerously high hoop stresses in the inner ring. Figure 3-34 illustrates an inner ring that cracked because of excessive interference fit.

Housing fits that are unnecessarily loose allow the outer ring to fret, creep, or even spin. Examples of fretting were seen in Figures 3-18 and 3-19. Lack of support to the outer ring results from excessive looseness as well as from faulty housing-bore contact. A cracked outer ring was shown in Figure 3-19.

Inadequate or Unsuitable Lubrication

Any load-carrying contact, even when rolling, requires the presence of lubricants for reliable operations. The curvature of the contact areas between rolling element and raceway in normal operation results in minute amounts of sliding motion in addition to the rolling. Also, the cage must be carried on either of the rolling elements or on some surface of the bearing rings, or on a combination of these. In most types of roller bearings, there are roller end faces which slide against a flange or a cage. For these reasons, adequate lubrication is even more important at all times. The term "lubrication failure" is too often taken to imply that there was no oil or

Figure 3-33. Wear due to "creep."

Figure 3-34. Axial cracks caused by an excessive interference fit.

grease in the bearing. While this happens occasionally, the failure analysis is normally not that simple. In many cases, a study of the bearing leaves no doubt in the examiner's mind that lubrication failed, but why the lubricant failed to prevent damage to the bearing is not obvious.

When searching for the reason why the lubricant did not perform, one must consider first its properties; second, the quantity applied to the bearing; and third, the operating conditions. These three concepts comprise the adequacy of lubrication. If any one concept does not meet requirements, the bearing can be said to have failed from inadequate lubrication.

Viscosity of the oil—either as oil itself or as the oil in grease—is the primary characteristic of adequate lubrication. The nature of the soap base of a grease and its consistency, together with viscosity of the oil, are the main quality points when considering a grease. For the bearing itself, the quantity of lubricant required at any one time is usually rather small, but sufficient quantity must constantly be available. If the lubricant is also a heat-removal medium, a larger quantity is required. An insufficient quantity of lubricant at medium to high speeds induces a temperature rise and usually a whistling sound. An excessive amount of lubricant produces a sharp temperature rise because of churning in all but exceptionally-slow-speed bearings. Conditions inducing abnormally high temperatures can render a normally adequate lubricant inadequate.

When lubrication is inadequate, surface damage will result. This damage will progress rapidly to failures that are often difficult to differentiate from primary fatigue failure. Spalling will occur and often destroy evidence of inadequate lubrication. However, if caught soon enough, one can find indications that will pinpoint the real cause of the short bearing life.

One form of surface damage is shown in stages in Figure 3-35, a, b, c, and d. The first visible indication of trouble is usually a fine roughening or waviness on the surface. Later, fine cracks develop, followed by spalling. If there is insufficient heat removal, the temperature may rise high enough to cause discoloration and softening of the hardened bearing steel. This happened to the bearing shown in Figure 3-36. In some cases, inadequate lubrication initially manifests itself as a highly glazed or glossy surface which, as damage progresses, takes on a frosty appearance and eventually spalls. The highly glazed surface is seen on the roller of Figure 3-37.

(a)   (b)   (c)   (d)

Figure 3-35. Progressive stages of spalling caused by inadequate lubrication.

Figure 3-36. Discoloration and softening of metal caused by inadequate lubrication and excessive heat.

Figure 3-37. Glazing caused by inadequate lubrication.

In the "frosty" stage, it is sometimes possible to feel the "nap" of fine slivers of metal pulled from the bearing raceway by the rolling element. The frosted area will feel smooth in one direction but have a distinct roughness in the other. As metal is pulled from the surface, pits appear, and so frosting advances to pulling. An example of pulling is seen in Figure 3-38.

Another form of surface damage is called smearing. It appears when two surfaces slide and the lubricant cannot prevent adhesion of the surfaces. Minute pieces of one surface are torn away and rewelded to either surface. Examples are shown in Figures 3-39 through 3-42. A peculiar type of smearing occurs when rolling elements slide as

Figure 3-38. Effect of rollers pulling metal from the bearing raceway.

Figure 3-39. Smearing on spherical roller end.

Figure 3-40. Smearing on cylindrical rollers caused by ineffective lubrication.

Figure 3-41. Smearing on cage pockets caused by ineffective lubrication.

Figure 3-42. Smearing on cylindrical outer raceway.

they pass from the unloaded to the loaded zone. Figure 3-43 illustrates the patches of skid smearing, one in each row. A lubricant that is too stiff also causes this type of damage. This is particularly likely to happen if the bearing is large.

Wear of the bearing as a whole also results from inadequate lubrication. The areas subject to sliding friction, such as locating flanges and the ends of rollers in a roller bearing, are the first parts to be affected. Figures 3-44 and 3-45 illustrate the damage done and the extent of wear.

Where high speeds are involved, inertial forces become important and the best lubrication is demanded. Figure 3-46 shows an advanced case of damage from high speed with inadequate lubrication. Inertia forces acting on the rolling elements at high speed and with sudden starting or stopping can result in high forces between rolling elements and the cage.

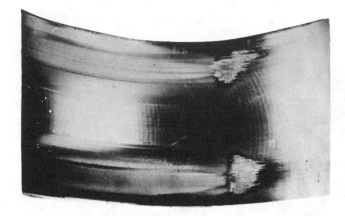

Figure 3-43. Skid smearing on spherical outer raceway.

Figure 3-44. Grooves caused by wear due to inadequate lubrication.

Figure 3-45. Grooves caused by wear due to inadequate lubrication.

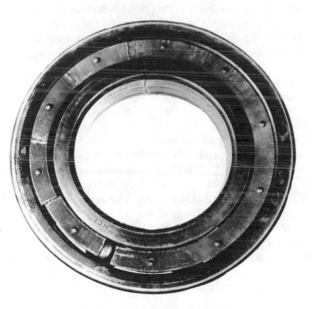

Figure 3-46. Broken cage caused by ineffective lubrication.

Figure 3-47 shows a large-bore tapered-roller bearing failure due to an insufficient amount of lubricant resulting from too low a flowrate in a circulating oil system. The area between the guide flange and the large end of the roller is subject to sliding motion. This area is more difficult to lubricate than those areas of rolling motion, accounting for the discoloration starting at the flange contact area. The heat generated at the flange caused the discoloration of the bearing and resulted in some of the rollers being welded to the guide flange.

Figure 3-47. Rollers welded to rib because of ineffective lubrication.

The foregoing defines the principal lubrication-related surface failures. Importance has been assigned to lubrication adequacy and those factors attacking it. Implied, then, is the need to know how to avoid surface failures. Briefly, these are the guidelines:

1. Sufficient elastohydrodynamic film prevents surface distress (glazing, pitting).
2. Good boundary lubrication guards against smearing and sliding surface wear.
3. Clean lubricants prevent significant wear of rolling surfaces.
4. Sufficient lubricant flow keeps bearings from overheating.

As long as the surfaces of the rolling element and raceway in rolling contact can be separated by an elastohydrodynamic oil film, surface distress is avoided. The continuous presence of the film depends on the contact area, the load on the contact area, the speed, the operating temperature, the surface finish, and the oil viscosity.

SKF research has developed a procedure for determining the required oil viscosity when the bearing size, load, and speed are known and when operating temperature can be reasonably estimated. Charts enabling calculation have been published in trade journals and are available in "A Guide to Better Bearing Lubrication," from SKF, or in Volume 1 of Practical Machinery Management for Process Plants—Improving Machinery Reliability—page 233.

When the elastohydrodynamic oil film proves suitable in the rolling contacts, experience shows it is generally satisfactory for the sliding contacts at cages and guide flanges. When, in unusual applications, viscosity selection must be governed by the sliding areas, experience has proven that the viscosity chosen is capable of maintaining the necessary elastohydrodynamic film in the rolling contacts.

Ineffective Sealing

The effect of dirt and abrasive in the bearing during operation was described earlier in this chapter under "Faulty Mounting Practices." Although foreign matter can enter the bearing during mounting, its most direct and sustained area of entry can be around the housing seals. The result of gross change in bearing internal geometry has been pointed out. Bearing manufacturers realize the damaging effect of dirt and take extreme precautions to deliver clean bearings. Not only assembled bearings but also parts in process are washed and cleaned. Freedom from abrasive matter is so important that some bearings are assembled in air-conditioned white rooms. Dramatic examples of combined abrasive-particle and corrosive wear, both due to defective sealing, are shown in Figures 3-48 and 3-49. Figure 3-50 shows a deep-groove ball bearing which has operated with abrasive in it. The balls have worn to such an extent that they no longer support the cage, which has been rubbing on the lands of both rings.

In addition to abrasive matter, corrosive agents should be excluded from bearings. Water, acid, and those agents that deteriorate lubricants result in corrosion. Figure 3-51 illustrates how moisture in the lubricant can rust the end of a roller. The corroded areas on the rollers of Figure 3-52 occurred while the bearing was not rotating. Acids forming in the lubricant with water present etch the surface as shown in Figure 3-53. The lines of corrosion seen in Figure 3-54 are caused by water in the lubricant as the bearing rotates.

Figure 3-48. Advanced abrasive wear.

Figure 3-49. Advanced abrasive wear.

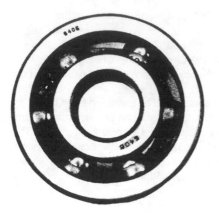

Figure 3-50. Advanced abrasive wear.

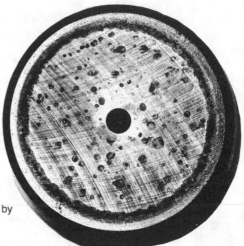

Figure 3-51. Rust on end of roller caused by moisture in lubricant.

Figure 3-52. Corrosion on roller surfaces caused by water in lubricant while bearing was not moving.

Figure 3-53. Corrosion on roller surface caused by formation of acids in lubricant with some moisture present.

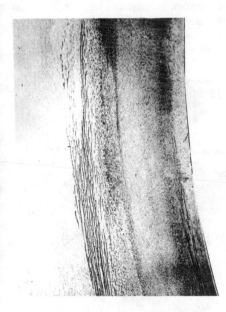

Figure 3-54. Corrosion streaks caused by water in the lubricant while the bearing rotated.

Vibration

Rolling bearings exposed to vibration while the shafts are not rotating are subject to a damage called false brinelling. The evidence can be either brightly polished depressions or the characteristic red-brown stain of fretting. Oxidation rate at the point of contact determines the appearance. Variation in the vibration load causes minute sliding in the area of contact between rolling elements and raceways. Small particles of material are set free from the contact surfaces and may or may not be oxidized immediately. The debris thus formed acts as a lapping agent and accelerates the wear. Another identification of damage of this type is the spacing of the marks on the raceway. The spacing of the false brinelling will be equal to the distance between the rolling elements, just as it is in some types of true brinelling. If the bearing has rotated slightly between periods of vibration, more than one pattern of false-brinelling damage may be seen.

A type of false brinelling with abrasive present is seen in Figure 3-55. There was no rotation between the two rings of the bearing for considerable periods of time, but while they were static they were subject to severe vibration. False brinelling developed with a production of iron oxide, which in turn acted as a lapping compound.

The effect of both vibration and abrasive in a rotating bearing is seen in the wavy pattern shown in Figure 3-56. When these waves are more closely spaced, the pattern is called fluting and appears similar to cases shown in the next section under "Passage of Electric Current Through the Bearing." Metallurgical examination is often necessary to distinguish between fluting caused solely by abrasive and vibration and that caused by vibration and passage of electric current.

Since false brinelling is a true wear condition, such damage can be observed even though the forces applied during the vibration are much smaller than those corresponding to the static carrying capacity of the bearing. However, the damage is more extensive as the contact load on the rolling element increases.

Figure 3-55. False brinelling caused by vibration with bearing stationary.

Figure 3-56. False brinelling caused by vibration in the presence of abrasive dirt while bearing was rotating.

False brinelling occurs most frequently during transportation of assembled machines. Vibration fed through a foundation can generate false brinelling in the bearings of a shaft that is not rotating. False brinelling during transportation can always be minimized and usually eliminated by temporary structural members that will prevent any rotation or axial movement of the shaft.

It is necessary to distinguish between false and true brinelling. Figures 3-57 and 3-58 are 100X photomicrographs of true and false brinelling in a raceway, respectively. In Figure 3-57 (true brinelling) there is a dent produced by plastic flow of the raceway material. The grinding marks are not noticeably disturbed and can be seen over the whole dented area. However, false brinelling (Figure 3-58), does not

Figure 3-57. Example of true brinelling—100x.

Figure 3-58. Example of false brinelling—100x.

involve flow of metal but rather a removal of surface metal by attrition. Notice that the grinding marks are removed. To further understand false brinelling, which is very similar to fretting corrosion, one should remember that a rolling element squeezes the lubricant out of its contact with the raceway, and that the angular motion from vibration is so small that the lubricant is not replenished at the contact. Metal-to-metal contact becomes inevitable, resulting in submicroscopic particles being torn from the high points. If protection by lubricant is absent, these minute particles oxidize and account for the red-brown color usually associated with fretting. If there is a slower oxidation rate, the false brinelling depression can remain bright, thereby adding to the difficulty in distinguishing true from false brinelling.

Passage of Electric Current Through the Bearing

In certain applications of bearings to electrical machinery there is the possibility that electric current will pass through a bearing. Current that seeks ground through the bearing can be generated from stray magnetic fields in the machinery or can be caused by welding on some part of the machine with the ground attached so that the circuit is required to pass through the bearing.

An electric current can be generated by static electricity, emanating from charged belts or from manufacturing processes involving leather, paper, cloth, or rubber. This current can pass through the shaft to the bearing and then to the ground. When the current is broken at the contact surfaces between rolling elements and raceways, arcing results, producing very localized high temperature and consequent damage. The overall damage to the bearing is in proportion to the number and size of individual damage points.

Figure 3-59 and the enlarged view Figure 3-60 show a series of electrical pits in a roller and in a raceway of a spherical roller bearing. The pit was formed each time the current broke in its passage between raceway and roller. The bearing from which this roller was removed was not altogether damaged to the same degree as this roller. In

Figure 3-59. Electric pitting on surface of spherical roller caused by the passage of a relatively large current.

fact, this specific bearing was returned to service and operated successfully for several additional years. Hence, moderate amounts of electrical pitting do not necessarily result in failure.

Another type of electrical damage occurs when current passes for prolonged periods and the number of individual pits accumulate astronomically. The result is the fluting shown in Figures 3-61 through 3-65. This condition can occur in ball or roller bearings. Flutes can develop considerable depth, producing noise and vibration during operation and eventual fatigue from local overstressing. The formation of flutes rather than a homogeneous dispersion of pits is not clearly explained but is possibly related to initial synchronization of shocks or vibrations and the breaking of the current. Once the fluting has started it is probably a self-perpetuating phenomenon.

Individual electric marks, pits, and fluting have been produced in bearings running in the laboratory. Both alternating and direct currents can cause the damage. Amperage rather than voltage governs the amount of damage. When a bearing is under radial load, greater internal looseness in the bearing appears to result in greater electrical damage for the same current. In a double-row bearing loaded in thrust, little if any damage results in the thrust-carrying row, although the opposite row may be damaged.

Figure 3-60. Electric pitting on surface of spherical outer raceway caused by the passage of a relatively large current.

Figure 3-61. Fluting on surface of spherical roller caused by prolonged passage of electric current.

Figure 3-62. Fluting on inner raceway of cylindrical roller bearing caused by prolonged passage of electric current.

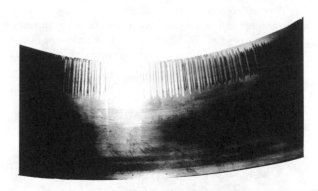

Figure 3-63. Fluting on outer raceway of spherical roller bearing.

Figure 3-64. Fluting on inner raceway of spherical roller bearing.

Figure 3-65. Fluting on outer raceway of self-aligning ball bearing caused by prolonged passage of relatively small current and vibration.

Troubleshooting Bearings

This section presents some helpful hints on bearing troubleshooting—what to look for, how to recognize the reason for the trouble, and practical solutions, wherever possible.

Observations of bearing trouble can be reduced to a few classifications, listed in the following order. For ease of relating them to conditions and solutions, they are coded A to G inclusive.

Observation or Complaint

A—Overheated bearing.
B—Noisy bearing.
C—Replacements are too frequent.
D—Vibration.
E—Unsatisfactory equipment performance.
F—Bearing is loose on shaft.
G—Hard-turning shaft.

The following table lists some typical conditions that will result in bearing failures. The first column numerically codes each typical condition (Nos. 1 to 54 inclusive). The third column is the observation or complaint code (A to G) to which the condition may apply.

TYPICAL CONDITIONS RESULTING IN BEARING FAILURES

EACH CONDITION COULD CAUSE ANY ONE OF THE COMPLAINTS LISTED OPPOSITE THE RESPECTIVE CONDITION IN COLUMN 3

CODE	CONDITION	COMPLAINT
1.	Inadequate lubrication (Wrong type of grease or oil)..........................	A-B-C-G
2.	Insufficient lubrication (Low oil level—loss of lubricant through seals)...............	A-B-C-G
3.	Excessive lubrication (Housing oil level too high or housing packed with grease).......	A-G
4.	Insufficient clearance in bearing (Selection of wrong fit).........................	A-B-C-E-G
5.	Foreign matter acting as an abrasive (Sand, carbon, etc.).......................	B-C-D-E-G
6.	Foreign matter acting as a corrosive (Water, acids, paints, etc.)...................	B-C-D-E-G
7.	Bearings pinched in the housing (Bore out of round)............................	A-B-C-D-E-G
8.	Bearings pinched in the housing (Housing warped).............................	A-B-C-D-E-G
9.	Uneven shimming of housing base (Distorted housing bore - possible cracking of base)	A-B-C-D-E-G
10.	Chips in bearing housing (Chips or dirt left in housing).........................	B-C-D-E-G
11.	High air velocity over bearings (Oil leakage).................................	C
12.	Seals too tight (Cup seals)	A-G
13.	Seals misaligned (Rubbing against stationary parts)............................	A-B-G
14.	Oil return holes plugged (Oil leakage).......................................	A
15.	Preloaded bearings (Opposed mounting).....................................	A-B-C-G
16.	Preloaded bearings (Two held bearings on one shaft)...........................	A-B-C-E-G
17.	Bearing loose on shaft (Shaft diameter too small).............................	B-C-D-E-F
18.	Bearing loose on shaft (Adapter not tightened sufficiently)......................	B-C-D-E-F
19.	Bearing too tight internally (Adapter tightened excessively).....................	A-B-C-E-G
20.	Split pillow block with uneven surfaces (Oil leakage)..........................	C
21.	Spinning of outer ring in housing (Unbalanced load)...........................	A-C-D-E
22.	Noisy bearing (Flat on roller or ball due to skidding)..........................	B-D-E
23.	Excessive shaft expansion (Resulting in opposed mounting)......................	A-B-C-E-G
24.	Excessive shaft expansion (Resulting in insufficient clearance in bearing)............	A-C-E-G
25.	Tapered shaft seat (Concentration of load in bearing).........................	C-D-E
26.	Tapered housing bore (Concentration of load in bearing).......................	C-D-E
27.	Shaft shoulder too small (Inadequate shoulder support—bending of shaft)..........	C-D-E-G
28.	Shaft shoulder too large (Rubbing against bearing seals).......................	A-B-C
29.	Housing shoulder too small (Inadequate shoulder support)......................	C-D-E-G
30.	Housing shoulder too large (Distortion of bearing seals).......................	B-C-G
31.	Shaft fillet too large (Bending of shaft)	C-D-E-G
32.	Housing fillet too large (Inadequate support)................................	C-D-E-G
33.	Insufficient clearance in labyrinth seals (Rubbing)............................	A-B-C-G
34.	Oil gauge breather hole clogged (Shows incorrect oil level).....................	A-C
35.	Shafts out of line (Linear misalignment)....................................	A-C-D-E-G
36.	Shafts out of line (Angular misalignment)...................................	A-C-D-E-G
37.	Constant oil level cups (Incorrect level)....................................	A-C
38.	Constant oil level cups (Located against rotation of bearing)...................	A-C
39.	Lockwasher prongs bent (Rubbing against bearing)............................	A-B-E-G
40.	Incorrect positioning of flingers (Rubbing against covers)......................	A-B-C-G
41.	Pedestal surface uneven (Bending of housing causing pinching of bearing)..........	A-C-D-E-G
42.	Ball or roller denting (Hammer blows on bearing).............................	B-C-D-E
43.	Noisy bearing (Extraneous conditions)......................................	B
44.	Lubricant leakage and entrance of dirt into bearing (Worn out seals)...............	C
45.	Vibration (Excessive clearance in bearing)..................................	D-E
46.	Vibration (Unbalanced loading)...	D-E

	TYPICAL CONDITIONS RESULTING IN BEARING FAILURES (Continued)	
CODE	**CONDITION**	**COMPLAINT**
47.	Hard turning shaft (Shaft and housing shoulders out of square with bearing seat).......	C-E-G
48.	Bearing loose on shaft (Knurling and center punching of shaft for bearing seat).......	A-F
49.	Discoloration of bearings (Use of blow torch to remove bearing).................	B
50.	Oversized shaft (Overheating and noise).................................	A-B-C-E-G
51.	Undersized housing bore (Overheating of bearing).........................	A-B-C-E-G
52.	Oversized housing bore (Overheating of bearing—spinning of outer ring)..........	A-B-C-D-E
53.	Enlarged housing bore (Excessive peening of non-ferrous housings)...............	A-B-C-D-E
54.	Noisy bearing (False brinelling)..	B

The following pages offer practical solutions to the trouble conditions which were observed. Column 1 refers back to the code of the *typical condition* listed in the foregoing table. Column 2 is the *reason for that condition*, and column 3 is the *practical solution*.

(text continued on page 128)

TROUBLE CONDITIONS AND THEIR SOLUTION

OVERHEATED BEARING
Complaint "A"

CODE TO TYPICAL CONDITION	REASON FOR CONDITION	PRACTICAL SOLUTION	
1	Wrong type of grease or oil causing break-down of lubricant.	Consult reliable lubricant manufacturer for proper type of lubricant. Check SKF Catalog instructions to determine if oil or grease should be used.	
2	Low oil level. Loss of lubricant through seal. Insufficient grease in housing.	Oil level should be just below center of lowest ball or roller in bearing. Using grease, lower half of pillow block should be ½ to ⅔ full.	
3	Housing packed with grease, or oil level too high . . . causing excessive churning of lubricant, high operating temperature, oil leakage.	Purge bearing until only lower half of housing is ½ to ⅔ full of grease. Using oil lubrication, reduce level to just below center of lowest ball or roller.	
4	Bearings selected with inadequate internal clearance for conditions where external heat is conducted thru shaft, thereby expanding excessively the inner ring.	Replacement bearing should have identical marking as original bearing for proper internal clearance. Check with SKF if bearing markings have become indistinct.	
7-8-9 41-51	Housing bore out of round. Housing warped. Excessive distortion of housing. Undersized housing bore.	Check and scrape housing bore to relieve pinching of bearing. Be sure pedestal surface is flat, and shims cover entire area of pillow block base.	

OVERHEATED BEARING — *Complaint "A"* (Continued)

CODE TO TYPICAL CONDITION	REASON FOR CONDITION	PRACTICAL SOLUTION	
12	Leather or composition seals with excessive spring tension or dried out.	Replace leather or composition seals with ones having reduced spring tension. Lubricate seals.	
13 33-40	Rotating seals rubbing against stationary parts.	Check running clearance of rotating seal to eliminate rubbing. Correct alignment.	
14	Oil return holes blocked—pumping action of seals cause oil leakage.	Clean holes. Drain out used oil—refilling to proper oil level with fresh lubricant.	BLOCKED PASSAGE
15	Opposed mounting.	Insert gasket between housing and cover flange to relieve axial preloading of bearing.	GASKET
16 23-24	Two "held" bearings on one shaft. Excessive shaft expansion.	Back off covers in one of the housings, using shims to obtain adequate clearance of outer ring, to permit free axial bearing motion.	SHIMS SHAFT EXPANSION
19	Adapter tightened excessively.	Loosen locknut and sleeve assembly. Retighten sufficiently to clamp sleeve on shaft but be sure bearing turns freely.	
21-52	Unbalanced load. Housing bore too large.	Rebalance machine. Replace housing with one having proper bore.	CLEARANCE
28	Rubbing of shaft shoulder against bearing seals.	Remachine shaft shoulder to clear seal.	RUBBING
34	Incorrect oil level. Result: no lubricant in bearing.	Clean out clogged hole to vent oil gauge.	OIL LEVEL IN GAGE CLOGGED VENT OIL LEVEL IN HOUSING

OVERHEATED BEARING — *Complaint "A"* (Continued)

CODE TO TYPICAL CONDITION	REASON FOR CONDITION	PRACTICAL SOLUTION	
35-36	Incorrect linear or angular alignment of two or more coupled shafts with two or more bearings.	Correct alignment by shimming pillow blocks. Be sure shafts are coupled in straight line — especially when three or more bearings operate on one shaft.	LINEAR MISALIGNMENT / ANGULAR MISALIGNMENT
37-38	Incorrect mounting of constant oil level cup. (Too high or too low.) Cup located opposite rotation of bearing permitting excessive flow of oil, resulting in too high oil level.	The oil level at standstill must not exceed the center of the lowermost ball or roller. Sketch illustrates correct position of constant level oil cup with respect to rotation. Better replace constant level oiler with sight gage.	1-STATIC OIL LEVEL 2-OPERATING OIL LEVEL
39	Prong rubbing against bearing.	Remove lockwasher — straighten prong or replace with new washer.	RUBBING
48	Knurling and center punching of bearing seat on shaft.	Unsatisfactory because high spots are flattened when load is applied, when fit is loose, metalize shaft and regrind to proper size.	
50	Bearing seat diameter machined oversize, causing excessive expansion of bearing inner ring, thus reducing clearance in bearing.	Grind shaft to get proper fit between inner ring of bearing and shaft.	
53	"Pounding-out" of housing bore due to soft metal. Result: enlarged bore . . , causing spinning of outer ring in housing	Rebore housing and press steel bushing in bore. Machine bore of bushing to correct size.	

NOISY BEARING
Complaint "B"

1	Wrong type of grease or oil causing break-down of lubricant.	Consult reliable lubricant manufacturer for proper type of lubricant. Check SKF Catalog instructions to determine if oil or grease should be used.
2	Low oil level. Loss of lubricant through seal. Insufficient grease in housing.	Oil level should be at center of lowest ball or roller in bearing, at standstill. Using grease, lower half of pillow block should be ½ to ⅔ full. *See Illustration—Complaint "A", Condition 2*
4	Bearings selected with inadequate internal clearance for conditions where external heat is conducted thru shaft, thereby expanding excessively the inner ring.	Replacement bearing should have identical marking as original bearing for proper internal clearance. Check with SKF if markings have become indistinct.
5	Foreign matter (dirt, sand, carbon, etc.) entering bearing housing.	Clean out bearing housing. Replace worn-out seals or improve seal design to obtain adequate protection of bearing.
6	Corrosive agents (water, acids, paints, etc.) entering the bearing housing.	Addition of a shroud and (or) flinger to throw off foreign matter.

NOISY BEARING — *Complaint "B"* (Continued)

CODE TO TYPICAL CONDITION	REASON FOR CONDITION	PRACTICAL SOLUTION
7-8 9-51	Housing bore out of round. Housing warped. Excessive distortion of housing. Undersized housing bore.	Check and scrape housing bore to relieve pinching of bearing. Be sure pedestal surface is flat, and shims cover entire area of pillow block base. *See Illustration—Complaint "A", Conditions 7-8-9-41-51*
10	Failure to remove chips, dirt, etc. from bearing housing before assembling bearing unit.	Carefully clean housing, and use fresh lubricant.
13 33-40	Rotating seals rubbing against stationary parts.	Check running clearance of rotating seal to eliminate rubbing. Correct alignment. *See Illustration—Complaint "A", Condition 13-33-40*
15	Opposed mounting.	Insert gasket between housing and cover flange to relieve axial pre-loading of bearing. *See Illustration—Complaint "A", Condition 15*
16-23	Two "held" bearings on one shaft. Excessive shaft expansion.	Back off covers in one of the housings using shims to obtain adequate clearance of outer ring to permit free axial bearing motion. *See Illustration—Complaint "A", Conditions 16-23-24*
17-18	Shaft diameter too small. Adapter not tightened sufficiently.	Metallize shaft and regrind to obtain proper fit. Retighten adapter to get firm grip on shaft.
19	Adapter tightened excessively.	Loosen locknut and sleeve assembly. Retighten sufficiently to clamp sleeve on shaft but be sure bearing turns freely. *See Illustration—Complaint "A", Condition 19*
22	Flat on ball or roller due to skidding. (Result of fast starting.)	Carefully examine balls or rollers, looking for flat spots on the surface. Replace bearing.
28	Rubbing of shaft shoulder against bearing seals.	Remachine shaft shoulder to clear seal. *See Illustration—Complaint "A", Condition 28*
30	Distortion of bearing seals.	Remachine housing shoulder to clear seal.
39	Prong rubbing against bearing.	Remove lockwasher—straighten prong or replace with new washer. *See Illustration—Complaint "A", Condition 39*
42	Incorrect method of mounting. Hammer blows on bearing.	Replace with new bearing. Don't hammer any part of bearing when mounting.
43	Interference of other movable parts of machine.	Carefully check every moving part for interference. Reset parts to provide necessary clearance.
49	Distorted shaft and other parts of bearing assembly.	Only in extreme cases should a torch be used to facilitate removal of a failed bearing. Care should be exercised to avoid high heat concentration at any one point so distortion is eliminated.

NOISY BEARING — *Complaint "B"* (Continued)

CODE TO TYPICAL CONDITION	REASON FOR CONDITION	PRACTICAL SOLUTION
50	Bearing seat diameter machined oversize causing excessive expansion of bearing inner ring, thus reducing clearance in bearing.	Grind shaft to get proper fit between inner ring of bearing and shaft.
52	Unbalanced load. Housing bore too large.	Rebalance unit. Replace housing with one having proper bore.
53	"Pounding-out" of housing bore due to soft metal. Result: enlarged bore . . . causing spinning of outer ring in housing.	Rebore housing and press steel bushing in bore. Machine bore of bushing to correct size.
54	Bearing exposed to vibration while machine is idle.	Carefully examine bearing for wear spots separated by distance equal to the spacing of the balls. Replace bearing.

REPLACEMENTS ARE TOO FREQUENT
Complaint "C"

1	Wrong type of grease or oil causing break-down of lubricant.	Consult reliable lubricant manufacturer for proper type of lubricant. Check SKF Catalog instructions to determine if oil or grease should be used.
2	Low oil level. Loss of lubricant through seal. Insufficient grease in housing.	Oil level should be at center of lowest ball or roller in bearing. Using grease, lower half of pillow block should be ½ to ⅔ full. *See Illustration—Complaint "A", Condition 2*
4	Bearings selected with inadequate internal clearance for conditions where external heat is conducted thru shaft, thereby expanding excessively the inner ring.	Replacement bearing should have identical marking as original bearing for proper internal clearance. Check with SKF if bearing markings have become indistinct.
5	Foreign matter (dirt, sand, carbon, etc.) entering into bearing housing.	Clean out bearing housing. Replace worn-out seals or improve seal design to obtain adequate protection of bearing.
6	Corrosive agents (water, acids, paints, etc.) entering the bearing housing.	Addition of a shroud and (or) flinger to throw off the foreign matter.
7-8-9 41-51	Housing bore out of round. Housing warped. Excessive distortion of housing. Undersized housing bore.	Check and scrape housing bore to relieve pinching of bearing. Be sure pedestal surface is flat, and shims cover entire area of pillow block base. *See Illustration—Complaint "A", Conditions 7-8-9-41-51*
10	Failure to remove chips, dirt, etc. from bearing housing before assembling bearing unit.	Carefully clean housing, and use fresh lubricant.

CODE TO TYPICAL CONDITION	REASON FOR CONDITION	PRACTICAL SOLUTION
11	Oil leakage resulting from air flow over bearings. (Example: forced draft fan with air inlet over bearings.)	Provide proper baffles to divert direction of air flow.
15	Opposed mounting.	Insert gasket between housing and cover flange to relieve axial pre-loading of bearing. *See Illustration—Complaint "A", Condition 15*
16 23-24	Two "held" bearings on one shaft. Excessive shaft expansion.	Back off covers in one of the housings, using shims to obtain adequate clearance of outer ring, to permit free axial bearing motion. *See Illustration—Complaint "A", Conditions 16-23-24*
17-18	Shaft diameter too small. Adapter insufficiently tightened.	Metallize shaft and regrind to obtain proper fit. Retighten adapter to get firm grip on shaft. *See Illustration—Complaint "B", Conditions 17-18*
19	Adapter tightened excessively.	Loosen locknut and sleeve assembly. Retighten sufficiently to clamp sleeve on shaft but be sure bearing turns freely. *See Illustration—Complaint "A", Condition 19*
20	Oil leakage at housing split. Excessive loss of lubricant.	If not severe, use thin layer of gasket cement. Don't use shims. Replace housing if necessary.
21-52	Unbalanced load. Housing bore too large.	Rebalance machine. Replace housing with one having proper bore. *See Illustration—Complaint "A", Conditions 21-52*
25-26	Unequal load distribution on bearing.	Rework shaft, housing, or both, to obtain proper fit. May require new shaft and housing.
27	Inadequate shoulder support causing bending of shaft.	Remachine shaft fillet to relieve stress. May require shoulder collar.
29	Inadequate support in housing causing cocking of outer ring.	Remachine housing fillet to relieve stress. May require shoulder collar.
28	Rubbing of shaft shoulder against bearing seals.	Remachine shaft shoulder to clear seal. *See Illustration—Complaint "A", Condition 28*
30	Distortion of bearing seals.	Remachine housing shoulder to clear seal. *See Illustration—Complaint "B", Condition 30*
31	Distortion of shaft and inner ring. Uneven expansion of bearing inner ring.	Remachine shaft fillet to obtain proper support.
32	Distortion of housing and outer ring. Pinching of bearing.	Remachine housing fillet to obtain proper support.
33-40	Rotating seals rubbing against stationary parts.	Check running clearance of rotating seal to eliminate rubbing. Correct alignment. *See Illustration—Complaint "A", Conditions 13-33-40*

REPLACEMENTS ARE TOO FREQUENT — *Complaint "C"* (Continued)

CODE TO TYPICAL CONDITION	REASON FOR CONDITION	PRACTICAL SOLUTION
34	Incorrect oil level. Result: no lubricant in bearing.	Clean out clogged hole to vent oil gauge. *See Illustration—Complaint "A", Condition 34*
35-36	Incorrect linear or angular alignment of two or more coupled shafts with two or more bearings.	Correct alignment by shimming pillow blocks. Be sure shafts are coupled in straight line—especially when three or more bearings operate on one shaft. *See Illustration—Complaint "A", Conditions 35-36*
37-38	Incorrect mounting of constant oil level cup. (Too high or low.) Cup located opposite rotation of bearing.	The oil level at standstill must not exceed the center of the lowermost ball or roller. Locate cup with rotation of bearing. Replace constant level oiler with sight gage. *See Illustration—Complaint "A", Conditions 37-38*
42	Incorrect method of mounting. Hammer blows on bearing.	Replace with new bearing. Don't hammer any part of bearing when mounting.
44	Excessively worn leather (or composition), or labyrinth seals. Result: lubricant loss; dirt getting into bearing.	Replace seals after thoroughly flushing bearing and refilling with fresh lubricant.
47	Shaft and housing shoulders and face of locknut out-of-square with bearing seat.	Remachine parts to obtain squareness.
50	Bearing seat diameter machined oversize, causing excessive expansion of bearing inner ring, thus reducing clearance in bearing.	Grind shaft to get proper fit between inner ring of bearing and shaft.
53	"Pounding-out" of housing bore due to soft metal. Result: enlarged bore . . . causing spinning of outer ring in housing.	Rebore housing and press steel bushing in bore. Machine bore of bushing to correct size.

VIBRATION

Complaint "D"

	REASON FOR CONDITION	PRACTICAL SOLUTION
5	Foreign matter (dirt, sand, carbon, etc.) entering bearing housing.	Clean out bearing housing. Replace worn-out seals or improve seal design to obtain adequate protection of bearing.
6	Corrosive agents (water, acids, paints, etc.) entering the bearing housing.	Addition of a shroud and (or) flinger to throw off foreign matter.
7-8 9-41	Housing bore out of round. Housing warped. Excessive distortion of housing. Undersized housing bore.	Check and scrape housing bore to relieve pinching of bearing. Be sure pedestal surface is flat, and shims cover entire area of pillow block base. *See Illustration—Complaint "A", Conditions 7-8-9-41-51*
10	Failure to remove chips, dirt, etc. from bearing housing before assembling bearing unit.	Carefully clean housing, and use fresh lubricant.
17-18	Shaft diameter too small. Adapter not tightened sufficiently.	Metallize shaft and regrind to obtain proper fit. Retighten adapter to get firm grip on shaft. *See Illustration—Complaint "B", Conditions 17-18*

CODE TO TYPICAL CONDITION	REASON FOR CONDITION	PRACTICAL SOLUTION
21-52	Unbalanced load. Housing bore too large.	Rebalance machine. Replace housing with one having proper bore. *See Illustration—Complaint "A", Conditions 21-52*
22	Flat on ball or roller due to skidding. (Result of fast starting.)	Carefully examine balls or rollers, looking for flat spots on the surface. Replace bearing.
25-26	Unequal load distribution on bearing.	Rework shaft, housing, or both, to obtain proper fit. May require new shaft and housing. *See Illustration—Complaint "C", Conditions 25-26*
27	Inadequate shoulder support causing bending of shaft.	Remachine shaft fillet to relieve stress. May require shoulder collar. *See Illustration—Complaint "C", Condition 27*
29	Inadequate support in housing causing cocking of outer ring.	Remachine housing fillet to relieve stress. May require shoulder collar. *See Illustration—Complaint "C", Condition 29*
31	Distortion of shaft and inner ring. Uneven expansion of bearing inner ring.	Remachine shaft fillet to obtain proper support. *See Illustration—Complaint "C", Condition 31*
32	Distortion of housing and outer ring. Pinching of bearing.	Remachine housing fillet to obtain proper support. *See Illustration—Complaint "C", Condition 32*
35-36	Incorrect linear or angular alignment of two or more coupled shafts with two or more bearings.	Correct alignment by shimming pillow blocks. Be sure shafts are coupled in straight line—especially when three or more bearings operate on one shaft. *See Illustration—Complaint "A", Conditions 35-36*
42	Incorrect method of mounting. Hammer blows on bearing.	Replace with new bearing. Don't hammer any part of bearing when mounting.
45	Excessive clearance in bearing, resulting in vibration.	Use bearings with recommended internal clearances.
46	Vibration of machine.	Check balance of rotating parts. Rebalance machine.
53	"Pounding-out" of housing bore due to soft metal. Result: enlarged bore . . . causing spinning of outer ring in housing.	Rebore housing and press steel bushing in bore. Machine bore of bushing to correct size.

UNSATISFACTORY PERFORMANCE OF EQUIPMENT

Complaint "E"

CODE TO TYPICAL CONDITION	REASON FOR CONDITION	PRACTICAL SOLUTION
4	Bearings selected with inadequate internal clearance for conditions where external heat is conducted thru shaft, thereby expanding excessively the inner ring.	Replacement bearing should have identical marking as original bearing for proper internal clearance. Check with SKF if bearing markings have become indistinct.
5	Foreign matter (dirt, sand, carbon, etc.) entering bearing housing.	Clean out bearing housing. Replace worn-out seals or improve seal design to obtain adequate protection of bearing.
6	Corrosive agents (water, acids, paints, etc.) entering the bearing housing.	Addition of a shroud and (or) flinger to throw off foreign matter.
7-8-9 41-51	Housing bore out of round. Housing warped. Excessive distortion of housing. Undersized housing bore.	Check and scrape housing bore to relieve pinching of bearing. Be sure pedestal surface is flat, and shims cover entire area of pillow block base. *See Illustration—Complaint "A", Conditions 7-8-9-41-51*
10	Failure to remove chips, dirt, etc. from bearing housing before assembling bearing unit.	Carefully clean housing, and use fresh lubricant.

CODE TO TYPICAL CONDITION	REASON FOR CONDITION	PRACTICAL SOLUTION
16 23-24	Two "held" bearings on one shaft. Excessive shaft expansion.	Back off covers in one of the housings, using shims to obtain adequate clearance of outer ring, to permit free axial bearing motion. *See Illustration—Complaint "A", Conditions 16-23-24*
17-18	Shaft diameter too small. Adapter not tightened sufficiently.	Metallize shaft and regrind to obtain proper fit. Retighten adapter to get firm grip on shaft. *See Illustration—Complaint "B". Conditions 17-18*
19	Adapter tightened excessively.	Loosen locknut and sleeve assembly. Retighten sufficiently to clamp sleeve on shaft but be sure bearing turns freely. *See Illustration—Complaint "A", Condition 19*
21-52	Unbalanced load. Housing bore too large.	Rebalance machine. Replace housing with one having proper bore. *See Illustration—Complaint "A", Conditions 21-52*
22	Flat on ball or roller due to skidding. (Result of fast starting.)	Carefully examine balls or rollers, looking for flat spots on the surface. Replace bearing.
25-26	Unequal load distribution on bearing.	Rework shaft, housing, or both, to obtain proper fit. May require new shaft and housing. *See Illustration—Complaint "C", Conditions 25-26*
27	Inadequate shoulder support causing bending of shaft.	Remachine shaft fillet to relieve stress. May require shoulder collar. *See Illustration- —Complaint "C", Condition 27*
29	Inadequate support in housing causing rocking of outer ring.	Remachine housing fillet to relieve stress. May require shoulder collar. *See Illustration—Complaint "C", Condition 29*
31	Distortion of shaft and inner ring. Uneven expansion of bearing inner ring.	Remachine shaft fillet to obtain proper support. *See Illustration— Complaint "C", Condition 31*
32	Distortion of outer housing and ring. Pinching of bearing.	Remachine housing fillet to obtain proper support. *See Illustration Complaint "C", Condition 32*
35-36	Incorrect linear or angular alignment of two or more coupled shafts with two or more bearings.	Correct alignment by shimming pillow blocks. Be sure shafts are coupled in straight line— especially when three or more bearings operate on one shaft. *See Illustration—Complaint "A", Conditions 35-36*
39	Prong rubbing against bearing.	Remove lockwasher—straighten prong or replace with new washer. *See Illustration—Complaint "A", Condition 39*
42	Incorrect method of mounting. Hammer blows on bearing.	Replace with new bearing. Don't hammer any part of bearing when mounting.
45	Excessive clearance in bearing, resulting in vibration.	Use bearings with recommended internal clearances.
46	Vibration of machine.	Check balance of rotating parts. Rebalance machine.
47	Shaft and housing shoulders, and face of locknut out of square with bearing seat.	Remachine parts to obtain squareness.
50	Bearing seat diameter machined oversize, causing excessive expansion of bearing inner ring thus reducing clearance in bearing.	Grind shaft to proper fit between inner ring of bearing and shaft.
53	"Pounding-out" of housing bore due to soft metal. Result: enlarged bore . . . causing spinning of outer ring in housing.	Rebore housing and press steel bushing in bore. Machine bore of bushing to correct size. If loads are not excessive, tighter fit in housing, without the use of the steel bushing, may correct the trouble.

CODE TO TYPICAL CONDITION	REASON FOR CONDITION	PRACTICAL SOLUTION

BEARING IS LOOSE ON SHAFT
Complaint "F"

CODE TO TYPICAL CONDITION	REASON FOR CONDITION	PRACTICAL SOLUTION
17-18	Shaft diameter too small. Adapter not tightened sufficiently.	Metallize shaft and regrind to obtain proper fit. Retighten adapter to get firm grip on shaft. *See Illustration—Complaint "B", Conditions 17-18*
48	Knurling and center punching of bearing seat on shaft.	Unsatisfactory because high spots are flattened when load is applied. When fit is loose, metallize shaft and regrind to proper size.

HARD TURNING OF SHAFT
Complaint "G"

CODE TO TYPICAL CONDITION	REASON FOR CONDITION	PRACTICAL SOLUTION
1	Wrong type of grease or oil causing break-down of lubricant.	Consult reliable lubricant manufacturer for proper type of lubricant. Check SKF Catalog instructions to determine if oil or grease should be used.
2	Low oil level. Loss of lubricant through seal. Insufficient grease in housing.	Oil level should be just below center of lowest ball or roller in bearing. Using grease, lower half of pillow block should be ½ to ⅔ full. *See Illustration—Complaint "A", Condition 2*
3	Housing packed with grease, or oil level too high . . . causing excessive churning of lubricant, high operating temperature, oil leakage.	Purge bearing until only lower half of housing is ½ to ⅔ full of grease. Using oil lubrication, reduce level to just below center of lowest ball. *See Illustration—Complaint "A", Condition 3*
4	Bearings selected with inadequate internal clearance for conditions where external heat is conducted thru shaft, thereby expanding excessively the inner ring.	Replacement bearing should have identical marking as original bearing for proper internal clearance. Check with SKF if bearing markings have become indistinct.
5	Foreign matter (dirt, sand, carbon, etc.) entering bearing housing.	Clean out bearing housing. Replace worn-out seals or improve seal design to obtain adequate protection of bearing.
6	Corrosive agents (water, acids, paints, etc.) entering the bearing housing.	Addition of a shroud and (or) flinger to throw off foreign matter.
7-8 9-41 51	Housing bore out of round. Housing warped. Excessive distortion of housing. Undersized housing bore.	Check and scrape housing bore to relieve pinching of bearing. Be sure pedestal surface is flat, and shims cover entire area of pillow block base. *See Illustration—Complaint "A", Conditions 7-8-9-41-51*
10	Failure to remove chips, dirt, etc. from bearing housing before assembling bearing unit.	Carefully clean housing, and use fresh lubricant.

HARD TURNING OF SHAFT — *Complaint "G"* (Continued)

CODE TO TYPICAL CONDITION	REASON FOR CONDITION	PRACTICAL SOLUTION
12	Leather or composition seals with excessive spring tension or dried out.	Replace leather or composition seals with ones having reduced spring tension. Lubricate seals. *See Illustration—Complaint "A", Condition 12*
13 33-40	Rotating seals rubbing against stationary parts.	Check running clearance of rotating seal to eliminate rubbing. Correct alignment. *See Illustration—Complaint "A", Condition 13-33-40*
15	Opposed mounting.	Insert gasket between housing and cover flange to relieve axial pre-loading of bearing. *See Illustration—Complaint "A", Condition 15*
16 23-24	Two "held" bearings on one shaft. Excessive shaft expansion.	Back off covers in one of the housings, using shims to obtain adequate clearance of outer ring, to permit free axial bearing motion. *See Illustration—Complaint "A", Conditions 16-23-24*
19	Adapter tightened excessively.	Loosen locknut and sleeve assembly. Retighten sufficiently to clamp sleeve on shaft but be sure bearing turns freely. *See Illustration—Complaint "A", Condition 19*
39	Prong rubbing against bearing.	Remove lockwasher. Straighten prong or replace with new washer. *See Illustration—Complaint "A", Condition 39.*
27	Inadequate shoulder support causing bending of shaft.	Remachine shaft fillet to relieve stress. May require shoulder collar. *See Illustration—Complaint "C", Condition 27*
29	Inadequate support in housing causing cocking of outer ring.	Remachine housing fillet to relieve stress. May require shoulder collar. *See Illustration—Complaint "C", Condition 29*
30	Distortion of bearing seals.	Remachine housing shoulder to clear seal. *See Illustration—Complaint "B", Condition 30*
31	Distortion of shaft and inner ring. Uneven expansion of bearing inner ring.	Remachine shaft fillet to obtain proper support. *See Illustration—Complaint "C", Condition 31*
32	Distortion of housing and outer ring. Pinching of bearing.	Remachine housing fillet to obtain proper support. *See Illustration—Complaint "C", Condition 32*
35-36	Incorrect linear or angular alignment of two or more coupled shafts with two or more bearings.	Correct alignment by shimming pillow blocks. Be sure shafts are coupled in straight line—especially when three or more bearings operate on one shaft. *See Illustration—Complaint "A", Conditions 35-36*
47	Shaft and housing shoulders, and face of locknut out of square with bearing seat.	Remachine parts to obtain squareness.
50	Bearing seat diameter machined oversize, causing excessive expansion of shaft and bearing inner ring, thus reducing clearance in bearing.	Grind shaft to get proper fit between inner ring of bearing and shaft.

(text continued from page 117)

Journal and Tilt-Pad Thrust Bearings*

It is not at all unusual to find unexpected bearing damage when machinery is opened for routine inspection, for turnaround maintenance, or for the purpose of troubleshooting vibration events. Very often the bearings are then blamed for faulty design or deficient manufacture when in fact an extraneous source is more likely the culprit.

This section will direct the analyzing engineer or technician on the basis of available facts to the most probable causes of bearing damage. Moreover, this section will help him decide what corrective action should be taken. This decision must be based on the nature and severity of the damage and on the significance of the type of damage.

Scoring Due to Foreign Matter or Dirt

Contamination of the lubricant includes:

1. "Built-in" dirt in crankcases, on crankshafts, in oil distribution headers, in cylinder bores, etc. present at the time of machine assembly.
2. Entrained dirt entering through breathers or air filters and particles derived from fuel combustion in internal combustion engines.
3. Metallic wear particles resulting from abrasive wear of moving parts.

"Dirt" may polish the surfaces of whitemetal-lined bearings, burnish bronze bearings, abrade overlays or other bearing linings, and score both bearing and mating surfaces, with degrees of severity depending upon the nature and size of the dirt particle or upon oil-film thickness and type of bearing material.

The effect of dirt on journal bearings is shown in Figure 3-66, which illustrates a whitemetal-lined bearing scored and pitted by "dirt." Similarly, Figure 3-67 shows a whitemetal-lined bearing with "haloes" caused by dirt particles.

Bearings in the condition shown in Figure 3-67 should be scrapped and new bearings fitted after journal, oil passages, and filters are cleaned. The oil should be changed, also.

Bearings in the condition shown in Figure 3-66 can be refitted after cleaning bearings and journal surfaces, provided any clearance increase due to wear can be tolerated.

Tilting-pad thrust bearings are similarly affected by dirt. In Figure 3-68, note concentric scoring of the thrust pad due to dirt entering the bearing at high speed. Figure 3-69 illustrates scoring by dirt entering the bearing at startup, while Figure 3-70 depicts the surface of the pad in Figure 3-69 at higher magnification. Figure 3-70 also shows irregular tracks caused by the passage of shotblast spherical steel particles.

Figure 3-66. Scoring caused by dirt.

Figure 3-67. Effects of con-taminated lube oil

Figure 3-68. Concentric scoring due to dirt entering at high speed.

Figure 3-69. Dirt-intrusion damage at low-speed operation.

Figure 3-70. Magnification of Figure 3-69. Irregular tracks were caused by spherical steel particles.

The damage shown in Figures 3-68 through 3-70 requires damaged pads to be scrapped. The lubricating system should be cleaned, the oil changed, and new pads fitted.

Wiping of Bearing Surfaces

A wiped bearing surface is where surface rubbing, melting, and smearing is evident. This is usually caused from inadequate running clearance with consequent surface overheating, from inadequate oil supply, or from both.

Damaged journal bearings are shown in Figure 3-71. This whitemetal-lined turbine bearing is wiped in both top and bottom halves because of inadequate

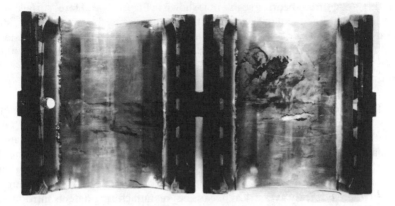

Figure 3-71. Wiping caused by tight clearance.

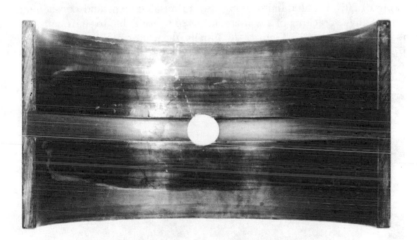

Figure 3-72. Wiping caused by barreled journal surfaces.

clearance. An overlay-plated-copper lead journal bearing wiped because of a barrel-shaped journal is illustrated in Figure 3-72.

Recommended action is as follows. If the journal bearing is lightly wiped, refit the bearing after cleaning the surface to remove any loose metal, provided the clearance can be tolerated. If it is cracked, fit a new bearing and increase the clearance when fitting. If vibration synchronous with shaft rotational frequency is present, check balance of rotor, alignment, coupling condition, etc. If vibration frequency is at half shaft speed or less, fit antiwhirl bearings.

A tilting-pad thrust-bearing segment is shown in Figure 3-73. Here, surface wiping of a whitemetal-lined pad in successive, thin layers is caused from excessive, steady load at startup. To remedy the situation, fit new pads. Improve lubrication at startup, possibly by using a prelubrication oil system. If possible, reduce loading at startup.

Fatigue Cracking

Fatigue cracking is caused by the imposition of dynamic loads in excess of the fatigue strength of the bearing material at operating temperature. The fatigue strength, especially of low-melting-point materials such as whitemetals and lead-base overlays, is greatly reduced at high temperatures, hence overheating alone may cause fatigue failure. Other causes of fatigue failure are overloading, cyclic out-of-balance loadings, high cyclic centrifugal loading due to overspeeding, and shafts that are not truly cylindrical because of manufacturing defects introduced by honing, filing, etc.

Journal bearings with fatigue defects are shown in Figures 3-74 through 3-77. Figure 3-74 illustrates fatigue cracking of a whitemetal-lined bearing due to shaft ridging caused by differential wear in the relief groove. Figure 3-75 shows fatigue cracking of 20% tin-aluminum lining due to misalignment and consequent edge loading. Fatigue cracking of a whitemetal-lined bearing due to shaft deflection and edge loading is depicted in Figure 3-76. Finally, the whitemetal-lined top-half turbine bearing in Figure 3-77 is partially fatigue cracked due to out-of-balance loading and an excessively wide cutaway section.

These journal bearings should be discarded. Investigate and, if possible, rectify the causes of misalignment, shaft deflection, and overloading. Increase the width of lands on the top half (Figure 3-77). Fit new bearings, possibly of stronger material if underdesign is indicated.

Figure 3-73. Damage caused by excessive steady load at startup.

Figure 3-74. Fatigue cracking due to shaft ridging.

Figure 3-75. Results of misalignment and high edge-loading.

Figure 3-76. Fatigue cracking due to shaft deflection and high edge-loading.

Figure 3-77. Fatigue cracking due to high rotor imbalance.

Corrosion

Corrosion of the lead in copper-lead and lead-bronze alloys, and of lead-base whitemetals, may be caused by acidic oil oxidation products formed in service, by ingress of water or coolant liquid into the lubricating oil, or by the decomposition of certain oil additives.

Removal of overlays by abrasive wear or scoring by dirt exposes the underlying lead in copper-lead or lead-bronze interlayers to attack, while in severe cases the overlays may be corroded.

The journal bearings shown in Figures 3-78 through 3-81 have been damaged by corrosion. In Figure 3-78, severe corrosion of the surface of an unplated copper-lead lined bearing is caused by the lead phase being attacked by acidic oil oxidation products. Figure 3-79 magnifies a fractured section of the bearing shown in Figure 3-78, indicating the corrosion depth of the lead layer. Corrosion of a marine turbine whitemetal bearing is illustrated in Figure 3-80. Water in the oil has caused a smooth, hard, black deposit of tin dioxide to form on the surface. Figure 3-81 captures

Figure 3-78. Corrosion caused by unserviceable lube oil.

Figure 3-79. Magnification of Figure 3-78.

Figure 3-80. Tin-dioxide formation attributed to water contamination of lube oil.

Figure 3-81. Sulphur corrosion of phosphor-bronze bushing.

"sulphur corrosion" of a phosphor bronze bushing. This damage was initiated by the decomposition of a lubricating oil additive and by gross pitting and attack of the bearing surface.

In all of these damage cases it will be necessary to scrap the bearings. Investigate the condition of the oil to ascertain the cause of corrosion. Eliminate water in the oil. In case of sulphur corrosion, change the bearing material from phosphor bronze to a phosphorous-free alloy such as lead-bronze, silicon bronze, or gunmetal. It would also be appropriate to consult with oil and bearing suppliers.

Cavitation Erosion

Cavitation erosion is an impact fatigue attack caused by the formation and collapse of vapor bubbles in the oil film. It occurs under conditions of rapid pressure changes during the crank cycle in internal combustion engines. The harder the bearing material the greater its resistance to cavitation erosion.

Impact cavitation erosion occurring downstream of the joint face of an ungrooved whitemetal-lined bearing is shown in Figure 3-82. Figure 3-83 is a section through the main bearing. It illustrates the mechanism of cavitation and modifications made to the groove to limit or reduce damage. In Figure 3-84, observe the result of discharge cavitation erosion in an unloaded half-bearing caused by rapid movement of the journal in its clearance space during the crank cycle. Also of interest is Figure 3-85, a set of diesel-engine main bearings showing cavitation erosion of the soft overlay while the harder tin-aluminum is unattacked.

The journal bearings shown in Figures 3-82 through 3-85 may be put back into service if cavitation attack is not very severe or extensive. Investigate increasing the oil-supply pressure, modifying the bearing groove, blending the edges or contours of grooves to promote streamline flow, reducing the running clearance, or changing to a harder bearing material.

Figure 3-82. Cavitation-erosion damage.

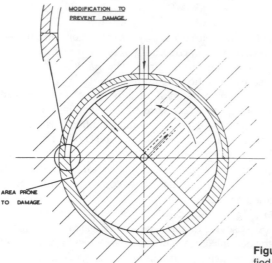

Figure 3-83. Section through modified main bearing.

Figure 3-84. Result of discharge-cavitation erosion.

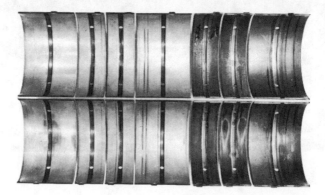

Figure 3-85. Cavitation erosion in diesel-engine main bearings.

"Black Scab" or "Wire Wool" Damage

A large dirt particle (probably not less than one mm across) carried into the clearance space by the lubricating oil and becoming embedded in the bearing may form a hard scab of material by contact with the steel journal or thrust collar. This scab will then cause very severe damage to the mating steel surface, which is literally machined away with the formation of so-called "wire wool." The action is self-propagating once it has started. The susceptibility to scab formation appears to depend upon the nature of the lubricant and the composition of the steel of the rotor shaft or collar. Steels containing chromium or manganese in excess of 1% appear to be particularly susceptible to scab formation, especially in high-speed machines with bearing rubbing speeds of over 20 meters (> 63 feet) per second.

Journal bearings exhibiting this type of damage are shown in Figures 3-86 and 3-87. In Figure 3-86, a whitemetal-lined compressor bearing shows "black scab" at its inception. In Figure 3-87, the 13% chromium-steel journal running in the bearing shown in Figure 3-86 exhibits severe machining damage.

Figure 3-86. Whitemetal-lined compressor bearing with early indication of "black scab."

Figure 3-87. Severe machining damage on 13% chrome journal.

Figure 3-88. Thrust pad with severe "black scab" formation.

Black-scab formation on a whitemetal-lined thrust pad is pictured in Figure 3-88. Black-scab or wire-wool damage renders the bearings unusable; they must be scrapped. Pay particular attention to cleanliness during assembly and avoid contaminating the bearing surface and oil passages with hard debris. Investigate changing the journal or collar surface material by sleeving with mild steel or by hard-chrome plating. Changing the bearing alloy is unlikely to be effective.

Pitting Due to Electrical Discharge

Electrical discharge through the oil film between the journal and bearing in electrical machinery or on the rotors in fans and turbines may occur from faulty insulation or grounding, or from the build-up of static electricity. This can occur at very low voltages and may cause severe pitting of bearing or journal surfaces, or both. In extreme cases damage may occur very rapidly. The cause is sometimes difficult to diagnose because pitting of the bearing surface is ultimately followed by wiping and failure, which may obscure the original pitting.

Fine hemispherical pitting and scoring of a whitemetal-lined generator bearing due to electrical discharge is evident in Figure 3-89. Figure 3-90 represents a close-up of more severe electrical discharge pitting of whitemetal.

Bearings have to be discarded if severely pitted or wiped. Examine and, if necessary, regrind the journal to eliminate pitting. Investigate grounding conditions of the machine, especially around insulation. Pay particular attention to fittings such as guards, thermocouple leads, water connections, etc. which may be bridging the insulation. Fit new bearings and run for only a short time, depending upon the period run prior to failure. Disassemble and examine.

Figure 3-89. Whitemetal-lined generator bearing with electrostatic-erosion damage.

Figure 3-90. Closeup of electrical discharge pitting of whitemetal bearing.

Damage Caused by Faulty Assembly

Under this heading are included:

1. Fretting damage due to inadequate interference fit in a weak housing.
2. Excessive interference fit, causing bearing-bore distortion.
3. Effect of joint-face offset or absence of joint-face relief in bearing. This causes overheating and damage in the region of the bearing split.
4. Fouling at crankshaft fillets due to incorrect shaft radii.
5. Misalignment or shaft deflection, causing uneven wear of bearing.
6. Entrapment of foreign matter between bearing and housing during assembly, causing bearing-bore distortion and localized overheating.
7. Effect of grinding marks on shaft.

The damage observed in Figure 3-91 represents fretting between the back of the bearing and the housing due to inadequate interference fit. Alternatively, this could also be due to a weak housing design. In Figure 3-92, observe overheating and fatigue at joint faces due to excessive interference fit, causing bearing-bore distortion. Joint face offset at assembly or a weak housing design could again cause this defect. Fretting at the joint faces of a connecting rod due to inadequate bolt tension is depicted in Figure 3-93. In Figure 3-94, observe uneven wear of an overlay-plated bearing due to misalignment. Fouling between shaft radius and the bearing-end face caused the overheating damage in Figure 3-95. Finally, Figure 3-96 illustrates severe scoring of a whitemetal-lined compressor bearing caused by shaft grinding marks.

Figure 3-91. Fretting caused by inadequate interference fit.

Figure 3-92. Overheating and fatigue damage caused by bearing-bore distortion.

Figure 3-93. Fretting caused by inadequate bolt tension.

Figure 3-94. Uneven wear caused by misalignment.

Figure 3-95. Excessive heating due to inadequate oil flow caused this damage.

Figure 3-96. Scoring caused by grinding marks in shaft.

In Figure 3-97, observe the back portion of a tin-aluminum bearing showing a flat spot on the steel shell. This was caused by debris entrapped between the shell and housing. Figure 3-98 depicts the bore of the bearing shown in Figure 3-97. Entrapped debris has deformed the bearing bore, resulting in flexure and fatigue damage. In Figure 3-99, angular misalignment of shaft or housing or dirt trapped behind the carrier ring caused damage to the pads (however, on one side only).

Whenever faulty assembly is suspected or confirmed, investigate and if necessary correct the interference conditions and housing design. Correct the misalignment, discard the bearings, and fit new bearings or pads.

Figure 3-97. Flat spot on steel shell caused by careless assembly.

Figure 3-98. Bearing-bore deformation resulting from the careless assembly in Figure 3-97.

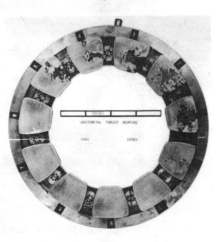

Figure 3-99. Angular misalignment or dirt trapped behind carrier ring caused damage to pads.

Inadequate Lubrication

Inadequate lubrication may be due to insufficient oil-pump capacity, inadequate header or oil passage diameters, incorrect grooving design, or accidental interruption of the oil supply.

In the case of machines provided with shaft-driven lubricating oil pumps only, the failure condition may originate with oil starvation during startup. This in turn could be caused by operating hand-priming pumps inadequately.

Depending upon the duration and severity of the lubrication failure, surface melting, wiping, or complete seizure of the bearing may occur.

Journal bearings damaged by inadequate lubrication are shown in Figures 3-100 through 3-102. Fatigue cracking caused by inadequate lubrication due to axial grooving in the loaded area is responsible for the damage in Figure 3-100. Figure 3-101 shows the effect of overheating and damage of a tin-aluminum bearing surface due to reduced oil flow. This was caused by omitting the drain hole at one end of the bearing, in effect causing the bearing cavity to be sealed and inhibiting through-flow. Finally, Figure 3-102 illustrates the seizure of a 20% tin-aluminum bearing due to lubrication failure.

Figure 3-100. Fatigue cracking caused by inadequate lubrication.

Figure 3-101. Overheating caused by insufficient oil flow.

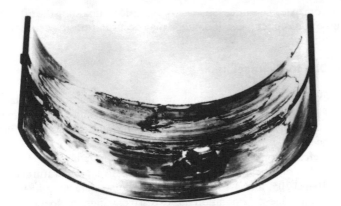

Figure 3-102. Tin-aluminum bearing seized due to lubrication failure.

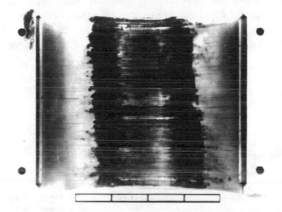

Figure 3-103. Surface damage caused by external vibration.

Recommended action is quite straightforward. Investigate and, if necessary, correct the groove design. Investigate all lubrication conditions. Scrap the bearings. Examine and, in cases of seizure, arrange for dye-penetrant checking of the journals. Fit new bearings.

Fretting Damage Due to Vibration

The damage we are talking about here is fretting damage in the bore as distinct from fretting damage elsewhere. Bearing operating surfaces may suffer fretting damage while the shaft is at rest. This could be caused by vibrations being transmitted to the machine from external sources, etc.

Figure 3-103 shows surface damage caused by external vibration while the journal was not rotating. Should this be the failure mode, eliminate the transmission of vibration from external sources, if feasible. Consider mounting the affected machine on springs or on rubber pads. If damage may occur in transit, secure the journal to prevent vibration.

Damage Due to Overheating

Lining extrusion and cracking may be due to overheating. A reduction in the strength of linings may result, so that the material yields and develops cracks because of the transmission of normal and shear forces through very thin oil films. Wiping does not necessarily occur under such conditions.

Surface deformation can be caused in an anisotropic material by thermal cycling.

Tilting-pad thrust bearings damaged by excessive temperature are shown in Figures 3-104, 3-105, and 3-106. Figure 3-104 depicts cracking of the whitemetal lining of a pad due to operation at excessively high temperatures. Note the displacement of whitemetal over the edge of the pad due to extrusion. Cracking and displacement of the whitemetal lining of a pad due to overheating under steady load conditions is shown in Figure 3-105. In Figure 3-106, observe thermal "ratcheting" of the whitemetal lining of a pad due to in-service thermal cycling through an excessive temperature range. In all three cases, it would be appropriate to investigate and, if possible, reduce maximum operating temperatures. Fit new or reconditioned pads.

Figure 3-104. Whitemetal liner damaged due to excessive temperature.

Figure 3-105. Overheating under steady load conditions caused white-metal cracking and extrusion.

Figure 3-106. Thermal "ratcheting" due to repeated cycling over an excessive temperature range.

Pivot Fatigue

Axial vibration can cause damage and fatigue of tilting-pad pivots. Both pivot and carrier ring may suffer damage by indentation or fretting. In some cases tiny hemispherical cavities may be produced. Damage may occur due to axial vibration imposed upon the journal, or may be caused by the thrust collar face running out of true.

In Figure 3-107, the pad carrier ring shows damage due to axial shaft vibrations. Similarly, Figure 3-108 depicts a pad pivot with hemispherical cavities caused by pivot fretting due to vibration.

Here, it will be necessary to recondition the carrier ring and fit new or reconditioned pads. Investigate and eliminate the cause of axial vibration.

Gear Failure Analysis

Someone once said that "Gears wear out until they wear in, and then they wear forever." The American Gear Manufacturers Association (AGMA) describes this mechanism more clearly as follows: "It is the usual experience with a set of gears on a gear unit . . . assuming proper design, manufacture, application, installation, and operation . . . that there will be an initial 'running-in' period during which, if the gears are properly lubricated and not overloaded, the combined action of rolling and sliding of the teeth will smooth out imperfections of manufacture and give the working surface a high polish. Under continued proper conditions of operation, the gear teeth will show little or no signs of wear."

Despite the truth of this observation, failures of gear teeth will occur as a result of certain extraneous influences. In many situations, early recognition may suggest a remedy before extensive damage takes place. High-speed gears on process

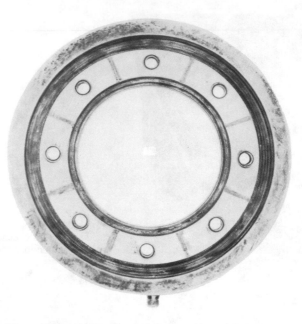

Figure 3-107. Pad carrier ring damaged by axial shaft vibration.

Figure 3-108. Pad pivot with hemispherical cavities caused by pivot fretting due to vibration.

machinery should therefore receive regular and thorough inspections as part of a well-managed predictive maintenance program.

The AGMA cites 25 modes of gear-tooth failure. Ku[4] thought it more logical to classify gear-tooth failure modes into two basic categories: strength-related modes and lubricant-related modes. Major modes of strength-related failure are, for instance, plastic flow and breakage. Examples of lubricant-related failure modes are rubbing wear, scuffing or scoring, and pitting.

Table 3-4 shows the principal gear failure modes and distress symptoms and relates them to their causes. Gear failure statistics are presented in Tables 3-5 and 3-6.[5]

Gear failure analysis is a complex procedure. It draws clues as to the most probable cause of failure from visual inspection, metallurgical analysis, tribological mechanisms, system dynamic analysis, design review of the casings, bearings, shafts, and seals, and, finally, the detailed analysis of the individual gears. A universally applicable approach to gear failure analysis is the subject of the following section.

Table 3-4
Gear Failure Modes and Their Causes

		Failure Mode											
		Fracture/Separation						Wear				Deformation	
		Tooth		Flank									
Failure Causes		Fracture (breakage)—overload	Fracture (breakage)—fatigue	Cracking *	Initial pitting	Destructive pitting *	Spalling *	Wear *	Scuffing *	Scoring	Corrosion	Plastic flow *	Warm flow *
Operating Conditions / Manufacture and Design	Manufacturing problems		●	●	●	●	●	●				●	
	Overload through misalignment	●											
	Frequent load cycles		●										
	Fatigue design		●		●	●	●						
	Operating Conditions (speed, loading)					●	●	●	●	●	●		
Lubrication	Viscosity							●	●	●	●		
	Quality							●	●	●	●		
	Quantity							●	●	●			●
	Contamination							●	●	●	●		

Legend: * In many cases leading to fatigue failure.

Table 3-5
Distribution of Stationary Gear Failure Causes[5]

Failure Cause	Distribution of Failure Cases, %
Vendor problems	36.0
Planning/design	12.0
Assembly	9.0
Manufacturing	8.0
Material of construction	7.0
Operating problems	47.0
Maintenance	24.0
Repair	4.0
Mishandling	19.0
Extraneous influences	17.0
Foreign bodies	8.0
Disturbances from the driving or driven side	7.0
Disturbances from electrical supplies	2.0

Table 3-6
Gear Failure Mode Distribution[5]

(Gear) Failure Mode	Approximate Distribution of Failure Cases, %
Forced fractures	50
Fatigue failures	16
Changes in load patterns	15
Incipient cracks	15
Distortions	4

Preliminary Considerations*

The traditional first step in gear failure analysis is to visually examine the failed parts. Gears, shafts, bearings, casings, lubricant system, and seals should be given attention. Data on the magnitude of the external loads applied by driving and driven equipment may be very helpful in differentiating between failure types. A correct evaluation of the failure type is essential in establishing the direction and depth of the remainder of the study.

---

*Based on "Introduction to Gear Failure Analysis," a paper given at the National Conference of Power Transmission, Chicago, Illinois, 1979, by P. M. Dean, MTI. By permission of the conference committee and the author.

In order to improve communication concerning gear failure problems, the AGMA published a standard in 1943 identifying some of the more frequently encountered failures. The latest revision of this standard, AGMA 110.XX, "Nomenclature of Gear Tooth Failure Modes," was published in September 1979. The photographs used in this section are taken (with permission) from this document[6] and are identified by the same captions used in the standard.

Analytical Evaluation of Gear Theoretical Capability

An important step when investigating a gear failure is to analytically evaluate the theoretical capabilities of the gears. In some cases these data will help differentiate between types or degrees of failure. More frequently, however, these data will lead the investigator toward essential bits of evidence. For example, in a case where pitting is discovered relatively early in the life of a gear set, a decision must be made: Will the gears correct themselves and the pits heal over, or will the pitting continue into destructive pitting, causing gear failure? If the investigator has made a good theoretical evaluation of the capabilities of the specific gears to withstand the known loads, he will be better guided toward the types of evidence needed. In this case, the theoretical loads may be within the capabilities of the gears, and only minor surface errors are being corrected. If the type of pits and distribution of the pitted areas support the normal-load/local-asperity thesis, the decision to run longer can be made with some confidence. On the other hand, the early discovery of pitting may have been fortuitous: The gears could be trying to say something. In one case, the pitted zones may indicate improper alignment to a degree that the internal pitting process could never correct. Even more subtle are cases wherein initial pitting cannot be explained by asperities or even by the anticipated loads. In such cases, a torsional dynamic analysis or highly instrumented measuring technique may be applied to determine if transient loads caused the observed pits.

The analytical techniques suggested here have been found to give good, realistic correlations. In cases where the calculated results agree with observed data, the investigator knows to look deeper into the problem for more subtle causes.

It is our intent to show the more generally accepted methods of establishing the surface stresses, bending fatigue stresses, flash temperature (for scoring), and lubricant film thickness (for wear). The details of applying these methods must remain beyond the scope of this text, but are covered in detail in the references noted. Key equations are shown as an aid in identification, and the terms of all equations may be found in Appendix A.

Metallurgical Evaluation

An extremely important aspect of a failure investigation is a thorough metallurgical analysis. The purpose of such an analysis is to determine if the gear did, in fact, have the physical properties anticipated in the design.

There are a number of potential defects that can only be found by sectioning the gear to obtain metallurgical specimens. Such defects include excessive grain size, nonmagnetic inclusions, seams, cracks and folds, undissolved carbides, excessive retained austenite, intermediate transformation products, grain boundary networks, excessive decarburization, banding and nonuniform transition from case to core, and excessive white layer (nitriding), depending on the type of heat-treatment process. Unfortunately, some gears are cut and finish-machined with much more attention than is given to them during heat treatment. Hence, many failures have been caused by improper metallurgy.

General Mechanical Design

Close attention must be given to evaluating the adequacy and accuracy of the mountings. A major effort in any investigation is to establish the degree, if any, to which misalignment may have contributed to failure. Misalignment can be caused either by manufacturing errors or by load-induced deflections in the shaft, bearings, or casings. In some cases, visual evidence may be found in the form of polished areas, or the lack of such areas, on contacting surfaces of the teeth, which may indicate operating alignment errors. In some cases it may be necessary to resort to deflection tests to determine the degree to which misalignment may occur under load.

In most gear failures, machine elements such as shafts, bearings, and seals will show signs of damage or will have failed. A study of these elements will help develop the clues necessary to reconstruct the failure sequence.

Lubrication

Proper lubrication is essential. The lubricant has two principal functions: to minimize rubbing friction and to carry off heat. If the lubricating film thickness is adequate and the lubricant is clean, wear will be minimized. The lubricant must sometimes provide protection against corrosion.

Some applications require lubricants having many additives so that the lubricant can adequately perform many tasks. Sometimes the lubricant itself fails. Its internal chemical compounds can break down with time, with heat, or by unforeseen reactions with chemicals from its ambience. Several types of gear failures can be traced back to lubricant failure.

Defects Induced by Other Train Components

Gearing is frequently part of a complete power train that includes a number of major components connected by shafts and couplings. In certain applications, such as in the petrochemical industry, compressor trains can be quite extensive, which in turn leads to very involved spring-mass systems. Such systems frequently have numerous potential problem areas of critical speeds and nodal points. The actual operating loads experienced by the various gears may be markedly different from the name-plate power ratings of the driving equipment.

In such applications, torsional dynamic analysis is essential during gear selection, and if failures occur, such analyses should be carefully reviewed or repeated.

Wear

Gear-tooth wear occurs as the surface material of the contacting areas is worn away during operation.[7] The type of wear that is under way can be determined by evaluating the:

1. Visual appearance.
2. Operating film thickness.
3. Particle size present in the oil.
4. Gear surface hardness.
5. Load operating profile.
6. Ambience.

It is convenient to group the more generally recognized types of wear into the following categories. These categories are based on appearance, cause, and effect.

1. Normal wear
   a. polishing
   b. moderate wear
   c. excessive wear
2. Abrasive wear
3. Corrosive wear

When evaluating wear, it is helpful to know the type of lubrication regime in which the gears are operating. At present, there are three generally recognized operating regimes:

Specific film thickness, which is the ratio of film thickness to surface asperity height, is a convenient index of the lubrication regime in effect. The risk of wear is evaluated in Table 3-7.

Table 3-7
Risk of Wear for Specific Ratios of Film to Asperity Thickness

Film Thickness Ratio (λ)	Operating Regime	Risk of Wear
up to 1.5	boundary (wear)	high
1.51 to 4.0	mixed	moderate
4.1 and over	full elastohydrodynamic	low

A typical equation to determine film thickness in helical gearing is:

$$h = \frac{2.65 \; \alpha^{.54}}{E^{.3} \; w^{.13}} \; \left(\eta_o \; \frac{\pi N_p}{30} \right)^{.7} \; \frac{(C \sin \phi_n)^{1.13}}{\cos^{1.56} \psi} \; \frac{m_G^{.43}}{(m_G + 1)^{1.56}}$$

An equation for specific film thickness is:

$$\lambda = \frac{2 \; h}{S_1 + S_2}$$

These equations, based on the work of Dowson and Higginson, are derived and discussed in Ref. 8. Terms are explained in Appendix D. A similar approach may be found in Ref. 9.

Since the film thickness is a function of the temperature of the oil within the mesh contact zone, there may be an uncertainty as to its actual value. It is helpful to plot both the film thickness and the specific film thickness for the range of temperatures that might exist within the contact zone (see Figure 3-109).

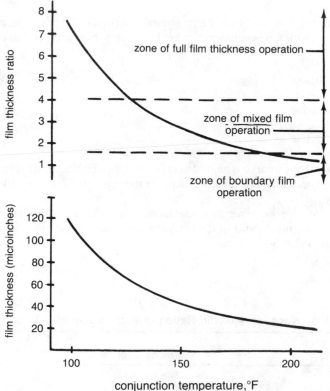

Figure 3-109. Film-thickness ratio and film thickness vs. conjunction temperature.

Normal Wear

An examination of gears operating within the "mixed" film-thickness operating regime (see Table 3-7) will usually show that they are operating with normal wear. Types of normal wear are:

Polishing. Gears that have run for a considerable period of time with good lubrication will usually show polishing. (see Figures 3-110 and 3-111). If the calculated film-thickness ratio is at the high end of the "mixed" regime, one should anticipate only polishing. A uniform polish, extending from end-to-end and from root to tip of the tooth, observed after an appreciable period of running, is an indication of a well-designed, correctly assembled and maintained gear set.

Polishing usually means a low rate of wear. Unless the gears are to have an extremely large number of mesh cycles during their anticipated life, low rates of polishing are usually of no concern. If the polishing, observed early in the life of the set, seems to indicate some wear, a heavier oil or a lower oil inlet temperature may be considered. These changes will tend to reduce wear by thickening the oil film.

Figure 3-110. Polishing. Run-in under operating conditions, this hardened hypoid pinion has developed a highly polished surface. Undoubtedly the extreme-pressure additives in the lubricant had a tendency to promote polishing on the tooth surfaces.

Figure 3-111. Polishing. This herringbone pinion, nitrided to 42 Rockwell C, shows a high degree of polish after 800 hours of operation with a mild E.P. lubricant.

Moderate wear. If the oil film is not sufficiently thick, wear may occur at a rate somewhat greater than in polishing. With this type of wear, a ridge at the operating pitch line of the teeth is noticed (see Figure 3-112). Metal is worn away in the addendum and the dedendum regions, where there is relative sliding. The operating pitch line, a zone of no sliding, then becomes evident, since it is higher than the remainder of the tooth.

Moderate wear can be controlled in some cases by an increase in film thickness, as discussed under "Polishing." This type of wear may also be a sign of dirt in the oil. In such a case the oil should be filtered to a particle size smaller than the film thickness. Moderate wear is not generally a problem unless its rate of progress causes the gear to run improperly because of vibration, breakage, or noise before the anticipated design life has expired.

Excessive wear. If the specific-film-thickness ratio is 1.0 or less, excessive wear may be anticipated, particularly for sets which will be subjected to many cycles (see Figure 3-113). If the wear rate is sufficiently high, so much material will be removed that the teeth will either break off due to thinning or run too rough to be satisfactory, long before the end of their anticipated life.

All of the modifications suggested under "Moderate Wear" may be applied to correct this problem. In addition, harder materials or a change in gear geometry may be needed.

If the gears that exhibit excessive wear are operating with a film-thickness ratio near or below 1.0, the wear can be explained on the basis of asperity contact. If, however, the specific film thickness is nearer 4, the search should concentrate on evidence of abrasive wear or the existence of high loads due to excessive vibratory or torsional activity.

Figure 3-112. Moderate Wear. Wear has taken place in the addendum and dedendum sections of the gear teeth. This causes the operating pitch line to be visible because little if any material has been removed.

Figure 3-113. Excessive Wear. Material has been uniformly worn away from the tooth surface causing a deep step. The tooth thickness has been decreased and the involute profile destroyed. Some slow-speed gears can operate on a profile of this nature until so much material has been worn away the tooth fails in beam bending fatigue.

Figure 3-114. Abrasive Wear. This herringbone pinion, nitrided to 42 Rc, shows abrasive wear from foreign particles imbedded in the mating gear.

Abrasive Wear

If dirt particles are present in the oil and their size is greater than the thickness of the lubricating film, abrasive wear will usually occur (see Figure 3-114). In most cases, abrasive wear can be reduced by attention to the cleanliness of the lubricant system. Tight enclosures, adequate seals, fine filters, and regular oil-sampling or changeout practices are required. An extreme case of abrasive wear is shown in Figure 3-115.

Figure 3-115. Abrasive Wear (Extreme Case). A large portion of the tooth thickness has been worn away due to an accumulation of abrasive particles in the lube-oil supply. Note the deep ridges and the end of the gear teeth that were not subjected to the abrasive action.

Corrosive Wear

Corrosive wear is induced on the gear-teeth surfaces as a result of a chemical action attacking the gear material (see Figures 3-116 and 3-117). The active factors attacking the materials may come from sources external to the casings of the gear set or from within the lubricant itself.

If the sources of the active factors are external to the gear casing and its lubricant system, they may enter through the breather, seals, or gaskets, in the casings, or through openings in the lubricant system. Once the method of entry is found, it is usually possible to devise means to limit further entry. In particularly hostile environments, a change in materials or special means of protecting the materials may be required.

The active factor may also result from a chemical breakdown within the lubricant itself. Lubricants containing extreme pressure additives sometime break down at high temperatures, forming very active chemical compounds. When such lubricants are used in the gear set, they should be checked at regular intervals to determine their continued suitability for service.

The original manufacturing practice itself can sometimes be the source of chemical attacks. Careful cleanup and neutralizing or passivating procedures must be undertaken after completing metal etching to remove grinding burns or chemical stripping to remove copper overlays used in case-carburizing processes.

The wear rate can sometimes be estimated by a series of measurements on specific teeth over an extended period of time. Unless wear is very pronounced, a single examination of a gear set may not be sufficient to determine if the rate of wear is so fast that the teeth will be unsatisfactory before the anticipated service life has expired.

Figure 3-116. Corrosive Wear. It is evident from this gear that considerable wear has taken place yet the surface still shows signs of being attacked chemically. Wear of this nature will continue until the gear surface is completely beyond usefulness.

Figure 3-117. Corrosive Wear. This AISI 9310 case-carburized and ground pinion shows the corrosive effects of running in a lubrication system that was contaminated with hydrogen sulfide (H_2S). The pinion has a 10 normal diametral pitch and was operating at a more than 10,000 ft/min pitch line velocity.

A gear set undergoing moderate wear (see Figure 3-112) which has been in service for some time may show the presence of a ridge at the operating pitch line near the center of the tooth. Any measurements made to establish either the amount or rate of wear should be made on the addendum or dedendum zones, away from the ridge at the operating pitch line.

Wear rate is a function of the load intensity, materials compatibility, material hardness, load-carrying characteristics of the lubricant, and operating time.

Even though all of the parameters that can be calculated—wear, scoring, and pitting—indicate a low probability of failure, some gears, such as those used in high-speed rotating machinery, accumulate so many mesh cycles during their normal life that even a low rate of wear will become appreciable in time. As a general rule, such gears should be made as hard as possible to keep to a minimum the wear from the small particulate matter in the oil and from the large number of starting cycles.

Scoring

Under certain combinations of operating load, speed, material-lubricant combinations, and operating temperatures, the oil film that is supposed to separate the active profiles of the meshing teeth may break down because of a high load temperature and allow metal-to-metal contact. This results in alternate welding and tearing actions called scoring. In some cases this may so deteriorate the active surfaces of the teeth that meshing performance is affected. The more generally recognized types of scoring may be grouped into the following categories. As in wear, these categories are based on appearance, cause, and effect:

1. Frosting
2. Moderate scoring
3. Destructive scoring

When evaluating scoring, it is helpful to analytically evaluate the degree of scoring risk inherent in the gear design. An intense effort, particularly in the field of aircraft gear design, has made possible a reasonable method of evaluating scoring risk in any given design. A typical equation for predicting flash temperature is as follows:

$$T_f = T_b + \frac{W_{te}^{.75}}{F_e} \left(\frac{50}{50\text{-}S} \right) \frac{Z_t(n_p)^{.5}}{P_d}$$

For a discussion covering the use of this equation and the evaluation of the individual terms, see Refs. 9 and 10.

Experience with this method of scoring evaluation indicates that scoring probability may be characterized as shown in Table 3-8.

For a given geometry of gears, scoring is strongly influenced by inlet temperatures, surface finish, and rubbing speed. In general, the flash temperature number at each

Table 3-8
Probability of Scoring Using AGMA Method

Calculated Flash Temperature Number	Degree of Risk
up to 275	low
275 to 350	moderate
over 350	high

5-10 points along the line of action should be determined. When tabulated, such an evaluation will appear as in Table 3-9. These data, when compared with those in Table 3-8, will give a good indication as to where, and often to what degree, scoring will likely occur.

Frosting. This is the mildest form of scoring. The affected areas of the tooth, usually the addendum, give the appearance of an etched surface (see Figures 3-118 and 3-119). When magnified, the area appears to consist of very fine pits, .001 inch or so

Table 3-9
Tabulation of Lubricant Flash Temperature
Number Along Gear Line of Action

Station	Roll Angle (Deg.)	Location	Flash Temperature Index
1	22.196	outside diameter	160.17
2	18.565		155.18
3	14.934		154.57
4	11.302	high point, single-tooth contact	162.95
5	12.474		160.88
6	13.646		158.16
7	14.818		154.92
8	15.989	low point, single-tooth contact	151.02
9	14.199		156.09
10	12.408		161.01
11	10.617	form diameter	163.82

Figure 3-118. Frosting. This nitrided helical gear shows frosting in its early stages where random patches are showing up. Close inspection shows small patches of frosting which follow along the machining marks where there are very small high and low areas. This frosting pattern shows no radial welding or tear marks.

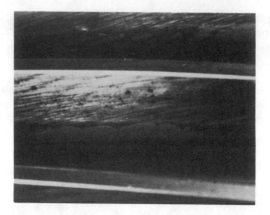

Figure 3-119. Frosting. This nitrided 4140 pinion shows early stages of frosting. Frosting appears in separate patches in the addendum and the dedendum areas of the teeth. Damage to the gear is not severe at this stage.

Figure 3-120. Moderate Scoring. This 5 diametral pitch 8620 case-carburized and ground spur gear shows scoring in the addendum and dedendum regions. Note also how scoring extends across some teeth and only part way on others.

deep. The frosted area usually marks the higher asperities or undulations that may exist on the surface of the teeth.

Frosting is frequently an indication of local errors in the heights of the tooth surfaces. If these areas can be worn away during continued running of the teeth without causing damage, the frosting will usually disappear.

Light to moderate scoring. This is a somewhat more intense form than frosting. In addition to the etched appearance, close examination shows fine scratches resulting from a sequence of welding and tearing-apart actions (see Figures 3-120 and 3-121).

Figure 3-121. Light to Moderate Scoring. The scoring action on this gear covers a large portion of the tooth surface. It has a frosty appearance and, upon close examination, it is evident that there has been metal-to-metal contact, a welding and a tearing apart due to rotation.

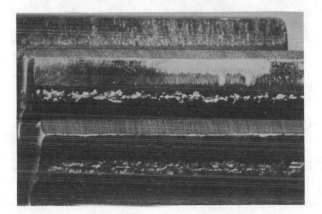

Figure 3-122. Destructive Scoring. Heavy scoring has taken place above and below the pitch line leaving the material at the pitch line high or proud. As a result of this, the pitch line pits away in an effort to redistribute the load. Usually, the gear cannot correct itself and ultimately fails.

Destructive scoring. This type of scoring results in a significant amount of material removal, and it may extend well into the addendum and the dedendum zones (see Figures 3-122 and 3-123). If the load is sufficiently concentrated at the pitch line, pitting may occur at such a rate that the tooth can never recover in its effort to redistribute the load.

Scoring, unlike tooth breakage or pitting, generally occurs soon after the gears are put into service, if it is going to occur at all. Thus, it may give early indications if a new gear design is of inadequate accuracy, if the gears have not been properly aligned (Figure 3-124), or if the design is deficient in some respect.

Figure 3-123. Destructive Scoring. This AISI 9310 ground aircraft spiral bevel gear shows destructive scoring.

Figure 3-124. Localized Scoring. This medium-hard (4340) marine gear which was hobbed and shaved shows scoring in a definite uniform pattern. In this case the waviness of the tooth surface contributed to the heavier scored patches superimposed on a more conventional uniform scored area.

If scoring is observed in local zones soon after a gear set has been put into service, the location and shape of the scoring patterns may indicate operating and assembly problems, which, if corrected, will eliminate further scoring (see Figure 3-125). Scoring is sometimes observed on gears that should have been given tip or profile modifications in their design (see Figure 3-126). In such cases, the addition of such modifications may correct scoring.

Since scoring is a local breakdown of the lubricating film, some control can be exercised through the lubricant itself. A change to a higher-viscosity lubricant or the use of extreme pressure additive lubricants may be considered. A lowering of the oil inlet temperature at the mesh and an increased flow of cool oil to reduce the gear-tooth temperature may also be effective.

Figure 3-125. Localized Scoring. This high-speed, high-load AISI 9310 case-carburized and ground helical gear shows localized scoring on the ends of the teeth. This failure was caused by a design oversight: when the gear casing reached operating temperature, differential thermal expansion caused a shift in alignment across the face of the gear. The resulting load distribution caused scoring.

Figure 3-126. Tip and Root Interference. This gear shows clear evidence that the tip of its mating gear has produced an interference condition in the root section. Localized scoring has taken place causing rapid removal in the root section. Generally, an interference of this nature causes considerable damage if not corrected.

Surface Fatigue

Surface fatigue results from repeated applications of compressive stress on the contacting areas of the teeth. This type of surface endurance failure is characterized by small pits in the surfaces of the teeth. These result from subsurface stresses beyond the endurance of the material.

It is convenient to group the more generally recognized types of surface fatigue into the following categories:

1. Pitting
   a. initial pitting
   b. destructive pitting
   c. spalling
2. Case crushing

The possibility of surface fatigue failure in the form of pitting or spalling can be predicted quite well by analytical techniques. The equations generally used for these predictions appear as given below and in Refs. 11 and 12.

The value of surface compressive stress number S_c is calculated as follows:

$$S_c = C_p ((W_T C_o/C_v) (C_s/d\ F) (C_m C_f/I))^{.5}$$

This is compared to an allowable value of surface compressive-stress number:

$$S_c' = S_{ac} (C_L C_H/C_T C_R)$$

These equations are discussed in detail in Ref. 7; the terms are defined in Appendix A. The term S_{ac} is the fatigue stress at which the material can be operated for 10^6 cycles. It is desirable to calculate a number of S_c values, one for each combination of load intensity and operating cycles at that given intensity.

When these two sets of data are plotted, the results appear as in Figure 3-127. In the particular case shown here, there is one point at which the load intensity for its number of operating cycles exceeds the capability of the material. When the gear set has run long enough to accumulate cycles at this load intensity, pitting can probably be observed. Pitting, which is a fatigue phenomenon, usually takes up to a few millions of load cycles before it can be observed, even though the load intensity is in the pitting range.

If a gear set is examined shortly before pitting has become clearly visible to the naked eye, it is usually desirable to make an estimate of the kind of pitting that is under way. Pitting tends to fall in one of two categories: "Initial pitting," a corrosive form of pitting which, when it has finished its task, will disappear; or "destructive pitting," which will continue as long as the gears are run.

If the applied stress is only slightly higher than the surface endurance limit of the material, the pits will tend to be small—just barely visible to the unaided eye. If, however, the applied stress is quite high relative to the capability of the material, the initial pits will be appreciably larger. In some cases, small pits may combine to form still larger ones if the high tooth loads are continued.

Initial pitting. This is a form of pitting that usually results in the correction of local areas, thus producing a gear set giving satisfactory operation (see Figure 3-128).

The initial contact in a new set will occur on the highest of the tooth asperities. If the load intensity on these areas is not unduly excessive, small pits will form and these areas will be corrected.

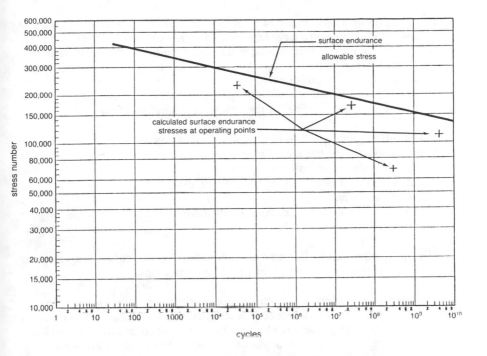

Figure 3-127. Allowable vs. calculated surface endurance stresses.

Figure 3-128. Initial Pitting. Pitting started on the outside end of the helix due to a small out-of-alignment condition. The pits were small in diameter. The pitting distress pattern progressively worked its way from the very end of the tooth toward the middle of the tooth. After a length of time, the fine pitting tapered out. Upon continued operation, the pitting stopped altogether and the pitting surface began to burnish over, indicating no new pits were being formed and that the load across the tooth had a tendency to be more evenly distributed. (This type of pitting is often called corrective pitting.)

Corrective pitting. This is a form of pitting in which local high spots are pitted and worn down to a point where a large enough area of the remainder of the active tooth surface will come into contact to reduce the load intensity to the point where the overall surface stress is less than that which will cause pitting. In this case, the initial pitted areas will usually polish over.

If a gear set is examined shortly after pitting has appeared, the pitting may be characterized as "initial pitting" if the following criteria are met:

1. The pits are very small—under .030 inch in diameter.
2. The areas showing pits are quite localized and can be logically interpreted as high spots due to manufacturing errors.
3. The analytical evaluations of the surface endurance stresses indicate that pitting should not occur.
4. The pitted or polished areas extend well along the tooth, indicating good gear alignment.

This type of pitting is usually found in gears having slight manufacturing errors, or in teeth that should have been given profile modifications.

Destructive pitting. Destructive pitting is characterized by large pits (see Figures 3-129) and is often seen first in the dedendum region of the driving gear member. In this form of pitting, the initial areas that pit have pits so deep they cannot carry load. As the pits form, the load shifts to adjacent areas which in turn pit, and which also can carry no load. The pits are at all times so large that the teeth cannot heal over with continued running. Thus, pitting proceeds over the tooth from zone to zone until the entire tooth is destroyed.

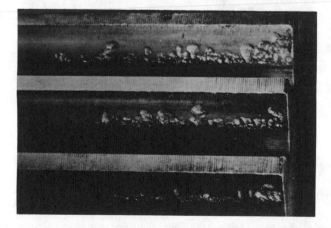

Figure 3-129. Destructive Pitting. Heavy pitting has taken place predominantly in the dedendum region. The pitted craters are larger in diameter than those denoted as corrective pitting. Some pitting has taken place in the addendum region as well.

Figure 3-130. Spalling. This hardened pinion shows an advanced stage of tooth spalling (not to be confused with case crushing). Material has been more or less progressively fatiguing away from the surface until a large irregular patch has been removed.

Figure 3-131. Spalling. This bronze worm gear shows an advanced stage of spalling in the loaded area of the tooth.

Spalling. This is a form of pitting in which large shallow pits occur (see Figures 3-130 and 3-131). Destructive pitting and spalling generally indicate that the gear set is loaded considerably in excess of the capability of the gear geometry and/or materials. Changes in material type and hardness may, in some cases, be sufficient to eliminate destructive pitting.

Failures from the Manufacturing Process

Just as some gear failures are "designed-in" to the gear through a lack of appreciation of the design requirements of the job, others may get "manufactured-in" by improper shop practices.

Grinding Cracks

Grinding cracks are an example of a manufactured-in problem. Grinding is a process that can produce an excellent gear—if it is done correctly. It does have the inherent danger that, through lack of proper attention to detail, localized areas of the teeth may suffer grinding burns. When feeds, speeds, grain size, wheel hardness, and coolant are not just right, intense localized hot spots will be developed by the wheel, and "burns" result. The localized heat will draw hardness of the steel, or in some cases quench cracks will form (see Figures 3-132 and 3-133). These cracks can become the focus of stress raisers, leading to tooth breakage.

Figure 3-132. Typical quench cracks in medium hard material.

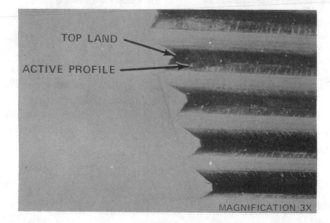

Figure 3-133. Grinding Cracks (3 × Size). Grinding cracks of the parallel type. These cracks can be seen by eye; however, they show up best when a flaw-detection procedure is utilized such as magna-flux or magna-flow. These cracks were found on a finished ground case-carburized AISI 9310 gear.

Burns can be prevented by good shop practice. They can be detected by chemical etching techniques which make the burned areas visible. The teeth may also be inspected with similar nondestructive techniques for grinding cracks.

Case Crushing

Case crushing is a subsurface fatigue failure which occurs when the endurance limit of the material is exceeded at some point beneath the surface. It is a fatigue-type failure, generally occurring in case carburized, nitrided, or induction-hardened gears.

The Hertzian stresses resulting from two teeth in contact occur along bands extending across the teeth. Both the width of the band of contact and the depth from the surface to the point of maximum shear stress can be calculated by methods given in Ref. 13. Accordingly, the maximum stress occurs at a point about .393 of the width of the band below the surface.

Figure 3-134 illustrates a section across a case-carburized and hardened gear tooth in its normal plane. Superimposed on this section is the results of a Tukon

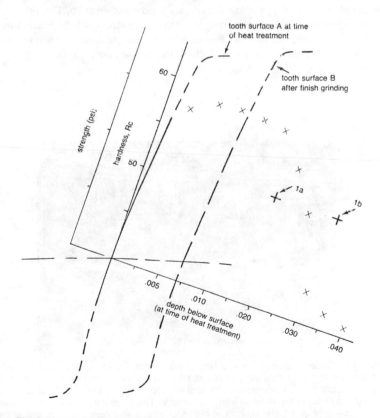

Figure 3-134. Relationship of subsurface strength to subsurface stress.

microhardness survey. Plotted perpendicularly to the surface near the pitch circle is depth to point of measurement (abscissa) versus hardness (ordinate). The X's are the values of measured hardness. In a good case-hardening job, the surface is only slightly less hard than the major part of the case because of decarburization.

The strength of the material is directly related to hardness; the maximum hardness, therefore strength, exists just below the surface and extends to a depth of about .020 inch, where it falls off to the core hardness and strength. The solid line "A" is the tooth surface at the time of the case-carburizing and hardening process.

When the tooth is run with a mating gear under load, a subsurface shear stress is developed. Its magnitude is plotted as point 1a to the same scale as that of the material strength. Since the stress is below the value that the steel can carry at this depth, the tooth will be satisfactory under this load.

Most case-carburized gears are ground to correct the almost inevitable errors that result from the distortions and changes in size that result from the heat-treating process. In some cases a highly variable amount of material must be ground from the teeth of an individual gear in order to clean it up to drawing-size tolerances. In such cases, the tooth surface may be ground down to the dashed line "B" in Figure 3-134.

When this part of the tooth is loaded up to the same value as before, the same stress will occur at the same depth below the surface, and is shown at 1b. This point now lies well inside the weaker core material, the prevailing stress is sufficiently intense to cause a subsurface shear failure. The failure can progress generally parallel to the surface and break through the surface where the stresses are right (see Figure 3-135), or progress deeper into the tooth, as in Figure 3-136.

Figure 3-135. Case Crushing. Several large longitudinal cracks, and several smaller ones, have appeared in the contact surface of this case-carburized bevel gear. The major cracks originated deep in the case-core structure and worked their way to the surface. Long chunks of material are about to break loose from the surface. The cracks should not be confused with normal fatigue cracks, which result from high root stress and normally form below the contacting face.

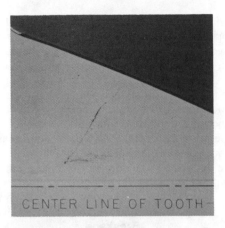

CENTER LINE OF TOOTH

Figure 3-136. Case Crushing Internal. This internal crack started just below the hardened case in the weaker core material. The material is AISI 9310; total case depth. 100 deep; origin of failure. 150 deep. The crack grew in both toward the case material in one direction and toward the opposite side root section in the other direction.

Figure 3-137. Random Fatigue Breaks. These breaks are not typical beam bending fatigue but instead originate high on the tooth flank and are somewhat uniformly dispersed across the face of the gear. Failures of this nature originate from stress raisers other than of the root fillet. In this installation, the fatigue failures were the result of grinding cracks on the tooth flanks during manufacture.

Random Failure

Sometimes only a part of a tooth will break out. In many such cases the crack did not originate in an area that is even close to the zone of the theoretical maximum (critical) bending stress within the tooth (see Figure 3-137).

Such failures may be generally attributed to the lack of proper control in the manufacturing process. The teeth may exhibit notches near the root sections, there may be cutter tears on the surface, the surfaces may have grinding burns, or the teeth may not have the proper profiles. Heat treatment may also be at fault. Improper heat treatment can produce high local stresses. The material may not be clean, being full of seams, inclusions, dirt, or other stress risers. All of these items are potential sources of local stress concentrations sufficient to cause failure (see Figure 3-138).

Local stresses due to pitting or spalling which are located away from the critical stress zones can also produce a focal point for a random failure.

A failure at the root of the tooth does not always result in an individual tooth breaking out. In some cases, the internal stress pattern will cause the fatigue failure to proceed down through the rim of the gear and into the web (see Figures 3-139 and 3-140). If this failure goes to its ultimate conclusion, an entire chunk containing several gear teeth will break out.

Breakage

Breakage is a type of failure in which part of a tooth, an entire tooth, or sometimes several teeth break free of the gear blank. It results either from a few load stresses greatly in excess of material capabilities or from a great many repeated stresses just above the endurance capability of the material.

It is convenient to group the more generally recognized breakage failure modes as follows:

1. Bending fatigue breakage.
2. Overload breakage.
3. Random fatigue breaks.

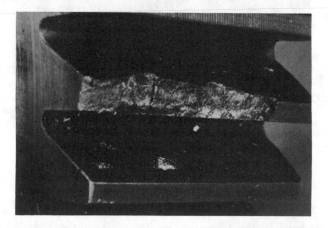

Figure 3-138. Random Fatigue Break. This AISI 4140 medium-hard gear ultimately failed from tooth fracture above the pitch line. The failure originated from large-diameter pits on the tooth surface, as can be seen on the bottom tooth. Note how a crack has propagated around the end of the gear tooth.

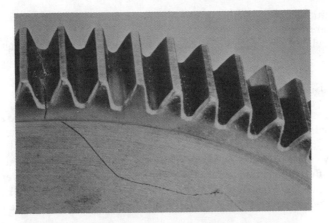

Figure 3-139. Rim & Web Failure. This hardened and ground spur gear fractured through the root section. The crack propagated from the root section down into the web of the gear. Failures of this nature are not uncommon on highly loaded thin-rim and web sections.

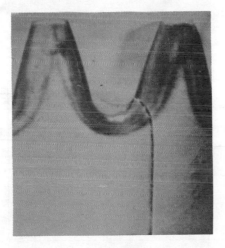

Figure 3-140. Rim Failure of a Coarse Pitch, Wide Face Width Industrial Gear. Note that the rim failure is not necessarily associated with the high root stresses but may be the result of high residual stresses in the gear wheel itself.

When evaluating a breakage failure, it is helpful to know the theoretical intensity of the bending stresses at the critical section of the tooth. It is also desirable to know the number of operating cycles at which each magnitude of bending stress will occur.

A typical equation for calculating the bending stress at the critical section of the tooth is:

$$S_t = ((W_t K_o)/K_v) \ (P_d/F) \ (K_s K_m/J)$$

A definition of each symbol is given in Appendix A. Details on the use of this equation are found in Ref. 11.

The calculated value of S_t is compared with an allowable value of S_t', determined as follows:

$$S_t' = S_{at}K_L/(K_TK_R)$$

This equation includes the effects of fatigue life and temperature, and allows for a factor-of-safety. The term S_{at} is the fatigue stress at which the material can be operated for 10^6 cycles.

In many cases, particularly where gears will accumulate a significant number of cycles at different load intensities, a plot showing the allowable stress versus fatigue cycles and the individual applied load stresses versus fatigue cycles makes it easy to determine which, if any, of the operating load points is a critical value (see Figure 3-141).

In the case of overload failures, the same equations apply, except that it is customary to use the yield stress for the value of S_{at}.

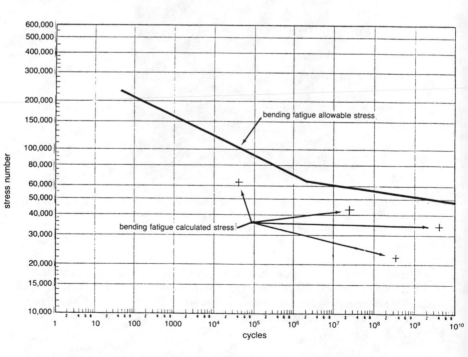

Figure 3-141. Allowable vs. calculated bending fatigue stresses.

Figure 3-142. Fatigue Failure. This tooth fracture shows the "eye" and "beach marks" that typify a subsurface fatigue failure.

Figure 3-143. Beam Bending Fatigue. This AISI 9310 case-carburized and ground aircraft power spur gear shows a fatigue crack caused by stresses at the root fillet.

Bending Fatigue Breakage

This failure mode is characterized by a fracture usually showing an "eye," beach marks, and an area of overload breakage (see Figure 3-142). The "eye" is the area where the failure started. The beach marks cover an area where the failure progresses step by step because of fatigue. These areas often show indications of fretting corrosion. After the fatigue failure has progressed well across the tooth, the remaining material will fail as a conventional fracture. Failure is the result of the continued application of load over many cycles.

The failure usually starts at the point of maximum bending stress or, if imperfections are present in the teeth, near a stress raiser (see Figure 3-143). The whole tooth or a major part may break away (see Figure 3-144).

Figure 3-144. Beam Fatigue Failure. This tooth fracture shows the type of "eye" and "beach marks" that typify a surface-oriented fatigue failure. The pinion was AISI 4340 hobbed and shaved.

This type of failure occurs because a local stress exceeds the fatigue capabilities of the material. In some gears, stress raisers are present because of manufacturing techniques. The most common of these are notches at the root fillets, cutter tears, inclusions, heat treatment and grinding cracks, and residual stresses. All of these can produce focal points from which fatigue cracks may originate.

Overload Breakage

In this type of failure, the fracture is characterized by a stringy, fibrous break, showing evidence of having been torn apart (see Figure 3-145). This type of failure is generally the result of an accidental overload far in excess of the anticipated design loads.

Lubricated Flexible/Coupling Failure Analysis

There are two basic types of shaft couplings employed throughout the petrochemical industry: couplings that are mechanically flexible, and those that rely on material flexibility, such as disc and diaphragm couplings. Designers and users of mechanically flexible couplings must contend with wear, which in turn is addressed by lubrication. Figure 3-146 shows the basic styles of couplings found in the petrochemical industry. By far the most widely used lubricated coupling is the gear coupling. When the gear coupling was invented in 1918, industry was changing from steam to electric power and from belt- to direct-driven machinery. Flexible shaft couplings permitted this change to direct shaft connection, and the gear coupling played an important role in this transition.

During the last ten years, the nonlubricated, high-speed, high-torque diaphragm coupling has gained increasing popularity. In spite of this development, a large

Figure 3-145. Overload Breakage. The break in this case-carburized AISI 9310 hardened and ground spur gear has a brittle fibrous appearance and the complete absence of any of the common beach marks associated with fatigue failure. One tooth failed from shock loading, and the force was then transmitted to successive teeth. As a result, several teeth were stripped from the rotating pinion.

Table 3-10
Causes of Mechanical Coupling Failures[5]

Failure Cause	Failure Distribution, %
Vendor Problems	26.0
Planning/design	13.0
Manufacturing	7.0
Material of construction	3.0
Assembly	3.0
Operating Problems	63.0
Mishandling	38.0
Maintenance	24.0
Repair	1.0
Extraneous Influences	11.0
Foreign bodies	4.0
Overloading	7.0

portion of major process compressor trains is still equipped with gear-style shaft couplings of various sizes and configurations. Their basic purposes are to transmit rotating power to a driven machine, to accommodate known or predictable misalignment of the connected shafts, and to protect both driving and driven machines from distress caused by misalignment.

As can be expected, 75% of all gear-coupling failures are due to improper or insufficient lubrication, as is true for all other lubricated couplings. The distribution of coupling failure causes is shown in Table 3-10; the distribution of failure modes is shown in Table 3-11.

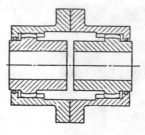

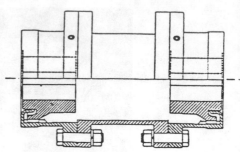

A: double-engagement gear-tooth coupling, grease (batch) lubricated; O-ring seal

B. same as A, but spacer-type; metal lip seal

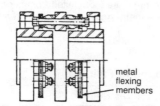

C: disc-type coupling with resilient components

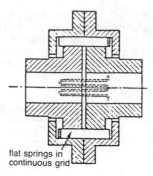

D: grid coupling

flat springs in continuous grid

metal flexing members

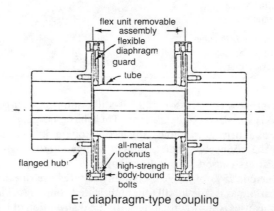

E: diaphragm-type coupling

flex unit removable assembly

flexible diaphragm

guard

tube

flanged hub

all-metal locknuts

high-strength body-bound bolts

Figure 3-146. Flexible shaft couplings used on process machinery.

Table 3-11
Distribution of Failure Modes of Mechanical Couplings[5]

Failure Mode	Distribution of Failure Cases, %
Forced fractures	60.0
Scuffing	18.0
Mechanical and corrosive surface damage	15.0
Incipient cracks	5.0
Bending, deformation	2.0

Table 3-12
Types of Typical Gear Coupling Failures[14]

Continuous Lube	*Standard or Sealed Lube*	*Excessive Misalignment*
1. Wear	1. Wear	1. Tooth breakage
2. Corrosive wear	2. Fretting	2. Scoring
3. Coupling contamination	3. Worm tracking	3. Cold flow
4. Scoring and welding	4. Cold flow	4. Wear
5. Worm tracking		5. Pitting

Gear-Coupling Failure Analysis

Few machinery components have received the amount of coverage in technical literature that has been devoted to gear couplings. No doubt, this is because lubricated couplings have frequently become the weak link in even well-managed process plants. By weak link, we are not implying that they are unreliable and prone to failure but that they need to be inspected and maintained periodically. The authors know of a well-managed petrochemical plant where during 15 years of running some 22 gear couplings on major centrifugal compressor trains, 3 forced outages were caused by gear-coupling failures. We leave it to our readers to decide whether they want to be satisfied with this record.

Machinery engineers are often fascinated by gear couplings because their "defect limit" is often difficult to determine. We believe from our experience with gear-coupling service representatives that this job is equally difficult for the expert. While it is relatively easy to judge the presence of "worm tracking," an assessment and recommendation in view of "slight wear" or even "corrosive attack" is much more difficult. Quite often the unavailability of a spare or the cost of change adds to the difficulty of making a decision after a gear-coupling inspection. K. W. Wattner of Texas Custom Builders showed clearly that wear is the predominant failure mode in gear couplings. His paper on failure analysis of high-speed couplings[14] lists the most common failure modes encountered in these couplings, as shown in Table 3-12. In short, assessing the various degrees of wear in gear couplings remains a challenge for even the most experienced machinery problem solver.

Gear-Coupling Failure Mechanisms

As gear-type couplings try to accommodate shaft misalignment, a sliding motion takes place inside the gear mesh. This motion is the main cause of coupling wear and failure. The sliding motion is composed of an oscillatory axial motion and a rocking motion that exists whenever the coupling is misaligned. The velocity of the oscillatory motion is usually called the sliding velocity. Figure 3-147 shows a gear-coupling half angularly misaligned.[15] The undulating shape of the developed gear-tooth path inside the hub is characteristic for coupling distress and failure caused by excessive misalignment.

Under proper alignment conditions and good lubrication, gear couplings can perform satisfactorily for many years. When those conditions become unfavorable, there are distress signals and forces that will adversely influence the coupled equipment before an actual coupling failure occurs. These distress signals may be subtle, but will nevertheless cause expensive downtime. Specifically there are axial forces generated by an increase in the coefficient of sliding friction and by bending moments which become excessive because of misalignment that is always present in even the most carefully assembled machinery train. The effects of misalignment are considerably more burdensome in gear couplings than in diaphragm couplings and may go so far as to weaken other major machinery components (Ref. 16).

The machinery problem solver is encouraged to get involved in regular coupling inspections. Records and photographs should be kept so that wear progress on gear couplings can be tracked. To facilitate this activity, refer to Table 3-13, which is presented in familiar matrix form.

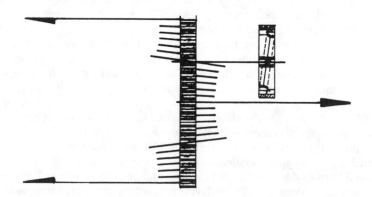

Figure 3-147. Developed plan of an angularly misaligned gear coupling half.[15]

Table 3-13
Gear Coupling Distress/Failure Analysis

Distress/failure Cause		Tooth fracture	Cracking	Pitting	Gouging △	Broken hub	Broken/sheared key	Flange fracture	Wear (adhesive)	Scuffing	Welding	Fretting/corrosive wear △3	Worm tracking	Shaft fretting	Cold flow (mesh)	Moisture/contamination	Bolt loosening	Leaking △	Overheating	Sleeve hunting	Sludging
Improper shrink fit	Design/Installation	①				①	①		⑤	④	③				②						
Random re-assembly		⑤	①	①					③	⑤	④	③	③		③						
High sliding velocity					①				③	④	④	③	③		③						
High misalignment								②	④	⑤	⑤	④	④					①			
Sealing △			⑤	⑤												①					
Insufficient bolt torque																	①				
Load requirements	Operating Conditions						②	①							①						
Vibration imposed by driving/driven machinery				③					②		②	②	②							①	
High ambient temperature										②									②		
Viscosity—low	Lubrication	②	②	③					①	①	①	①	①						①		
Quality*/Filtration		③	③	④					①	①	①	①	①								①
Penetration factor △		④	②																		
Quality—loss of lubrication		③	③	④					①	①	①	①	①			②		②	①		

Notes: (#) Distress/failure mode—cause probability ranking
*Resistance to centrifuging △
△ Low-speed/straight-tooth couplings
△ Friction oxidation/wear oxidation/chafing/bleeding/cocoa/red mud
△ Batch lubricated couplings

In all machinery component failure analysis activities, it is necessary to understand the design and functional peculiarities of the component and of the machinery system it is associated with. This is especially true when analyzing gear-coupling failures or distress signs. The following example will prove this point.

In 1972, Petronta had to shut down their 8500-hp propylene compressor after experiencing high lateral vibrations on the compressor. The compressor train configuration is similar to that in Figure 9-21. Upon inspection of the high-speed coupling, it was discovered that large portions of the hub gear teeth were broken. The coupling, a spacer type B, Figure 3-146, had been in service for 15 years and had received inspections combined with regreasing operations at two-year intervals. Because the failure was judged extremely serious, it was decided to have the failed coupling analyzed by a reputable engineering laboratory. Their failure analysis report read as follows:

1.0 Introduction (narrative deleted)

2.0 Investigation Procedures and Results
  2.1 Visual Examination
    2.1.1 The submission, as received, consisted of a male and female splined component and numerous spline segments which had separated from the male component. Some segments were jammed in the bottom of the female spline, necessitating trepanning out of a retainer insert to facilitate their removal.
    2.1.2 All surfaces were lightly rusted and dry (i.e. no trace of grease) and a dry, finely divided, red rust-like deposit was present at the bottom of the female spline. With two exceptions, the fracture faces on the male spline segments left integral with the hub were either in a circumferential plane or at an acute angle with the spline end faces. Where circumferential fracture faces were observed, most, and in some cases all, of the entire spline was absent and the surfaces were generally scalloped and heavily fretted (Figure 3-148). In all cases the scalloping had undercut the spline end-face radii at the bases of the individual teeth. The acute angled fracture faces were predominantly brittle, and characteristic "sunburst" brittle fracture patterns were seen originating at the loaded flanks on the root radii of several splines. Occasional chevron marks in the fracture surfaces indicated that most of the cracks had propagated in the direction opposite to rotation.
    2.1.3 The broken pieces of male splines had fracture surface markings which corresponded with neighboring pieces or with the fracture surfaces of male splines left integral with the hub. However, in two instances beach-like markings were observed on pieces where the fracture front was emerging at the unsupported radial/circumferential spline end-face surface at an acute angle. These pieces did not have common surfaces with the integral male splines. Many of the pieces had "sunburst" fracture patterns focusing at the loaded spline flank root radii.
    2.1.4 Examination of the female component of the coupling indicated that none of the splines were fractured or visibly cracked, but heavy fretting was

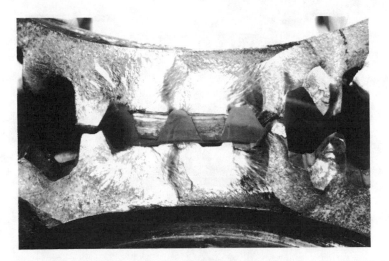

Figure 3-148. Mating fracture faces on a multispline segment showing a heavily fretted scalloped fracture and some "sunburst" fractures (magnified 1.4x).

observed on the load-bearing flanks (Figure 3-149). In the majority of cases the fretted patterns were not symmetrical across the areas in contact with the male splines; in most instances pronounced steps were visible, with up to four distinct depths of fretting present on one spline. The depths of the fretting varied from less than 0.001 in. (the highest fret marks were hardly detectable) to 0.060 in., and when viewed as a whole, the splines exhibited a definite progressive pattern from spline to spline in areas with the same depth of fretting.

2.1.5 Upon reassembly of all the pieces of the male coupling, it was apparent that the individual male splines were fretted to different degrees in much the same manner as their female counterparts. Some of the broken pieces exhibited very little fretting, while the splines still integral with the coupling exhibited the most severe fretting (Figure 3-150). Observation of the reassembled male and female portions of the coupling showed a geometrical similarity between the fretting wear patterns. The patterns on both components were mapped, the male spline pieces being in alphabetical sequence A through T in Figure 3-151.* From this exercise it was shown that the depth and shape of fretting marks on the individual male spline pieces corresponded exactly with mating areas on the female coupling, and that the depths of the fretting marks inflicted on the areas of the female coupling in contact with any particular piece of fractured male coupling were constant from spline to spline. In general, four distinct depths of fretting were observed (Figure 3-151).

(text continued on page 167)

---

*The Metallurgist apparently omitted the letter L.

Figure 3-149. Severe fretting on the female splines. Note the even depth and the diminishing pattern following the trapped male spline segment (magnified 0.5x).

Figure 3-150. Integral male spline and a multispline segment showing the changed dimensions resulting from fretting (magnified 2x).

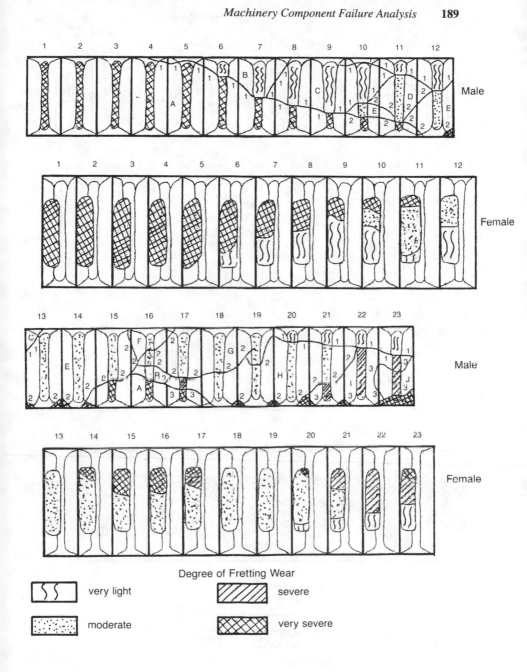

Figure 3-151. Diagram showing fracture pattern and sequence of fracture of male splines and well as the wear pattern on the corresponding female splines. The numbers along the cracks indicate the fracture phase during which the cracks occurred.

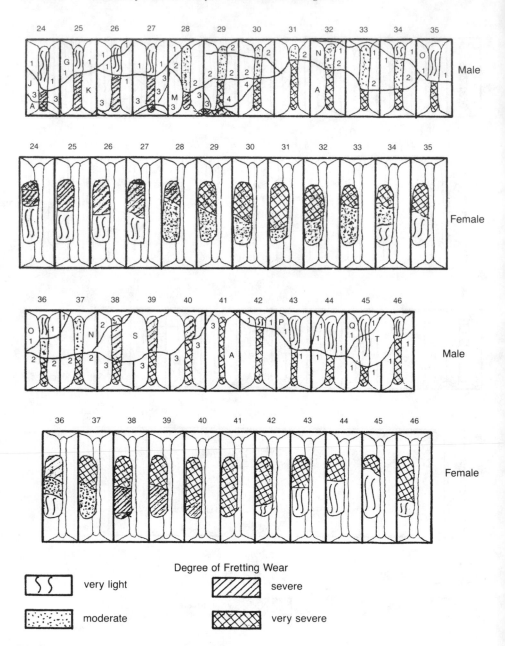

Figure 3-151. Continued.

(text continued from page 187)

2.1.6 Three pieces of the male spline (indexed as R, S, and T on figure 3-151) were missing, but the fretting depths inflicted by them on the corresponding position of the female coupling fell into one of the four main categories.

2.2 Metallographic Examination

2.2.1 Representative sections from individual segments of spline from the failed male components of the coupling were prepared for metallographic examination.

2.2.2 The structure on all sections consisted of coarse-tempered martensite free from significant inclusions and structural defects. Grain size was predominantly #4—ASTM E112. Brittle transgranular fracture progression was observed at the "sunburst" fracture sites. No evidence of possible nucleation sites was found at the focal point of the "sunbursts."

2.2.3 Examination immediately adjacent to surfaces showing visual evidence of fretting (i.e. the load-bearing flanks on the splines and the fretted scallop-shaped fracture faces (Figure 3-148) showed no significant structural deformation. A uniform oxide film, 0.0005 in. thick, was noted on the scalloped surfaces, and fracture progression appeared to be by a brittle transgranular mechanism.

2.3 Hardness Surveys

2.3.1 A hardness traverse was made along the axis of one of the transverse male spline microspecimens, starting at the crown and finishing at the center of the hub ruling section. The readings ranged from a high of 330 HV/313 BHN (Hardness Vickers/Brinell Hardness Number) to a low of 309 HV/293 BHN. There was a slight tendency for the lower figures to occur at the larger ruling sections.

2.3.2 Tukon microhardness surveys normal to fretted fracture faces and normal to a "sunburst" nucleus failed to show any appreciable change in hardness at the fretted surface or at the sunburst nucleus.

2.4 Chemical Analysis

2.4.1 Drillings from the hub of the male coupling were spectrograhically analyzed. The analysis showed that the metal was a low-alloy steel of the SAE 4140 type (ref. Chemical Laboratory Report #3873).

2.5 Scanning Electron Microscope Examination

2.5.1 Examples of "sunburst" fracture faces, scalloped areas, and some coarse "beachmark"-type progression fractures were submitted to the Research Foundation for S.E.M. (Scanning Electron Microscope) examination, and analysis. The results were inconclusive, presumably due to the age of the fracture. The characteristics of brittle, ductile, or fatigue fractures normally observed at high magnifications i.e. faceted cleavage, ductile dimple and striations, were not readily identified. A small area on the heavily fretted scalloped fracture showed striation-type markings, but the validity of this observation is questionable because of the presence of an oxide layer on this surface and the visually apparent fretting, both of which would mask any true fatigue striations.

3.1 Visual and metallographic examination indicated that failure of the splined coupling occurred in three or four major phases by a brittle fracture mechanism. Four distinct depths of fretting predominated, and since the shallowest fretting was less than 0.001 in. and the heaviest of the order of 0.060 in. it is surmised that breakup occurred at significant intervals of time since the last greasing operation. In general, the fretting depths were constant from spline to spline on multispline segments that had separated from the male component, suggesting rapid fracture progression. Also, several "sunburst"-type fraction patterns would occur on adjacent splines on each segment, indicating multifracture initiation points as opposed to a single crack front propagating through the segments. "Sunburst" fracture patterns are characteristic of high-energy brittle cleavage cracks radiating outward from a critical size nucleus. The majority of the male spline fracture faces exhibited this phenomenon, with the apparent nuclei being close to the root radius on the fretted load-bearing flanks of the splines.

3.2 Brittle cleavage failures of the type observed are normally associated with a combination of the following three conditions or factors.
   a) a notch "flaw" (this is essential)
   b) low operating temperatures
   c) a rapid change in strain rate

On a gear-type coupling, condition c would arise at startup of the compressor or perhaps during a speed change if it is a variable-speed unit. Temperature effects (condition b) cannot be discussed because the time of failure and therefore the conditions prevailing at that time are not known. A "flaw" (condition a) can be mechanical (e.g. machining nicks, indentations, notches caused by fretting or wear, etc.), metallurgical (e.g. fatigue cracks, inclusions, etc.), or chemical (e.g. alloy segregation, corrosion, etc.). The nuclei of the brittle "sunburst" fractures were closely scrutinized, using visual, metallographic, S.E.M. and hardness survey techniques, for evidence of flaws. Fretting damage was the only flaw found, and some of this occurred after the initial fracture. There was no evidence to suggest a metallurgical flaw such as small fatigue cracks, but the possibility of some fatigue-initiated brittle fracture, with the fatigue portion being subsequently masked by fretting, cannot be completely discounted. The complete absence of any grease in the coupling would favor the development of fretting. The two-year regreasing cycle was presumably proven to be quite adequate, and therefore the complete absence of any grease would suggest that an abnormal condition was prevailing. Any misalignment in the coupling would favor both the fretting and a change in strain application rates.

3.3 The material was SAE-4140-type steel, heat treated to within the desired hardness range, but the relatively coarse structure noted metallographically would tend to favor brittle-fracture progression.

4. Conclusions

4.1 Failure of the splined coupling occurred by a brittle fracture mechanism in several discrete stages.

4.2 Operating conditions since its last inspection were abnormal.

4.3 Material quality was as specified.

This failure analysis report was presented because it is an excellent example of accomplished metallurgical failure analysis. What then was the root cause of this coupling failure? The answer lies in Figure 3-151: The undulating wear pattern is quite apparent. What the failure analyst described as "operating conditions" was severe misalignment accompanied by a gradual loss of the grease charge.

Determining the Cause of Mechanical Seal Distress

It has been estimated that at least 30% of the maintenance expenditures of large refineries and chemical plants are spent on pump repairs. In turn, each dollar spent on pump repairs contains a 60- or 70-cent outlay for mechanical seals.

Seal failure reduction is therefore a priority assignment for mechanical technical service personnel in the petrochemical industry. But unless troubleshooting procedures are systematic and thorough, the root cause of a failure may remain hidden, resulting in costly repeat failures. Only when the analysis includes the entire pumping circuit, its service conditions, and the physical and thermodynamic properties of the fluid in contact with the seal can the analyzing engineer or technician expect to be successful.

However, careful examination of the failed seal parts will help determine whether the problem lies in seal selection or installation, the liquid environment, or pump operation, for instance. Careful examination is possible only if the entire seal is available, in a condition as undisturbed as possible. Component inspection for wear, erosion, corrosion, binding, fretting, galling, rubbing, excessive heat, etc. will point to further clues and lead to the root causes of most mechanical seal problems. Tables 3-14 and 3-15 will allow the reader to relate predominant failure modes to the most probable failure cause.

Typical failure causes are best explained with the help of pictures from the files of a major manufacturer of mechanical seals.*

Troubleshooting and Seal-Failure Analysis

Causes of Seal Failure

A mechanical seal has failed when leakage becomes excessive. Here is a summary of the common causes of seal failure:

- Mishandling of components. This includes allowing the seal's components to become dirty, chipped, scratched, nicked, or otherwise damaged before or during assembly.
- Incorrect seal assembly. This includes improper placement of the seal components in the seal cavity.

(text continued on page 196)

---

*Courtesy Durametallic Company. Reprinted by permission.

Table 3-14
Mechanical Seal Failure Analysis

		Cause	Swelling	Extruding	Cracking	Corrosion	Burning/overheating	Hardening	Cracking hard face	Breakage	Corrosion	Uneven/widened wr. track	Chipping	Pitting—carbon	Grooving—hard face	Erosion—carbon	Heat checking	Breakage	Corrosion	Hardening	Coking	Clogging—I.D.
			Secondary Seal						Seal Faces									Bellows				
Seal Selection Strategy — **Service Condition**		Temperature			①		①	①									①				①	
		Pressure (PV)		①									③	③								
		Corrosives	①			①					①							①	①			
		Abrasives										④			②							
		% Solids																				①
		Coking											②	②	①						①	②
		Flashing (VP)											③	③								
		Poor lubricant					④	④										②				
		High viscosity											①	①								
	Selection	Face material combination					④	④			①		①	①	①		①				①	
		"O" ring compatibility	①		②	①	②	②														
		Flush design					③	③	①				③		②	①	②	②				
		Quench design											②	④	③						②	③
Operation and Maintenance		Misalignment—external										②										
		Misalignment—internal										①										
		Vibration—axial																①				
		Vibration—radial																				
		Installation procedure		②					②	①		③						④				
		Run-out																				
		Loss of flush					③	③	③	④												
		Thermal shock								②												
		Frequent stop/start													⑤			③				
		Pump cavitation																				
		Secondary																⑤		①		④

Table 3-14 (continued)

Legend: (#) Denotes failure mode/cause probability ranking

Breakage—springs	Corrosion—	Fretting—Drive/Sleeve	Clogging—Springs/Retainer	Rubbing—Internal Diameter	Comments and Notes—See Table 3-15	Notes
				④	Select high temperature seal if required	(1)
						(1)
①	①				Proper material selection to resist corrosion/stress corrosion cracking	(1)
		②	①	①		
					Note	(1)
	②					
	②				Choose harder face combinations	(1)
					Elastomer compatibility should be checked	(1) (2)
					Note	(1) (3)
						(1)
					Check coupling (external) alignment	
	①	①		①	Note	(4)
	②					
				②	Note	(5)
	③			⑤		(6)
				③	Note	(7)
	④					
	⑤		②		Look for primary failure mode	(8)

Table 3-15
Mechanical Seal Failure Analysis—Notes

Note #

1. See Ref. 17 for seal selection strategy.
2. If consistent secondary sealing element deterioration is experienced, conversion to metal bellows might be indicated. If experiencing O-ring deterioration through high/low temperature, chemicals, and pressure, check standard compatibility tables.
3. Flush designs should preferably be developed in cooperation with seal *and* pump vendors.
4. Internal alignment of bearings, gland plate, and stuffing box. If experienced frequently, pump repair and installation practices may need review.
5. Identify source of vibration. Take corrective action, i.e. rotor balancing, realignment.
6. Consult one of several excellent texts on seal installation and maintenance. Develop appropriate assembly check lists.
7. Shaft run-out must be reduced.
8. Certain failure modes may occur as a consequence of, or secondary to, another seal defect, i.e. clogging of bellows caused by failure of seal faces or secondary sealing element.

(text continued from page 193)

- Incorrect seal selection. This includes selecting the wrong materials of construction or an incorrect design for the combination of pressures, temperatures, speeds, and fluid properties.
- Improper startup. This includes failing to pressurize a double seal before starting a pump or inadvertently running a seal dry.
- Inadequate environmental controls. This includes lack of adequate bypass, external flush, or temperature controls.
- Fluid contamination. This includes the presence of harmful solid particles in the seal cavity fluid.
- Poor equipment conditions. This includes excessive shaft run-out, deflection, or vibration.
- Worn-out seal. The seal may have completed a satisfactory life cycle for the particular seal face materials selected.

Troubleshooting Mechanical Seals

This section lists specific maintenance problems, their likely causes, and recommended solutions. When using this list for troubleshooting, read the entire list all the way through to find the problem that most closely describes the actual circumstances.

Whenever possible, the seal installation site should be inspected while the seal is still in operation. Much can be learned from observing operating temperature, environmental controls, equipment vibration, and other factors.

Problem. Seal spits and sputters in operation.
Causes. Product is vaporizing and flashing across the seal faces.

Solution.
- Take steps to maintain product in a liquid condition in the seal cavity. Check seal housing pressure, temperature, and vapor pressure of product.
- Have seal manufacturer check seal design. What's the competition doing? Review the experience history of other users and compare.
- Check environmental controls.

Problem. Seal leaks.
Causes.
- Gland bolts may be loose.
- Gland gasket may be defective. TFE gasket may have cold-flowed.
- Seal faces may not be flat. Overtight gland bolts or pipe strain may cause warpage of insert.
- Shaft packing or insert mounting may have been nicked or pinched during installation.
- Carbon insert may be cracked, or the face of the insert may have been chipped during installation.
- Seal faces may have been scored by foreign particles.
- Liquid may be leaking under equipment shaft sleeve.
- Excessive equipment vibration.
- Faulty equipment.

Solution. Recheck and correct gland bolt torque. Disassemble seal, analyze problem, and repair or replace as indicated. Check environmental controls. Check condition of equipment.

Problem. Seal squeals during operation.
Causes. Inadequate amount of liquid at sealing faces.
Solution.
- Install or enlarge bypass lines or external flush lines.
- Install a bushing in the bottom of the seal housing to increase seal cavity pressure.
- Vent the stuffing box.
- Check and adjust the pump suction.

Problem. Carbon dust accumulates on outside of gland ring.
Causes. Inadequate amount of liquid at sealing faces. Liquid film is flashing and evaporating between the seal faces, leaving residue that grinds away the carbon. Pressure may be too high for the type of seal and product.
Solution. Contact the manufacturer. Review industry experience and check into available configurations and materials.

Seal-Failure Analysis

Seal-failure analysis is a powerful tool for the diagnosis, correction, and prevention of major problems encountered in seal maintenance.

There is an important difference between seal-failure analysis and troubleshooting. In troubleshooting, the focus is on immediate problems. In seal-failure analysis, the focus is on specific symptoms within a failed seal. The purpose of seal-failure analysis is not only to correct a specific failure, but to correct the destructive conditions that gave rise to the failure.

There are four basic questions in seal-failure analysis:
1. What does the damage look like? Is the damage chemical, mechanical, or thermal?
2. How does the damage affect seal performance?
3. What does the damage indicate about a seal's past history?
4. What corrective steps can be taken to eliminate various types of damage from recurring?

Seal failure due to chemical action. There are several kinds of destructive chemical action to which a seal is subject. Four of the most common of these are:
1. Overall Chemical Attack

   Symptoms. Chemical attack leaves the parts appearing dull, honeycombed, flaky, or starting to crumble or break up. See Figure 3-152. Weight and material hardness readings taken on the damaged parts will be substantially lower than readings on the original parts.

Figure 3-152. Chemical attack produces overall corrosion of seal components.

Causes. Chemical attack is corrosion caused by using the wrong materials of construction for the chemical environment. If double seals have been used, the pressurizing system may be failing or the buffer fluid may be contaminated.
Action.
- Obtain a complete chemical analysis of the product being sealed and upgrade the seal's materials of construction to meet chemical requirements.
- Neutralize the corrosive environment by using double seals, or flush a single seal from an external source with a clean, compatible fluid.

2. Fretting Corrosion
   Symptoms. Fretting corrosion is one of the most common types of corrosion encountered in mechanical seals. It not only causes leakage at secondary seals but damages the sleeve directly beneath the secondary seal. This area will appear pitted or shiny bright. See Figure 3-153.
   Causes. Fretting corrosion results from a constant back-and-forth movement of secondary seals over a shaft sleeve or from constant vibration of the shaft packing over a shaft sleeve. These actions remove the passive oxide metal coating that would normally protect the sleeve from corrosion.
   Action.
   - Eliminate the causes of excessive vibration of the secondary seals. This can be achieved by making sure that the shaft run out, shaft deflection, and axial end play on the equipment are held within limits cited in the manufacturer's

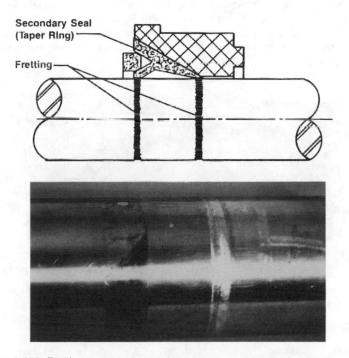

Figure 3-153. Fretting corrosion arises from vibration under dynamic secondary seal.

literature. Also see Volume 3 in this series, entitled *Machinery Component Maintenance and Repair.*

- Apply protective coatings of hard-facing alloys, chrome oxide, or aluminum oxide directly under the area where secondary seals slide.
- Upgrade the base material of the shaft or sleeve to a material that does not depend upon passive or protective coatings for corrosion resistance.
- Replace any TFE V-rings, wedge rings, or taper rings with elastomer O-ring secondary seals. Elastomer O-ring secondary seals are less susceptible to fretting corrosion because they are better able to absorb minor internal axial shaft movement.
- Replace a pusher-type seal with a nonpusher-type seal such as a metal or elastomer bellows seal having static secondary seals.

3. Chemical Attack on O-rings

Symptoms. Swollen O-rings or O-rings that have taken a permanent set preventing axial movement of the sliding seal face are likely to have been damaged by chemical attack. Chemical attack can leave O-rings hardened, bubbled, broken, or blistered on the surface, or with the appearance of having been eaten away. See Figure 3-154.

Causes. The most likely causes of chemical attack are incorrect material selection or the loss or contamination of the seal buffer fluid.

Action.

- Make a complete chemical analysis of the product and reevaluate the O-ring material selection. Frequently, the presence of trace elements, originally overlooked when specifying the seals, will be at fault. If a suitable material cannot be found, create an artificial environment by flushing the seal from an external source, or use a seal design that incorporates a static secondary seal component.

Figure 3-154. Chemical attack of O-rings.

4. Leaching

Symptoms. Leaching normally causes a minor increase in seal leakage and a large increase in wear of the carbon faces. Ceramic and tungsten carbide faces that have been leached will appear dull and matted, Figure 3-155, even though no coating is present on them. Hardness readings on such seal faces will indicate on the Rockwell A-scale a decrease of 5 points or more from the original value.

Figure 3-155. Leaching.

Causes. Leaching occurs through chemical attack on the binder that holds the base material together in carbon, metal, and ceramic materials. This type of attack can be up to several thousandths of an inch deep and can damage the seal parts beyond repair. Sodium hydroxide and hydrofluoric acid solutions, for example, will leach out the free silica binders in ceramic seal rings and result in an excessive rate of wear on the carbon seal faces. If this wearing is allowed to continue, it will dislodge aluminum oxide particles from the ceramic seal face, cause further abrasive damage between the seal faces, and shorten the life of the seal.

Action. Two procedures may be followed:

- Upgrade the base material of the seal. Use high-purity (99.5%) aluminum oxide ceramic for any applications involving sodium hydroxide or hydrofluoric acid solutions. Upgrade cobalt-bound tungsten carbide materials, which leach in the presence of water and other mild chemicals, to a nickel binder. Replace reaction-sintered Silicon Carbide 1 with alpha-sintered Silicon Carbide 2.
- Use a seal arrangement that will provide a buffer fluid at the seal faces, such as a single seal with a flush from an external source or a double seal with a suitable buffer fluid system.

Seal failure due to mechanical action. Six types of seal failure caused by faulty mechanical action are listed.

1. Face Distortion

Symptoms. Excessive leakage at the seal frequently indicates face distortion. Distorted seal faces themselves will show a nonuniform wear pattern. Sometimes a nonuniform wear pattern is difficult to detect, but when a seal face is lightly polished on a lapping plate, high spots will clearly appear at two or more points if it is distorted. See Figure 3-156.

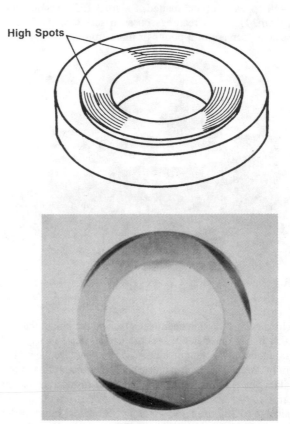

Figure 3-156. Distortion of the seal faces creates uneven wear, causing leakage.

Causes. A number of factors can be responsible for seal face distortion:
- Improper assembly of seal parts may cause nonuniform loads at two or more points around the seal face. This frequently occurs with rigidly mounted or clamp-style seal faces, because uneven torquing of the gland nuts may transmit deflections directly to the seal faces.
- Improper cooling will induce thermal stresses and distortions at the seal faces.
- Improper finishing or processing of the seal parts may leave them with high spots at several points around the seal faces.
- Face distortion can be caused by improper gland support, resulting from debris or deposits left in the gland or by physical damage to the metal in the gland.
- Poor surface finish at the face of the seal housing, due to corrosion or mechanical damage, can also cause face distortion.

Action.
- Relap the seal faces to remove the distortion.
- Consider the use of flexibly mounted stationary seal faces to compensate for any gland distortion.
- Readjust the gland by positioning the gland nuts finger-tight and then tighten them evenly with a torque wrench to an appropriate value.
- Check gland dimensions. Clean and check the seal housing face and gland finish.

2. Seal Face Deflection

Symptoms. Seal face deflection may be indicated by uneven wear at the seal face similar to that encountered with face distortion. The wear pattern, however, for seal face deflection is continuous for 360 degrees around the seal faces and is concave or convex. A convexed seal face usually results in abnormally high leakage rates, while one that is concaved usually results in excessive seal face torque and heat. See Figure 3-157. Seals exhibiting either condition will generally not be stable under cyclic pressure conditions.

Causes. Seal face deflection may arise from:
- Improper stationary seal face support.
- Swelling of secondary seals.
- Operation beyond the pressure limits of the seals.
- Inadequate balancing of hydraulic and mechanical loads on the primary seal faces.

Action.
- Check the seal design's operating limits and consult Durametallic Corporation to determine if another seal design is necessary.
- Consider flexible mounting of the stationary seal.
- Replace carbon seal faces with seal faces of materials having a higher modulus of elasticity—like bronze, silicon carbide, or tungsten carbide. These will have greater resistance to hydraulic and mechanical bending loads.

3. Tracking

Symptoms. The wear pattern of the nose on the larger seal face surface is wider than the nose.

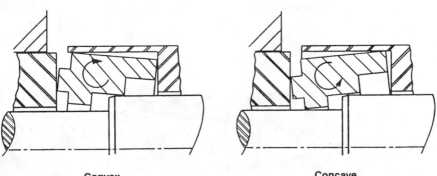

Convex   **Concave**

Figure 3-157. Face deflection of seal occurs as a continuous 360° pattern.

Causes. This condition can be caused by misalignment of the two seal faces, by equipment in poor operating condition, or by deflection of seal or equipment components under excessive pressure.

Action.
- Check and adjust the centering of the seal in the seal housing, equipment condition, or operating conditions.

4. Extrusion

Symptoms. O-rings or other secondary seals show deformation from being squeezed through the close clearances around the primary seal faces. Frequently these O-rings or secondary seals will appear cut or, in some cases, peeled. See Figure 3-158.

Causes. Excessive temperatures, pressures, or chemical attack soften the O-ring or cause excessive mechanical stresses on the O-ring for the given clearance.

Action.
- Check O-ring clearances for the application. Replace O-ring if necessary. See Figure 3-159.
- Check the chemical compatibility and temperature limits of the secondary seals. Replace if indicated.
- Install anti-extrusion rings if necessary.
- Check and correct the condition of the equipment.

5. Erosion

Symptoms. Seal face may be eaten away or washed out in one localized area. See Figure 3-160. Erosion will commonly occur on a stationary seal face until seal

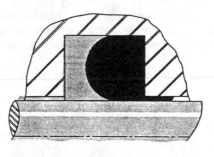

Figure 3-158. Extruded O-rings appear cut or peeled due to squeezing past very small clearances.

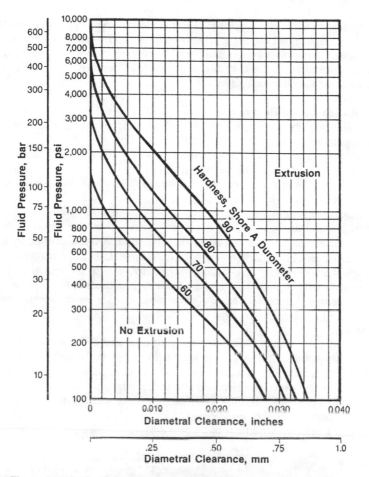

Figure 3-159. Maximum allowable clearance for O-ring secondary seals.

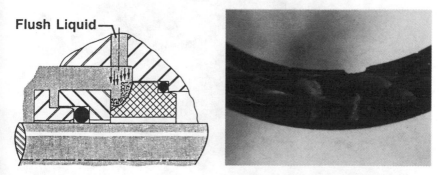

Figure 3-160. Erosion caused by excessive flushing or abrasives.

face distortion or breakage occurs. Erosion most often takes place in seal faces of carbon-graphite, but can arise in other materials under severe conditions.

Causes. Excessive flush rates or a flush fluid contaminated with abrasive particles will cause erosion. Either condition will result in a "sandblast" effect on a localized area of the stationary seal face.

Action.

- Reduce the seal flush rate.
- Eliminate abrasives from the seal flush fluid by using filters or cyclone separators.
- Replace carbon seal faces with those of an erosion-resistant material such as bronze, tungsten carbide, or silicon carbide.
- Relocate the seal flush or install a shroud to protect the stationary seal face from the direct flow of the flush.

6. Excessive Drive Pin Wear

Symptoms. Premature wear of drive pins or drive slots. See Figure 3-161.

Causes. Premature wear can occur on drive mechanisms due to heavy loads or large degrees of movement between the drive mechanism and other wear surfaces. High wear rates may also occur with relatively little movement if the drive mechanism is not properly lubricated. Such may be the case, for example, in drive mechanisms that operate in dry environments or in environments containing abrasive particles.

Action.

- Check the condition of the equipment, and limit shaft end play, shaft deflections, or any out-of-squareness of the shaft with respect to the seal housing.
- Check the piloting and centering of the seal components.
- Incorporate hardened drive pins into the seal design.
- Consider seal designs, such as double seals, that will put the drive mechanism in a better lubricating environment.
- Check pressure limitations of the seal design.

Figure 3-161. Excessive wear on drive pins, drive dents, and drive slots.

Seal failure due to heat buildup. There are seven types of seal failure caused by heat buildup:

1. Heat Checking

   Symptoms. Heat checking is indicated by the presence of fine to large cracks that seem to radiate from the center of the seal ring. See Figure 3-162. These cracks act as a series of cutting and scraping edges against carbon graphite and other seal face materials.

   Causes. High friction heat at the seal faces caused by:
   - Lack of proper lubrication.
   - Vaporization at the seal faces.
   - Lack of proper cooling.
   - Excessive pressures or velocities.

   Action.
   - Check operating conditions and make sure that they are within the prescribed application limits for the seal.
   - Confirm that adequate cooling and flow are available at the seal faces to carry away seal-generated heat.
   - Make sure the seal has not been mechanically overloaded. A thrust bearing or thrust collar in the equipment may have become damaged or inoperative, thereby creating excessive seal face loads.
   - Check, and if necessary, improve the cooling and lubrication at the seal faces.
   - Consult knowledgeable manufacturers for recommendations on reducing seal generated heat.

Figure 3-162. Radial cracks in metal or ceramic rings indicate heat checking.

2. Vaporization

   Symptoms. Any popping, puffing, or blowing of vapors at the seal faces is evidence of vaporization. Vaporization does not frequently cause catastrophic failure, but it usually shortens seal life. Inspection of the seal faces reveals signs of chipping at the inside and outside diameters and pitting over the entire area. See Figure 3-163.

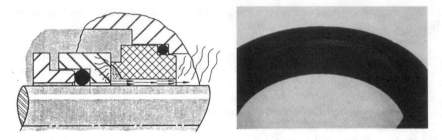

Figure 3-163. Vaporization shortens seal life and impairs seal performance.

Causes. Vaporization often occurs when heat generated at the seal faces is not adequately removed, and the liquid between the seal faces boils or flashes. Vaporization can also be caused by operating the seal too near the vapor pressure of the product in the seal cavity. Other operating conditions that will bring about vaporization include:

• Excessive pressure for a given seal.
• Excessive seal face deflection.
• Inadequate cooling and lubrication of the seal.

Action.

• Improve circulation and cooling at the seal faces. Check seal flush or cooling.
• Make sure the pressure in the seal cavity is well above the vapor pressure of the product. Excessive temperatures or low pressures can cause product flashing. Review the vapor-temperature curves for the product.
• Check the seal design to see if it is being used within its pressure and speed limits.
• Consult the manufacturer for recommendations on reducing seal-generated heat.

3. Blistering

Symptoms. Blistering, Figure 3-164, is characterized by small circular sections that appear raised on the carbon seal faces. Sometimes this condition is best detected by lightly lapping the seal faces and viewing with an optical flat. These blisters separate the two seal faces during operation and cause high leakage rates. Blistering usually occurs in three stages:

Stage 1: Small raised sections will appear at the seal faces.
Stage 2: Cracks will appear in the raised sections, usually in a starburst pattern.
Stage 3: Voids in seal face.

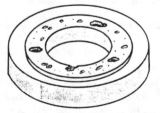

Figure 3-164. Voids remain in carbon seal face as a result of blistering.

Causes. The causes of blistering are still debated. The most accepted explanation is that high viscosity fluids, such as SAE #10 (ISO grade 22) oils penetrate the carbon seals over an extended period. When the seals heat up, the oil is suddenly forced out of the pores. Blistering commonly occurs in seals that are started and stopped frequently in applications involving high viscosity fluids.

Action.
- Reduce the viscosity of the fluid in the seal cavity either by substituting a new seal cavity fluid or by increasing the fluid temperature.
- Eliminate frequent starts and stops of equipment containing mechanical seals.
- Substitute a nonporous seal face material such as tungsten carbide, silicon carbide, or bronze for the carbon.
- Check cooling and circulation to the seal faces. Improper cooling and circulation will make seals more susceptible to blistering.

4. Spalling

Symptoms. Spalling is similar to blistering, but occurs on surfaces away from the seal face, such as the outside diameter and back side of the seal. See Figure 3-165.

Causes. Spalling, like blistering, often occurs in conditions of thermal stress in a carbon seal. Unlike blistering, however, spalling seems to occur with virtually any fluid and is the result of sealing fluid being suddenly driven off when the seal is over-heated. In most cases, spalling is due to dry running of the seal. Heavily spalled parts are a good indication that equipment was allowed to run dry for an extended period of time.

Action.
- To avoid dry running of a mechanical seal, a pressure or load switch should be added to the equipment, or an alternative sealing method, such as a double seal, should be used.

5. O-Ring Overheating

Symptoms. Overheated, elastomer O-rings harden, crack, and become very brittle. Secondary seals of TFE will harden and tend to discolor, becoming bluish-black or brown and showing signs of cold flowing. O-rings may deform to the shape of the secondary seal cavity.

Causes. Overheating is generally due to lack of cooling or to lack of adequate flow in the seal cavity. It can also be the result of selecting materials which are not designed to withstand the excessive temperatures of an application.

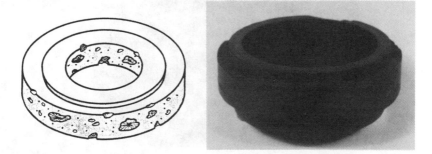

Figure 3-165. Spalling is similar to blistering but occurs on surfaces other than seal face.

Action. If overheating of O-rings is noted:

- Check cooling and flow in the seal cavity area—including the flush lines—for blockage, and inspect heat exchangers for buildup of scale.
- Apply additional cooling.
- Consider an elastomer with higher temperature limits.
- Consider a tandem seal arrangement incorporating high-temperature secondary seals.

6. Oxidation and Coking

   Symptoms. Oxidation and coking leave a varnish, a lacquer, or an abrasive sludge on the atmospheric side of the seal. See Figure 3-166. This can cause rapid wear of the seal faces or hangup in both pusher and nonpusher types of mechanical seals.

   Causes. Coking is the result of oxidation or chemical breakdown of the product to form heavy residues.

Figure 3-166. Varnish or abrasive sludge settles on atmospheric side of mechanical seal.

Action.
- Apply a steam purge to the atmospheric side of the pusher and nonpusher mechanical seals to carry away sludge or abrasive debris.
- Flush the seal from a clean, cool external source.
- Apply cooling to the seal cavity by using a seal housing water jacket, or apply a water-cooled or air-cooled heat exchanger.
- Replace carbon faces with hard seal faces that withstand the abrasive action.
- Consider replacing heat transfer fluid with a fluid having a higher coking temperature limit. In general, hydrocarbons should be cooled to below 250°F (121°C) in the seal cavity to prevent coking and oxidation. This temperature-limit varies greatly depending on the fluid handled. The oxidation or coking limits for some heat transfer fluids, for example, are above 350°F (177°C).

Summary of Mechanical Seal Failure Analysis

Perhaps the majority of the seal troubles encountered in modern process plants can be summed up in the following words: heat, abrasives, misalignment, faulty installation, faulty specification of conditions, errors in material selection, off-design operation of pumps, and carelessness. All of these causes of seal failure can be avoided or eliminated by careful design, installation, maintenance, and pump operation. To solve a seal problem, the troubleshooter or analyst must examine the evidence for clues as to the root cause of the deficiency. Using a systematic approach, he is helped by observing the failure symptoms, but will direct his energies towards treating and curing only the root cause.

While failure analysis is not always straightforward and exact, it will nevertheless provide an intelligent and repeatable approach to the solution of seal maintenance problems. The following four sequential steps highlight the procedure:

Step 1. Carefully observe the characteristics of the problem. A seal design is not always at the root of a problem.

Step 2. Carefully examine possible solutions to the problem. Past experience, feedback from equipment manufacturers, and consultation with a knowledgeable seal expert will be valuable in formulating a list of potential remedies.

Step 3. Choose an appropriate remedy and take corrective action. Selecting the best one will require analysis of cost, availability of hardware, and future economic benefits.

Step 4. Follow up on the problem-solving efforts.

Lubricant Considerations

Machinery maintenance in process plants uncovers a seemingly endless variety of lubricated component failures. Frequently it appears that the lubricant or the lubrication mechanism contributed to these defects and failures. A closer look, however, often reveals that the lubricant is seldom at fault if good lubrication practices have been observed.[18]

The role of the lubricant in machinery component failures can only be determined through objective analysis. Frequent failure causes have been design- and application-related deficiencies, material and manufacturing defects, and unfavorable operating and maintenance conditions. In the last category, inadequate lubricant quality is one of the possible failure causes. Experience shows that many potential or actual machinery component defects were thus caused by user-induced problems. Therefore, failure incidents attributable to contaminated lubricant usage could at best have been deferred, but never totally prevented, even if "superior" lubricants had been used.

The following section will discuss lubrication analysis as a vital part of the root-cause assessment of lubricated machinery component failures.

Lubrication Failure Analysis

Detailed examination of the lubricating oil or grease and of the lubrication system is as important as the analysis of, say, a failed roller bearing. It would just not make sense to only look at part of the picture. Failures which, at first sight, could be attributed to application-related deficiencies or operating conditions often result from dirt and water in the oil, dirty filters, acidic oil, insufficient oil flow, the use of an improper grade of lubricant—in other words, neglect of the good lubrication practices to which we have alluded earlier (Refs. 19 and 20).

Even where the lubricant and its circulating system are not at fault (in a given bearing failure, for instance), evidence from the type of wear particles in the oil and consideration of the oil properties can often help point out the root causes of failures. However, it is not usually possible to determine "remaining bearing life" from a wear-particle analysis.

Lubricant life, if examined in the context of life expectancy, will be limited by three major factors: oxidation, thermal decomposition, and contamination. All three factors will equally influence the lubricant life characteristic, as shown in Figure 3-167, where t_D, indicates the time at which the lube oil is no longer serviceable and should be discarded. In other words, linear deterioration of lube oil properties decreases remaining service life exponentially. A similar point is made in Figure 3-168, which illustrates the rapid deterioration of mineral oils exposed to high service temperatures and/or high oxygen concentrations.

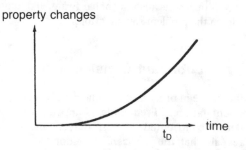

Figure 3-167. Lube-oil life curve.[18]

Research conducted in Japan (Ref. 21) states that about ten percent of steam-turbine failures are due to problems initiated by lube-oil degradation. Besides the many factors affecting the service life of turbine oil, this research looked also at the size and operating conditions of the turbines investigated. These factors included:

1. Output of the steam turbine.
2. Steam temperature.
3. Residence time of oil in the oil tank (V/Q in minutes).
4. Heat transferred to the oil (Q•r•C• t/V).

Where:

V = Quantity of turbine oil charged, liters
Q = Main oil-pump flowrate, liters/minute
r = Turbine-oil specific gravity, kilograms/liter
C = Turbine-oil specific heat, kilocalories/kilogram, kcal/kg- °C
t = Tank to cooler-outlet oil temperature difference, °C

The findings of this research were graphically represented and show the effect of unit size on the original lube-oil life remaining (Figure 3-169) and the effect of transferred heat on the original lube oil life remaining. (Figure 3-170).

In this context it is noteworthy that the life of most lubricants can be significantly extended by a conscientiously applied program of oil conditioning, or on-stream purification (Ref. 22).

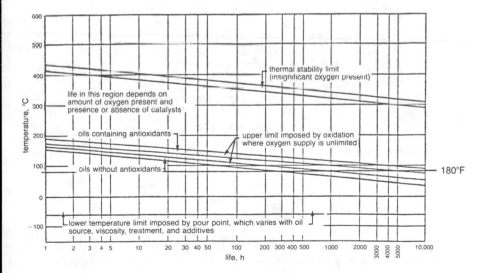

Figure 3-168. Temperature limits for mineral oils.[8]

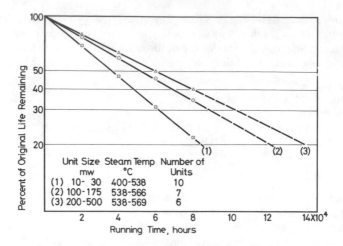

Figure 3-169. Effect of turbine size on lube-oil life remaining (—measured; ---extrapolated).[21]

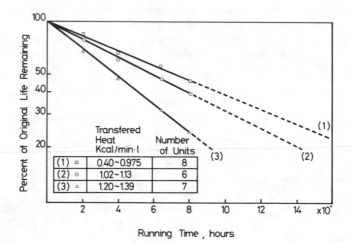

Figure 3-170. Effect of heat transferred on lube-oil life remaining (—measured; ---extrapolated).[21]

Why Lube Oil Should Be Purified

In the past two decades the benefits of oil purification have been viewed primarily in the areas of reducing oil consumption and improving machinery reliability, thereby contributing to plant operating cost reductions (Ref. 23).

While these benefits are self-evident, it is becoming increasingly important to focus on methods of reducing plant emissions. Oil purification, on-site and on-stream, has now moved to the forefront because of environmental pressures and oil-disposal related concerns. In addition to social cost justification, economic benefits accrue from the avoidance of potential steep fines from the environmental regulatory agencies.

The viability of on-stream oil purification has been documented by a world-scale chemical plant that has had an inventory of 22,000 U.S. gallons of lube oil in service for more than ten years, with no plans to replace it (Ref. 24).

As regards machinery operation and maintenance, it should be noted that lube oils of any grade or specification generally suffer from three common sources of contamination: dirt; hydrocarbon, gas, or other process dilutants; and water intrusion.

Of these, the first one, dirt, is usually filterable; hence, it can be readily controlled. However, dirt is often catalyzed into sludge if water is present. Experience shows that if water is kept out of lube oil, sludge can be virtually eliminated.

The second contaminant, process-dependent dilution, is seen in internal combustion engines and gas compressors where hydrocarbons and other contaminants blow past piston rings or seals and are captured within the lube and/or seal oils. Dilution results in reduced viscosity, lower flash points, and noticeable reduction of lubrication efficiency.

The last one, water, is perhaps both the most elusive and vicious of rotating machinery enemies. In lube oil, water acts not only as a viscosity modifier but also actively erodes and corrodes bearings through its own corrosive properties and the fact that it dissolves acid gases such as the ones present in internal combustion engines. Moreover, water causes corrosion of pumps, and promotes rust formation on cold steel surfaces where it condenses.

In some systems, water promotes biological growth that, in itself, fouls oil passages and produces corrosive chemicals. Water-contaminated lube oils are far more detrimental than is generally known. Because of the elusive nature of and the many misunderstandings associated with water contamination, this segment concerns itself with water, its effects, and various means for removing it from a lube-oil system.

Forms of Water Contamination Vary

In oil systems associated with process machinery, water can, and often will, exist in three distinct forms: free, emulsified, and dissolved. Before examining the effects of water contamination, it may be useful to more accurately define these terms.

Free water is any water that exists in excess of its equilibrium concentration in solution. This is the most damaging form for water to be in. Free water is generally separable from oil by natural settling due to gravity.

Emulsified water is a form of free water that exists as a colloidal suspension in the oil. Due to the electrochemical reactions and properties of the oil/water mixture in a particular system, some or all of the water that is in excess of the solubility limit forms a stable emulsion and will not separate by gravity even at elevated temperatures. In this respect, emulsified water behaves as dissolved water, but it has the damaging properties of free water and modifies the apparent viscosity of the lubricant.

Dissolved water is simply water in solution. Its concentration in oil is dependent upon temperature, humidity, and the properties of the oil. Water in excess of limits imposed by these condition is free water. The equilibrium concentration of water in typical lube oils is given in Figure 3-171. Dissolved water is not detrimental either to the oil or the machinery in which it is used.

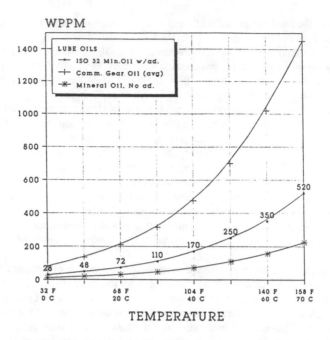

Figure 3-171. Water in oil solubility chart, at various reservoir temperatures.

For corrosion to occur, water must be present. Free water, in particular, will settle on machinery surfaces and will displace any protective surface oil film, finally corroding the surface. Emulsified water and dissolved water may vaporize due to frictional heat generated as the lube oil passes through bearings. Very often, though, the water vapors recondense in colder pockets of the lube-oil system. Once recondensed, the free water continues to work away at rusting or corroding the system (Ref. 25).

Larger particles generated by corrosion slough off the base metal surface and tend to grind down in the various components making up the lube system, i.e., pumps, bearings, control valves, and piping. The mixing of corrosion products with free and emulsified water in the system results in sludge formation that, in turn, can cause catastrophic machinery failures. Suffice it to relate just one of many examples of water-related damage to major machinery.

In one documented catastrophic failure of a steam turbine at a medium-sized U.S. refinery, the initial problem was unrelated to lubrication and led to coupling distress and severe unbalance vibration. When the coupling bolts sheared, the steam turbine was instantly unloaded and the resulting overspeed condition activated a solenoid dump valve. Although the oil-pressurized side of the trip piston was thus rapidly depressured, the piston stem refused to move and the turbine rotor sped up and disintegrated. The root cause of the failure to trip was found to be water contamination of the turbine control oil. Corrosion products had lodged in the trip cylinder and, although enveloped in control oil, the compression spring pushing on the trip piston had been weakened by the presence of water (Ref. 26).

Next, it should be pointed out that water is an essential ingredient for biological growth to occur in oil systems. Biological growth can result in the production of acidic ionic species and these enhance the corrosion effects of water. By producing ionic species to enhance electrochemical attack of metal surfaces, biological activity extends the range of corrodible materials beyond that of the usual corrodible material of construction, i.e., carbon steel.

While corrosion is bad enough in a lube-oil system, erosion is worse as it usually occurs at bearing surfaces. This occurs through the action of minute free water droplets explosively flashing within bearings due to the heat of friction inevitably generated in highly loaded bearings.

Additive loss from the lube-oil system is another issue to contend with. Water leaches additives such as anti-rust and anti-oxidant inhibitors from the oil. This occurs through the action of partitioning. The additives partition themselves between the oil and water phase in proportions dependent upon their relative solubilities. When free water is removed from the oil by gravity, coalescing or centrifuging, the additives are lost from the oil system, depleting the oil of the protection they are designed to impart.

The severity of the effects of water on bearing life due to a combination of the above is best illustrated by Figure 3-172 (Ref. 2), where we see that bearing life may be extended by 240% if water content is reduced from 100 wppm (weight parts per million) to 25 wppm. However, if water content is permitted to exist at 400 wppm, most of which will be free water, then bearing life will be reduced to only 44% of what it could be at 100 wppm.

Ideal Water Levels Difficult to Quantify

Manufacturers, operating plants, and the U.S. Navy all have their own water condemnation limits as illustrated in Table 3-16. Extensive experience gained by a multinational petrochemical company with a good basis for comparison of competing water removal methods points the way here. This particular company established that ideal water levels are simply the lowest practically obtainable and should always be kept below the saturation limit. In other words, the oil should always have the ability to take up water rather than a propensity to release it.

Methods Employed to Remove Water

Centrifuges have been used for decades. They operate on the principle that substances of different specific gravities (or densities) such as oil and water can be separated by centrifugal force. Centrifuges achieve a form of accelerated gravity settling, or physical separation. At a given setting, centrifuges are suitable for a narrow range of specific gravities and viscosities. If they are not used within a defined range, they may require frequent, difficult readjustment. They will not remove entrained gases such as hydrogen sulfide, ethane, propane, ethylene, etc., or air. Although centrifuges provide a quick means to separate high percentages of free water, they are maintenance intensive because they are high-speed machines operating at up to 30,000 rpm. More importantly, they can only remove free water to 20 wppm above the saturation point in the very best case, and none of the dissolved or emulsified water. In fact, centrifuges often have a tendency to emulsify some of the water they are intended to remove.

Coalescers are available for lube-oil service and have found extensive use for the dewatering of aircraft fuels. Unfortunately, coalescers remove only free water and tend

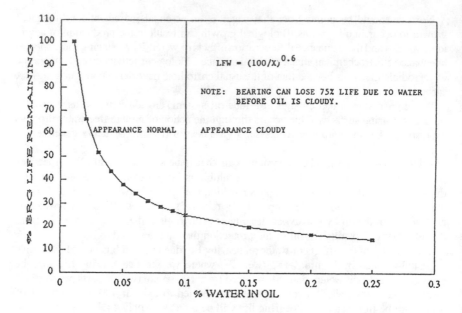

$$LFW = (100/X)^{0.6}$$

NOTE: BEARING CAN LOSE 75% LIFE DUE TO WATER BEFORE OIL IS CLOUDY.

APPEARANCE NORMAL   APPEARANCE CLOUDY

Effect of Water in Oil on Bearing Life Based on 100% Life at 100 ppm Water in Oil

% water	ppm water (X)	Life factor (LFW)	% life decrease	% life remaining
0.0025	25	2.297	−129.740	229.740
0.010	100	1.000	0.000	100.000
0.020	200	0.660	34.025	65.975
0.030	300	0.517	48.272	51.728
0.040	400	0.435	56.472	43.528
0.050	500	0.381	61.927	38.073
0.060	600	0.341	65.872	34.128
0.070	700	0.311	68.887	31.113
0.080	800	0.287	71.283	28.717
0.090	900	0.268	73.242	26.758
0.100	1000	0.251	74.881	25.119
0.150	1500	0.197	80.305	19.695
0.200	2000	0.166	83.428	16.572
0.250	2500	0.145	85.504	14.496
0.500	5000	0.096	90.436	9.564
1.000	10000	0.063	93.690	6.310
2.000	20000	0.042	95.837	4.163

Note: Less than 1000 ppm (.1%) water in oil is not detectable by appearance—past this point oil appears cloudy

Figure 3-172. Bearing life increases with low water-in-oil concentrations. (Adapted from "How Dirt and Water Affect Bearing Life," Timken Bearing Co., *Machine Design*, July 1986.)

Table 3-16
Water Condemnation Limits

Guidelines Issued by User or Manufacturer	Max, Allowable Water Content Quoted	Sampling Frequency	Purifier Operation
Chemical Plant, Texas	4,000 wppm	Monthly	Intermittent
Consulting Firm	2,000 wppm (steam turbines only)	—	—
Refinery, Europe	1,000 wppm (BFW) and cooling water pumps	Monthly	Intermittent
Refinery, Kentucky	1,000 wppm	—	As Required
U.S. Navy (MIL-P-20632A)	500 wppm	—	Continuous
Consulting Firm	500 wppm gas turbines only	—	—
Chemical Plant, Canada	200	Visual	As Required
Utility, Michigan	100	Monthly	Continuous
Chem. Plant, Texas	100	6 months	Intermittent
Chem. Plant, Louisiana	100	Monthly	Intermittent
Refinery, Texas	100	—	Intermittent
Consulting Firm	100	Weekly	As Required
Utility, Europe	100	—	Continuous
Refinery, Texas	100	—	Continuous
Utility, Michigan	100	—	—
Turbine Mfr., Japan	100	—	—
Compressor Mfr., Pennsylvania	40	—	—
Refinery, Louisiana	—	—	Intermittent

to be maintenance intensive. More specifically, a coalescer is a type of cartridge filter that operates on the principle of physical separation. As the oil/water mixture passes through the coalescer cartridge fibers, small dispersed water droplets are attracted to each other and combine to form larger droplets. The larger water droplets fall by gravity to the bottom of the filter housing for drain-off by manual or automatic means. Since coalescers, like centrifuges, remove only free water, they must be operated continuously to avoid long-term machinery distress. The moment they are disconnected, free water will form and begin to cause component damage. Again, because they are based on a physical separation principle, they are only efficient for a narrow range of specific gravities and viscosities.

As mentioned, coalescers are used in thousands of airports throughout the world to remove water from jet fuel. Water, of course, freezes at high altitudes. Refining of jet fuels is closely controlled to rigid specifications, which allows successful water removal by this method. There are several disadvantages to coalescers. They are only efficient over a narrow range of specific gravities and viscosities. They do not remove dissolved water, which means they must be operated continuously, and it is expensive to change elements.

Filter/dryers are also cartridge-type units that incorporate super-absorbent materials to soak up the water as the wet oil passes through the cartridges. They remove free and emulsified water, require only a small capital expenditure, and are based on a very simple technology. However, they do not remove dissolved water, and their operation might be quite costly because the anticipated use rate of cartridges is highly variable due to the changing water concentrations. The amount of water contamination at any given time would be difficult, if not impossible, to predict. Additionally, high cartridge use creates a solid-waste disposal problem.

Although vacuum oil purifiers have been used since the late 1940s, low-cost versions generally tend to suffer reliability problems. Therefore, the user should go through a well-planned selection process and should use a good specification. For a properly engineered vacuum oil purifier, see Figures 3-173 and 3-174. Good products will give long, trouble-free service (Ref. 27).

A vacuum oil purifier operates on the principle of simultaneous exposure of the oil to heat and vacuum while the surface of the oil is extended over a large area. This differs from the other methods we have discussed in that it is a chemical separation rather than a physical one. Under vacuum, the boiling point of water and other contaminants is lowered so that the lower boiling point constituents can be flashed off. Typical operating conditions are 170°F (77°C) and 29.6″ Hg (10 Torr). Because water is removed as a vapor in a vacuum oil purifier, there is no loss of additives from the oil system. The distilled vapors are recondensed into water to facilitate rejection from the system. Noncondensibles such as air and gases are ejected through the vacuum pump.

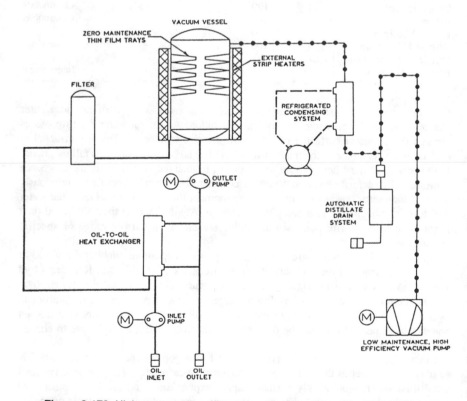

Figure 3-173. High vacuum oil purifier schematic. (Courtesy: Allen Filters, Inc., Springfield, Missouri, U.S.A.)

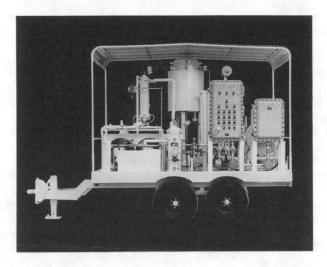

Figure 3-174. Mobile high vacuum oil purifier. (Courtesy: Allen Filters, Inc., Springfield, Missouri, U.S.A.)

Vacuum Oil Purifiers More Closely Examined

As illustrated in Figure 3-173, the typical components of a vacuum oil purifier are an inlet pump; a filter typically rated at 5 microns; some method of oil heating such as electric heaters, steam, hot water, or heat transfer fluid; a vacuum vessel; and a vacuum source such as a mechanical piston vacuum pump or water eductor. Vacuum oil purifiers may or may not incorporate a condenser depending upon the application. A discharge pump is employed to return the oil to the tank or reservoir, and an oil-to-oil heat exchanger may be employed for energy conservation (Ref. 28).

Vacuum oil purification is applied across a broad spectrum of industries: power generation and transmission, automotive, aluminum, refining and petrochemicals, steel, mining, construction, plastic injection molding, metalworking, and food processing (Ref. 29).

Vacuum oil purification is the only extended range method capable of removing free, emulsified, and dissolved water. Since vacuum oil purifiers can remove dissolved water, they can be operated intermittently without the danger of free water forming in the oil. Furthermore, they are the only method of oil purification that will simultaneously remove solvents, air, gases, and free acids.

In virtually all instances, a cost justification study by medium and large users of industrial oils will favor well-engineered vacuum oil purifiers over centrifuges, coalescers, and filter/dryers. Cost justification is further influenced in favor of vacuum oil purifiers by bottom-line analyses that look at the cost of maintenance labor and parts consumption.

Sound Alternative Available for On-Stream Purification

Not every application needs, or can justify, a vacuum oil purifier. Due to monetary constraints, some users are willing to give up flexibility, but not effectiveness. Other users may have troublesome machines that are continuously subject to free water contamination and require an inexpensive, dedicated dehydration device operated continuously to purify the lube-oil system. These users may be well served by air sparging (Ref. 30), or by the very latest method of oil purification, which operates on the chemical separation principle of air stripping. It is important to recognize that this chemical separation principle, too, removes free, dissolved, and emulsified water. It, therefore, ranks as a viable alternative and close second to vacuum oil purification as the preferred dehydration method, particularly for smaller systems (Ref. 30).

Ambient air stripping units are intended for light-duty application on small or medium-sized reservoirs. They are specifically designed for dedicated use on individual machines for water removal only. No operator attention is required, and such units are simple to install. Since they are compact and lightweight, they may be set on top of a lube-oil reservoir. They are available at low initial cost—in the U.S., in the $16,000 to $26,000 range (1996)—and are extremely easy to maintain. These modern, self-contained stripping units remove water at or above atmospheric pressure in the vapor phase and, therefore, conserve oil additives. Units similar to the one shown in Figures 3-175 and 3-176 are capable of removing free, emulsified, and dissolved

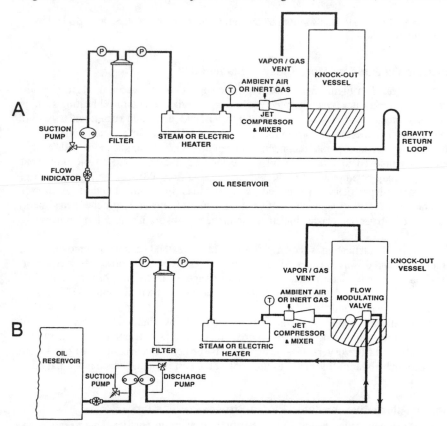

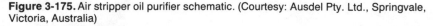

Figure 3-175. Air stripper oil purifier schematic. (Courtesy: Ausdel Pty. Ltd., Springvale, Victoria, Australia)

water to well below the saturation levels, yielding a product that is, in most cases, as good as fresh lubricant. Such units can reduce water concentration to below 15 wppm and, as can be seen in Figures 3-175 and 3-176, the technology is rather straightforward.

Operating Parameters for State-of-the-Art Stripping Units

As depicted in Figure 3-175A, air stripping units draw oil from the bottom of an oil reservoir by a motor-driven gear pump, which comprises the only moving part of the unit. The oil is then forced through a filter that removes particulates and corrosion products, and then to a steam or electric heater for temperature elevation. From there the oil goes to a very efficient mixer/contactor (jet pump) where ambient air, or low-pressure nitrogen, is aspirated into the wet oil mixture. The air is humidified by the water in the oil during its period of intimate contact, and this is the method by which the oil is dehydrated. Since even relatively humid air can absorb even more moisture when heated, ambient air is usually a suitable carrier gas for this water stripping process. The wet air is then vented to atmosphere, while the oil collects in the bottom of a knock-out vessel. In Figure 3-175A, a gravity or pump-equipped return loop allows the now dehydrated oil to flow back to the reservoir. Figure 3-175B depicts an air stripping unit with a pressurized oil return arrangement.

The choice of air versus nitrogen depends on the oxidation stability of the oil at typical operating temperatures between 140°F (60°C) and 200°F (93°C). The choice is also influenced by flammability and cost considerations.

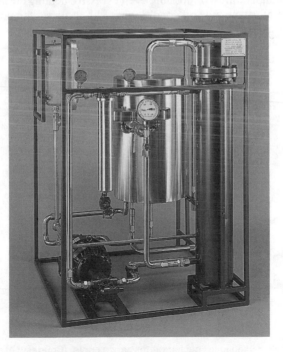

Figure 3-176. Air stripper compact design permits dedicated use on lube-oil reservoirs. (Courtesy: NuVision Systems, Houston, Texas)

Vacuum oil purifiers and modern self-contained stripping units produce comparable oil quality, and are generally preferred over the less effective and maintenance intensive physical separation methods. In general, the cost of a self-contained stripping unit is only a fraction of the cost of a vacuum oil purifier, and operating and maintenance costs are markedly lower due to its much simpler concept and construction.

The comparable performance of vacuum oil purification versus air stripping is illustrated in Figures 3-177 and 3-178. Following a performance test conducted by a major multinational oil/chemical company, the results were plotted as shown in Figure 3-177. The test conditions were vacuum oil purifier circulating flow rate of 375 U.S. gallons per hour on a 75 U.S. gallon oil sample that had been contaminated with 2% water (20,000 wppm), 162°F (72°C) average operating temperature, and 29.96″ Hg (1 Torr) vacuum level. The exact test conditions were then computer simulated for an air stripping device, and the results plotted as shown in Figure 3-178.

Six Lube-Oil Analyses Are Required

Research efforts spearheaded years ago by the utilities industry have led to optimized lube-oil analysis methods for steam-turbine lube oils. Ref. 31 recommends testing for color, foreign solids, neutralization number, viscosity, and water content. However, more recent studies indicate that these tests alone are not sufficient to determine early on if oxidation has progressed to an undesirable degree.[32] Lube-oil oxidation can result from prolonged exposure to atmospheric oxygen, high bearing and reservoir temperatures, or possibly even excessive heating during processing in vacuum-dehydrator-type lube-oil conditioners with incorrect temperature settings.

The plant described in Ref. 22 opted for an analysis program which checks for appearance, water, flash point, viscosity, total acid number, and additive content. These tests are briefly described as follows.

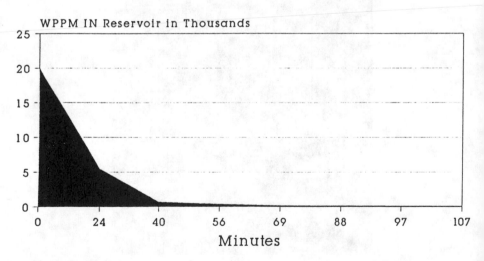

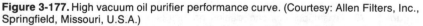

Figure 3-177. High vacuum oil purifier performance curve. (Courtesy: Allen Filters, Inc., Springfield, Missouri, U.S.A.)

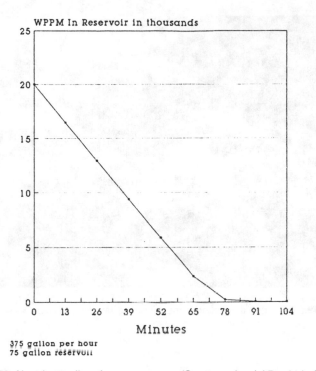

WPPM In Reservoir in thousands

Minutes

375 gallon per hour
75 gallon reservoir

Figure 3-178. Air stripper oil performance curve. (Courtesy: Ausdel Pty. Ltd., Springvale, Victoria, Australia)

Appearance Test

The test for appearance is purely visual. If free water is noted during this test, none of the remaining tests are run. If the oil is very dark, the additive test may not give sufficiently accurate results and may warrant different handling.

Testing for Dissolved Water

To determine the amount of water dissolved in the sample, automated Karl Fischer titration equipment is used and ASTM testing method D-1744 is followed. Figure 3-179 shows this testing apparatus.

Typical machinery reliability guidelines allow a maximum dissolved water content of 40 wppm in premium grade ISO-32 lube oil. This value was chosen because it generally leaves a reasonable margin for additional water contamination, and thus allows sufficient elapsed time, before saturation occurs and free water starts collecting in a given lube-oil reservoir. Free water is highly undesirable in major turbomachinery lube-oil systems because it catalyzes sludge formation.

We strongly disagree with condemnation limits as high as 2,000 wppm water as perhaps inadvertently advocated by some lube-oil analysis laboratories by reporting water content as "pass," or "less than 0.2%." On many occasions, the catastrophic failure of steam

Figure 3-179. Automated titration equipment used to test for dissolved water (Karl Fischer method).

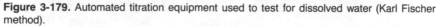

turbines has been attributed to the presence of free water in the oil. As long as the oil is contaminated with free water, relay valves and pistons may stick and overspeed trip bolts may seize. This is due to certain equipment design features which allow water separation in feed lines, valves, pistons, and trip bolts. Even the occasional overspeed trip testing and exercising of moving parts cannot eliminate these risks as long as wet oil is introduced into the turbine lube and governing system.[20] While it may be true that properly designed and maintained lube-oil systems should not contain free water but should, instead, allow for gravity-settling and subsequent removal by alert operators, neither of these requirements can be depended upon in the "real world" process plant.

The guideline value of 40 wppm of dissolved water gave this plant an adequate cushion before free water started forming in the lube-oil reservoirs.

Flash-Point Testing

Flash-point testing is important whenever there is risk of light hydrocarbons dissolving in the lubricant. Using ASTM procedure D92, flash points are automatically determined by the testing equipment shown in Figure 3-180, an apparatus which employs the familiar Cleveland open cup principle. Self-imposed guidelines required the turbine lube oil in this plant to have a minimum flash point in excess of 190°C (374°F) to be considered fully satisfactory. If the oil had been diluted by a lighter hydrocarbon, the flash-point would be lowered. The test was not performed on samples containing more than one percent water.

Figure 3-180. Laboratory testing apparatus allows rapid determination of flash point (Cleveland open cup method).

Viscosity Test

The steam-turbine lube oil is measured at 38°C (100°F) on a kinematic viscosimeter, Figure 3-181, and the centistoke reading converted to Saybolt Universal Seconds (SUS) for easy reference in this non-metric plant environment. As is the case with flash-point readings, viscosity readings will fall if oil is diluted by light hydrocarbons and rise if it is contaminated by heavier components or becomes oxidized. Viscosity readings ranging from 27.1 − 37.6 cSt (140 − 194 SUS) are considered acceptable at this temperature.

Excessively low viscosity reduces oil-film strength and therefore the ability to prevent metal-to-metal contact. Other functions such as contaminant control and sealing ability may also be reduced. Excessively high viscosity impedes effective lubrication. Contaminants which cause the oil to thicken may cause accelerated wear and/or corrosion of lubricated surfaces and leave harmful deposits.

This test is performed per ASTM D-445; ASTM D-2161 is used to convert the cSt measurements to SUS units.

Figure 3-181. Kinematic viscosimeter for testing according to ASIM D-445.

Total Acid Number

Acid formation can result from high-temperature or long-term service. A titroprocessor, Figure 3-182, is used to determine the acid number according to ASTM D-664. The total acid number should not exceed 0.3 if the steam-turbine lube oil is to qualify for further long-term service.

Oil acidity can be considered a serviceability indicator. Acidity increases primarily with progressive oxidation, although on nonturbine applications products of combustion or sometimes even atmospheric contaminants can cause an increase in lube-oil acidity. The total acid number is a measure of the quantity of acidic material in the lube oil, which for test purposes must be neutralized by titration with a basic material.

Determination of Additive Content

Premium steam-turbine lube oils contain a phenolic oxidation inhibitor which may become depleted after long-term operation of lube oils contaminated by water or exposed to high operating temperatures. Oxidation inhibitors minimize the formation of sludge, resins, varnish, acid, and polymers.

New oil contains approximately 0.6 weight percent of this important additive, and depletion to less than 0.2 weight percent requires action. Additive content is, therefore, monitored with the infrared spectrophotometer shown in Figure 3-183. The presence of additives shows up in the infrared spectrum as low transmissibility,

Figure 3-182. Titroprocessor determines total acid number.

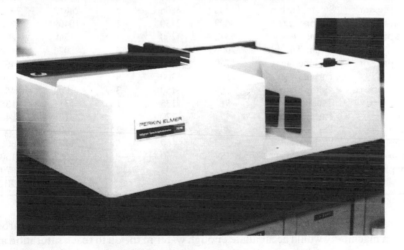

Figure 3-183. Infrared spectrophotometer measures amount of oxidation inhibitor contained in sample.

with λ around approximately 3660 cm^{-1}, where the frequency v is the reciprocal of the wave length λ.

The analysis procedure for this test is relatively straightforward and follows the instrument manufacturer's guidelines employing the test sample and a known control sample. The equipment incorporates a plotter which records the absorption points in the infrared spectrum as a function of wavelength and, therefore, frequency.

Periodic Sampling and Conditioning Routines Implemented

As indicated earlier, the plant described in Ref. 22 has implemented a rigorous program of periodically analyzing the condition of 36 lube-oil reservoirs. Although all of these reservoirs contain ISO Grade 32 turbine oil, only 29 are associated with steam-turbine-driven machinery. The remaining seven reservoirs perform lubrication duty on *motor*-driven pumps and compressors, which does not necessarily make them immune from the intrusion of atmospheric moisture condensation. Table 3-17 gives an overview of the various systems, the principal equipment types with which they are associated, and respective working capacities of the reservoirs.

Table 3-17
Self-Contained Lube-Oil Systems Covered by Lube-Oil
Analysis and Conditioning Program at a Large Olefins Plant

		Oil Reservoir Working Capacities				
		Capacity		No. of	Total Capacity	
Unit	Driver	Liters	Gallons	Units	Liters	Gallons
Compressor	Steam Turbine	21423	5660	1	21423	5660
Compressor	Steam Turbine	14406	3806	2	28812	7612
Compressor	Motor-Gear	10409	2750	1	10409	2750
Compressor	Motor-Gear	4542	1200	1	4542	1200
Fan	Steam Turbine	2006	530	4	8024	2120
Compressor	Turbine-Gear	1760	465	1	1760	465
Pump	Turbine-Gear	1135	300	3	3405	900
Compressor	Motor-Gear	568	150	3	1704	450
Compressor	Expander	511	135	1	511	135
Fan	Turbine-Gear	454	120	7	3178	840
Pump	Turbine-Gear	303	80	2	606	160
Pump	Turbine-Gear	227	60	5	1135	300
Pump	Turbine	189	50	3	567	150
Pump	Motor	132	35	2	264	70
				36	86340	22812

At program inception, the arbitrary assumption was made that steam-turbine-driven machinery would accumulate enough water in the oil to reach saturation after perhaps two months of operation, whereas it would take motor-driven machinery four months before free water would collect in the lube-oil reservoir. Accordingly, a schedule was prepared for the two trailer-mounted lube-oil conditioners, calling for operation at predefined locations for time periods ranging from perhaps one eight-hour shift on a small pump reservoir to as much as three or four days on a very large compressor lube-oil reservoir. These operating times were also calculated on a highly arbitrary basis, with the intention simply to reduce an assumed water contamination level of 500 wppm to as low as perhaps 20 wppm. The vacuum lube-oil conditioners, also called vacuum dehydrators, would be able to remove this amount of water in one pass and send the clean oil back into the reservoir. Assuming again that during this period no significant amount of water enters the

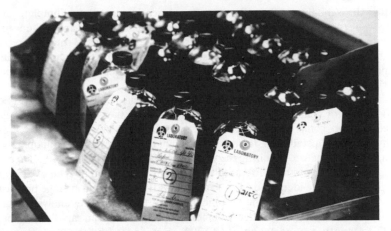

Figure 3-184. Lube-oil samples collected during the first two or three working days of each month.

reservoir, it would take some six or eight complete passes of all the lube oil through the vacuum dehydrator before the desired water concentration level would be reached. In other words, for a 14,000-liter (approximately 3,600-gallon) lube-oil reservoir, it would take the 1,400-liter (approximately 360-gallon)-per-hour vacuum dehydrator 10 hours for one complete pass. The rule of thumb would thus be to leave the unit hooked up for 60–80 operating hours. A more rigorous method is available in advanced calculus texts for the more analytical souls who feel inclined to set up a differential equation to accurately calculate the whole thing.

Much was learned from these initially scheduled lube-oil reclamation procedures. Spot checking showed that the vacuum dehydrator was often hooked up to a lube-oil reservoir which at the time still had acceptable contamination levels. Both time and effort to further purify the oil were thus wasted, since achieving even lower contamination readings was academic, at best. Also, it was learned that, for organizational and logistical reasons, sending lube-oil samples to outside laboratories could be costly, time-consuming, or involve too many people. Instrumentation acquisition was justified based on the anticipated sampling frequency.

In late 1981, the lube-oil sampling and reconditioning program was modified. During the first two or three working days of each month, all 36 lube-oil reservoirs are sampled and the containers tagged as shown in Figure 3-184. The on-site laboratory immediately tests all samples for water content only and reports its findings to the mechanical planner responsible for scheduling the lube-oil conditioning tasks. The planner now singles out those reservoirs which contained excessive amounts of dissolved, or perhaps even free, water and arranges for lube-oil purification to start first on the reservoirs with the highest water concentrations. Only the reservoirs which violate the water contamination limit are purified during this particular month. The duration of hookup to one of the two vacuum dehydrators is based on experience and convenience, invoking again the rough rule-of-thumb given earlier.

After the scheduled lube-oil reservoir purification tasks are complete on the excess-water reservoirs, their contents are again sampled and taken to the laboratory. This time all six standard analyses are performed and routinely reported in tabular form. Nonroutine analyses are requested by the plant's machinery engineers as required. These would normally include analyses for all six properties described earlier.

Water contamination levels are graphically represented in Figure 3-185. "Before purification" (or dehydration) vs. "after purification" water contamination levels are plotted for two compressor lube-oil reservoirs, CC-700 and CC-801 (Ref. 28). In this hypothetical installation, the guideline might be to allow water contamination to reach 250 wppm before requiring the vacuum dehydrator to be hooked up to a given lube-oil reservoir, and to purify until the residual water content is below 25 wppm. This plant may have identified CC-700 and CC-801 as requiring average purification intervals of 30 days. Using the graphical method of presenting analysis data, the technical service person, planner, or laboratory technician can rapidly spot that the month-to-month contamination of CC-700, as evidenced by the "before purification" plot, seems to remain steady, whereas the "after purification" residual contamination seems to have dropped. From this observation he may conclude that the vacuum dehydrator hookup time could be somewhat reduced while still achieving a residual contamination level of 20 wppm. Observing the "before" vs. "after" trend of CC-801, the reviewer may conclude that successive vacuum dehydrator hookup intervals can be increased in cold-weather operation, for example.

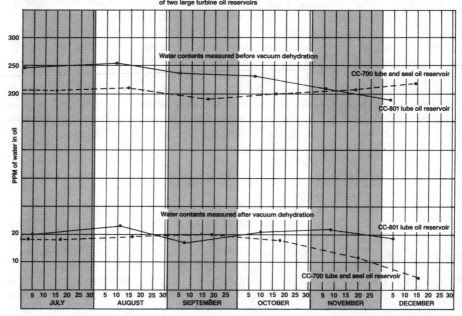

Figure 3-185. Graphical representation of "before vs. after purification" of two compressor lube-oil reservoirs.

Calculated Benefit-to-Cost Ratio

The benefits of on-stream reclamation or purification of lube oil have been described in numerous technical papers and other publications. Monetary gains are almost intuitively evident from the data given earlier, and further elaboration does not appear warranted. Suffice it to say that even a gold-plated, four-wheel, canopied vacuum dehydrator would cost less than two hours of unscheduled downtime brought on by contaminated lube oil in a major turbocompressor.

Ref. 22 mentions a detailed investigation of annual costs and savings for the in-house analysis portion of their plant-wide turbine lube-oil reconditioning and analysis program. Costs consisted of laboratory technician wages and expendables such as glassware and chemicals. Higher charges for outside contract laboratory work, labor, and electric power were saved by eliminating "precautionary reclamation" practiced in the many instances where delayed reporting or logistics problems deprived the plant of timely feedback. Their net annual savings substantially exceeded the cost of acquiring supplemental laboratory instrumentation (i.e. instrumentation specifically required for in-house analysis of lube oil and not otherwise needed by their laboratory). A benefit-to-cost ratio approaching 1.8, with discounted cash flow returns exceeding 100%, was reported.

Knowledgeable sources have calculated that for most lubrication systems using more than 200 liters (approximately 50 gallons) of lubricant, oil analysis generally proves more profitable than a routine time/dump program (Ref. 29). Similar findings have been reported by large-scale users of hydraulic oil whose reconditioning efforts have proven successful and profitable (Ref. 33).

There are many reasons why thoughtful engineers should make an effort to put in place lube-oil preservation and waste-reduction programs: economy, environmental concerns, and just plain common sense, to name a few. A conscientiously implemented program of lube-oil analysis and reconditioning can rapidly pay for itself through lube-oil savings and reductions in machinery failure frequency.

Lube-oil analysis techniques are relatively easy to understand, and automated laboratory equipment makes the job more precise and efficient than it was a few decades ago. Employing these techniques in conjunction with a well-designed lube-oil conditioner is considerably better than selling, burning, or otherwise disposing of lube oil in a modern plant environment.

Wear-Particle Analysis*

We have often had occasion to compare the merits of on-stream lube-oil purification and analysis with those of wear-particle analysis. The two approaches differ not only in substance, but also in philosophy. Where the former is totally aimed at *prevention* of component degradation by proactive techniques, the latter is entirely *predictive* in nature. In other words, wear-particle analysis defines the origin and severity of distress after the fact, whereas on-stream lube-oil purification and analysis maintains lubricants in a state of purity intended to ward off the development of distress.

Nevertheless, wear-particle analysis—also known as ferrography—can be a powerful weapon in the predictive maintenance arsenal. Unlike its predecessors (particle counting

---

*Source: Predict(SM) Monitoring Service, Cleveland, Ohio. Adapted by permission.

and spectrography), wear-particle analysis produces detailed information about machinery condition. Particles that collect in lubricants are examined and analyzed for shape, composition, size distribution, and concentration. From these particle characteristics, operating wear modes within machinery can be determined, as shown in Figure 3-186. This allows the prediction of imminent machine distress, enabling users to take corrective action.

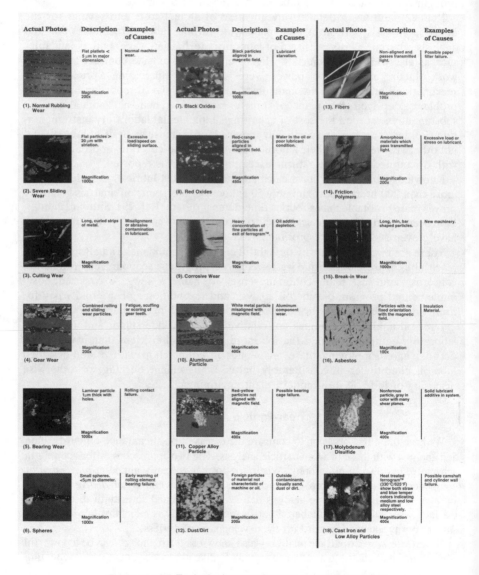

Figure 3-186. Typical microscopic wear particles observed.

Wear-particle analysis is a machine condition monitoring program for lubricated equipment. It provides accurate insights into the condition of a machine's lubricated components, simply by examining particles contained in the lubricant. It identifies wear-related failures at an early stage, providing the full benefits of effective machine condition monitoring.

All particles found in lubricated systems are not wear particles. The strength of this technique is that it enables analysts to distinguish between the wear particles that provide the earliest indication of abnormal wear, and the large quantities of particulate matter typically present in any operating equipment.

Wear-particle analysis is conducted in two stages. The first stage involves monitoring and trending wear particles, the single most important parameter in the early detection of abnormal wear. Once abnormal wear is indicated, the diagnostic stage of wear-particle analysis comes into play. This detailed microscopic analysis identifies the wear mechanisms causing particle formation, the sources of the particles, and hence the wearing components.

Every lubricated wear surface generates particles in normal operation. These normal wear particles are the product of benign wear. As Figure 3-187 shows, a lubricated system in normal operation experiences a steady increase in small particle concentration until an oil change.

After an oil change this concentration is reduced to zero. A gradual build-up of small particles in this system begins as soon as the equipment is returned to service. Most importantly, when abnormal wear begins there is no sharp, instantaneous increase in small particles present in the system.

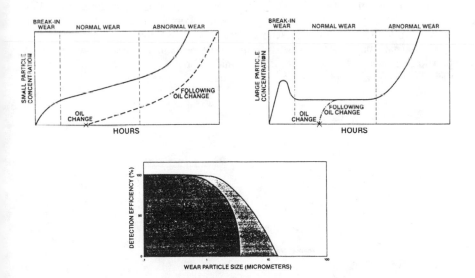

Figure 3-187. A lubricated system in normal operation experiences a steady increase in small particle concentration until an oil change.

Large particles, on the other hand, reach a dynamic equilibrium in a normally operating system. This means that large particles are formed and removed from the system at the same rate. After an oil change, the large particle concentration returns almost immediately to equilibrium. When abnormal wear begins, there is a dramatic increase in the formation of large particles.

This dramatic increase in large particle concentration provides the first indication that a system is experiencing abnormal wear.

Particles caused by specific wear modes have distinctive characteristics that reveal the wear mechanism at work. Analysts can classify wear particles by both type and metallurgy. They also have a thorough working knowledge of the equipment being monitored and the metallurgy of its components. This gives them the ability to "see" inside operating equipment, identify abnormal wear conditions, and pinpoint the wearing components without taking the equipment out of service. Since wear-particle analysis provides the earliest indication of abnormal wear, the results are obtained with sufficient lead time to correct these problems, or schedule timely maintenance activities before secondary damage occurs.

The fundamental difference between traditional lube-oil analysis and wear-particle analysis is that oil analysis is a technique for determining *lubricant* condition, while wear-particle analysis diagnoses *machine* condition. At best, the "wear-metals analysis" (spectrographic analysis) component of oil analysis provides an incomplete profile of wear metals in a system. This is because the instruments used are insensitive to particles larger than 8 microns.

Since the onset of abnormal wear is first revealed by the increased concentration of wear particles larger than this, spectrographic oil analysis has limited capability to provide useful information about machine condition. But, don't forget our earlier acquaintance, the six-step lube-oil analysis. It provides useful information on the condition of the lubricant even before the inception of wear. This allows for the correction of an oil deficiency, that, if undetected, might lead to component wear. Obviously, then, the six-step lube-oil analysis represents an "early warning" tool.

Wear-metal analysis is conducted using instruments that embody the operating principles shown in Figure 3-188. Typical microscopic wear particles, together with their description and typical origins, are illustrated in Figure 3-186.

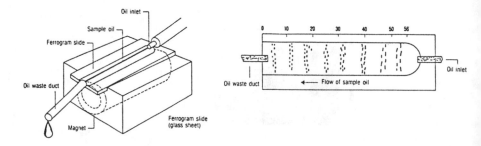

Figure 3-188. Ferrography device for examining wear particles in oil, and principle.

Oil debris monitoring is sometimes achieved with ferromagnetic concentration sensors*, with a claimed capability of detecting particles ranging in size from 1 micron to more than 1,000 microns. These relatively compact devices trap and then release ferromagnetic particles with an electromagnet. A ferromagnetic concentration sensor measures the effect of debris on a sensing coil mounted in a flange on the tubing. It thereby provides a real-time measurement of the wear debris being picked up by the lubricating fluid, without in any way impeding the flow of the fluid.

The data signal from each sensor is typically stored in the user's computer. Operators can then monitor this information periodically or after an alarm threshold has been exceeded.

Grease Failure Analysis

Process machinery management should include an occasional thorough grease analysis. Most plants will have access to customer lube-oil and grease analysis laboratories maintained by the major oil suppliers.

Figure 3-189 shows the result, albeit inconclusive, of a typical grease analysis. This analysis was requested after coupling grease failure on a large turbo train had been suspected.

Temperature limits for greases are shown in Figure 3-190. In many cases grease life will be controlled by volatility or migration. This cannot be depicted simply, as it varies with pressure and the degree of ventilation. However, in general, the limits might be slightly below the oxidation limits.

Contamination of oils and greases is the most prevalent cause of premature component wear and system failures. Contaminants may be:

1. Built-in: residual sand, metal chips from machining operations, thread sealants, weld spatter, or other foreign particles that enter during fabrication.
2. Generated: pieces of metal (from wear of moving parts), elastomeric particles abraded from seals or packings, and byproducts from oil degradation.
3. Introduced: airborne dirt, oil absorbents, and water, which enter the system through seal clearances, breather caps, and other reservoir openings.

Although introduced contaminants usually have the most detrimental effect on the life of both lubricants and equipment, overgreasing and mixing of incompatible greases (Table 3-18) are not far behind. When lubricating greased bearings in such traditional users as electric motors, be absolutely certain that the drain plug is removed while applying a new charge of premium grease.

It is important to note that a very large percentage of bearings are shipped with a protective charge of synthetic polyurea grease. This grease is totally incompatible with the lithium-base grease typically applied by the majority of users or purchasers of machinery.

---

*Source: Sensys, Napean, Ontario, Canada K2E 6T8.

Major Oil Limited

82 06 09

R 235   Used Grease
                                         C-52 Coupling

Maintenance Department

Dear Sir:

Inspections obtained on the small sample of grease from C-52 coupling from which oil had run out when shutdown, are as follows:

Ash, sulphated %	1.58
Spectrographic Analysis of Sulphated Ash	
Major Component Compounds of:	Mo, Si, Zn, Ca
Minor Component Compounds of: (High)	Pb
(Low)	Na, Fe
Water, %	Trace
Penetration (unwk)	369
(wk)	388

The spectrographic analysis of the ash shows both molybdenum and lead which are foreign to both Galena Tramo EP and Ronex EP2. The penetration of the sample is also softer than we would expect from Tramo EP or Ronex EP2. This would indicate that the grease, in the coupling, has been contaminated with another grease or is not Tramo EP or Ronex EP2. There was insufficient sample to determine the oil viscosity which would have helped to identify the grease in the coupling.

A portion of the sample was dispersed in solvent and the residue examined under a microscope. It showed little, if any, iron wear but did contain some blue-black, non-magnetic particles which may be molybdenum disulphide.

Yours very truly,

R. W. Wilson

Research Technologist

Figure 3-189. Coupling grease analysis.

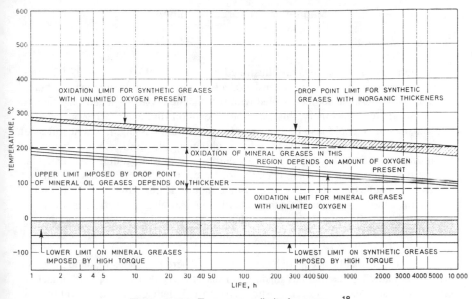

Figure 3-190. Temperature limits for greases.[18]

Table 3-18
Results of Grease Incompatibility Study*

	Aluminum complex	Barium	Calcium	Calcium 12-hydroxy	Calcium complex	Clay	Lithium	Lithium 12-hydroxy	Lithium complex	Polyurea
Aluminum complex	X	I	I	C	I	I	I	I	C	I
Barium	I	X	I	C	I	I	I	I	C	I
Calcium	I	I	X	C	I	C	C	B	C	I
Calcium 12-hydroxy	C	C	C	X	B	C	C	C	C	I
Calcium complex	I	I	I	B	X	I	I	I	C	C
Clay	I	I	C	C	I	X	I	I	I	I
Lithium	I	I	C	C	I	I	X	C	C	I
Lithium 12-hydroxy	I	I	B	C	I	I	C	X	C	I
Lithium complex	C	I	C	C	C	I	C	C	X	I
Polyurea	I	I	I	I	C	I	I	I	I	X

B =	Borderline compatibility
C =	Compatible
I =	Incompatible

---

*From Meyers, E. H., "Incompatibility of Grease," presented at the National Lubrication Grease Institute's 49th Annual Meeting.

Magnetism in Turbomachinery*

Scope of Magnetism

Simple forms of magnetism are illustrated in Figure 3-190. In Figure 3-190A, the source of magnetism is shown to consist of a magnetically hard component containing a magnetic residual source coupled with soft magnetic parts that carry the field to an air gap. In turbomachinery, this gap would be an oil film in a bearing or seal. The magnetism source could also be from an electric current passing around a soft steel part as shown in Figure 3-190B. In order to obtain a field similar to that caused by the residual source of Figure 3-190A, the current "I" must be a direct current. This current can originate from some external source or it can occur because of internal currents of an electromagnetic or electrostatic nature. If this current is alternating in nature, then the field in the air gap is alternating. An alternating field can also be produced if the air gap should change due to mechanical motions or vibration by affecting an otherwise direct field. This condition may exist in the situation as shown in Figure 3-190C.[34]

Magnetic fields and currents exist almost everywhere in various amounts; however, a combination of factors has made them particularly damaging to turbomachinery. Specifically, there has been frosting, spark track damage, and even welding on shaft journals and collars. Coupling and gear teeth have become pitted and welded. The critical factors found in turbomachinery attendant with this damage are: (1) high operating speed, (2) large steel cross sections in casings, piping, etc.; (3) a source of magnetism, either electromagnetic or residual, (4) extremely close running gaps, and (5) in the case of sudden or disastrous events, an upsetting incident. This latter can consist of compressor surge, high vibration, temporary loss of lubrication or shock, as well as other causes.[35]

Methods to reduce damaging currents have been under investigation in recent decades. Many of these are well-known in the industry and others are being developed for the special situation of turbocompressors. Simply stated, they consist of either removing damaging magnetic fields or impeding the conductivity of the current path in the affected component. One straightforward measure is to demagnetize components and structures to remove residual magnetism that may produce damaging fields. This will be discussed later on with respect to effective cycling demagnetization.

Another method is to reduce the magnetic field by insertion of nonmagnetic components in critical magnetic flow paths.[36]

A third method to reduce residual magnetism is to raise the temperature of the components to above the Curie temperature and hold it there until all the magnetic dipoles have been normalized to a nonmagnetic state. This is not usually possible with precision machinery.

Damaging currents have been reduced or eliminated as a factor in electrical machinery by placing insulation at the bearing housing or pedestal and maintaining it

---

*From information provided by Mr. Paul I. Nippes, Magnetic Products and Services, Inc., Holmdel, New Jersey 07733.

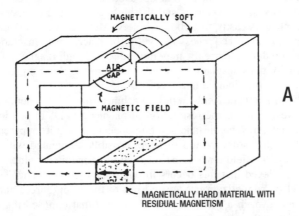

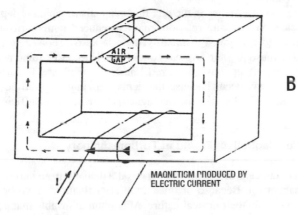

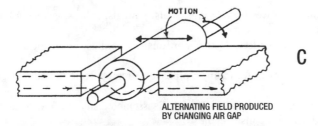

Figure 3-190. Simple magnetic circuits. A, B, C.

in an insulated state to block current conduction through the bearing. For various reasons, such as system dynamics and the high temperature of operation, insulation cannot be applied as easily to turbomachinery. Methods are being employed at this time; however, their effectiveness has yet to be determined. Insulation, which may be effective in reducing currents, may not reduce through magnetic flux to the shaft unless unusually thick insulation is used. This through flux may generate localized currents in the bearing-journal interface.

A property of magnetic fields in turbomachinery not fully understood but continuing under investigation is the restoration of magnetic fields in previously demagnetized components. In some instances, a component has had its apparent external field reduced. Then after some period of time, possibly associated with mechanical or thermal cycling, the field tends to be restored to its previous value, or some portion thereof. It is believed that this occurs through one of two mechanisms. The first is that apparent demagnetization occurred not to the true source, but resulted from inducing magnetism in a soft magnetic field zone that opposed that in the original hard zone, giving the surface the appearance of reduced magnetism. Shortly thereafter, the magnetism on the soft zone disappeared and the original hard magnetism projected through.

Another possible cause lies in the fact that magnetism, which is depicted as a form of oriented electron spin and rotation, may be reduced temporarily by altering the electron spins and rotation. Presumably there is an overpowering internal atomic force to restore these to their original conditions. Thus, after some period of time or after "seasoning" effects have occurred, there is a tendency toward original field restoration. This may occur because the demagnetizing field did not line up in the axis of the residual field and/or the demagnetizing field has insufficient demagnetizing strength.

The Extent of Magnetism Found in Turbomachinery

The primary concern in turbomachinery is the damage from currents that are generated by magnetism. Because of this, three classifications have been assigned to components based on their critical nature. Maximum allowable magnetic field levels for the three classifications are:

- Three gauss or lower for bearings, seals, journals, collars, gears, and other oil film surfaces.
- Six gauss or lower for all other portions of bearings, seals, journals, collars, and gears.
- Eight gauss or lower for casings, pipes, etc., remote from the oil film surfaces.

Levels are for components in a fully disassembled state in free air as measured by a calibrated hall probe instrument.

These values are acceptable as consistent unipolar and possibly dipolar magnetization. Multi poles, even within these levels, may still give problems; so, it is wise to remove the poling effects. Larger fields are sometimes permitted when it is clear that they will not cause destructive currents and when there is limited downtime or excessive cost of demagnetization. Obviously, this does represent a certain risk.

Unfortunately, it is necessary to evaluate the magnetic fields in a disassembled state. It is in the assembled state that the highest level of magnetism occurs, providing capabilities of generating damaging currents. Fully satisfactory methods of measuring such internal fields have yet to be developed.

An interim measure developed by John Sohre and recommended for turbomachinery users is to monitor the alternating field levels at partings of casings and bearing housings while the unit is running, using an ordinary telephone pick-up and tape recorder unit. Numerous analyses have been made of such tapes with respect to magnetic field frequencies and strengths; however, no positive relationship has yet been found with respect to bearing damage. Still, there is the obvious condition in which an increase in magnetic field is expected to manifest itself as a higher recorded level of audio frequency noise. Efforts are also being made to relate static and other types of audio noise disturbances to current discharge through an oil film or a similar discharge. Whenever this noise occurs on the tape, it may be possible to evaluate in the laboratory if it is reasonable to expect that a condition of increased field strength does occur.[37, 38]

The three different classifications listed above are covered here with respect to residual fields found on measured components prior to demagnetization in the open air.

Critical Components. These components are those that experience damage from currents either of a gradual or sudden nature requiring shutdown of the train. First and foremost among these are bearing pads, sleeves, and seals. Obviously, when these are nonmagnetic, only the earth's field strength can be measured on the free, open-air unit. Even when these are made of magnetic materials, fields are generally low, indicating that they are usually a conduit for either the magnetic field or electric current. In some instances, hardened buttons or rockers produce high field levels, sometimes in the range of 15 to 20 gauss; however, in absence of these types of damage items, the pads rarely have fields in excess of 6 gauss. Current damage can occur on nonmagnetic components as readily as on magnetic items as is indicated from spark tracks and frosting noted on bronze bearing pads as well as aluminum bearings.

Examples of damage to turbomachinery bearings sleeve can be seen in Figures 3-191, 3-192, and 3-193. However, bearing currents are not confined to turbomachinery and have been found on ball and roller bearings as well. An example of this is shown in Figure 3-194, which is a typical ball roller bearing used in a radial load thrust application. Damage occurs in lines rather than in random discharge patterns noted on bearing pads and sleeves. This would indicate that a timing of the race and ball rotation is critical to the time when the current flows.

Other critical components are the mating parts for those mentioned above, namely, the shaft journal (Figure 3-195) and seal areas as well as thrust collars. The journals have rarely been found to have field levels in excess of 10 gauss; however, high through shaft fields have been found. Sometimes an occasional high magnetic local spot in the range of 15 to 20 gauss appears as if someone put a magnetic clamp or similar device onto the shaft. The thrust collar, on the other hand, is often machined on a magnetic chuck and may show very strong regions of alternate north and south poles across its face. This magnetism must be removed; otherwise, a strong set of local eddy currents occur.

(*text continued on page 246*)

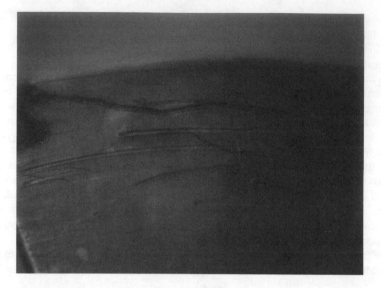

Figure 3-191. Typical spark tracks on a thrust shoe.

Figure 3-192. Thrust shoe severely damaged by electrical shaft currents (frosted area). The wedge shaped area at leading edge is the only remaining original surface. The babbitt is completely gone and spark-erosion has completely progressed into the steel backing at the white triangular area at the trailing edge OD. (0.085″ metal has been spark eroded at this point.) The dark area is oil-cake. Bearing was close to failure.

Figure 3-193. Damage to a tilt-pad bearing by short-circuited homopolar-generation loop *within* the bearing.

Figure 3-194. Ball bearing inner race damaged by shaft currents. The surface of the small grooves shows fine frosting.

Figure 3-195. Typical spark frosting on approximately one-half the circumference of a journal bearing and seal area.

(*text continued from page 243*)

Couplings and gears are also critical items. These were found to have field strengths varying up to 50 gauss. Usually high fields in these items have resulted from lack of proper demagnetization following magnetic particle inspection testing or from use of magnetic chucks during machining.

Primary Components. These are areas that may originate or focus magnetism causing magnetic currents to flow resulting in damage to components of the "critical" category mentioned above. Shafts have been found to have the greatest potential for conducting magnetism and retaining high residual levels. These are usually high alloy steel with an accompanying hard magnetism capable of high residual energy storage. The ability of shafts to retain magnetism has been very apparent. Some that have been magnetic particle inspection tested retained levels as high as 30 gauss even though they were supposedly demagnetized. Also, shafts have been found to have high residual magnetism after having operated for some period of time, even though they were reportedly installed in a demagnetized state. Where there has been a sudden crash accompanied by excessive bearing currents, shafts have always been found to have high levels of residual magnetism, usually oriented axially. Magnetic fields have reached hundreds of gauss in the open air, the highest levels being located at the corners at the very ends of the shaft.

Also in this classification are items mounted on the shaft such as wheels, blades, and shrouds, all of which may carry through the character of the shaft magnetism or which may show an individual-type magnetism having various magnetic polarities and strengths as if these had been magnetized during earlier inspection or machining.

Magnetic fields on these items reach 25 to 30 gauss in the open air condition. Figure 3-196 is convincing evidence of the magnitude of magnetic fields on critically important equipment. Nozzle rings, diaphragms, and other interstage elements are similar to the shaft-mounted components. These are usually of a high alloy steel and are often found to have very high levels of magnetism in the open air reaching up into the 50 gauss range. While they sometimes carry the character of the casing in which they are mounted, more often these have fields that are circumferentially oriented and occasionally have poles, especially indicative of testing or machining with magnetic chucks. Usually these elements are in halves, and placing them together produces an extreme field at the parting.[39]

Also in the primary classification are the seal and bearing housings. These generally have low residual fields because they are usually of a mild steel; however, since this is not always the case, they may require demagnetizing. They constitute a significant conduit for through current or flux from the casing into the bearing pads, seals, and thrust units.

Finally, the main casing and appendages to this main casing should be considered. These almost always provide a through conduit for magnetizing fields and may have residual field retention as well. Magnetic fields have been found to reach as high as 25 gauss in casings, especially at the corners and on dowels or bolts. This magnetism can register fields upwards of 100 gauss. These latter items can usually be demagnetized locally without difficulty. Sometimes the casings and the bearing housing partings have experienced current arcing damage. Because this does not require a shut-

Figure 3-196. Steam turbine rotor with residual magnetic field sufficiently strong to hold nails and small mirrors. (Source: Sohre Turbomachinery, Inc., Ware, Massachusetts 01082)

down, it is not regarded as critical damage. This has been observed in several instances, particularly on turbine casings.[40]

Secondary Components. Next are the secondary areas. These areas are conduits or remote sources for residual magnetism. They include the bases on which the train is mounted, plus the structural steel and supports as well as piping and the interconnected valves, heat exchangers, and other items that constitute a magnetically coupled system. These items will probably not experience damage, but their role can be most instrumental in producing or supporting troublesome magnetic fields. It is rare that the magnetism in these elements, when disassembled and laid out as individual pieces, can be detected in any great amount. Still, in an assembled state, they have the potential to create a very high field. Residual magnetism is usually produced within these elements while welding is in progress, either directly on them or where welding current flows to ground through them, or as enclosed by a magnetic loop. This latter situation may occur when the welder has pulled the welding electrode cable through a closed magnetic loop, of which the compressor train is one of the components, and when the ground return current path is through the structure to the ground electrode outside this loop. In either of these cases, the exact source of magnetism cannot easily be located, and it is necessary to measure the field when the compressor and its structure and piping interconnections are made in the fully assembled or a semi-assembled state. A properly demagnetized machinery train requires checking the field and demagnetizing as the final assembly is being made. This is in addition to individual component demagnetization, which has already been covered. Magnetism found in piping and around structures varies considerably. Levels can be as high as 100 gauss. To make these measurements requires a certain degree of skill in locating the magnetic path, which in the fully assembled state will produce very high magnetic fields.[41]

Present State-of-the Art Demagnetizing. Since the mid 1980s leading consulting engineers such as Paul Nippes and John Sohre have used all of the various types of demagnetizing equipment discussed here. First, there are the low voltage, high current units. These include magnetic particle inspection power supplies, direct current welders, and DC generators that are usually a part of an M-G set. All units use a heavy, single-conductor-insulated cable normally 2/0 or 4/0 in size. Because of the large diameter, these cables are very stiff and are extremely difficult to position or to coil around the suspected residual magnetic source. Major problems exist with polarity reversal and the limited range of current steps. Because of the high current involved, it is not practical to have an automatic reversing switch in order to change polarity, a necessary feature for proper down cycling and demagnetization. Furthermore, the very small steps occasionally needed at the bottom to remove residual magnetism, especially in closed magnetic loops or minor configurations, usually cannot be obtained with these devices. Thus, a certain amount of residual magnetism may remain in the equipment.

There is one advantage to the high current, low voltage source, in that current can be passed directly through the components. While this has a theoretical advantage, there are many practical and ultimate drawbacks in doing this. Some units have been demagnetized by passing thousands of amps through the center of the shaft, alternately reducing these down to hundreds of amps with very successful demagnetization of all full-ring components. However, on segmented shrouds, or where discontinuities occurred in casings such as inlet and outlet piping or feet, quite strong north and south polarities develop at the edges of the discontinuities. This is a magnetic condition that under certain circumstances could result in magnetic current problems.

Next to be considered is a common coil in a circular or rectangular shape using alternating current as a source. This coil can be either a high voltage, low current coil of the conventional type, or it can be a low voltage, high current coil. These coils, fed from an AC source, are very successful in demagnetizing thin wall components such as bearing housings, pads, sleeves, coupling components, bolts, and nuts. However, when the steel section exceeds a half-inch in thickness and certainly when it exceeds one inch, the lack of penetration of the alternating field into the component prevents soaking in of the 60 hertz cycling field. Often what is believed to have been demagnetized, has not been, and strong internal fields continue to exist. Also with the AC supply, when the current to the coil is interrupted, there is a good probability that high residual fields are induced. On the other hand, gradual removal of the component from the field, or a gradual reduction of its voltage supply, will down cycle the field and proper demagnetizing will occur in thin-walled elements.

One type of demagnetizer, which has been used because of its portability and suitability for placing of the coil, is the hand-held variable current unit. This has been used successfully by skilled operators in reducing magnetism of components and assembled parts. It has the advantage of having a light cable, approximately ⁵⁄₁₆ of an inch in diameter, which can be placed into or around most areas requiring demagnetization. It operates from a 120 volt AC source.

Figure 3-197 shows the automatic demagnetizing device developed to replace hand-held variable current units. Here, a small diameter multi turn coil is used and, therefore, one can place the coil in otherwise inaccessible areas. Furthermore, the unit depicted in Figure 3-197 has programmed cycling with polarity reversal. By following the detailed instructions supplied with this equipment, a technician should be able to position the demagnetizing coil in such a way that automatic cycling of the equipment will reduce residual magnetism in a satisfactory manner. If, for some reason a trial location and cycling event did not succeed in reducing the field, other coil positioning or cycling conditions may be tried until demagnetizing has succeeded. The equipment has a manual variable current mode; however, it is not recommended that this be used except to determine initial settings for the automatic phase or for placing a bucking field into a piece of equipment. This latter item is performed with the full understanding that it does not constitute proper demagnetization, but is a stopgap measure for expediency, and that a certain amount of risk is involved in leaving the equipment in this state. The extent of existing or residual magnetism is monitored with the devices shown in Figure 3-198.

Controlling Magnetism and Shaft Currents. As was stated earlier, the effects of magnetism can be gradual or suddenly disastrous to the point at which a decision to shut down may have to be made. This decision is usually unannounced and unexpected, and there is severe damage to the shaft, bearings, and seals. A very high level of residual magnetic field is found in the damaged unit. No method can be proposed to assure control or limitation of this event.[42]

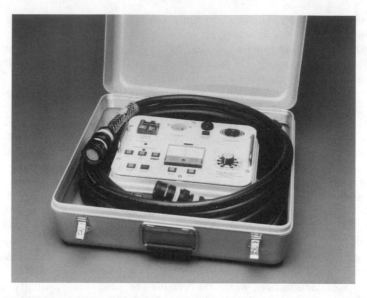

Figure 3-197. "Auto-Degauss" traveling unit. (Source: Magnetic Products and Services, Inc., Holmdel, New Jersey)

Figure 3-198. Magnetism of critical and/or large rotating machinery can be measured with these dual monitors.

On the other hand, the gradual buildup of magnetic effects is determined either by a phone pick-up with tape recorder measurements being made at selected locations or by the level of brush current determined from a reliable shaft-mounted brush such as that developed by John Sohre (Figures 3-199 and 3-200).* A very positive indica-

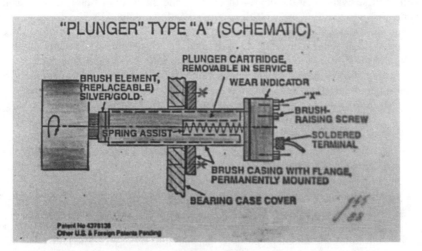

Figure 3-199. Typical arrangement of Sohre Turbomachinery, Inc. TYPE-A "plunger type" brush being used on a shaft end or collar. The brush can also be used radially, running on the shaft OD.

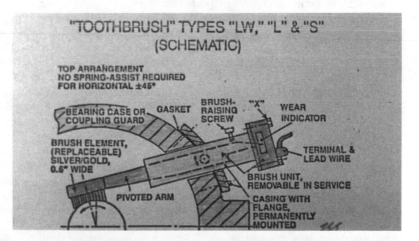

Figure 3-200. Typical arrangement of Sohre Turbomachinery, Inc. "toothbrush" type brushes running on shaft OD. These brushes can also be run against a shaft end or collar when equipped with spring assist.

---

*Sohre Turbomachinery, Inc., Ware, Massachusetts 01082.

tor, though not necessarily caused by currents, is a change in shaft position, whether dropping at the journal pads or moving axially towards the thrust-bearing loaded position. One further positive determination for gradual deterioration is the extent of damage to the bearings and seals, as well as couplings and gears, at each time of disassembly when, at equal intervals, the extent of damage is increased. It is important in all of these events that the degree of damage and the level of fields, as well as the phone pick-up recording, be well-documented and maintained in files for later comparison in evaluation of the condition of equipment. This will then provide a decision point for action. In any case, once magnetism is determined to be high by using a reliable gaussmeter with a hall probe, then the need for demagnetizing is clear. Following this, efforts should be made to maintain the equipment in a demagnetized state. Some of the suggestions that may be put into action are discussed below.

1. Existing Installations

   For the many units now installed and being manufactured, the prospect of residual magnetism causing problems is very real. Units should be maintained in a demagnetized state. Every effort should then be made to control and prevent remagnetization.

   - All components upon receipt following purchase, repair, or testing should be entirely free of residual magnetism.
   - Thorough deep-soaked demagnetization should be conducted on any component following magnetic particle inspection or on any component discovered to have high level of magnetic fields.
   - Welding on the compressor, turbine, or its piping should be controlled very carefully. The ground clamp and electrode cables should both be strung along the same path to the work area. Then the ground clamp should be connected to the same metal piece that is to be welded.
   - All components should have ground straps interconnected to the structure or station ground grid. The ground grid should have a ground resistance of less than 3 ohms. Also, lightning rods and other tall structures should have cables firmly interconnected to the ground grid. They further must be routed so they are not near nor do they link magnetic circuits such as closed-loop piping between the compressors or turbine to heat exchangers, condensors, boilers, etc. The goal is to provide a low-impedance discharge path for atmospheric discharge current but in such a way that component magnetization cannot occur.
   - Reliable brushes should be applied to the shafts to drain away electrostatic charge and to shunt persistent electromagnetic currents around bearing and seal surfaces. Brushes must be continuously conducting. This may require the use of a properly designed wire bristle brush (Figure 3-201). Close initial monitoring of the brush currents and voltages is necessary to assure that there is no compounding of magnetic fields due to the internal current paths. If this occurs, then brushes should be removed until components are demagnetized and there is assurance that remagnetizing currents are arrested.

Figure 3-201. Examples of Sohre Turbomachinery, Inc. sacrificial metal fiber brush elements. Extensive government testing has proved sacrificial metal fiber brush elements are far superior in performance and maintenance to any other shaft grounding device.

2. Future Installations

Remagnetization and occurrence of bearing damage is expected to continue unless significant measures are taken to correct the problems. Some of these measures are:

- Insulation of all bearings, seals, couplings, and other components through which damaging currents now flow.
- Installation of permanent brushes to shunt currents around the affected components.
- Selection of materials for the equipment that is magnetically soft rather than magnetically hard.
- Installation of coils in units at the time of original manufacture that can be used to effect demagnetization without having to disassemble the equipment.
- Installation of sensors at the time of original manufacture to detect shaft currents and internal magnetic fields and thus provide for continuous monitoring.

Additional Considerations. Magnetic problems have received much attention because of forced downtime and physical damage to components. Other effects, neglected up to now, deserve consideration. In addition to the quest for higher productivity, elimination of wasted energy and improved efficiency are eagerly sought. Preliminary considerations indicate that elimination of circulating eddy currents due to residual magnetism could constitute considerable energy savings. These eddy currents can exist in a machine even though the circulating current paths may not damage components. Demagnetization and prevention of future remagnetization are required for their elimination.

References

1. Good, W. R., "A Few Observations About Paper Machine Lubrication," SKF *Ball Bearing Journal* No. 202, p. 8.

2. Walp, H. O., "Recognition of the Causes of Bearing Damage as an Aid in Prevention," SKF Industries. 1958.

3. *Handbook of Loss Prevention.* Editors: Allianz Versicherungs-AG. Translated from the German by P. Cahn-Speyer. Berlin-Heidelberg-New York: Springer-Verlag, 1978, p.33.

4. Ku, P. M., "Gear Failure Modes—Importance of Lubrication and Mechanics," ASLe *Transactions,* Vol. 19, 1976, p. 239.

5. *Handbook of Loss Prevention,* Editors: Allianz Versicherungs-AG. Translated from the German by Cahn-Speyer. Springer Verlag, Berlin, 1978.

6. AGMA 110.xx, The American Gear Manufacturers Association, Nomenclature of Gear-Tooth Failure Modes (1979).

7. AGMA 215.01, The American Gear Manufacturers Association, (R 1974) Information Sheet for Surface Durability (Pitting) of Spur, Helical, Herringbone and Bevel Gear Teeth (1966).

8. EHD Lubricant Film Thickness Formulas for Power Transmission Gears (L. S. Akin), Journal of Lubrication Technology—Transactions of the ASME, July 1974, pgs. 426-436.

9. AGMA 217.01, The American Gear Manufacturers Association, Information Sheet—Gear Scoring Design Guide for Aerospace Spur and Helical Power Gears (1965).

10. AGMA 411.xx, The American Gear Manufacturers Association, Gear Scoring Design Guide for Aerospace Spur and Helical Power Gears (undated).

11. AGMA 225.01 The American Gear Manufacturers Association, Information Sheet for Strength of Spur, Helical, Herringbone and Bevel Gear Teeth (1967).

12. AGMA 226.01, The American Gear Manufacturers Association, Information Sheet, Geometry Factors for Determining the Strength of Spur, Helical, Herringbone and Bevel Gear Teeth (1970).

13. Dudley, D. W., "Practical Gear Design," McGraw-Hill Book Co., 1954.

14. Wattner, K. W., "High Speed Coupling Failure Analysis", Proceedings of 4th Turbomachinery Symposium, Texas A & M University, College Station, TX, 1975, pp. 143-148.

15. Calistrat, M. M., "Some Aspects of Wear and Lubrication of Gear Couplings," ASME Paper No. 74-DET-94, Presented at the Design Engineering Technical Conference, New York, N.Y., Oct. 5-9, 1974.

16. Bloch, H. P., "How to Uprate Turboequipment by Optimized Coupling Selection," *Hydrocarbon Processing,* Jan. 1976, pp. 87-90.

17. Bloch, H. P., "Selection Strategy for Mechanical Shaft Seals In Petrochemical Plants", Proceedings of 38th ASME Petroleum Mechanical Engineering Workshop and Conference, Philadelphia, Pa., Sept. 12-14, 1982, pp. 17-24. Also reprinted in *Hydrocarbon Processing,* January, 1983.

18. Neale, M. J., *Tribology Handbook,* Halsted Press, New York-Toronto, 1973.

19. Bloch, H. P., "Optimized Lubrication of Antifriction Bearings for Centrifugal Pumps," ASLE Paper No. 78-AM-1D-2, Presented at the 33rd Annual Meeting, Dearborn, Michigan, April 17-20, 1978.

20. Bloch, H. P., "Criteria for Water Removal from Mechanical-Drive Steam-Turbine Lube Oils," ASLE Paper No. 80-A-1E-1, Presented at the 35th Annual Meeting, Anaheim, California, May 5-8, 1980.

21. Watanabe, H., and C. Kobayashi, "Degradation of Lube Oils—Japanese Turbine Lubrication Practices and Problems," *Lubrication Engineering,* Vol. 34, August 1978, pp. 421-428.

22. Bloch, H. P., "Results of a Plant-Wide Turbine Lube-Oil Reconditioning and Analysis Program," Presented at the 38th Annual Meeting of the American Society of Lubrication Engineers, Houston, Texas, April 25-28, 1983.

23. Bloch, H. P., J. L. Allen, and T. Russo, "Cost Justification and Latest Technology for the On-Stream Purification of Turbomachinery Lube Oil," Presented at the International Turbomachinery Maintenance Congress, Singapore, October 9-11, 1990.

24. Allen, J. L., "On-Stream Purification of Lube Oil Lowers Plant Operating Expenses," *Turbomachinery International,* July/August 1989, pp. 34, 35, 46.

25. Symposium on Steam Turbine Oils, ASTM Special Publication No. 211, September 17, 1956.

26. Bloch, H. P., and J. R. Carroll, "Preventive Maintenance Can Be More Effective than Predictive Programs," *Oil and Gas Journal,* July 30, 1990, pp. 81–86.

27. Adams, M. A., and H. P. Bloch, "Vacuum Distillation Methods for Lube Oils Increase Turbomachinery Reliability," Proceedings of the 17th International Turbomachinery Symposium, Texas A&M University, Dallas, Texas, November 8–10, 1988.

28. Bloch, Heinz P., "Vacuum Oil Conditioner Removes Contaminants from Lubricating Oil," *Chemical Processing,* March 1982, pp. 84–85.

29. Jacobson, D. W., "How to Handle Waste Oil," *Plant Engineering,* June 10, 1982, pp. 107–109.

30. Bloch, Heinz P., "Reclaim Compressor Seal Oil," *Hydrocarbon Processing,* October 1974, pp. 115–118.

31. "Oelbuch," Anweisung fuer Pruefung, Ueberwachung und Pflege der im Elektrischen Betrieb verwendeten Oele und synthetischen Fluessigkeiten mit Isolier- und Schmiereigenschaften/Herausgeber: VDEW. 4. Auflage, 1963, Frankfurt: Verlags- und Wirtschaftsgesellschaft der Elektrizitaetswerke (VWEW).

32. Grupp, H., "Moderne Untersuchungsverfahren Fuer Schmieroele und Hydraulikfluessigkeiten von Dampfturbinen," *Der Maschinenschaden,* 55(1982) Heft 2, pp. 84–87.

33. Sullivan, J. R., "In-Plant Oil Reclamation—A Case Study," *Lubrication Engineering,* Volume 38, July 1982, pp. 409–411.

34. Sohre, J. S. and Paul I. Nippes, "Electromagnetic Shaft Currents in Turbomachinery—An Update," Presented August 19, 1980 at the 25th Symposium on *Safety in Ammonia Plants and Related Facilities,* American Institute of Chemical Engineers, Portland, Oregon (also printed in abbreviated form in Volume 23, Ammonia Plant Safety, a CEP Technical Manual published by the American Institute of Chemical Engineers).
35. Sohre, J. S. and Paul I. Nippes, "Electromagnetic Shaft Currents and Demagnetization on Rotors of Turbines and Compressors," presented December 1978, at the 7th Turbomachinery Symposium, Texas A & M University, College Station, Texas.
36. Sohre, J. S., "Shaft Currents in Rotating Equipment: Electromagnetic and Moving-Particle Induced," an informal review of literature. October 19, 1971.
37. Sohre, J. S., "Are Magnetic Currents Destroying Your Machinery?" *Hydrocarbon Processing,* April 1979, pp. 207–212.
38. Sohre, J. S., "Shaft Currents Can Destroy Turbomachinery," *Power,* March 1980, pp. 96–100.
39. Schier, V., "Selbsterregte Unipolare Gleichstroeme in Maschinenwallen," Elektrotechnische Zeitschrift, November 12, 1965 (in German). A translation of this article was prepared by the Associated Electrical Industries, Ltd., Power Group, Research Laboratory, Trafford Park, Manchester 17, United Kingdom. Translation #3925 "Self-excited Homopolar Direct Currents in the Shafts of Machines."
40. Oearcem, C. T., "Bearing Currents—Their Origin and Prevention," *The Electrical Journal,* August 1927. Vol. XXIV, No. 8, pp. 372–6.
41. Rosenberg, L. T., "Eccentricity, Vibration, and Shaft Currents in Turbine Generators," *AIEE Transactions,* April 1955, pp. 38–41.
42. Wilcock, D. F., "Bearing Wear Caused by Electric Current," *Electric Manufacturing,* February 1949, pp. 108, 111.

Chapter 4

Machinery Troubleshooting

Machines should work;
people should think
IBM Pollyanna Principle[1]

Proof is often no more than lack
of imagination in providing an
alternative explanation
De Bono's 2nd Law[2]

If failure analysis represents the post mortem of a failed machinery component, then troubleshooting—as an extension of failure analysis—encompasses all root cause determination activities. Often, the troubleshooting process will not stop there but will lead to problem elimination. Two basic cases will illustrate the principle: One, a machinery manufacturer experiences consistent fatigue failures of a drive shaft during prototype tests of his newly designed equipment. Two, a machinery owner experiences similar failures after years of trouble-free operation. In the first case, failure analysis will uncover the cause of the failure, and remedial action through a change of design parameters can be achieved relatively fast. In the second case, failure analysis will be part of a process generally called troubleshooting. Here, once failure analysis is completed and, say, "fatigue fracture," due to low-cycle torsional vibration has been determined as the immediate cause-and-effect relationship, troubleshooting begins. What caused torsional vibration suddenly to appear? The experienced engineer will say: "But it does not usually work this way! How about diagnostics? Is that not a part of troubleshooting?" A justifiable question, because quite often machinery engineers are called upon to solve problems associated with machinery performance deficiencies as well as to assess machinery health. Typical process machinery performance deficiencies, or simply problem symptoms, are listed in Table 4-1.

Table 4-1
Typical Process Machinery Performance Deficiencies

SYMPTOMS Abnormal Performance Parameters	Compressors	Blowers	Fans	Pumps	Turbines	Electric Motors	Engines	Gears	Lubrication Systems
Change in Efficiency	•			•	•				
No Delivery				•					•
Insufficient Capacity	•	•	•	•	•		•		•
Abnormal Pressure	•	•	•	•					•
Excessive Power Requirement	•	•	•	•					
Excessive Leakage	•	•		•	•			•	•
Excessive Noise	•	•	•	•	•	•	•	•	
Overheating	•	•		•	•	•	•	•	
Knocking	•			•			•		
Excessive Discharge Temperature	•	•			•		•		
Fails to Start	•	•	•	•	•	•	•		•
Fails to Trip					•	•	•		
Lack of Power					•	•	•		
Excessive Steam/Fuel Consumption					•		•		
Failure of Automatic, Control, or Safety Devices	•	•	•	•	•	•	•	•	•
Freezing	•			•					
Abnormal or Fluctuating Speed					•	•	•		
Stoppage	•	•	•	•	•	•	•	•	•
Vibration—Excessive	•	•	•	•	•	•	•	•	

The appearance of one or more of these symptoms will require the machinery engineer to answer the following questions:

1. If the machine is operating, must it be shut down to reduce consequential damage? How serious is the problem? How fast must we react to it? Is the problem increasing, is it holding, or has it decreased?
2. If the machine is available for maintenance, must it be opened for inspection and repair?
3. What components and their failure mode and effect chains might be causing the symptoms to appear, and what parts and facilities must be on hand to reduce downtime?

How does the experienced troubleshooter go about answering these vital questions? In the following pages, we will explore several competing approaches and demonstrate methods that have been applied successfully.

Competing Approaches

Organizational approaches. Quite often the organizational approach to trouble-shooting does not help. The structuring of organizations in process plants to provide "fixes" for symptoms and failures will not lead to conclusive identification and elimination of the true cause. Instead of doing something about the *cause* of a machinery problem, the "maintenance" approach of just correcting symptoms or effects is employed. Later in this chapter, in a section about organizing for successful failure analysis and troubleshooting, the shortcomings of the purely organizational, very conventional approach will be addressed in detail.

Learning to live with the problem. "The problem is no problem any more because we have learned to live with it," one of the authors overheard somebody say in connection with a machinery problem. We are all familiar with the thought behind it. If it is an ongoing problem but only occurs, say, every three months or so, we will have adapted to it with all kinds of contingent actions: "When it appears, we do this—and that is quite normal!" It goes without saying that lack of proper motivation is the reason for this approach, but another reason is a lack of understanding of machinery failure modes.

Well-meaning approach. With this approach, the troubleshooting effort is well intended but ends up creating new problems in the course of curing the old ones. It represents failure to compare the advantages and disadvantages of a solution.

Another characteristic of this approach is the failure to follow up on recommendations made once the failure cause has been determined. We often fail to translate the correction of a problem into preventive action in areas where the same problem might occur. Parochial attitudes sometimes will deter us from doing so.

Mr. Machinery approach. This approach actually belongs in the categories of organizational approaches and the well-meaning approach. It is the approach taken when sole responsibility of a plant's machinery fate rests on one single person—Mr.

Machinery. His batting average is usually good. How does one know whether or not it can be improved? Quite often he is in charge because he is an efficient firefighter. He can make the trouble go away. He is used to calling the shots. He is the machinery troubleshooter by definition. How successful might he be?

Throughout this text we will show how failure analysis and troubleshooting must be a cooperative effort, with very little room for Mr. Machinery going it alone.

Approaches caused by wrong thinking patterns. Machinery failure analysis and troubleshooting is really the process of *understanding* what causes machinery to fail or malfunction. "Understanding is thinking"[2], and as one successful troubleshooter once said, troubleshooting is knowing when to think and when to act. If failure analysis and troubleshooting at times are not successful, it may be that our thinking patterns do not allow us to be efficient and successful.

Our thinking process is impeded by two factors: one is an extraneous factor; the other is a factor determined by the thinking process itself. The first factor is the recognition that machinery component life can be either predictable or not (we alluded to this fact before, and we have addressed the need to recognize this when undertaking failure analysis and troubleshooting of process machinery). The second factor is the subject of the following discussion.

We should begin by saying that the same processes that make the mind so effective a thinking device are also responsible for its mistakes.[2] Consequently, there is this first mistake:

The monorail mistake. Because machinery vibration signature analysis has worked so well in the inspection of jet engines, it is easy to assume the same would be true for major unspared process machinery such as centrifugal compressors and turbines. However, where in one case there are literally hundreds of identical machines, in the other case, among process machines, really no two are the same. Whereas, in one case, preliminary diagnostics of one jet engine could be validated in a number of others, with heavy process machinery this is not possible. So consequently, miracles, in terms of troubleshooting results, are expected from the process plant machinery engineer equipped with all the paraphernalia of vibration analysis. The monorail mistake simply involves going directly from one idea to another in an inevitable manner, ignoring all qualifying factors.[2] Another example: "This is a vibration severity chart published by experts, and according to it, this machine runs with excessive vibration!" The qualifying factor being neglected is the fact that all machines are not created equal. If one cannot think of any qualifying factors in machinery troubleshooting, it is easy to make the monorail mistake.

The magnitude mistake. This is the second major thinking mistake that arises directly from the way the mind works. The mind moves from one idea to the other in what seems a valid way if only the "names" of the ideas are looked at. Repair reports on machinery are full of magnitude mistakes. We see "normal wear," "failed from excessive misalignment," etc. Magnitude mistakes are unlikely to occur in situations where the engineer has had direct experience. The machinery engineer has to ask the questions: "How much has it worn—in what time?" "What was the amount of misalignment?" etc.

The misfit mistake. You are walking down the street and recognize the general appearance of someone you know very well. You quicken your pace to meet the person, only to find a total stranger. This is the misfit mistake, because the idea of what something is does not fit reality. One recognizes certain features but does not wait until all the features are registered before jumping to a conclusion. The more familiar the situation, the quicker the jump to recognition. "Heard this noise before and it turned out to be. . . ." Misfit mistakes are easy to make in troubleshooting; in fact, the troubleshooting matrices presented in this text provide a good idea as to how many misfit mistakes can be made if this danger is not recognized.

The must-be mistake. This mistake is also called the "arrogance mistake." There may be nothing wrong with the way information has been put together to reach a conclusion, but arrogance fixes this conclusion so that no change or improvement is possible. When new evidence in the process of machinery troubleshooting emerges, the must-be mistake prevents us from changing our conclusions. This mistake goes so far as shutting out the possibilities of a completely new viewpoint which has not even been generated. "Don't confuse me with facts—I have made up my mind!"

The miss-out mistake. This mistake arises when someone considers only part of a situation and yet reaches conclusions that are applied to the whole situation. Quite often this will happen to the troubleshooter relying on second-hand information. For instance, in the hunt for input into the background of a thrust bearing failure of a large process compressor, an investigator could be misled by the fact that there is no record of a recent process upset which caused the machine to be liquid slugged. We have all been in the situation where we were left with the vague impression that there must be a fuller picture somewhere—fuller than the one we have. When we sense this in a machinery troubleshooting situation, we have several choices:

1. To reject the conclusion that is put forward because we are somehow convinced that it is based on only part of the whole picture.
2. To reject the conclusion that has been offered because we dislike it and, consequently, claim that it must be based on only part of the picture.
3. To accept the conclusion with reservations, but still look for the whole picture.
4. To accept the conclusion because we like it, and elect that the picture is, after all, complete.
5. To conclude that the remainder of the picture is not really there, since we cannot find it and that we have the whole picture.

The Professional Problem Solver's (PPS) Approach

We have been moving toward the PPS approach to failure analysis and troubleshooting by discussing the impediments to the thinking process. The last "mistake" seems to be the most critical one, as selection plays an important role in all information processing. There are three elements of selection involved in problem solving and troubleshooting[3]:

1. *Selective encoding.* One must choose elements to encode from the often numerous and irrelevant bits of information. The trick is to select the right elements and eliminate the wrong ones.
2. *Selective combination.* There may be many possible ways for the encoded elements to be combined or otherwise integrated. The trick is to select the right way of combining them. This process is usually based on a thorough knowledge of the interrelated functions of the various machinery parts and the effect of adverse conditions. An important insight in this context is that most machinery troubles have several causes which combine to give the observed result. A single-cause failure is a very rare occurrence.
3. *Selective comparison.* New information must be related to older pieces of information. There are any number of analogies or relations that might be drawn. The trick is to make the right comparisons.

These three ingredients of our mental process are commonly called insight. They have one factor in common: *selection.* However, there are other ingredients of a professional problem solver's approach to failure analysis and troubleshooting.

Prior knowledge. Even apparently simple problems often require prior knowledge. Consider the performance deficiencies listed in Table 4-1. It just does not make sense to try to diagnose these symptoms without at least some knowledge of what their causes could be.

Prior knowledge is based on experience, and if our experience has been the right kind, we will have built up certain basic troubleshooting axioms. Good examples are:

On lubrication:

1. Lubrication- or tribology-related problems on machinery often arise because deflections increase with size, while lubricant film thicknesses generally do not.
2. Temperatures have a very significant effect on lubricated components directly and indirectly because of differential expansions and thermal distortions. Therefore, we must always check temperatures, steady temperature gradients, and temperature transients when troubleshooting lubricated components.

On "trouble spots":

1. Trouble is likely to occur at interfaces, i.e. tight clearances, so we should expect fit corrosion and rubs. Cavitation will occur where an interface exists between vapor bubbles and a liquid, and between the liquid and an adjacent wall.[4] Only when both interfaces are present at the same time (see Figure 4-1), does cavitation occur.
2. Components that are designed to move relative to each other, but that are not in fact moving, will seize up.
3. Experienced maintenance people know to match-mark an assembly before taking it apart, to make sure that they can put it back together. But handling similar jobs a number of times can lead to overconfidence. Disregarding a simple precaution such as match-marking can then cause improper reassembly and serious operating problems after startup.

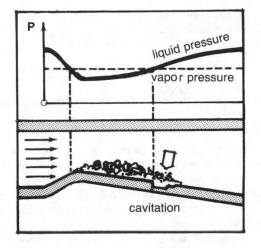

Figure 4-1. Cavitation event (modified from Ref. 5).

On problem solving:

1. Problems are best solved one at a time.
2. Problems have causes. Identify the cause before trying to solve the problem.
3. Before doing something, know what you want to accomplish.
4. Problems occur. It's better to plan for them than to be surprised.

And, on the lighter side: By definition, when you are investigating the unknown, you do not know what you will find.[1]

Executive processes. These are the processes involving the planning, monitoring, and performance evaluation in failure analysis and troubleshooting. To start with, one must first study the problem carefully, define it with regard to identity, location, time, and extent, and modify the perception of the problem as new information becomes available. Another important executive process involves monitoring the solution process—keeping track of what has been done, is being done, and still needs to be done—and then switching strategies if progress is too slow.

Motivation. Really challenging problems often require a great deal of motivation on the part of the problem solver. Successful problem solvers are often those who are simply willing to put in the necessary effort.

Style. People approach problems with different styles of thinking and understanding. Some tend to be more impulsive, and others more reflective. The most successful problem solvers are the ones who manage to combine both styles, for certain points in the troubleshooting process require following impulses and other points require reflection or rigorous calculation.

In summary, the professional problem solver's approach to failure analysis and troubleshooting involves the coordinated application of good selection process, prior knowledge, executive processes, motivation, and style.

The Matrix Approach to Machinery Troubleshooting

In the preceding chapters it became evident that process machinery trouble-shooting must contain certain diagnostic steps. We will encounter machinery trouble situations where the cause is known (no diagnostics required), where the cause is suspected, and where the cause is unknown. A responsible troubleshooter will not walk away from any of these situations—including the one where the cause is known. The following will cover diagnostics and address the approach to the situations where the cause is unknown or suspected.

Diagnostics is the method of reaching a diagnosis. Diagnosis, in turn, is simply another word for "recognition," as, for instance, the recognition of a disease.[5] In preceding sections we have alluded to certain aspects of machinery diagnostics. We should be able to differentiate between two basic forms: one, where diagnostics or failure analysis is performed after the failure has occurred; and two, where diagnosis is part of "machinery health monitoring" efforts, as shown in Table 4-2, i.e. where we attempt to recognize a machine's health or condition by assembling all pertinent performance and condition data. This approach is also referred to as condition monitoring, which plays an important role in predictive maintenance.

A good application of this form of diagnostics is in connection with health monitoring of large diesel engines and gas-engine compressors. Here, exhaust-gas temperature-monitoring programs have been used as an inexpensive and general machinery health indicator, but seldom have these programs provided adequate early warning of impending failure. Consequently, sophisticated diagnostic techniques were developed in the form of the engine analyzer. Typical diagnostic displays obtained with an engine analyzer are shown in Figures 4-2 and 4-3. Few other process machines allow the extent of failure diagnosis that can be achieved with this type of monitoring on these particular machines. Figure 4-4 and the inset text in Figure 4-5 give a good idea of what is possible in other fields with respect to failure and malfunction diagnosis.[6]

Closer to home, automatic fault diagnosis is available for motor shutdown trip conditions on large pump drives, as shown in Table 4-3.[7]

Frankly, as much as we would like to subscribe to the more predictive, on-line type diagnosis, its undeniable disadvantage is a frequently difficult-to-define cost/benefit ratio.

It has been said that 99% of all machinery failures are preceded by some nonspecific malfunction sign. The corollary is that with the exception of reciprocating compressors, internal combustion engines, and integrated electrical systems, most process machinery systems will give little warning of *specific* individual component distress prior to total failure. It therefore stands to reason that we have to be prepared to apply both post-mortem and on-line diagnostics in our failure-fighting strategy.

Table 4-2
Process Machinery Diagnostics

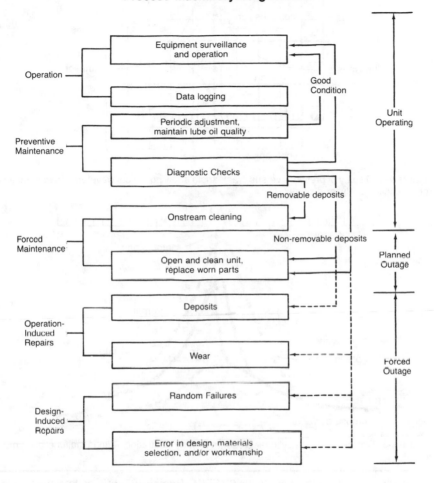

We thought it a good idea to look to the medical world for suitable analogies pertaining to our topic; after all, we borrowed the terms diagnosis and diagnostics from there. Three analogies come to mind:

1. *Symptoms.* Like the human body, a working process machine has certain "vital signs" that reveal to the experienced troubleshooter its internal state of health. Like the physician, the machinery analyst must rely primarily on his senses to obtain the symptomatic data he needs. Unlike the physician however, he very seldom has had those ten years or so of university training on the same model. The machinery troubleshooter must above all use his general experience and

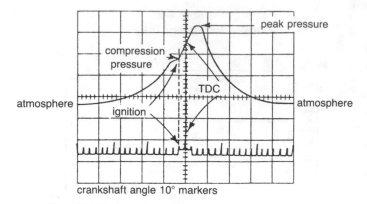

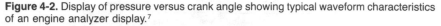

Figure 4-2. Display of pressure versus crank angle showing typical waveform characteristics of an engine analyzer display.[7]

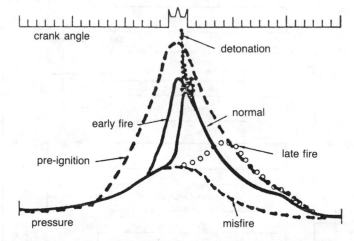

Figure 4-3. Two-cycle engine cylinder pressure display illustrating various malfunction symptoms.[8]

. . . the "Forward Facing Crew Cockpit" (FFCC-Boeing 767) affords total on-screen information and control: The flight monitoring system projects visible to the pilot onto the screen at any time engine operating data and flight systems conditions. Additionally, it indicates "Maintenance work due" for the ground maintenance crew. Computers, able to store some 200 malfunction conditions, warn the flight crew of approaching problems and indicate what measures are required for failure elimination.

The onboard computer is able to search within seconds for the root cause along the "trouble chain", and—above all—it does this quietly: Three quarters of all usual audible signals are now silenced in the computer cockpit . . . (From *Aviation Week Space Technology*)

Figure 4-4. The computer cockpit.

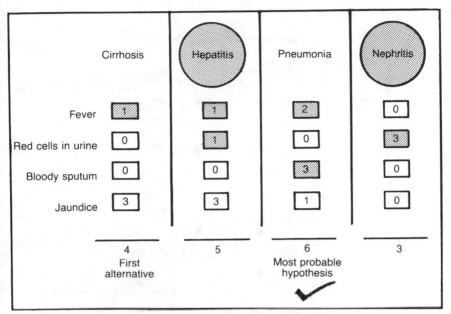

In a simplified example, the computer examines a patient with fever, blood in the urine, bloody sputum from the lungs, and jaundice. The program adds together numbers that show how much each symptom is related to four possible diagnoses—cirrhosis of the liver, hepatitis, pneumonia, and nephritis (kidney disease)—and picks pneumonia as top contender. The runner-up in score is hepatitis. But because hepatitis has one symptom not shared with pneumonia (blood in the urine), the computer chooses cirrhosis as first alternative. This process, called partitioning, focuses the computer's attention on groups of related diseases.

Figure 4-5. Computerized medical diagnosis (modified from Ref. 6).

his five senses. These, unaided by suitable instrumentation, will often tell him that "something" is wrong, without being specific. Consequently, in machinery troubleshooting as in medical diagnostics one has to wait for significant changes in a disease or distress situation to obtain from a given quantity of nonspecific symptoms specific ones that allow a valid diagnosis. Examples of nonspecific symptoms are shown in Table 4-4. Quite often the machinery troubleshooter must work with nonspecific symptoms and act timely or the specific symptom will show itself in a forced shutdown or even a wreck.

The following typical case will serve as an example of nonspecific machinery trouble symptoms. The patient: An 800-hp, 320-rpm induction motor driving a balanced-opposed reciprocating two-stage process gas compressor. The symptoms: Operating technicians have noticed the following gradual change in the motor's operating behavior:

- Unusual axial vibrations at 0.1 IPS (usual level: 0.05 IPS (inches/second peak-to-peak velocity).

Table 4-3
Motor Fault Diagnosis Display[7]

U/B	O/L	TRIPPED RELAY INDICATION				U/C	ON	PROBABLE CAUSE OF TRIP
---	---	START	RTD	G/F	TRIP	---	---	---
	•	•			•		•	START TIMER EXCEEDED • locked rotor • too many starts • start timer set too low
	•				•		•	OVERLOAD • excessive overloads during running • mechanical jam, rapid trip • short circuit
			•		•		•	HIGH TEMPERATURE (RTD OPTION) • blocked ventilation • cooling fins blocked • high ambient temperature
•	•	•			•		•	BLOWN FUSE START • check fuses
•					•		•	EXCESSIVE UNBALANCE • blown fuses • single phase supply • loose wiring connection • short between motor windings
				•	•		•	GROUND FAULT • motor winding to case short • wiring touching metal ground
•	•			•	•		•	GROUND FAULT • motor winding to case short • wiring touching metal ground
					•		•	MEMORY LOCKOUT • memory still locked out after power is reapplied
						•	•	UNDERCURRENT ALARM • pump loss of suction • pump closed discharge valve • blower loss of airflow • conveyor overload warning

Table 4-4
Nonspecific Symptoms in Diagnostics

Machinery Trouble Diagnosis	Medical Diagnosis
Increased Temperature	Fever
Noise	Looking Ill
Vibration	Skin Rash
High Bearing Temperature	Coated Tongue

- Discrete rubbing sound from inside the motor.
- Rubbing sound decreasing, disappearing with decreasing load.
- Ampere reading fluctuating ± 15 around 86% of full electrical load.

What is the troubleshooter to do who gets called upon the scene? Will he wait until further, more *specific* symptoms appear? Will he have the motor uncoupled from its driven machine in order to perform an idle test run and perhaps some electrical tests for rotor bar looseness? Is there perhaps a bearing problem in connection with water-contaminated oil that should be investigated? And, finally, are there even enough distress symptoms to warrant any action at all?

These and other thoughts will go through the troubleshooter's head as he looks at the symptoms before him.

2. *Diagnosis* in machinery troubleshooting and in the medical profession is far from an exact science. Respected specialists in medicine will examine the same set of symptoms and arrive at different conclusions,[9] most likely because there are often too many nonspecific distress symptoms.

3. *Computerized Diagnosis.* Computer diagnosis in medicine is already functioning well.[6] In a way this is not surprising, as it is applied, again, to only one model and not to a diverse population of process machinery. The approach, however, is interesting in our context. It functions as described in Figure 4-5.

We are now ready to look at the information needed to perform machinery troubleshooting and diagnostics. Here are the requirements:

1. List distress or *trouble symptoms.* Start with the ones that are most easily perceptible:
   a. *Taste/odor:* gas, acid leakage, etc.
   b. *Touch:* overheating, vibration, etc.
   c. *Sound:* excessive noise, knocking, piston slap, rubbing, detonation, annunciator alarms, etc.
   d. *Sight:*
      1. *Direct sight or observation:* vapor, fume, and fluid leakage, vibration, loosening, smoke, fire, etc.
      2. *Indirect sight or observation:*
         a. Changes (up/down) in indicator readings:
            - Pressure
            - Temperature
            - Flow
            - Position
            - Speed
            - Vibration
         b. Changes (up/down) in performance:
            - Pressure ratios
            - Temperature ratios
            - Power demand
            - Product loss
            - Efficiencies

    e. Internal inspection results.

    f. Failure analysis (post-mortem) results.

2. List *possible cause or hypothesis:* Begin by listing the machine's major components, systems, and auxiliaries. It is here where the troubleshooter's intimate knowledge of the machinery in his care finds it application.

3. Indicate *tie-in or relationship* of symptom to cause. Make use of experience by asking: What can and usually does go wrong with this component? What are the symptoms associated with this distress or malfunction? A typical aid for this effort will be a fault chart such as shown in Figure 4-6.

4. Indicate symptom/cause *probability ranking.* For instance, if port plugging, piston-ring sticking, excessive ash, and varnish and carbon deposits are

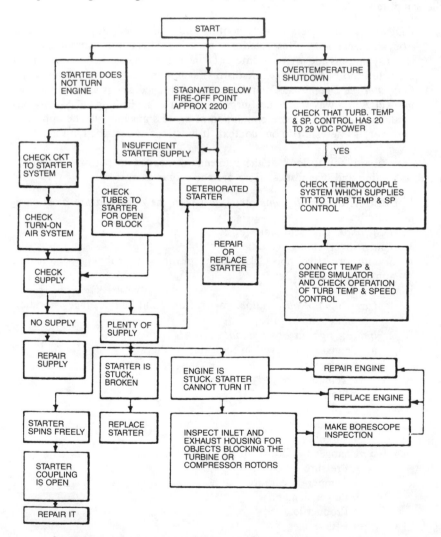

Figure 4-6. Gas-turbine troubleshooting (adapted from ref. 10).

symptomatic of overlubrication in internal combustion engines, which symptom is most indicative of overlubrication and which one less so? It is a good idea to rank relative probabilities of symptom/cause relationships in machinery troubleshooting, as shown in Figure 4-5. Later, in Chapter 12, we discuss the natural development of this approach in terms of expert system applications.

5. Indicate remedial action—*remedy*. Quite often remedial action is obvious and commonplace. Typical actions taken after the most probable cause of a trouble has been identified are checking, inspection, adjustment, lubrication, cleaning, balancing, replacing, and, of course, analysis and follow-up.

The following sections show troubleshooting guides in matrix form accompanied by appropriate comments. Test these against your own troubleshooting efforts, and keep in mind that in some cases the ultimate development could well be computerized machinery fault diagnosis.* Where possible, relevant failure statistics are presented for various types of process machinery. This will help in problem symptom and cause probability ranking or partitioning.

Troubleshooting Pumps

The centrifugal pump is the workhorse of petrochemical-plant pumping applications. As far as process pumps are concerned, Table 4-5 gives the reader, by way of a statistic, an indication of the most frequently encountered troubles.

Tables 4-6 and 4-7 represent the late Igor Karassik's troubleshooting guide for the most common on-line distress situations.

---

* Current machinery operating and maintenance manuals have troubleshooting information of varying quality. We believe suitably formatted troubleshooting information should be made a specified requirement for new process machinery purchases.

Table 4-5
Typical Yearly Repair Summary, Centrifugal Pumps

Centrifugal pumps installed: 2560
Centrifugal pumps operating at any given time: 1252 (average)

Total pumps repaired in 1979:	768
Pumps repaired at site location:	382
Pumps repaired at own shops:	267
Pumps repaired at outside shops:	119

Failure Causes	Distribution (%)
Mechanical seals	34.5
Bearing distress	20.2
Vibration events	2.7
Packing leakage	16.3
Shaft problems/couplings	10.5
Case failure/auxiliary lines	4.8
Stuck	4.3
Bad performance	2.5
Other causes	4.2
Total	100.0

An internationally recognized authority on pumps and their applications until his death in 1995, Mr. Karassik believed that, while no list of pump troubles can ever be complete, it makes sense to use checklists in diagnosing centrifugal pump troubles. The checklist approach shown in Tables 4-6 and 4-7 thus correlates observed symptoms with possible causes of trouble. In Reference 11, Igor Karassik refers to these situations as "field troubles," and no doubt his check charts for centrifugal pump problems are among the most comprehensive troubleshooting guides for centrifugal pumps.

Table 4-6
Check Chart for Centrifugal Pump Problems

Symptom*	Possible Cause of Trouble* (Each number is defined in Table 1)
1. Pump does not deliver liquid	1, 2, 3, 5, 10, 12, 13, 14, 16, 21, 22, 25, 30, 32, 38, 40
2. Insufficient capacity delivered	2, 3, 4, 5, 6, 7, 7a, 10, 11, 12, 13, 14, 15, 16, 17, 18, 21, 22, 23, 24, 25, 31, 32, 40, 41, 44, 63, 64
3. Insufficient pressure developed	4, 6, 7, 7a, 10, 11, 12, 13, 14, 15, 16, 18, 21, 22, 23, 24, 25, 34, 39, 40, 41, 44, 63, 64
4. Pump loses prime after starting	2, 4, 6, 7, 7a, 8, 9, 10, 11
5. Pump requires excessive power	20, 22, 23, 24, 26, 32, 33, 34, 35, 39, 40, 41, 44, 45, 61, 69, 70, 71
6. Pump vibrates or is noisy at all flows	2, 16, 37, 43, 44, 45, 46, 47, 48, 49, 50, 51, 52, 53, 54, 55, 56, 57, 58, 59, 60, 61, 67, 78, 79, 80, 81, 82, 83, 84, 85
7. Pump vibrates or is noisy at low flows	2, 3, 17, 19, 27, 28, 29, 35, 38, 77
8. Pump vibrates or is noisy at high flows	2, 3, 10, 11, 12, 13, 14, 15, 16, 17, 18, 33, 34, 41
9. Shaft oscillates axially	17, 18, 19, 27, 29, 35, 38
10. Impeller vanes are eroded on visible side	3, 12, 13, 14, 15, 17, 41
11. Impeller vanes are eroded on invisible side	12, 17, 19, 29
12. Impeller vanes are eroded at discharge near center	37
13. Impeller vanes are eroded at discharge near shrouds or at shroud/vane fillets	27, 29
14. Impeller shrouds bowed out or fractured	27, 29
15. Pump overheats and seizes	1, 3, 12, 28, 29, 38, 42, 43, 45, 50, 51, 52, 53, 54, 55, 57, 58, 59, 60, 61, 62, 77, 78, 82
16. Internal parts are corroded prematurely	66
17. Internal clearances wear too rapidly	3, 28, 29, 45, 50, 51, 52, 53, 54, 55, 57, 59, 61, 62, 66, 77
18. Axially-split casing is cut through wire-drawing	63, 64, 65
19. Internal stationary joints are cut through wire-drawing	53, 63, 64, 65
20. Packed box leaks excessively or packing has short life	8, 9, 45, 54, 55, 57, 68, 69, 70, 71, 72, 73, 74
21. Packed box sleeve scored	8, 9
22. Mechanical seal leaks excessively	45, 54, 55, 57, 58, 62, 75, 76
23. Mechanical seal: damaged faces, sleeve, bellows	45, 54, 55, 57, 58, 62, 75, 76
24. Bearings have short life	3, 29, 41, 42, 45, 50, 51, 54, 55, 58, 77, 78, 79, 80, 81, 82, 83, 84, 85
25. Coupling fails	45, 50, 51, 54, 67

Table 4-7
Possible Causes of Trouble

Suction Troubles

1. Pump not primed
2. Pump suction pipe not completely filled with liquid
3. Insufficient available NPSH
4. Excessive amount of air or gas in liquid
5. Air pocket in suction line
6. Air leaks into suction line
7. Air leaks into pump through stuffing boxes or through mechanical seal
8. Air in source of sealing liquid
9. Water seal pipe plugged
10. Seal cage improperly mounted in stuffing box
11. Inlet of suction pipe insufficiently submerged
12. Vortex formation at suction
13. Pump operated with closed or partially closed suction valve
14. Clogged suction strainer
15. Obstruction in suction line
16. Clogged impeller
17. Suction elbow in plane parallel to the shaft (for double-suction pumps)
18. Two elbows in suction piping at 90° to each other, creating swirl and prerotation
19. Selection of pump with too high a Suction Specific Speed

Other Hydraulic Problems

20. Speed of pump too high
21. Speed of pump too low
22. Wrong direction of rotation
23. Reverse mounting of double-suction impeller
24. Uncalibrated instruments
25. Impeller diameter smaller than specified
26. Impeller diameter larger than specified
27. Impeller selection with abnormally high head coefficient
28. Running the pump against a closed discharge valve without opening a by-pass
29. Operating pump below recommended minimum flow
30. Static head higher than shut-off head
31. Friction losses in discharge higher than calculated
32. Total head of system higher than design of pump
33. Total head of system lower than design of pump
34. Running pump at too high a flow (for low specific speed pumps)
35. Running pump at too low a flow (for high specific speed pumps)
36. Leak of stuck check valve
37. Too close a gap between impeller vanes and volute tongue or diffuser vanes
38. Parallel operation of pumps unsuitable for the purpose
39. Specific gravity of liquid differs from design conditions
40. Viscosity of liquid differs from design conditions
41. Excessive wear at internal running clearances
42. Obstruction in balancing device leak-off line
43. Transients at suction source (imbalance between pressure at surface of liquid and vapor pressure at suction flange)

Mechanical Troubles—General

44. Foreign matter in impellers
45. Misalignment
46. Foundation insufficiently rigid
47. Loose foundation bolts
48. Loose pump or motor bolts
49. Inadequate grouting of baseplate
50. Excessive piping forces and moments on pump nozzles
51. Improperly mounted expansion joints
52. Starting the pump without proper warm-up
53. Mounting surfaces of internal fits (at wearing rings, impellers, shaft sleeves, shaft nuts, bearing housings, etc.) not perpendicular to shaft axis
54. Bent shaft
55. Rotor out of balance
56. Parts loose on the shaft
57. Shaft running off-center because of worn bearings
58. Pump running at or near critical speed
59. Too long a shaft span or too small a shaft diameter
60. Resonance between operating speed and natural frequency of foundation, baseplate or piping
61. Rotating part rubbing on stationary part
62. Incursion of hard solid particles into running clearances
63. Improper casing gasket material
64. Inadequate installation of gasket
65. Inadequate tightening of casing bolts
66. Pump materials not suitable for liquid handled
67. Certain couplings lack lubrication

Mechanical Troubles—Sealing Area

68. Shaft or shaft sleeves worn or scored at packing
69. Incorrect type of packing for operating conditions
70. Packing improperly installed
71. Gland too tight, prevents flow of liquid to lubricate packing
72. Excessive clearance at bottom of stuffing box allows packing to be forced into pump interior
73. Dirt or grit in sealing liquid
74. Failure to provide adequate cooling liquid to water-cooled stuffing boxes
75. Incorrect type of mechanical seal for prevailing conditions
76. Mechanical seal improperly installed

Mechanical Troubles—Sealing Area

77. Excessive radial thrust in single-volute pumps
78. Excessive axial thrust caused by excessive wear at internal clearances or by failure or, if used, excessive wear of balancing device
79. Wrong grade of grease or oil
80. Excessive grease or oil in anti-friction bearing housings
81. Lack of lubrication
82. Improper installation of anti-friction bearings such as damage during installation, incorrect assembly of stacked bearings, use of unmatched bearings as a pair, etc.
83. Dirt getting into bearings
84. Moisture contaminating lubricant
85. Excessive cooling of water-cooled bearings

Of course, even Karassik's checklists could be further expanded by observing the symptoms of bearing distress with corresponding possible causes, or mechanical seal distress could be tabulated together with possible contributing causes. Similarly, vibration symptoms could be contrasted with causes, or stuffing box packing deterioration diagnosed from a symptom vs. cause comparison matrix.

On the simplified centrifugal pump symptom vs. cause comparison matrix, Table 4-8, the numbers listed in columns A through H indicate the probability; the lower the number, the higher the probability.

If we look at Table 4-8 to determine the most probable cause for insufficient pressure generation (Symptom "D"), we find that we should look at possible causes in this sequence:

1. Noncondensibles (air) in liquid.
2. Pump speed too low.
3. Wrong direction of rotation.
4. Total system head lower than design head of pump—pump is "running out."
5. Viscosity too high.
6. Two or more pumps in parallel operation and unsuitable.
7. Internal obstructions in lines ahead of pressure gauges.
8. Internal wear, i.e., wear rings worn.

If more than one symptom is observed, the checking sequence may be established by obtaining the average ranking number by cause. If the pump also works intermittently, alternately losing and regaining its prime after startup, the sequence of probable cause would be that shown in Table 4-9. Tables 4-10 through 4-14 have been compiled with the same partioning approach in mind.

However, before we leave the subject of pump troubleshooting, we should direct our attention to the more elusive problems and causes of pump distress.

Selection-Related Problems

Centrifugal pump impellers will usually perform well over a wide range of flows and pressures. However, impellers designed for conditions of low $NPSH_R$ (net positive suction pressure required at the suction eye vane tips) may suffer from recirculation when operating at low-flow conditions. Recirculation can occur at both the impeller inlet and outlet, and at very low capacities will usually be present at both. Operation at lower than design flows means operation at reduced efficiency; a higher proportion of the power input will be converted into frictional heat.

Figure 4-7 illustrates a section through a single-suction impeller with fluid recirculation vortices occurring near the periphery.[12] This internal recirculation will often cause significant reductions in seal and bearing life. Moreover, operation at low flow will result in higher bearing loads, increased shaft deflection (Figure 4-8), and potential fatigue failure of pump shafts.[13]

(*text continued on page 289*)

Table 4-8
Troubleshooting Guide—Centrifugal Process Pumps

	Symptoms				Symptoms			

		Symptoms							Symptoms				
D	**Insufficient Disch. Pressure**				**Short Bearing Life**							**E**	
C	**Intermittent Operation**				**Short Mech. Seal Life**						**F**		
B	**Insufficient Capacity**				**Vibration and Noise**					**G**			
A	**No Liquid Delivery**				**Power Demand Excessive**				**H**				

<table>
<thead>
<tr><th></th><th>Possible Causes</th><th>#</th><th>A</th><th>B</th><th>C</th><th>D</th><th>E</th><th>F</th><th>G</th><th>H</th><th>#</th><th>Possible Remedies</th></tr>
</thead>
<tbody>
<tr>
<td rowspan="3">**Suction Problems**</td>
<td>Pump Is Cavitating (Symptom For Liquid Vaporizing In Suction System)—Horizontal Pumps</td>
<td>1</td><td>2</td><td>1</td><td>1</td><td></td><td></td><td>9</td><td>1</td><td></td><td>1</td>
<td>* Check NPSHa/NPSHr Margin
* If Pump Is Above Liquid Level, Raise Liquid Level Closer To Pump
* If Liquid Is Above Pump, Increase Liquid Level Elevation</td>
</tr>
<tr>
<td>Insufficient Immersion Of Suction Pipe Or Bell (Vertical Turbine Pumps)</td>
<td>2</td><td>1</td><td>1</td><td>1</td><td></td><td></td><td></td><td>1</td><td></td><td>2</td>
<td>* Lower Suction Pipe Or Raise Sump Level
* Increase System Resistance</td>
</tr>
<tr>
<td>Pump Not Primed</td>
<td>3</td><td>1</td><td></td><td>2</td><td></td><td></td><td></td><td></td><td></td><td>3</td>
<td>* Fill Pump & Suction Piping Complete With Liquid
* Eliminate High Points In Suction
* Remove All Non-Condensibles (Air From Pump, Piping And Valves)
* Eliminate High Points In Suction Piping
* Check For Faulty Foot Valve Or Check Valve</td>
</tr>
<tr>
<td rowspan="3">**Hydraulic System**</td>
<td>Non-Condensibles In Liquid</td>
<td>4</td><td></td><td>2</td><td>3</td><td>1</td><td></td><td></td><td></td><td></td><td>4</td>
<td>* Check For Gas/Air Ingress Through Suction System/Piping
* Install Gas Separation Chamber</td>
</tr>
<tr>
<td>Supply Tank Empty</td>
<td>5</td><td>3</td><td></td><td></td><td></td><td></td><td></td><td></td><td></td><td>5</td>
<td>* Refill Supply Tank</td>
</tr>
<tr>
<td>Obstructions In Lines Or Pump Housing</td>
<td>6</td><td></td><td>9</td><td></td><td>7</td><td></td><td></td><td>7</td><td></td><td>6</td>
<td>* Inspect And Clear</td>
</tr>
<tr>
<td></td>
<td>**Possible Causes**</td>
<td>**#**</td><td>**A**</td><td>**B**</td><td>**C**</td><td>**D**</td><td>**E**</td><td>**F**</td><td>**G**</td><td>**H**</td><td>**#**</td>
<td>**Possible Remedies**</td>
</tr>
</tbody>
</table>

(table cont. on next page)

Table 4-8 (cont.)

Symptoms						Symptoms					

D Insufficient Disch. Pressure						Short Bearing Life **E**					
C Intermittent Operation						Short Mech. Seal Life **F**					
B Insufficient Capacity						Vibration and Noise **G**					
A No Liquid Delivery						Power Demand Excessive **H**					

Possible Causes	#	A	B	C	D	E	F	G	H	#	Possible Remedies
Strainer Partially Clogged	7		3							7	* Inspect and Clean
Pump Impeller Clogged	8	8	8					5		8	*Check For Damage And Clean
Suction or/& Dischrg. Valve(s) Closed	9	9								9	* Shut Down & Open Valves
Viscosity Too High	10		7		5			4		10	*Heat Up Liquid To Reduce Viscosity * Increase Size Of Discharge Piping To Reduce Pressure Loss * Use Larger Driver Or Change Type Of Pump * Slow Pump Down
Specific Gravity Too High	11							2		11	* Check Design Specific Gravity
Total System Head Lower Than Design Head Of Pump	12				4		11	3		12	* Increase System Resistance To Obtain Design Flow * Check Design Parameters Such As Impeller Size etc.
Total System Head Higher Than Design Head Of Pump	13	6	5	4			10	2		13	* Decrease System Resistance To Obtain Design Flow * Check Design Parameters Such As Impeller Size, etc.
Unsuitable Pumps In Parallel Operation	14	7	6		6					14	* Check Design Parameters
Improper Mechanical Seal	15						1			15	* Check Mechanical Seal Selection Strategy
Possible Causes	#	A	B	C	D	E	F	G	H	#	Possible Remedies

(left margin labels: Hydraulic System; Mechanical System)

Table 4-8 (cont.)

Symptoms					Symptoms			
D Insufficient Disch. Pressure					Short Bearing Life **E**			
C Intermittent Operation					Short Mech. Seal Life **F**			
B Insufficient Capacity					Vibration and Noise **G**			
A No Liquid Delivery					Power Demand Excessive **H**			

Possible Causes	#	A	B	C	D	E	F	G	H	#	Possible Remedies
Speed Too High	16								1	16	* Check Motor Voltage—Slow Down Driver
Speed Too Low	17	4	4		2					17	* Consult Driver Troubleshooting Guide
Wrong Direction Of Rotation	18	5			3				6	18	* Check Rotation With Arrow On Casing—Reverse Polarity On Motor
Impeller Installed Backward (Double Suction Imp.)	19		10						12	19	* Inspect
Misalignment	20					1	2	4	7	20	* Check Angular And Parallel Alignment Between Pump & Driver
Casing Distorted From Excessive Pipe Strain	21					2	3	5		21	* Check For Misalignment * Check Pump For Wear Between Casing And Rotating Elements * Analyze Piping Loads
Inadequate Grouting Of Base	22							6		22	* Check Grouting & Regrout If Required
Bent Shaft	23					3	4	7	8	23	* Check Deflection (Should Not Exceed 0.002″). Replace Shaft & Bearings If Necessary
Internal Wear	24			8					9	24	* Check Impeller Clearances
Possible Causes	#	A	B	C	D	E	F	G	H	#	**Possible Remedies**

(table continued on next page)

Table 4-8 (cont.)

Symptoms						Symptoms					

D	**Insufficient Disch. Pressure**			**Short Bearing Life**	**E**
C	**Intermittent Operation**			**Short Mech. Seal Life**	**F**
B	**Insufficient Capacity**			**Vibration and Noise**	**G**
A	**No Liquid Delivery**			**Power Demand Excessive H**	

	Possible Causes	#	A	B	C	D	E	F	G	H	#	Possible Remedies
Mechanical System	Mechanical Defects Worn, Rusted, Defective Bearings	25					5	8	10	25		* Inspect Parts For Defects—Repair Or Replace. Use Bearing Failure Analysis Guide * Check Lubrication Procedures
	Unbalance—Driver	26					5	7	9		26	* Run Driver Disconnected From Pump Unit—Perform Vibration Analysis
	Unbalance—Pump	27					4	6	3		27	* Investigate Natural Frequency
	Motor Troubles	28					6	8	10	11	28	* Consult Motor Troubleshooting Guide
	Possible Causes	**#**	**A**	**B**	**C**	**D**	**E**	**F**	**G**	**H**	**#**	**Possible Remedies**

Note: For vibration & short bearing life related causes, refer to special sections on bearing failure analysis and vibration diagnostics.

Table 4-9
Establishment of Average Ranking Numbers by Combining Probable Causes

Possible Cause	Symptom (C)	(D)	Sequence (#)
1. Pump is cavitating	1		1
2. Pump not primed	2		2
3. Non-condensibles in liquid	3	1	2*
4. Speed too low		2	2
5. Wrong direction of rotation		3	3
. . . and so forth.			

* Average = (3 + 1)/2

Table 4-10
Troubleshooting Guide—Vertical Turbine Pumps

	Symptoms					Symptoms				
	C Insufficient Disch. Pressure					**Vibration**		**D**		
	B Insufficient Capacity					**Abnormal Noise**		**E**		
	A Liquid Delivery					**Power Demand Excessive**		**F**		
Possible Causes	#	A	B	C		D	E	F	#	**Possible Remedies**
Pump Suction Interrupted (Water Level Below Bell Inlet)	1	1							1	* Check sump level
Low Water Level	2		8						2	* Check water level
Cavitation Due to Low Submergence	3						7		3	* Check submergence
Vortex Problems	4					8			4	* Install vortex breaker shroud
Suction or Discharge Recirculation	5					7			5	* Establish design flows
Operation Beyond Max. Capacity Rating	6					6	6		6	* Establish proper flow rate
Entrained Air	7			6					7	* Install separation chamber
Suction Valve Closed	8	2							8	* Shut down or open valve
Suction Valve Throttled	9		7						9	* Open valve
Strainer Clogged	10	4	6						10	* Inspect & clean
Impeller Plugged	11	3		4					11	* Pull pump & clean
Impeller or Bowl Partially Plugged	12		4				8		12	*Pull pump & clean
Impellers Trimmed Incorrectly	13		2	2				3	13	* Check for proper impeller size
Improper Impeller Adjustment	14					9		2	14	* Check installation/repair records
Impeller Loose	15	7	3	3					15	* Pull pump & analyze
Impeller Rubbing On Bowl Case	16						5		16	* Check lift
Wear Rings Worn	17			5					17	* Inspect during overhaul
Possible Causes	#	A	B	C		D	E	F	#	**Possible Remedies**

Left margin labels: Hydraulic System (rows 1–13), Mechanical System (rows 14–17)

Table 4-10 (cont.)

Symptoms								Symptoms	
C Insufficient Disch. Pressure					Vibration				**D**
B Insufficient Capacity					Abnormal Noise				**E**
A Liquid Delivery					Power Demand Excessive				**F**
Possible Causes	#	A	B	C	D	E	F	#	Possible Remedies
Shaft Bent	18				5		5	18	* Pull pump & analyze
Shaft Broken, or Unscrewed	19	6						19	* Pull pump & analyze
Enclosing Tube Broken	20				4			20	* Pull pump & analyze
Bearings Running Dry	21				2			21	* Provide lubrication
Worn Bearings	22				6			22	* Pull pump & repair
Column Bearing Retainers Broken	23				3			23	* Pull pump & analyze
Wrong Rotation	24	5	9	8				24	* Check rotation
Speed Too Slow	25		1	1				25	* Check rpm
Speed Too High	26						1	26	* Check rpm
Misalignment Of Pump Assembly Through Pipe Strain	27				4		4	27	* Inspect
Leaking Joints	28		4	5				28	* Inspect
Pumping Sand, Silt Or Foreign Material	29						6	29	* Check liquid pumped
Motor Noise	30					1		30	* Check sound level
Motor Electrical Imbalance	31				1			31	* Perform phase check
Motor Bearing Problems	32				2			32	* Consult motor/bearing troubleshooting guide
Motor Drive Coupling Out of Balance	33				3			33	* Inspect
Resonance: System Natural Frequency Near Pump Speed	34				10			34	* Perform vibration analysis
Possible Causes	#	A	B	C	D	E	F	#	**Possible Remedies**

Mechanical System

Table 4-11

Fire Pump Troubleshooting Guide[28]

Symptoms

#	Possible Causes	Disch. Press. Too Low For gpm. Discharge	Insufficient Water Discharge	Pump Loses Suction After Starting	Disch. Press. Not Constant For Same gpm	Too Much Power Required	Pump Is Noisy or Vibrates	No Water Discharge	Pump Unit Will Not Start	Pump or Driver Overheats	Excessive Leakage at Stuffing Box
1	Suction lift too high.	•	•	•	•		•	•			
2	Foot valve too small, partially obstructed or of inferior design causing excessive suction head loss.	•	•	•	•						
3	Air drawn into suction connection.	•	•	•	•		•				
4	Air drawn into suction connection through leak.	•	•	•	•		•	•			
5	Suction connection obstructed.	•	•	•				•			
6	Air pocket in suction pipe.	•	•	•				•			
7	Hydraulic cavitation from excessive suction lift.	•	•				•				
8	Well collapsed or serious misalignment.					•	•		•	•	

Suction System

• Check this cause (no probability ranking assigned).

Table 4-11 (cont.)

Symptoms

#	Possible Causes	Disch. Press. Too Low For gpm. Discharge	Insufficient Water Discharge	Pump Loses Suction After Starting	Disch. Press. Not Constant For Same gpm	Too Much Power Required	Pump Is Noisy or Vibrates	No Water Discharge	Pump Unit Will Not Start	Pump or Driver Overheats	Excessive Leakage at Stuffing Box
9	Stuffing box too tight or packing improperly installed, worn, defective, too tight, or incorrect type.	•			•	•	•		•	•	•
10	Water-seal or pipe to seal obstructed.	•	•	•	•					•	
11	Air leak into pump through stuffing box.	•	•	•	•						
12	Impeller obstructed.	•	•			•	•	•		•	
13	Wear rings worn.	•	•			•					
14	Impeller damaged.	•	•				•				
15	Wrong diameter impeller.	•	•			•				•	
16	Actual net head lower than rated.	•	•								
17	Casing gasket defective, permitting internal leakage. (Multi-stage pumps only.)	•	•			•					
18	Bad pressure gage on top of pump casing.	•									
19	Incorrect impeller adjustment. (Vertical pumps only.)		•			•			•	•	
20	Impellers locked.								•		

Pump

Category	No.	Possible Cause
Pump	22	Pump shaft or shaft sleeve scored, bent, or worn.
	23	Pump not primed.
	24	Seal ring improperly located in stuffing box, preventing water from entering space to form seal.
Driver/Pump	25	Excess bearing friction due to wear, dirt, rusting, failure, or improper installation.
	26	Rotating element binds against stationary element.
	27	Pump and driver misaligned. Shaft running off center because of worn bearings or misalignment.
	28	Foundation not rigid.
	29	Engine cooling system obstructed. Heat exchanger or cooling water system too small. Cooling pump faulty.
Driver	30	Faulty driver.
	31	Lack of lubrication.
	32	Speed too low.
	33	Wrong direction of rotation.
	34	Speed too high.
	35	Rated motor voltage different from line voltage, i.e., 220- or 440-volt motor on 208- or 416-volt line.
	36	Faulty electric circuit, obstructed fuel system or obstructed steam pipe, or dead battery.

Table 4-12
Troubleshooting Guide—Rotary Pumps

Symptoms							**Symptoms**				

		D			**Excessive Wear**						
		C	**Starts, But Loses Prime**				**Excessive Heat**			E	
		B	**Insufficient Capacity**				**Vibration and Noise**			F	
		A	**No Liquid Delivery**				**Excessive Power Demand**			G	

	Possible Causes	#	A	B	C	D	E	F	G	#	**Possible Remedies**
Hydraulic System	Suction Filter Or Strainer Clogged	1	1	1				1		1	* Clean strainer or filter.
	Pump Running Dry	2	2			2	1	4		2	* Reprime
	Insufficient Liquid Supply	3		2	4	3		3		3	* Look for suction restriction or low suction level.
	Suction Piping Not Immersed in Liquid	4	4		3					4	* Lengthen suction pipe or raise liquid level
	Liquid Vaporizing In Suction Line	5		4	2			5		5	* Check NPSH. Check for restriction in suction line.
	Air Leakage Into Suction Piping Or Shaft Seal	6		3	1			6		6	* Tighten & seal all joints. * Adjust packing or repair mechanical seal.
	Solids Or Dirt In Liquid	7				1				7	* Clean system. * Install filtration.
	Liquid More Viscous Than Designed For	8							1	8	* Reduce pumped medium viscosity. * Reduce pump speed. * Increase driver HP
	Excessive Discharge Pressure	9		5		7		9	2	9	* Check relief valve or by-pass setting. * Check for obstruction in discharge line.
Mechanical Problems	Pipe Strain On Pump Casing	10			5		4	8	5	10	* Disconnect piping and check flange alignment.
	Coupling, Belt Drive, Chain Drive Out Of Alignment	11			6		3	2	4	11	* Realign
	Rotating Elements Binding	12			4		2	7	3	12	* Disassemble & Inspect.
	Possible Causes	#	A	B	C	D	E	F	G	#	**Possible Remedies**

Table 4-12 (cont.)

Symptoms										Symptoms
D Excessive Wear										
C Starts, But Loses Prime					Excessive Heat **E**					
B Insufficient Capacity					Vibration and Noise **F**					
A No Liquid Delivery					Excessive Power Demand **G**					
Possible Causes	#	A	B	C	D	E	F	G	#	Possible Remedies
Internal Parts Wear	13		7						13	* Inspect & replace worn parts
Speed Too Low	14		6						14	* Check driver speed
Wrong Direction Of Rotation	15	3							15	* Check & reverse if required
Possible Causes	#	A	B	C	D	E	F	G	#	Possible Remedies

Mechanical Problems

Table 4-13
Troubleshooting Guide—Reciprocating (Power) Pumps

Symptoms											Symptoms
D Excessive Wear-Liquid End					Excessive Wear—Power End **E**						
C Short Packing Life					Excessive Heat—Power End **F**						
B Insufficient Capacity					Vibration and Noise **G**						
A No Liquid Delivery					Persistent Knocking **H**						
Possible Causes	#	A	B	C	D	E	F	G	H	#	Possible Remedies
Excessive Suction Lift	1	1								1	* Decrease suction lift—Consider charge or booster pump
Not Enough Suction Pressure Above Vapor Pressure	2	2								2	* Remove all non-condensibles (air) from pump, piping, & valves * Eliminate high points in suction piping * Check for faulty foot valve or check valve
Non-Condensibles (Air) In Liquid	3	3	1		5			3		3	* Install gas separation chamber
Possible Causes	#	A	B	C	D	E	F	G	H	#	Possible Remedies

Suction System

Table 4-13 (cont.)

	Symptoms					Symptoms					
	D Excessive Wear-Liquid End					Excessive Wear—Power End **E**					
	C Short Packing Life					Excessive Heat—Power End **F**					
	B Insufficient Capacity					Vibration and Noise **G**					
	A No Liquid Delivery					Persistent Knocking **H**					

	Possible Causes	#	A	B	C	D	E	F	G	H	#	Possible Remedies
Suction System	Cylinders Not Filling	4		2		2			4		4	* Attempt to prime pump. * Install foot valve at bottom of suction pipe if pump has to lift * Vortex in supply tank. Consider charge or booster pump
	Low Volumetric Efficiency	5		3							5	* Liquid with low specific gravity or high discharge pressure compressing & expanding in cylinder
	Supply Tank Empty	6	6								6	* Refill supply tank
Operating Conditions	Obstruction In Lines	7	4						5		7	* Inspect & clean
	Abrasives or Corrosives In Liquid	8			4	1					8	* Establish proper maintenance intervals * Consider design change
	Relief or Bypass Valve(s) Leaking	9	4								9	* Inspect & repair
Valves/Cylinders/Pistons/P. Rods	Pump Speed Incorrect	10	7					5			10	* Check for slipping belts. Consult belt drive & driver TSG * Check driver speed
	One or More Cylinder(s) Not Operating	11	5								11	* Stop & prime all cylinders
	Pump Valve(s) Stuck Open	12	6								12	* Check for damage & clean
	Broken Valve Springs	13	8		3				6		13	* Inspect & replace
	Worn Valves, Seats, Liners, Rods, &/or Plungers	14	5	9		4					14	* Inspect & repair
	Possible Causes	#	A	B	C	D	E	F	G	H	#	**Possible Remedies**

Table 4-13 (cont.)

Symptoms						Symptoms					

D	**Excessive Wear-Liquid End**			**Excessive Wear—Power End E**		
C	**Short Packing Life**			**Excessive Heat—Power End F**		
B	**Insufficient Capacity**			**Vibration and Noise G**		
A	**No Liquid Delivery**			**Persistent Knocking H**		

	Possible Causes	#	A	B	C	D	E	F	G	H	#	Possible Remedies
Valves/Cylinders/ Pistons/P. Rods	Improper Packing Selection	15			1						15	* Develop selection guidelines
	Scored Rod or Plunger	16		10							16	* Regrind or replace
	Misalignment of Rod or Plunger	17			2						17	* Check & realign. Tolerance: Eccentricity in box within 0.003″ max.
	Loose Piston or Rod	18								1	18	* Inspect & repair
Crosshead Bearings	Loose Cross-Head Pin or Crank Pin	19								2	19	* Adjust or replace * Check for proper clearances
	Worn Cross-Heads or Guides	20			3		4				20	* Adjust or replace
	Other Mechanical Problems: Worn, Rusted, Defective Bearings	21					1	3	1	3	21	* Inspect part for defect limits & repair or replace * Consult bearing failure analysis guide * Check lubrication procedures
	Inadequate Lubrication Conditions	22					2	1	2		22	* Establish lubrication schedule * Review design assumptions for crankcase lubrication, etc.
Loading Drive	Liquid Entry Into Power End	23					5	2			23	* Replace packing & parts affected
	Gear Problems	24					4		7	4	24	* Consult gear TSG†
	Driver Troubles	25							8		25	* Consult appropriate TSG†
	Overloading	26					3				26	* Check design assumptions.
	Possible Causes	**#**	**A**	**B**	**C**	**D**	**E**	**F**	**G**	**H**	**#**	**Possible Remedies**

Table 4-14
Troubleshooting Guide—Liquid Ring Pumps

Possible Causes	Symptoms				
	Noise Excessive	Vibration Up	Overheating Temperature Up	Capacity Down	Power Consumption Up
Speed too low				③	
Suction line leakage				①	
Service liquid temperature too high*			①	②	
Service liquid level too high	①	②		④	①
Service liquid insufficient			②	⑤	
Coupling misalignment	②	①	③		②

*Insufficient heat rejection (cooler make-up liquid).
Note: The numbers indicate what to check first, or symptom-cause probability rank.

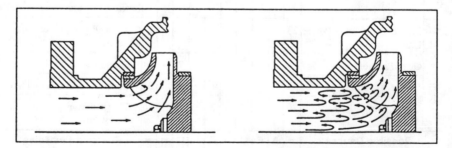

Figure 4-7. Single-suction impeller at (a) normal flow and (b) low flow conditions. At low flow conditions, internal recirculation takes place.

(*text continued from page 274*)

Design-Related Problems

Although statistically less prevalent, design problems do occur and may have to be addressed. Lack of hydraulic balance may overload thrust bearings, as discussed in Chapter 10 (see Figure 10-3). Here, the pump manufacturer had calculated a load of 1,000 lbs at zero-flow conditions. Frequent bearing failures were explained when field tests showed actual loads in the vicinity of 2,600 lbs. Since ball bearing life changes as the third power of the load ratio, this 2.6-fold load increase would reduce the probable bearing life to 1/17 of the pump manufacturer's anticipated bearing life. The problem was solved by redesigning the pump impeller.[14]

Older pumps that were originally designed for packing as a means to contain the pumpage may be prone to experience frequent mechanical seal failures. The root cause of the problem may well be related to shaft slenderness; braided packing acted as a stabilizing bushing. Insertion of a suitably dimensioned graphite bushing has often proved to be beneficial.

And then there are pumps with less than optimum internal gap or clearance values.[15, 16] Incorrect internal dimensions may cause pump component breakage, including impellers, fasteners, bearings, and mechanical seals. Correct gap values are given in Figure 4-9.

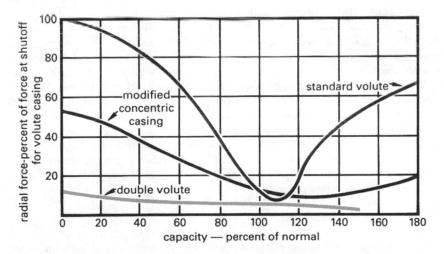

Figure 4-8. Radial forces on impellers in pumps of three different casing styles.

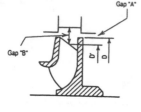

Type	Gap "A"	Gap "B" Percentage of Impeller Radius		
		Minimum	Preferred	Maximum
Diffuser	50 mils	4%	6%	12%
Volute	50 mils	6%	10%	12%

Figure 4-9. Gap recommendations as a percentage of impeller radius for different casing styles.

Installation Problems

A surprisingly large number of installation-related deficiencies continue to plague literally thousands of centrifugal pump users worldwide. The importance of not allowing pump piping to exert undue stresses on nozzles and baseplates is simply not appreciated by many maintenance workers, reliability professionals, and operators. When it is not possible for a *single* worker to *manually* push pump suction or discharge piping into position to mate with the pump flanges, pipe stress is excessive and should be corrected before inserting and torquing up the flange bolting. In other words, regardless of flange and nozzle size, it is never good practice to pull piping into place by means of chainfalls or other mechanical assistance devices. Figures 4-10 and 4-11 clearly show how piping-induced stresses are prone to create internal pump misalignment sufficient to cause the destruction of even the best mechanical seals.

Along the same lines, the customary practice of installing an entire pump-and-driver set on a common baseplate and then attempting to level and grout the total assembly on the concrete foundation does not represent "best practice." A reliability minded engineer will insist on installing equipment baseplates by themselves. This will greatly improve the probability of achieving truly level installations. A much-desired side benefit will be improved grouting access and full grout support under the entire baseplate. Figure 4-12[17] depicts grouting in progress. Finally, the detrimental effects of marginal alignment accuracy between driving and driven shafts should prompt pump owners to insist on good alignment. The lower diagonal line in Figure 4-13 represents reasonable and readily achievable alignment accuracies for centrifugal pumps,[18] while Figure 4-14 shows the estimated time to failure of rotating machinery due to misalignment.[19]

Assembly-Related Problems

A surprisingly large number of pump bearings are being assembled without too much attention given to acceptable fits and tolerances. The shop repair guidelines mentioned in Chapter 13 (Table 13-2) show bearing bore and shaft diameter mini-

Figure 4-10. Stationary seal ring damaged by piping-induced internal deflection of centrifugal pump.

Figure 4-11. Seal faces damaged by piping-induced contact with pump shaft.

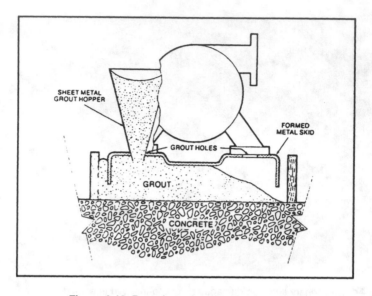

Figure 4-12. Pump baseplate grouting in progress.

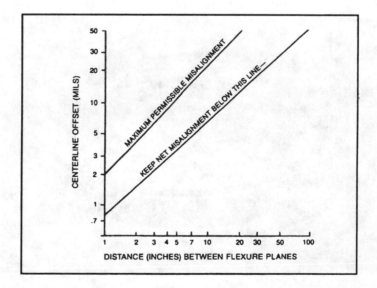

Figure 4-13. Allowable misalignment for centrifugal pumps in 1200–3600 RPM range.

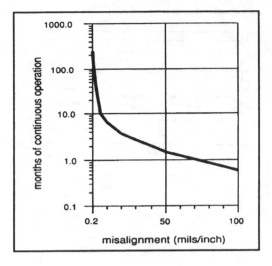

Figure 4-14. Estimated time to failure as a function of shaft misalignment.

mum and maximum values recommended for centrifugal pump radial bearings. What is perhaps less well-known is the fact that mating shafts and radial bearings, which respectively are at the high and low ends of their allowable tolerances, are likely to cause high operating temperatures. More specifically, oil-ring lubricated bearings will usually receive enough oil for adequate lubrication, but rarely will this amount be sufficient to preclude hot bearing housings if interference fits are at the high end of the apparent allowable interference spectrum. Hot bearing housings invite operators to pour liberal amounts of water on the assembly; this causes the bearing outer ring to contract and bearing internal clearances to vanish, and rapid failures are now inevitable. But even if the operator resists the impulse to provide this detrimental means of supplemental cooling, increased oil temperatures will lead to accelerated oxidation of the lubricant.

Angular contact bearings require even more care. Mounted back-to-back and often provided with a small axial preload gap, these bearings are often intended for shaft interference fits on the order of 0.0001″ to 0.0005″ only. Sets of these bearings are frequently supplied in centrifugal pumps complying with API 617, a standard specification issued by the American Petroleum Institute.[20] Figure 4-15 shows that installing two of these bearings back-to-back with a radial interference of 0.003″ will produce an almost insignificant preload of approximately 22 lbs (10 kg), whereas an interference fit of 0.0007″ would result in a mounted preload of 200 lbs (90 kg). A much more significant preload would be created by bearing inner ring temperatures higher than bearing outer ring temperatures. With such temperature differences in centrifugal pumps and electric motors typically approaching 50°F (31°C), actual preload values can easily exceed 2,000 lbs (900 kg). Worse yet, if the operator should

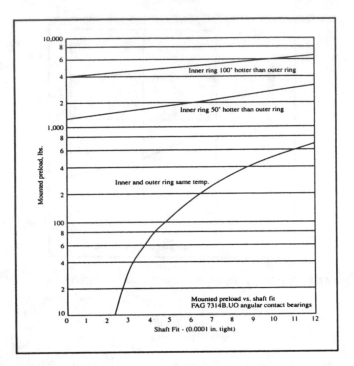

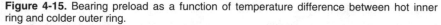

Figure 4-15. Bearing preload as a function of temperature difference between hot inner ring and colder outer ring.

decide to "cool" the bearing housing, a 100°F (55°C) temperature difference might result. At this time, the preload value will increase exponentially and the bearing is in imminent danger of destruction.

As so often in this reliability improvement game, plain attention to detail, that is, careful measurement and avoiding the traditional quick-fix solution, will pay dividends.

Improving Seal Life

Several Scandinavian pulp and paper plants routinely achieve in excess of 42 months mean-time-between-failure (MTBF). At the very same time in 1993, a number of North American paper mills reported only a dismal four months MTBF for their hundreds of mechanical seals. While the failure ratios between North America and Scandinavia are less severely biased in such industries as the HPI, there are still untold improvement options available to engineers and technicians who refuse to see their jobs as "business as usual."

A single ⅛-inch-diameter hole drilled as shown in Figure 4-16, will safeguard and dramatically extend the lives of many mechanical seals by venting air trapped during shop work or gases which might accumulate in stand-by pumps. Of course, operator

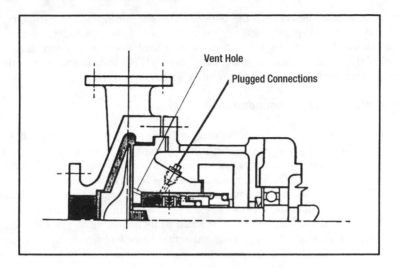

Figure 4-16. Vent hole connecting seal chamber and impeller region increases mechanical seal life.

diligence and conscientious venting of centrifugal pump seal housings could accomplish the same thing.

Another highly advantageous seal life extension program starts by identifying pumpage that contacts seal faces under pressure/temperature conditions at least 15°F away from the initial boiling point (IBP). Many of these services, and especially those containing massive amounts of solids, are candidates for dead ended stationary seals and steep-tapered housing bores. Figure 4-17 depicts one such seal; note the

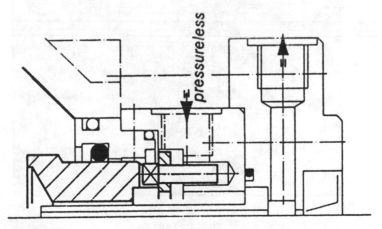

Figure 4-17. Stationary mechanical seal in steep-tapered seal housing for slurry pump in services allows "dead-ending," that is, no external flush needed to prevent vaporization of fluid at seal faces.

hard face, located adjacent to the impeller and the spring-loaded stationary face anchored to the seal gland. Stationary seals are able to accommodate greater shaft deflections and higher peripheral speeds than conventional mechanical seals. That many of the most dependable and reliable high MTBF installations use stationary seals in steep-tapered seal housings should come as no surprise.

Lubrication: Old, But Misunderstood

A rather large number of factors influences lubricating oil degradation and, consequently, pump bearing life. If your centrifugal pumps are equipped with rolling element bearings, there is little doubt that medium viscosity turbine oils (ISO Grade 68) will perform better than the lighter oils originally specified by many pump manufacturers. But, by far, the most frequent cause of lube-oil-related failure incidents is water and dirt contamination. With only 20 ppm water in pure mineral oil, bearing race and rolling element fatigue life is reduced by an incredible 48 percent. Although the fatigue life reduction is less pronounced with inhibited lubricants, there are always

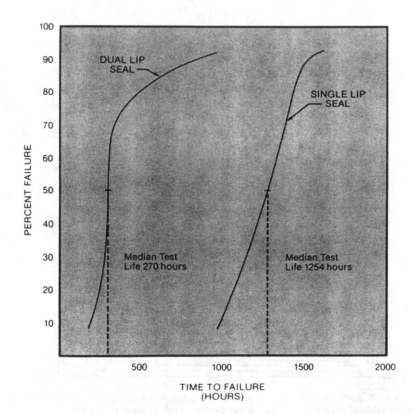

Figure 4-18. Life expectancy of elastomeric lip seals.

compelling reasons to exclude dirt and water from pump bearing housings.[21] Judging from Figure 4-18, lip seals are a poor choice for centrifugal pump installations demanding high reliability.[22, 23] Face seals similar to the ones shown in Figure 4-19 represent superior, "hermetic" sealing and should be given serious consideration.[24, 25]

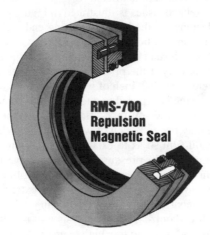

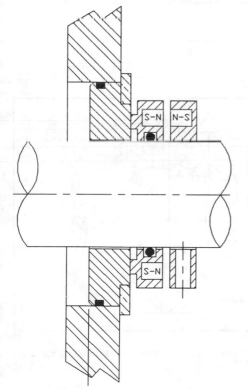

Figure 4-19. Magnetic face seal using repulsion principle allows for "hermetic sealing" of bearing housings.

On a related subject, have you explained to your operators and maintenance personnel that a full-bottle oiler is no guarantee of adequate lubrication? The height of the beveled tube determines the level of oil in the bearing housing, and all too often there will be costly misunderstandings. However, there are at least two considerably more elusive problems involving bottle oilers.

The first of these lube-related issues is illustrated in Figure 4-20. With a relatively viscous oil and close clearance at the bearing housing end seal (point "S," circled), an oil film may exist between seal bore and shaft surface. Good lube oils have a certain film strength and under certain operating conditions, this sealing film near the bearing end cap may break only if the pressure difference bearing housing interior-to-surrounding atmosphere exceeds ⅜ inch of water column.

If, now, the bearing housing is exposed to a temperature increase of a few degrees, the trapped vapors—usually an air-oil mix—floating above the liquid oil level will expand and the pressure may rise to ¼ inch of water column. While this would not be sufficient to rupture the oil film so as to establish equilibrium between atmosphere and bearing housing interior, the pressure buildup is nevertheless sufficient to depress the oil level from its former location near the center of a bearing ball at the 6 o'clock position to a new level now barely touching the extreme bottom of the lowermost bearing rolling element. At that time, the bearing will overheat and the lube oil in contact with it will carbonize. An oil analysis will usually determine that the resulting blackening of the oil is due to this high temperature degradation.

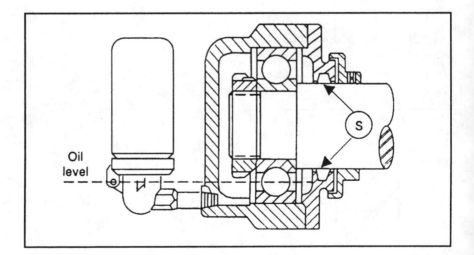

Figure 4-20. Bottle oilers may malfunction unless suitably large bearing housing vents are provided.

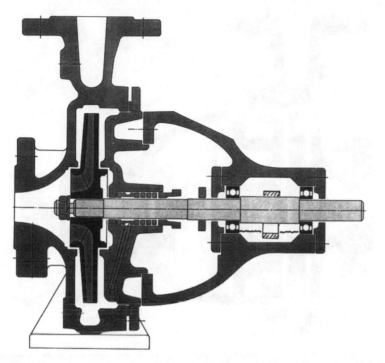

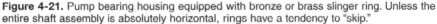

Figure 4-21. Pump bearing housing equipped with bronze or brass slinger ring. Unless the entire shaft assembly is absolutely horizontal, rings have a tendency to "skip."

The second of the elusive oil-related problems often causes the contents of bottle oilers to turn grayish in color. This one is primarily observed on ring-oil lubricated rolling element bearings.

Suppose you have very precisely aligned the shafts of pump and driver; nevertheless, shims placed under the equipment feet in order to achieve this precise alignment caused the shaft system to slant 0.005″ or 0.010″ per foot of shaft length. As a consequence, the brass or bronze oil slinger ring shown in Figure 4-21 will now exhibit a strong tendency to run "downhill." Thus bumping into other pump components thousands of times per day, the slinger ring gradually degrades and sheds numerous tiny specks of the alloy material. The specks of metal cause progressive oil deterioration and, ultimately, bearing distress.

Pump users may wish to pursue one of two time-tested preventive measures. First, use properly vented bearing housings or, better yet, hermetically sealed (Figure 4-19) bearing housings without oiler bottles. The latter are offered by some pump manufacturers and incorporate bull's-eye-type sight glasses to ascertain proper oil levels.

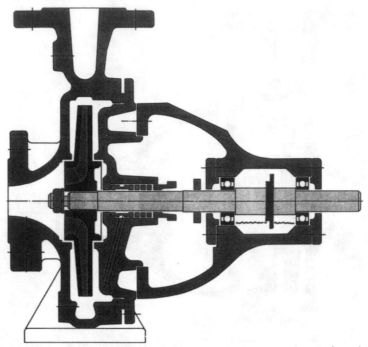

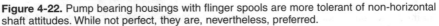

Figure 4-22. Pump bearing housings with flinger spools are more tolerant of non-horizontal shaft attitudes. While not perfect, they are, nevertheless, preferred.

The second preventive measure would take into account the need for radically improved pump and driver leveling during shaft alignment or, even more desirable, apply flinger spools (Figure 4-22). Of course, oil mist lubrication or direct oil injection into the bearings would represent an altogether more dependable, long-term satisfactory lube application method for centrifugal pumps.[26]

Other Practices Responsible for Reduced Pump Reliability

Among the most wasteful and costly practices leading to premature distress in pump antifriction bearings is water cooling. When applied to bearing housings, cooling water has been demonstrated to cause the bearing outer ring to contract. The resulting increase in bearing preload will often reduce the bearing life expectancy. Similar distress may be introduced by moisture condensation attributable to cooling of air present in bearing housings. Deletion of cooling water is feasible if ISO Grade 68 or 100 mineral oils are chosen. Even better results have been obtained in installations using oil mist lubrication or properly compounded synthetic oils of the diester and/or polyalpha-olefin variety. Many refineries have deleted cooling water from pumps with fluid temperatures around 740°F.[27] Similarly, a large number of pump users have come to realize that water-cooled pump support pedestals represent a serious liability at worst, and an unnecessary operating expense at best. Unforeseen corrosion incidents have led to the collapse of pedestals, in turn causing disastrous pump fires. Alignment of driver and driven shafts with predetermined offset values have proven fully satisfactory in every case.

Assembly-induced bearing problems also deserve our attention. Rolling element bearings furnished with plastic cages can suffer damage when overheated in defective heaters. They will be destroyed when heated by open flames, a practice observed on the assembly floor of a pump manufacturer in the 1980s! Do your maintenance/technical work forces know which code letter identifies plastic cages? (Answer: P, for SKF bearings.) Do they know that the letter directly behind the bearing identification number identifies the load angle and that dimensionally interchangeable angular contact bearings using the same number, but differing in just this one letter, may have allowable load ratings differing by a factor of 8-to-1? Compelling reasons for developing a purchase specification for your rolling element bearings and even more compelling reasons for workforce training!

Troubleshooting Centrifugal Compressors, Blowers, and Fans

Centrifugal compressors, blowers, and fans form the basic make-up of unspared machinery equipment in petrochemical process plants. Single, unspared centrifugal compressor trains support the entire operation of steam crackers producing in excess of 1,000,000 metric tons (approximately 2.2 billion lbs) of ethylene per year. When plants in this size range experience emergency shutdowns of a few hours' duration, flare losses alone can amount to $400,000 or more. Timely recognition of centrifugal compressor troubles is therefore extremely important. Table 4-15 shows what is most likely to happen to a centrifugal compressor or blower.

Table 4-16 presents a troubleshooting guide for centrifugal compressors and their lubrication systems. Similar guides may be constructed for other types of centrifugal machinery such as fans and blowers. Since Table 4-16 cannot possibly be all-encompassing, refer to special sections in this book addressing such related topics as vibration-causing syndromes and bearing, gear, and coupling failure analysis and troubleshooting.

Table 4-15
Typical Distribution of Unscheduled Downtime Events for Major Turbocompressors in Process Plants

Cause of Problem	Estimated Frequency	Estimated Hrs/Event	Average Downtime Events/Yr.	Hrs/Yr
Rotor/shaft	22%	122	44	54
Instrumentation	21%	4	42	2
Radial bearings	13%	28	26	7
Blades/impellers	8%	110	16	18
Thrust bearings	6%	22	12	3
Compressor seals	6%	48	12	6
Motor windings	3%	200	06	12
Diaphragms	1%	350	02	7
Miscellaneous causes	20%	70	40	28
All causes	100%		2.00	137 hours

Approximate number of shutdowns per train per year: 2

Table 4-16
Troubleshooting Guide—Centrifugal Compressor and Lube System

| | Symptoms | | | | | | Symptoms | | | |

D	Low Lube Oil Pressure
C	Loss Of Disch. Pressure
B	Compressor Surges
A	Excessive Vibration

| Excessive Brg. Oil Drain Temp. E |
| Units Do Not Stay In Alignment F |
| Water In Lube Oil G |

	Possible Causes	#	A	B	C	D	E	F	G	#	Possible Remedies
Rotor/Bearing System	Excessive Bearing Clearance	1	13							1	* Replace bearings
	Wiped Bearings	2					7			2	* Replace bearings * Determine & correct cause
	Rough Rotor Shaft Journal Surface	3					9			3	* Stone or restore journals * Replace shaft
	Bent Rotor (Caused By Uneven Heating or Cooling)	4	8							4	* Turn rotor at low speed until vibration stops, then gradually increase speed to operating speed. * If vibration continues, shut down, determine & correct the cause
	Possible Causes	#	A	B	C	D	E	F	G	#	Possible Remedies

Table 4-16 (cont.)

	Symptoms							Symptoms		
	D Low Lube Oil Pressure									
	C Loss Of Disch. Pressure					**Excessive Brg. Oil Drain Temp. E**				
	B Compressor Surges					**Units Do Not Stay In Alignment F**				
	A Excessive Vibration					**Water In Lube Oil G**				
Possible Causes	**#**	**A**	**B**	**C**	**D**	**E**	**F**	**G**	**#**	**Possible Remedies**
Operating In Critical Speed Range	5	9							5	* Operate at other than critical speed
Build-Up of Deposits On Rotor	6	10	4						6	* Clean deposits from rotor * Check balance
Build-Up Of Deposits in Diffuser	7		3						7	* Mechanically clean diffusers
Unbalanced Rotor	8	11							8	* Inspect rotor for signs of rubbing * Check rotor for concentricity, cleanliness, loose parts * Rebalance
Damaged Rotor	9	12							9	* Replace or repair rotor * Rebalance rotor
Loose Rotor Parts	10	15							10	* Repair or replace loose parts
Shaft Misalignment	11	5							11	* Check shaft alignment at operating temperatures * Correct any misalignment
Dry Gear Coupling	12	6							12	* Lubricate coupling
Worn or Damaged Coupling	13	7							13	* Replace coupling * Perform failure analysis
Liquid "Slugging"	14	14							14	* Locate & remove the source of liquid * Drain compressor casing of any accumulated liquids
Operating In Surge Region	15	16							15	* Reduce or increase speed until vibration stops * Consult vibration analysis guide
Insufficient Flow	16		1						16	* Increase recycle flow through machine
Possible Causes	**#**	**A**	**B**	**C**	**D**	**E**	**F**	**G**	**#**	**Possible Remedies**

Row group labels (left margin): Rotor/Bearing System · Coupling · Operating Conditions

Table 4-16 (cont.)

	Symptoms					Symptoms				
D		Low Lube Oil Pressure								
C		Loss Of Disch. Pressure				Excessive Brg. Oil Drain Temp. E				
B		Compressor Surges				Units Do Not Stay In Alignment F				
A		Excessive Vibration				Water In Lube Oil G				
Possible Causes	#	A	B	C	D	E	F	G	#	Possible Remedies
Operating Conditions Change In System Resistance Due to Obstructions or Improper Inlet or Disch. Valve Positions	17		2						17	* Check position of inlet/discharge valves * Remove obstructions
Compressor Not Up To Speed	18			1					18	* Increase to required operating speed
Excessive Inlet Temperature	19		2						19	* Correct cause of high inlet temperature
Leak In Discharge Piping	20			3					20	* Repair leak
Vibration	21						7		21	* Refer to "A" in symptom column
Sympathetic Vibration	22	4							22	* Adjacent machinery can cause vibration even when the unit is shut down, or at certain speeds due to foundation or piping resonance. A detailed investigation is required in order to take corrective measures
Assembly Improperly Assembled Parts	23	1							23	* Shut down, dismantle, inspect, correct.
Loose or Broken Bolting	24	2							24	* Check bolting at support assemblies. * Check bed plate bolting * Tighten or replace * Analyze
Support System Piping Strain	25	3						1	25	* Inspect piping arrangements and proper installation of pipe hangers, springs, or expansion joints
Possible Causes	#	A	B	C	D	E	F	G	#	Possible Remedies

(table continued on next page)

Table 4-16 (cont.)

	Symptoms					Symptoms				
	D **Low Lube Oil Pressure**									
	C **Loss Of Disch. Pressure**				**Excessive Brg. Oil Drain Temp. E**					
	B **Compressor Surges**				**Units Do Not Stay In Alignment F**					
	A **Excessive Vibration**				**Water In Lube Oil** **G**					
	Possible Causes	**#**	**A**	**B**	**C**	**D**	**E**	**F**	**G**	**#**	**Possible Remedies**
---	---	---	---	---	---	---	---	---	---	---	---
Support System	Warped Foundation or Bedplate	26						2		26	* Check for possible settling of the foundation support * Correct footing as required * Check for uneven temperatures surrounding the foundation causing
Lube Oil System	Faulty Lube Oil Pressure Gauge or Switch	27				1				27	*Calibrate or replace
	Faulty Temperature Gauge or Switch	28						2		28	* Calibrate or replace
	Oil Reservoir Low Level	29				2				29	* Add oil
	Clogged Oil Strainer/Filter	30				5				30	* Clean or replace oil strainer or filter cartridges
	Relief Valve Improperly Set or Stuck Open	31				8				31	* Adjust relief valve * Recondition or replace
	Incorrect Pressure Control Valve Setting on Operation	32				9				32	* Check control valve for correct setting and operation
	Poor Oil Condition/Gummy Deposits On Bearings	33					4			33	* Change oil * Inspect and clean lube oil strainer or filter * Check and inspect bearings * Check with oil supplier to ascertain correct oil species being used
	Possible Causes	**#**	**A**	**B**	**C**	**D**	**E**	**F**	**G**	**#**	**Possible Remedies**

Table 4-16 (cont.)

	Symptoms					Symptoms				
D	Low Lube Oil Pressure									
C	Loss Of Disch. Pressure					Excessive Brg. Oil Drain Temp. E				
B	Compressor Surges					Units Do Not Stay In Alignment F				
A	Excessive Vibration					Water In Lube Oil G				
Possible Causes	#	A	B	C	D	E	F	G	#	Possible Remedies
Inadequate Cooling Water Supply	34					5			34	* Increase cooling water supply to lube oil cooler * Check for above design cooling water inlet temperature
Fouled Lube Oil Cooler	35					6			35	* Clean or replace lube oil cooler
Operation at a Very Low Speed Without The Auxiliary Oil Pump Running (If main L.O. pump is shaft driven)	36				7				36	* Increase speed or operate aux. lube oil pump to increase oil pressure
Bearing Lube Oil Orifices Missing or Plugged	37				11				37	* Check to see that lube oil orifices are installed and are not obstructed * Refer to lube oil system schematic diagram for orifice locations
Oil Pump Suction Plugged	38				3				38	* Clear pump suction
Leak In Oil Pump Suction Piping	39				4				39	* Tighten leaking connections * Replace gaskets
Failure Of Both Main & Auxiliary Oil Pumps	40				6				40	* Repair or replace pumps
Oil Leakage	41				10				41	* Tighten flanged or threaded connections * Replace defective gaskets or parts
Clogged Or Restricted Oil Cooler Oil Side	42					1			42	* Clean or replace cooler
Possible Causes	#	A	B	C	D	E	F	G	#	Possible Remedies

Lube Oil System

(table continued on next page)

Table 4-16 (cont.)

	Symptoms					Symptoms			

D Low Lube Oil Pressure
C Loss Of Disch. Pressure
B Compressor Surges
A Excessive Vibration
Excessive Brg. Oil Drain Temp.**E**
Units Do Not Stay In Alignment **F**
Water In Lube Oil **G**

Possible Causes	#	A	B	C	D	E	F	G	#	Possible Remedies
Inadequate Flow Of Lube Oil	43				3			43		* Refer to "D" in symptom column * If pressure is satisfactory, check for restricted flow of lube oil to the affected bearings
Water in Lube Oil	44				8			44		* Refer to "G" in symptom column
Leak In Lube Oil Cooler Tube(s) or Tube Sheet	45						1	45		* Hydrostatically test the tubes and repair as required * Replace zinc protector rods (if installed) more frequently if leaks are due to electrolytic action of cooling water
Condensation in Oil Reservoir	46						2	46		* During operation maintain a minimum lube oil reservoir temperature of 120 deg. F to permit separation of entrained water * When shutting down, stop cooling water flow to oil cooler * Commission lube oil conditioning unit * Refer to lube oil management guide
										NOTE: Vibration may be transmitted from the coupled machine. To localize vibration, disconnect coupling and operate driver alone. This should help to indicate whether driver or driven machine is causing vibration.
Possible Causes	#	A	B	C	D	E	F	G	#	Possible Remedies

Lube Oil System

Table 4-17
Distribution of Basic Failure Causes—Reciprocating Compressors

Basic Failure Cause	Distribution of Incidents (%)
Valves	41.0
Piston Rings	14.0
Cylinders	1.0
Pistons	3.0
Piston Rods	10.0
Packings	10.0
Lubricator/System	18.0
Crossheads	1.0
Crankshafts	1.0
Bearings	1.0
Controls	1.0

Troubleshooting Reciprocating Compressors

Valves and piston rings, for obvious reasons, are frequently the weak link of reciprocating compressor operation. A typical statistic taken from a five-year operating history of three Canadian refineries with comparable compressor populations is shown in Table 4-17.

Valves.* A successful compressor valve application requires the careful analysis of the nature, properties, and quality of the gas that is compressed, and dimensions and specifications of the compressor, and the compressor's operating conditions. Then the proper-size valve is selected, with an acceptable compromise between efficiency and durability.

There are many different types of compressor valves manufactured by different manufacturers. The following is a list of the more representative valve designs.

1. Single- and multi-ring valves.
2. Plate valves.
3. Ring-plate valves.
4. Channel valves.
5. Feather-strip valves.

1. The *single- and multi-ring valve* is the valve most commonly used in reciprocating process gas compressors. The valve has one or more concentric valve rings that are generally spring loaded separately, as shown in Figure 4-23.

---

*Courtesy IIC, 364 Supertest Road, Downsview (Toronto), Ontario, M3J 2M2, Canada.

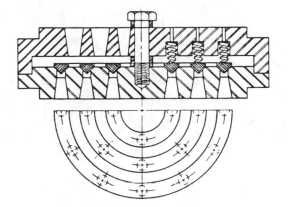

Figure 4-23. Pyramid ring valve.

A ring-type valve using rings with a contoured cross section, such as the IIC pyramid ring valve, is designed to lower the resistance to gas flow and improve efficiency.[29]

Attention to the distribution of the spring loads on the different valve rings is important in order to achieve proper timing in the opening and closing of the multiple valve rings.

2. *The plate valve* illustrated in Figure 4-24 uses a slotted disc for the valve element and springs for damping. This Hoerbiger design generally has an evenly distributed spring load over the entire valve area. Achieving proper damping without adversely affecting the flow efficiency is generally difficult with this design, as strong spring load is usually required for proper damping.

   The successful use of relatively thick damping plates, spring loaded separately from the Hoerbiger valve plate, is an attempt to achieve a fast opening of the valve plate (lightly spring loaded), with sufficient damping to avoid valve slamming.

3. The *ring-plate valve* is a hybrid between the multi-ring valve and the plate valve. Generally it uses two concentric, slotted ring plates as the valve elements. It is a well-designed valve, but it is somewhat limited in cushioning or damping. The major manufacturer of this type of valve is Cooper-Bessemer.

4. The *channel valve* in Figure 4-25 is designed by Ingersoll-Rand. It consists of a valve seat, a guard, channels with leaf springs, and guides to hold the channels in place. Sometimes a lapped, stainless-steel seat plate is used to improve wear and to facilitate repair.

   Each channel, complete with its leaf spring, operates individually by closing over a corresponding slot-shaped port. The distribution of spring load on the different channels inside the same valve is very important to achieve proper timing in the opening and closing of the channels.

   A small volume of gas is trapped between the channel and its leaf spring. This gives a pneumatic cushioning effect so that the channel can ease to a stop when opening.

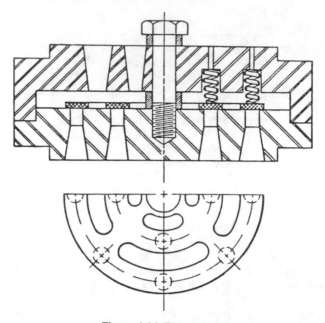

Figure 4-24. Plate valve.

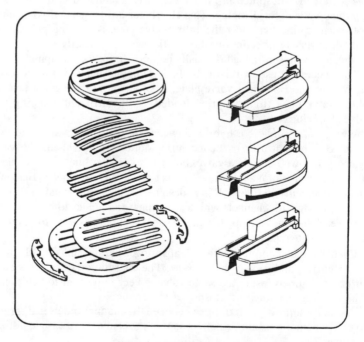

Figure 4-25. Channel valve.

5. *The feather-strip valve* in Figure 4-26 is very simple in design. Manufactured by Worthington, this valve consists of a seat, a guard, and several feather strips. These feather strips are made of light strips of steel whose flexing action permits the passage of gas. The valve seats merely by increasing contact from each end of the strip to the center, not by impact. It is generally an efficient valve, with a large seat-port area, lift area, and guard-port area. Since the feather strips are not cushioned, they are subjected to comparatively high mechanical wear.

Valve failures causes and their remedies.* Valve failures can generally be classified into three categories:

1. Mechanical wear and fatigue.
2. Foreign matters in the gas stream.
3. Abnormal action of valve elements and compressor-valve equipment.

Mechanical wear and fatigue. Mechanical wear in a compressor valve normally occurs where the valve components contact each other. Prolonged wearing of the moving valve element guides can cause sloppy valve action, cocking, and poor seating of the valve elements, leading to leakage, poor performance, and eventual failure. Wear on components should be checked and repaired regularly.

The moving valve elements are also subjected to cyclical, mechanical, and thermal stresses in their normal operation. The strength of the valve components must be considered in each application. Wear on valve parts can be minimized by lubricating the valve parts affected in a particular application. Synthetic lubricants will often outperform mineral oils in this service.

Mechanical fatigue load can be minimized by proper damping or cushioning of the moving valve parts. Proper cylinder cooling and interstage cooling will minimize heat stresses on the valve components.

Wear and fatigue on compressor valves can be sharply aggravated by the presence of foreign matter in the gas stream and by abnormal mechanical valve action.

Foreign matter in the gas stream. Foreign matter may include dirty gas, liquid carryover, coking or carbon formation, and corrosive chemicals.

1. *Dirty gas.* These are generally materials left inside the gas stream by an improper filtration system. There may be rust; fine, sandy, or abrasive particles; loose particles from the gas passages leading to the compressor; and sometimes even remnants from previous valve failures left inside the compressor. These materials accelerate wear on valve components such as the rings, the plates, and the guides. Foreign matter between the coils of the springs can cause rapid spring failures. This has a direct effect on the performance and service life of other valve parts.
2. *Liquid carryover.* Liquid may be carried into the gas stream in chemical processes, or it can be formed from condensation inside the interstage cooler of

---

* Modified from Reference 13.

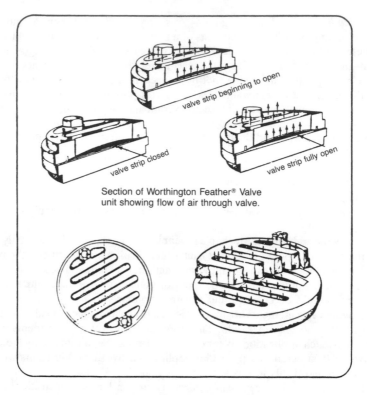

valve strip beginning to open

valve strip closed

valve strip fully open

Section of Worthington Feather® Valve
unit showing flow of air through valve.

Figure 4-26. Feather-strip valve.

the compressor system. In many instances, liquid is present in the first compression stage because excessively cold jacket cooling water promotes condensation or "sweating" of the suction gas stream. Liquid slugs are particularly hard on valves. Because of their high bulk modulus slugs of liquid can break the valve seats. The presence of liquid in the gas stream can dilute the lubricants used inside the compressor, causing accelerated wear on mechanical components.

Moisture traps should be used in the piping to avoid liquid damage. They should be periodically drained, together with the interstage separators. Piping should be designed such that there are no low spots where liquid may collect, and a sight glass or similiar devices can be used for visual inspection.

3. *Coking and carbon formation.* Excessive heat can cause coking of the lubricating oil and polymerization of gases containing unsaturates. These contaminants can reduce compressor efficiency and cause serious valve problems. A spray of steam or a suitable solvent in the form of a mist may be injected periodically into the compressor to clean the cylinders and the valves. Again, consideration should be given to dibasic ester-type synthetic lubricants which greatly resist coking and have very low coefficients of friction.

4. *Corrosive chemicals.* Corrosion can cause leakage at valve seating surfaces and can weaken the valve body to cause valve failures. It can be minimized or eliminated by using different valve materials or by using chemical scrubbers to remove or reduce the corrosive agents before they enter the compressor. Also, removing the moisture or heating up the gas before it enters the compressor can reduce or eliminate the formation of corrosive agents such as acids.

Corrosion takes place gradually, usually over several months. Upon failure of a valve, all the valves in the machine should be checked and repaired. Generally, it is good practice to keep a complete set of spare valves in stock in order to minimize downtime.

Abnormal actions of valve elements. These include slamming, fluttering of the valve elements, disturbances caused by the gas flow pattern, and the effects of resonance and pressure pulsations in suction and discharge pipes.

During normal compressor operation, the discharge valves are more susceptible to *slamming* during opening when the piston is pushing the gas out of the cylinder. This can be reduced by using either damping springs and/or pneumatic cushioning to slow down the valve elements as they move off the seat toward the guard. In other situations, slamming can also occur to compressor valves upon closing.

When opening and closing a valve, the inertia of the valve element has to be overcome. If the inertia of the valve element is high, the finite time required for it to open or close is comparatively longer than required for one with a lower inertia. With high inertia, the valve will close late and the backflow of gas, instead of the valve springs, will push the valve closed, causing the valve elements to be slammed against the seat. Slamming can usually be identified by a mottled appearance where the valve element contacts the seat. It can frequently be heard outside the compressor as clattering.

This problem can be solved by reducing the lift and/or increasing the spring rate to regulate the valve timing; or, where conditions allow, by changing the valve element to a lighter-weight material. A phenolic or fiber-reinforced plastic valve element is generally much lighter than a similar steel unit and, by its nature, more shock absorbent. However, these materials are limited by temperature and strength in their applications.

Valve fluttering occurs when the flow of gas passing through the valve is insufficient to lift the valve element fully to the guard containing the valve springs. The valve element will then oscillate between the seat and the guard. This can accelerate spring wear, and if the valve element strikes the seat and guard several times, the wear and fatigue of it is accelerated. Also, valve slamming is likely to occur when the valve is closing. Fluttering is generally identifiable when no definite pattern or impact trace of the valve springs exists on the valve elements, or when the valve elements appear to have spun. Fluttering may be eliminated by lowering the lift and/or by using lighter springs.

The flow pattern into and out of the compressor cylinder can be affected by obstructions in the compressor gas passages, and sometimes by the orientation of the valves inside the compressor. These problems are rare, but they are also less apparent and more difficult to troubleshoot.

A problem related to the flow pattern is the vibration of valve unloaders on the suction valves, activated by the drafting effect of the gas flowing into the cylinder. This is usually caused by unloader parts protruding across the direction of the gas stream, causing it to be pulled into vibration. The vibration may eventually cause some of the unloader fingers to break off, and leakage can occur in the suction valves due to unseated valve elements.

The solution is to use, where possible, a stronger retention spring on the unloader and/or to redesign the parts of the unloader protruding into the gas flow.

Resonance and pressure pulsations in the compressor gas passages can affect the timing in opening and closing the compressor valves. These problems can be corrected by changing the piping of the compressor and by using settling chambers, surge bottles, and pulse dampers. The application of surge bottles and pulse dampers is a science in itself and is beyond the scope of this text; however, their purpose is simply to prevent pressure pulses from traveling back into the compressor cylinder, thus unseating the valves and affecting the compressor. Sometimes a properly sized orifice installed in the compressor piping is all that is needed to cure the problem. The following section will describe suitable approaches to troubleshooting piping-system-related resonance and pressure pulsations.

General troubleshooting. Very few pieces of petrochemical process machinery lend themselves to early symptom-cause identification as do reciprocating compressors. This becomes evident when considering how the neglect of subtle performance changes can result in costly wrecks. For instance, high discharge temperatures can result from the simple primary cause of insufficient cooling water. Not responding to this cause will lead to an overheated cylinder and ultimately to such final events as piston seizure, ring breakage, and piston cracking. Similarly, not responding to the audible symptom of "knocking in cylinder," where the most probable cause is incorrect piston-to-head clearance will ultimately result in piston failure or rod breakage with attendant damage to the crosshead. A detailed analysis of the symptom-cause-failure chain would be in order. Table 4-18 will serve this purpose. Far more detailed information can be gleaned from the Bloch-Hoefner text *Reciprocating Compressors—Operation and Maintenance,* Houston, TX: Gulf Publishing Company, 1996.

When troubleshooting reciprocating compressors, the most important symptoms to watch for are unusual sounds and changes in pressures, temperatures, and flowrates. Consequently, the primary troubleshooting tools are our five senses, two pressure gauges, two temperature indicators, and a flowmeter. Generally, flowmeters are not available for each individual stage of compression, but considering that what goes in the front end comes out the back end, flow measurement at one stage along the way is sufficient in cases where no interstage inlets or knockouts are involved.

Compressor temperatures and pressures are basic to design calculations and help determine compressor health. The difference between the observed and calculated temperatures, ΔT, should be more or less constant from day to day. The actual observed or calculated temperature may vary; when suction temperature increases, so does discharge temperature. If the compression ratio across a cylinder increases, its discharge temperature also increases. Comparing calculated and actual discharge temperatures provides a yardstick with which to determine operating deviations.

Calculations help. The procedure for analyzing reciprocating compressors is to measure the temperatures and pressures across each cylinder when the unit is known to be in good operating condition. The discharge temperature is calculated from the compression ratio, suction temperature, and k-value of the gas. The k-value, a physical property of the gas, is the ratio of the specific heats. Figure 4-27 shows the necessary equations and k-values for some of the more common process gases.

The observed and calculated discharge temperatures are seldom exactly the same. There might be variations in pressure and temperature indicator readings as well as in gas properties. However, results should be comparable and differences between actual and calculated discharge temperatures reasonably constant from day to day if the compressor is operating properly. When the temperature difference increases, a problem with the compressor valves or piston rings can be suspected (see Table 4-18).

Discharge temperature is the most powerful tool in analyzing the unit for potential problems. The troubleshooter should be able to rely on a record of observed discharge temperatures and pressures when the valves and piston rings are known to be in good, but not necessarily new, condition.

Readings must be made with reliable gauges. With the information recorded, it is not necessary to calculate discharge temperature every day, only to have the raw data available when a problem is suspected. One should then go back and analyze the actual and calculated conditions.

Diagnostics. Other tools for diagnosing reciprocating compressor troubles are lube-oil analyses for metal content, vibration readings, and Beta analyzer-type diagnostics. In lube-oil analyses, particular emphasis is on finding the metals that are

---

$$T_2 = T_1 \cdot \left(\frac{P_2}{P_1}\right)^{\frac{K-1}{K}}$$

where for a single cylinder:
T_1 = Inlet Temperature, °Rankine (°F + 460)
T_2 = Outlet Temperature, °R
P_2 = Outlet Pressure, psia
P_1 = Inlet Pressure, psia

K = Ratio of Specific Heats, $MC_p/MC_v = \dfrac{MC_v - 1.986}{MC_v}$

K = 1.4 for a perfect diatomic gas

K-Values of Common Gases
Air, N_2, O_2, H_2 ~1.4
CO_2 1.28
Propane   1.15
Natural Gas   Ranges from 1.1 to 1.25

Figure 4-27. Discharge temperature calculations for reciprocating compressors.

Table 4-18
Troubleshooting Guide—Reciprocating Compressor

	Possible Causes	Noise Vibration — Knocking	Noise Vibration — Vibration	Pressure — Discharge Pressure Up	Pressure — Discharge Pressure Down	Pressure — Inter-Cooler Press. Up	Pressure — Inter-Cooler Press. Down	Pressure — Discharge Temp. Up	Temperature — Outlet Cooling Water Temp. Up	Temperature — Overheating Valves	Temperature — Overheating Cylinder	Temperature — Overheating Frame	Flow — Capacity Down	Int. Inspection Result — Carbon Deposits Abnormal	Int. Inspection Result — Piston Rings Cyl. Wear Up	Int. Inspection Result — Valve Wear/Breakage Up
Valves	L.P. Valves Wear/Breakage				②	①		①	③	③	②		①	④		⑤
	H.P. Valves Wear/Breakage		③		①		①									
	L.P. Unloading System Defective				②	②		②	④	④	⑧		②	⑤		⑦
	H.P. Unloading System Defective						②									
Pistons/Cylinders	L.P. Piston Rings Worn				④	⑤							⑦	⑨		
	H.P. Piston Rings Worn						③								⑥	
	Piston Rod Nut Loose	④														
	Piston Loose	⑥														
	Head Clearance Too Small	②														

Category		⑤											
Frame	Bearing Clearance Too High	⑤											
	Flywheel Or Pulley Loose	⑦	②							①		①	
	Crosshead Clearance Too High	③											
Support/Cooling/Lubrication	Cooling Water Qnty. Too Low			④	①		④						
	Cylinder Lubrication Inadequate	⑨	⑥	⑦			⑥			①		①	
	Frame Lubrication Inadequate	①						①					
	Cylinder Lubrication Excessive					⑧				②		⑧	
	Lubricating Oil Incorrect Spec.	⑩		⑧			⑦			①		②	②
	Foundation/Grouting Inadequate	⑧	④										
	Piping Support Inadequate		①										
Piping System	Resonant Pulsations (Suction or Discharge)											⑨	
	Suction Filter Dirty/Defective				③				③	③	⑤	⑤	
	Suction Line Restricted				④						④		

(table continued on next page)

Note: The numbers indicate what to check first, or the probability ranking.

Table 4-18 (cont.)

Possible Causes		Knocking	Vibration	Discharge Pressure Up	Discharge Pressure Down	Inter-Cooler Press. Up	Inter-Cooler Press. Down	Discharge Temp. Up	Outlet Cooling Water Temp. Up	Overheating Valves	Overheating Cylinder	Overheating Frame	Capacity Down	Carbon Deposits Abnormal	Piston Rings Cyl. Wear Up	Valve Wear/Breakage Up
System Leakage Excessive	Piping System				3								5			
System Demand Exceeds Compressor Capacity	Piping System				5											
Discharge Pressure Too High	Operating Conditions	11	7	2				3	5	1	1	3		6		
Discharge Temperature Too High	Operating Conditions													7		
Intercooler Fouled	Operating Conditions					4		6	6							
Liquid Carry-Over	Operating Conditions														3	3
Dirt/Corrosion Products Into Cyl.	Operating Conditions														4	4
Cylinder Cooling Jackets Fouled	Operating Conditions							5	2		5			10		
Running Unloaded Too Long	Operating Conditions									2						
Speed Incorrect	Operating Conditions		5		6						3	2	6	8		

NOTE: The numbers indicate what to check first, or the probability ranking.

in the bearings and running gear. Increases in metals contents from the previous analysis indicate the beginning of wear of these parts; we can frequently predict the point when wear becomes critical.

Vibration analysis is usually applied to rotating equipment such as centrifugal compressors. It can also be useful for reciprocating compressors. Vibration analysis can alert us to coupling misalignment, even with relatively low-speed reciprocating compressors. Some troubleshooters have used vibration measurements in the analysis of piping fatigue failures and fractures caused by inherent system resonance and pulsations. Some of these problems are caused by first or second orders of running speed, exciting resonant frequencies in pipe runs. Usually, if the vibration frequency is higher than twice the running speed of the machine, acoustic pulsations are suspected.

Beta analyzer indicator card readings allow us to directly recognize what is happening inside a compressor cylinder. This method of analysis, further explained in Chapter 5, can be extremely helpful in solving problems related to valve losses and piston-ring leakages. Such analytical equipment can be expensive, but there are companies supplying this service on either a contract or a one-time basis. However, we must alway remember that from a diagnostic point of view, a one-time reading is no substitute for recorded historical operating data. After all, the most decisive feature of a trouble symptom is its upside/downside *change*.

This concludes our discussion of driven process machinery troubleshooting. As an introduction to our discussion of drivers we include Table 4-19, a troubleshooting guide for V-belt drives. Coupling failure analysis and troubleshooting was discussed earlier in Chapter 3.

Table 4-19
Troubleshooting Guide—V-Belt Drive

	Symptoms					Symptoms				
	C Repeated Belt Fracture					Rapid Belt Wear D				
	B Bearings Overheating					Checked Or Cracked Belts E				
	A Belt Squeal					Belts Turn Over In Sheave F				
	Possible Causes	**#**	**A**	**B**	**C**	**D**	**E**	**F**	**#**	**Possible Remedies**
Wear and Tear	Rubbing Belt Guard	1				1			1	* Check clearance
Wear and Tear	Sheave Diameter too Small	2				3			2	* Check sheave side walls
Wear and Tear	Mismatched Belts	3				5			3	* Replace with matched setor banded belts
Wear and Tear	Slippage	4	3	2		6	1		4	* Check tension
Wear and Tear	Replacing One Belt Only in Multibelt Drive	5				8			5	* Replace as complete set
Wear and Tear	Misalignment	6				10			6	* Realign
	Possible Causes	**#**	**A**	**B**	**C**	**D**	**E**	**F**	**#**	**Possible Remedies**

Table 4-19 (cont.)

		#	A	B	C		D	E	F	#	
	Symptoms						**Symptoms**				
	C Repeated Belt Fracture						**Rapid Belt Wear** **D**				
	B Bearings Overheating						**Checked Or Cracked Belts** **E**				
	A Belt Squeal						**Belts Turn Over in Sheave** **F**				
	Possible Causes	**#**	**A**	**B**	**C**		**D**	**E**	**F**	**#**	**Possible Remedies**
Wear and Tear	Insufficient Contact Arc	7	2							7	* Increase center distance
	Belt Bottoming In Grooves	8	4							8	* Replace sheaves or belt
	Small Sheaves	9		4			3			9	* Redesign or use cog belt
	Backward Bending	10					4			10	* Use cog belt
	Overtensioned Drive	11		1						11	* Recheck tension
Design/Installation	Sheaves Too Far From Bearing	12		3						12	* Move sheaves closer to bearing block
	Sheaves Too Small	13		4			3			13	* Redesign drive
	Improper Installation	14			2			11	5	14	* Check if belt pried over sheave
	Misplaced Slack	15			3					15	* Keep slack on one side
	Extended Or Improper Storage	16					7			16	* Replace
Operating Conditions	Overloads	17	1				4		2	17	* Redesign drive
	Overheating	18					9			18	* Improve operating environment * Use cog belt
	Shock Loads Or Vibration	19			1				3	19	* Use banded belts
	Foreign Materials In Grooves	20			4				4	20	* Improve belt shielding
	High Ambient Temperatures	21						2		21	* Provide adequate ventilation * Use cog belts
	Worn Sheave Grooves	22					2			22	* Check Sheave side walls
	Broken Tensile Cords	23							1	23	* Check if belt was pried over sheave
	Possible Causes	**#**	**A**	**B**	**C**		**D**	**E**	**F**	**#**	**Possible Remedies**

Note: The numbers indicate what to check first, or the probability ranking.

Troubleshooting Engines

Internal combustion engines, such as large four- or two-cycle integral gas engine compressors, gas and diesel engine drivers, also belong to the diverse population of petrochemical process plant machinery. These services range as high as 16,000 BHP with brake-mean-effective-pressure ratings of up to 250 psi for 20-cylinder engines. Usually, this equipment consists of high-performance, finely tuned machines that must be installed with extreme care and maintained equally thoroughly. A typical duty for an integral gas engine compressor would be hydrogen gas compression service using refinery off-gas as fuel. Other typical applications are diesel engines as drivers for cooling tower pumps and emergency generators.

During the years 1971–1974, the Hartford Steam Boiler Inspection and Insurance Company compiled the failure statistics shown in Table 4-20. The equipment covered consisted of a population of 180 engines, including diesel and dual-fuel engines, spark-ignited gas and gasoline engines, and internal combustion engines driving reciprocating compressors. Table 4-21 shows a breakdown of failures by cause and of parts initially failed.

Earlier we alluded to the fact that large internal combustion engines, along with other reciprocating machinery, usually respond well to monitoring and on-line diagnostics. Here, as in the case of reciprocating compressors, final dramatic failures will result because subtle initial distress symptoms were either not recognized or not acted upon. Consider bearing failures, which head the list in both the number of engine failures and cost. Again, just the "middle" link in a chain of symptoms and causes often beginning with lubrication deficiency (see Chapter 3) and ending with a crankshaft failure. Consequently, paying attention to bearing wear and failure symptoms will eliminate a major portion of crankshaft failure causes.

Engine bearing wear or damage, although directly related to crankshaft alignment, is also the result of many other factors. Table 4-22 represents an engine troubleshooting chart that shows the following causes for bearing wear and failure symptoms as itemized on page 324.

Table 4-20
Distribution of Large Internal Combustion Engine Failures
and their Cost by Failure Modes[30]

Failure Mode	Distribution (%)	$-Value of Total (%)
Breaking/tearing	48.3	46.7
Cracking	26.1	19.4
Deforming	1.1	0.4
Scoring	5.6	8.2
All others	18.9	25.3
Totals	100.0	100.0

Table 4-21
Distribution of Large Internal Combustion Engine Failures and their Cost by Cause and Parts Failed First[30]

Failure Cause	Distribution	
	Incidents (%)	$—Value of Total (%)
Cause undetermined	28.9	17.2
Lubrication problems	12.8	17.1
Cooling problems (loss of, insufficient, inadequate)	10.6	6.1
Fatigue	10.0	5.2
Overheating (except from overload)	5.6	3.5
Material, defective	5.6	3.5
Misalignment	2.2	13.9
Overspeeding	2.2	8.4
Accumulation of liquid in cyl.	2.2	2.4
Cracking, progressive	2.2	1.1
Not otherwise classified	2.2	1.0
Vibration	1.7	6.3
Maintenance, inadequate	1.7	1.3
Properties, change in metallurgical	1.7	0.6
Wear (age or service)	1.7	0.4
Repair, improper	1.1	1.4
Device, loosening of locking	1.1	0.6
Workmanship, poor (owner)	1.1	0.6
Freezing, low ambient	1.1	0.1
All others	4.3	9.3
	100.0	100.0

Initial Part to Fail		
Bearings, all	24.4	33.5
Piston, piston rings	19.4	17.3
Cylinder, head, block, liner	16.7	8.3
Crankshaft	6.1	24.2
Valves	5.6	3.2
Rod, connecting, piston	4.4	3.3
Manifold	4.4	2.2
Lube-oil system	2.2	0.8
Gears	2.2	0.6
Cam shaft	1.7	0.6
Frame, cast	1.7	0.4
Couplings	1.7	0.2
Turbocharger rotor	1.1	1.1
Control, pressure, temperature	1.1	0.3
All others	7.3	4.1
	100.0	100.0

[1] Improper crankshaft alignment and bearing fit. Some remedial actions are:
   a. Inspect and adjust alignment and bearing clearances, i.e. "crush."
   b. Take crankshaft web deflection readings periodically for early detection of misalignment problems.
   c. Inspect foundation for any movement between engine base and grout.
   d. Correct a loose-grout problem as soon as possible. Use an oil-resistant polymeric grout.
   e. Ascertain that foundation bolt torque values are adequate.

[2-7] [9] [11] Fuel quality: overload/unsteady, rough-running-related causes.

[8] Inadequate oil filtration. Some remedial actions are:
   a. Obtain periodic lab analysis of crankcase oil to check for dilution and oxidation.
   - Lube-oil dilution on diesel engines, for instance, results from:
   - High-pressure tube fittings leaking between fuel-injection pumps and spray nozzles.
   - Clogged camshaft trough drains.
   - Wear of fuel-injection pump pushrod bushings.
   - Loose pushrod in the fuel injector.
   - Incorrectly tightened fittings on injection pumps and fuel return piping.
   b. Review procedure for keeping duplex oil strainers clean and available for changeover.

[10] Incorrect metallurgy. Refer to Chapter 3.

[12] Lack of lubrication. Remedial actions could be:
   a. Inspect oil pump drive.
   b. Monitor oil supply pressure.

[13] Incorrect oil viscosity. Remedial actions could be:
   a. Review manufacturer's recommendations.
   b. Obtain periodic lab analysis.

[14] Coolant leakage into lube oil. Remedial action is shown in 8a.

The foregoing will have given the reader an impression of the complexity of an engine troubleshooting system. We recommend, therefore, that Table 4-21 be referred to as a first guide only.

Troubleshooting Steam Turbines

Failure statistics. Steam turbines in the petrochemical industry range from smaller than 10 hp, stand-by pump drives, to larger than 60,000 hp, unspared process gas compressor drives. Table 4-23 shows a representative statistic of failure cases and their major causes and affected components.

Similarly, Table 4-24 shows failure-mode distributions from the same source. Accordingly, rubbing through axial and radial contact can be caused by:

1. Thermal distortion of rotor and casing.
2. Alignment changes—external and internal.
3. Unbalance.

(text continued on page 329)

Table 4-22
Engine Troubleshooting Guide

	Possible Cause	Hammering/Knocking	Pre-Ignition	Detonation	Misfiring	Overheating	Soot Formation (Exhaust)	Valve Leaking	Port Plugging (*)	Piston Blow-By	Bearing Wear	High Oil Consumption	Short Oil Life	Short Filter Life	High Maintenance	Power Loss	Varnish/Sludge	Ash Deposits	Carbon Deposits	Spark Plug Fouling	Piston Ring Sticking	Cylinder/Liner Wear	Valve Failure	Bearing Failure	Piston Seizure	Cylinder Hd. Cracking	Turbocharger Failure
						Monitoring/Surveillance Results											Internal Inspection Results										
Valves	Valve Angle—Incorrect					14		2										11		5			2				
	Adjustment—Incorrect			7		15	3	5															5				
	Metallurgy—Incorrect							3															3				
Piston/Rings	Ring Size—Incorrect					16				7				10		2		2		6	7				3		2
	Ring Groove—Worn									9			12								9	17					
	Oil Ctrl. Ring—Worn		8									1	13														
	Oil Ctrl. Slot—Plugged									10										11	10						
	Side Clearance—Excessive																				5						
	Piston/Ring Design Inadequate												14				7	3		7		6			4	3	

Cylinder	Liner Distortion					17				8											8	3				
	Surface Finish Inadequate																					2				
	Metallurgy—Incorrect																					4				
	Break-In Procedure—Incorrect																					1				
Ignition System	Timing—Incorrect		3	2	1	11	4				4					3		4		2		8		4	1	2
	Ignition Elements—Faulty				2	9												5		3						
	Spark Plug Gap—Incorrect				3													6		4						
Air/Fuel System	Compression Ratio—Too High		6	5	4	5					6											11		6		
	Air/Fuel Ratio—Incorrect		10	9	6	7	1						9	4		1		8		9		12				
	Intake Air Temp.—Too High		4	3		4					5		5				3					9		5		
	Filtration—Inadequate									3	9		15	2				12			3	15	8	9		
	Intake Air Restriction					12																				
	Scavenging—Inadequate		9	8	5	6			3			3		5		6		7	2	8						
	Fuel Quality—Inadequate				7	3							4		4		9					7				
	Fuel—Wet		2	1		8					2			13		4								2		
	Fuel—Unstable HC		1								3													3		

(*) Two-Cycle Engine

(table continued on next page)

Table 4-22 (cont.)

System	Possible Cause	\<-- Symptoms: Monitoring/Surveillance Results -->															\<-- Internal Inspection Results -->										
		Hammering/Knocking	Pre-Ignition	Detonation	Misfiring	Overheating	Soot Formation (Exhaust)	Valve Leaking	Port Plugging (*)	Piston Blow-By	Bearing Wear	High Oil Consumption	Short Oil Life	Short Filter Life	High Maintenance	Power Loss	Varnish/Sludge	Ash Deposits	Carbon Deposits	Spark Plug Fouling	Piston Ring Sticking	Cylinder/Liner Wear	Valve Failure	Bearing Failure	Piston Seizure	Cylinder Hd. Cracking	Turbocharger Failure
Cooling System	Cylinder Temp.—Too Low						2	1	4	6		4	3	9				10			6	10					
Cooling System	Cooling—Inadequate		5	4		2																	1		2	1	
Cooling System	Cooling Water Temp. Too Low												7	11													
Cooling System	Coolant Leakage Into Lube Oil										14		6				5							14			
Lube Oil System	Oil Quality—Inadequate							6	6	5		5	1	1	1		2	9		10	5		6	15			
Lube Oil System	Oil Viscosity—Incorrect					13					13											14	7	13			
Lube Oil System	Oil Filtration—Inadequate							7	1	2	8	2	16	8	2		8				2	16		8			
Lube Oil System	Overlubrication		7	6						1				6	3			1	1	1	1						1

Table 4-22 (cont.)

Possible Cause	Hammering/Knocking	Pre-Ignition	Detonation	Misfiring	Overheating	Soot Formation (Exhaust)	Valve Leaking	Port Plugging (*)	Piston Blow-By	Bearing Wear	High Oil Consumption	Short Oil Life	Short Filter Life	High Maintenance	Power Loss	Varnish/Sludge	Ash Deposits	Carbon Deposits	Spark Plug Fouling	Piston Ring Sticking	Cylinder/Liner Wear	Valve Failure	Bearing Failure	Piston Seizure	Cylinder Hd. Cracking	Turbocharger Failure
Lube Oil System																										
Lack of Lubrication					1		4			12		2									18	4	12			
Lubricator Failure								5																		
Oil Drain Interval Excessive									4				7			1										3
Contamination												10	3			6				4						
Oil Temp. Too Low												8				4										
Metallurgy—Incorrect										10													10			
Crankshaft Alignment/Brg. Fit.										1													1			
Misc.																										
Loose Parts	1												12								13					
Overload					10			2		7					5								7			

(*) Two-Cycle Engine

Table 4-23
Distribution of Steam Turbine Failure
Incidents by Cause and Affected Components[15]

Failure Causes	Distribution of Incidents (%)	Components	Distribution of Incidents (%)
Vendor Problems	64.1	Rotor blades	29.0
Planning, design, and calculation	16.5	Bearings	16.7
		Radial bearings (12.5)	
		Thrust bearings (4.2)	
Assembly	16.0		
Technology	10.6	Shaft seals, balance pistons	15.6
Manufacturing	8.7	Rotors with discs	10.3
Material	8.0	Casings with baseplates,	
		screws	9.8
Repair	4.3	Strainers, fittings	4.0
Operational Problems	15.3	Control	4.0
Surveillance	10.6	Guide blades and	
		diaphragms	3.4
Maintenance	4.7	Gears, transmissions	2.4
External Influences	20.6	Pipelines	0.8
Foreign bodies	7.2	Other parts	4.0
From the electrical grid	4.1		
Others	9.3		
	100		100

Table 4-24
Steam Turbine Failure Mode Distribution[15]

Failure Mode	(%) of Incidents
Rubbing	23.0
Fatigue and creep failure	18.5
Damage to bearings	14.6
Thermal stress cracking	11.7
Sudden failure	9.3
Incipient cracks	8.0
Mechanical surface damage	5.4
Corrosion and erosion	3.3
Shaft bending	2.4
Wear	2.3
Abrasion	1.5
	100

(text continued from page 323)

4. Damage to axial or radial bearings.
5. Off-design condition leading to thermally-induced dimensional changes.

Rotor blade failures can be caused by:
1. Dynamic overloading.
2. Loosening of blade seating.
3. Seating design deficiencies.

Bearing failures are discussed in Chapter 3. Thermal stress cracking occurs primarily in cast-steel components in the region of highest temperatures and at points of changes in wall thickness. Thermal stress cracking of turbine rotors is a rare occurrence. Other components affected by thermal stress cracking are trip and throttle valve casings, steam chests, and the wheel-chamber sections of high-pressure casings.

Table 4-25 represents a troubleshooting guide for general-purpose steam turbine drivers. Table 4-26 is a troubleshooting guide for a commonly used hydraulic governor, the TG-10 governor made by Woodward Governor Co. Finally, Figure 4-28 is a stair step-type diagnostic guide for governor-system troubles on larger steam turbines. *(text continued on page 336)*

Table 4-25
Troubleshooting Guide—Steam Turbines

Symptoms										Symptoms	
D Speed Increases as Load Decreases					**Governor Not Operating / E Excessive Speed Variation**						
C Insufficient Power					**O.S.T.† on Load Changes F**						
B Slow Start-Up					**O.S.T. at Normal Speed G**						
A Turbine Fails to Start					**Leaking Glands H**						
Possible Causes	**#**	**A**	**B**	**C**	**D**	**E**	**F**	**G**	**H**	**#**	**Possible Remedies**
Too Many Hand Valves Closed.	1			2		9	5			1	* Open additional hand valves.
Nozzles Plugged Or Eroded	2		6	7						2	* Remove nozzle pipe plugs and hand valves. Inspect nozzle holes. Clean as required.
Dirt Under Carbon Rings	3								1	3	* Steam leaking under carbon rings may carry scale or dirt which will foul rings. Remove, inspect, replace
Worn or Broken Carbon Rings	4								2	4	* Replace with new rings.
Possible Causes	**#**	**A**	**B**	**C**	**D**	**E**	**F**	**G**	**H**	**#**	**Possible Remedies**

(Left margin label spanning table: **Casing/Rotor System**)

†Overspeed trip.

Table 4-25 (cont.)

		Symptoms					Symptoms				
		D Speed Increases as Load Decreases					**Governor Not Operating / Excessive Speed Variation** **E**				
		C Insufficient Power					**O.S.T.† on Load Changes** **F**				
		B Slow Start-Up					**O.S.T. at Normal Speed** **G**				
		A Turbine Fails to Start					**Leaking Glands** **H**				
Possible Causes	**#**	**A**	**B**	**C**	**D**	**E**	**F**	**G**	**H**	**#**	**Possible Remedies**
Shaft Scored	5								3	5	* Shaft surface under carbon rings must be smooth to prevent leakage. Observe proper run-in procedures.
Leak-Off Pipe Plugged	6								4	6	* Make sure all condensate is draining.
Throttle Valve Travel Restricted	7		1	1	1	3	2			7	* Close the main admission valve and disconnect throttle linkage. Valve lever should move freely from full open to full closed.
Throttle Assembly Friction	8	4			2	3	3			8	* Disassemble throttle valve. Inspect for freedom & smoothness of movement of all parts. * Inspect valve stem for straightness and for build-up of foreign material.
Valve Packing Friction	9				3	4	4			9	* Inspect valve packing for excessive compression. If it is compressed to the point where "drag" on valve stem exists, replace.
Throttle Valve Loosening	10					5				10	* Some "floating" valve assemblies have critical end play. Replace or repair valve & stem as necessary.
Possible Causes	**#**	**A**	**B**	**C**	**D**	**E**	**F**	**G**	**H**	**#**	**Possible Remedies**

Left margin label: Trip and Throttle Valve

†Overspeed trip.

(table continued on next page)

Table 4-25 (cont.)

Symptoms									Symptoms	
D Speed Increases as Load Decreases					**Governor Not Operating / Excessive Speed Variation** **E**					
C Insufficient Power					**O.S.T.†** on Load Changes **F**					
B Slow Start-Up					O.S.T. at Normal Speed **G**					
A Turbine Fails to Start					Leaking Glands **H**					
Possible Causes	**#**	**A**	**B**	**C**	**D**	**E**	**F**	**G**	**H**	**#**	**Possible Remedies**
Throttle Valve & Seats Cut or Worn	11			4	5					11	* Remove valve assembly & check valve & seats for wear & steam cutting.
Trip Valve Set Too Close to Operating Speed	12					1	3			12	* Consult operating manual for proper trip-speed setting.
Trip Valve Does Not Open Properly	13	3		6						13	* Assure trip levers are properly latched.
Dirty Trip Valve	14							2		14	* Inspect & clean.
Steam Strainer Plugging	15	2	2	8						15	* Inspect & clean from all foreign matter.
Oil Relay Governor Set Too Low	16			3						16	* Refer to operating manual for speed adjustment & speed range limits.
Governor Droop Adjustment Needed	17					1				17	* An increase in internal droop setting will reduce variation or hunting.
No Governor Control At Start-Up	18		5			8				18	* Check for proper direction of rotation. * Check for proper governor speed range. * Consult governor trouble-shooting guide
Governor Lubrication Problem	19					2				19	* Low governor oil level, dirty, contaminated, or foamy oil, will cause poor governor response. * Drain, flush, and refill with proper spec. oil.
Possible Causes	**#**	**A**	**B**	**C**	**D**	**E**	**F**	**G**	**H**	**#**	**Possible Remedies**

†Overspeed trip.

Table 4-25 (cont.)

Symptoms						Symptoms			
D Speed Increases as Load Decreases					**Governor Not Operating / E** Excessive Speed Variation				
C Insufficient Power					**O.S.T.† on Load Changes F**				
B Slow Start-Up					**O.S.T. at Normal Speed G**				
A Turbine Fails to Start					**Leaking Glands H**				

Possible Causes	#	A	B	C	D	E	F	G	H	#	Possible Remedies
Inlet Steam Pressure Too Low or Exhaust Pressure Too High	20	1	4	5						20	* Use accurate gauges to check steam pressure at turbine inlet and the exhaust pressure close to the exhaust flange. * Low inlet pressure may be the result of auxiliary control equipment being too small, improper pipe sizing, excessive piping lengths, etc.
Light Load & High Inlet Steam Pressure	21					6	6			21	* The tendency for excessive speed variation is quite high when reserve capacity is available and steam pressure is high. * Open hand valves or install smaller throttle valve & body.
Load Higher Than Turbine Rating	22		3	4						22	* Ascertain the actual load requirements of the driven equipment.
Rapidly Changing Load	23					7				23	* Rapidly changing load can cause governor hunting. * Check turbine application.
High Starting Torque Of Driven Equipment	24									24	*Check the required starting torque. * Ascertain turbine is not overloaded.
Excessive vibration	25							1		25	* See vibration diagnostics and note below.
Possible Causes	#	A	B	C	D	E	F	G	H	#	Possible Remedies

Note: For vibration & short bearing life-related causes consult special sections on bearing failure analysis and vibration diagnostics.

†Overspeed trip.

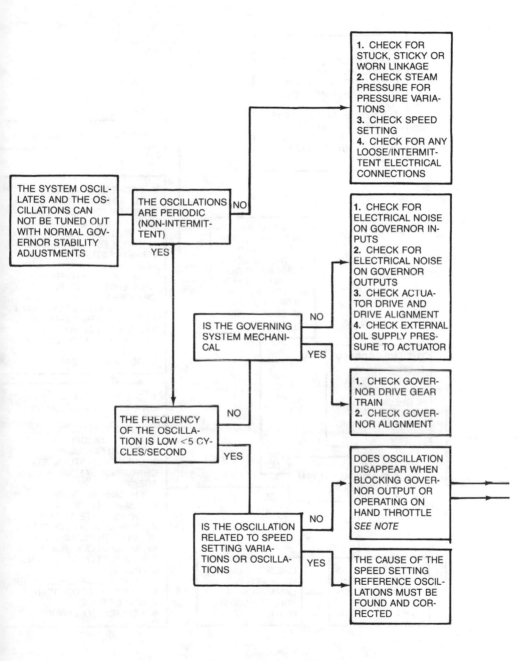

Figure 4-28. Steam-turbine governor troubleshooting guide.

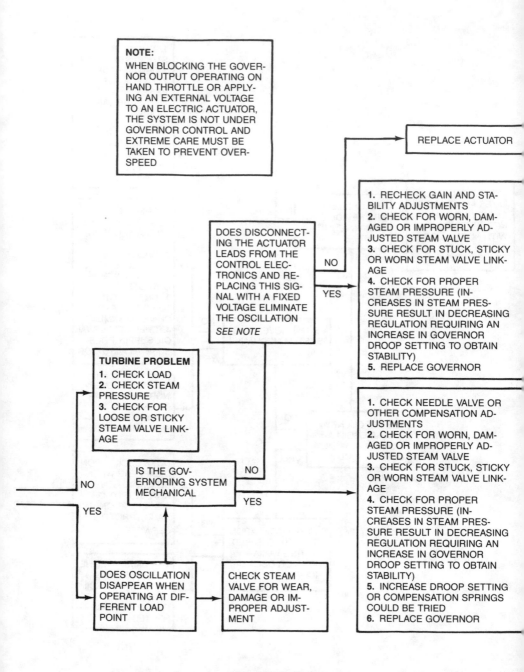

Figure 4-28. Continued.

Table 4-26
Troubleshooting Guide—TG-10 Governor

Symptoms

- **A** Engine/Turbine Hunts and Surges
- **B** Governor Output Jiggles
- **C** Governor Unstable/Oscillations
- **D** Engine/Turbine Cannot Obtain Rated Speed
- **E** Gov'nr. Does Not Start/Control
- **F** Gov'nr. Starts But Remains at Maximum

	Possible Causes	#	A	B	C	D	E	F	#	Possible Remedies
Governor Adjustment	Speed Setting Too Low	1				1			1	* Increase governor speed setting.
Governor Adjustment	Speed Setting Too High	2						1	2	* Reduce speed setting until governor controls, then adjust for desired speed.
Governor Adjustment	Insufficient Droop Adjustment	3	4						3	* Reposition droop adjusting lever to increase droop
Governor Adjustment	Too Much Droop	4			2				4	* Re-position the droop adjusting lever to increase droop.
Governor Adjustment	Governor Speed Range Is Incorrect For This Application	5				3			5	* Check speed range of governor.
Design/Installation/Maintenance	Wrong Governor Drive Rotation	6					1		6	* Check governor drive. * Reverse pump parts for different rotation.
Design/Installation/Maintenance	Worn Flyweight Pins	7		2					7	* Check flyweights and pins for smooth movement. * Replace both parts if one is defective.
Design/Installation/Maintenance	Pump Drive Pin Broken	8					3		8	* Disassemble pump housing to check pin & replace if required.
Design/Installation/Maintenance	Dirt/Contamination In Governor Oil	9	2		4				9	* Drain, flush, and refill with fresh oil of proper specification.
Design/Installation/Maintenance	Low Oil Level	10	1						10	* Add oil to level visible in the oil level indicator.
	Possible Causes	**#**	**A**	**B**	**C**	**D**	**E**	**F**	**#**	**Possible Remedies**

Table 4-26 (cont.)

Symptoms:

- A — Engine/Turbine Hunts and Surges
- B — Governor Output Jiggles
- C — Governor Unstable/Oscillations
- D — Engine/Turbine Cannot Obtain Rated Speed
- E — Gov'nr. Does Not Start/Control
- F — Gov'nr. Starts But Remains at Maximum

	Possible Causes	#	A	B	C	D	E	F	#	Possible Remedies
Design/Installation/Maintenance	Improper Alignment Of The Governor Coupling	11		1					11	* Check and realign as required
Design/Installation/Maintenance	Drive Key Missing Or Improperly Installed	12					2		12	* Check drive installation
Design/Installation/Maintenance	Binding Terminal Shaft Linkage	13	3						13	* Realign linkage as necessary.
Design/Installation/Maintenance	Insufficient/Incorrect Terminal Shaft Travel	14			1	2			14	* Check linkage. Review recommended travel from no load to full load.*
Design/Installation/Maintenance	High Steam Valve Gain	15			3				15	* Ascertain steam valve is not too large for this application.
	Possible Causes	#	A	B	C	D	E	F	#	Possible Remedies

(text continued from page 329)

Troubleshooting Gas Turbines

Although gas turbines are not as wide-spread in petrochemical plants as steam turbines, they have unique features that make them ideally suited as drivers for a number of refinery and chemical plant services. Gas turbines are able to operate on a wide variety of fuels, including low-grade fuels, and under severe environment conditions. Gas turbines have been found to experience more frequent failures than steam turbines. Figure 4-29 gives the distribution of primary failure causes for industrial gas turbines from 1970 to 1979. Figure 4-30 shows the component damage distribution for the same population of machines.

A typical gas turbine consists of three major systems: the compressor section, the combustor, and the turbine. For effective troubleshooting, a good understanding of these systems and their interaction is necessary. The following gas-turbine distress or failure modes must be understood for successful trouble diagnosis:

1. Air inlet system and filtration.
   a. Combustion air problems.
      1. Compressor fouling.

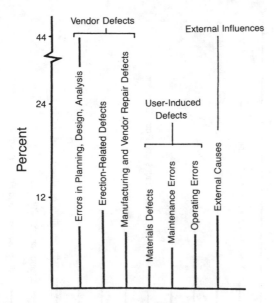

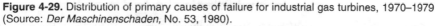

Figure 4-29. Distribution of primary causes of failure for industrial gas turbines, 1970–1979 (Source: *Der Maschinenschaden,* No. 53, 1980).

2. Compressor erosion.
3. Cooling air blockage.
4. Rotor unbalance.
5. Locking of turbine blade roots.
6. Hot corrosion or sulfidation.
2. Fuel system and treatment.
   a. Start-up and lightup failures.
   b. Tripping due to hydrate formation and freezing.
   c. Hydrocarbon liquid carryover.
   d. Combustor fouling.
3. Rotor system.
   a. Rubbing in hot-gas path.
   b. Bearing and seal failures.
4. Turbine and exhaust.
   a. Governor malfunction.
   b. Vibration problems.
5. Control system.
   a. Instrument air problems.

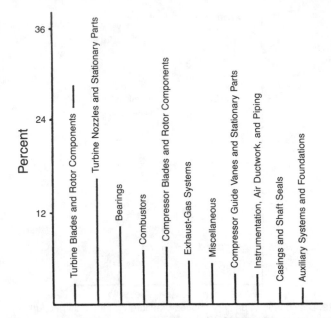

Figure 4-30. Distribution of gas-turbine component damage, 1970–1979 (Source: *Der Maschinenschaden,* No. 53, 1980).

    b. Loss of electrical continuity.
    c. Failure of air intake hoar-frost monitoring and deicing system.
6. Lube-oil system.
    a. Filter problems.
    b. Leakage.
    c. Pressure control problems.
7. Coupling and load system.
    a. Misalignment.
    b. Coupling distress.
    c. Gearbox and ancillary problems.
8. Environment.
    a. Power outage.
    b. Fuel supply.
    c. Operator error.
    d. Changes in load demand.

Finally, Table 4-27 is an appropriate gas-turbine troubleshooting guide.

Troubleshooting Electric Motors

Electric motors are the most commonly encountered prime movers in petrochemical process services. In the typically found two- or four-pole induction design, they range in size from fractional horsepower to more than 20,000 BHP in, for instance, refinery cat plant air blower service. Electric motor troubles appear in two forms: mechanical rotor/bearing difficulties and electrical, winding and supply-system problems. However, it has been estimated that more than 90% of electrical problems, mainly winding failures, can again be traced back to pre-occurring mechanical distress. Typical mechanical failure modes are overloading, flooding, and blockage of cooling air, single phasing, and bearing failures. Some of these failures, in turn, happen because of lack of maintenance or for reasons external to the motor itself. As an example, most single-phasing failures happen because of lack of maintenance on exactly those protective devices installed to guard against this danger.

Consider the case of a 3500-hp, four-pole, pipe-ventilated, squirrel-cage motor driving a centrifugal process compressor. The motor suffered a winding failure and caught fire. The fire was successfully extinguished and the motor disconnected and removed to a motor repair shop. The following failure diagnosis was made:

1. *Results of examination:*
   a. The fire in the motor was caused by a winding failure at that point where the windings emerge from the slots in the stator.
   b. There was a buildup of oil-impregnated dust, coke and fibrous material on the windings that fueled the fire.
   c. That portion of the windings normally encased in the slots of the stator showed widespread signs of overheating, evidenced by the fact that the insulating enamel powdered off at the slightest touch. That this same enamel was relatively intact in the fire-damaged area leads us to conclude that the deterioration was due to internal overheating.
2. *Most probable cause identification:*
   a. Multiple starts, i.e. starting the machine or subjecting it to inrush current levels more frequently than designed for.
   b. Operating the machine in a sustained overload condition. This is most unlikely here, since the machine has accurate relay protection schemes applied to prevent this, and since the compressor is assumed to be operating within design.
   c. Inadequate cooling air: fouled filters, fan failure, or operating the machine without fans.
   d. Fouled windings, i.e. the cooling air is unable to dissipate the normal heat buildup, which is caused by inadequate preventive maintenance (P.M.) on the motor to ensure clean windings and filter system.
3. *Conclusions:*
   a. It is suspected that this machine failure was caused by a combination of 2a, 2c, and 2d.

(text continued on page 342)

Table 4-27
Troubleshooting Guide—Gas Turbine

Symptoms

Category	Possible Cause	Combustion Noise (Erratic)	Vibration (Up)	Bearing Pressure (Up)	Bearing Pressure (Down)	Cooling Air Pressure (Up)	Cooling Air Pressure (Down)	Bleed Chamber Pressure (Erratic)	Fuel Pressure (Up)	Fuel Pressure (Down)	P_2/P_1 (Down)	P_3/P_4 (Down)	Bearing Temperature (Up)	Wheel Space Temperature (Up)	Exhaust Temperature (Up)	Exhaust Temperature (Down)	Exhaust Temperature Variance (Up)	T_2/T_1 (Up)	T_3/T_4 (Down)	Mass Flow (Down)	Comp. Eff. (Down)	Therm. Eff. (Down)
Compressor	Blade Damage		②					②			③							②		③	③	
Compressor	Bearing Failure		①		①								①									
Compressor	Filter Clogging										②											
Compressor	Surging		③	①				①					②							④	②	
Compressor	Fouling		③								①							①		①	①	
Combustor	X-Over Tube Failure								③ OR ③								③					
Combustor	Liner Cracking/Loosening								④ OR ④								④					
Combustor	Fouling	①							② OR ②							①	①					
Combustor	Clogging	②							①						①		②					

Header groups: Pressure/Pressure Ratio (Bearing Pressure, Cooling Air Pressure, Bleed Chamber Pressure, Fuel Pressure, P_2/P_1, P_3/P_4); Temperature/Temperature Ratio (Bearing Temperature, Wheel Space Temperature, Exhaust Temperature, Exhaust Temperature Variance, T_2/T_1, T_3/T_4); Efficiencies (Mass Flow, Comp. Eff., Therm. Eff.).

Possible Cause	Combustion Noise	Vibration	Pressure/Pressure Ratio								Temperature/Temperature Ratio								Efficiencies	
			Bearing Pressure		Cooling Air Pressure	Bleed Chamber Pressure	Fuel Pressure		P_2/P_1	P_2/P_4	Bearing Temperature	Wheel Space Temperature	Exhaust Temperature		Exhaust Temperature Variance	T_2/T_1	T_3/T_4	Mass Flow	Comp. Eff.	Therm. Eff.
	Erratic	Up	Up	Down	Down	Erratic	Up	Down	Down	Down	Up	Up	Up	Down	Up	Up	Down	Down	Down	Down
Turbine Blade Damage		②															②			②
Nozzle Distortion		④								①		③					③			③
Bearing Failure		①		①							①									
Cooling Air Failure					①						②	①								
Fouling		③										②					①			①

(text continued from page 339)

b. *Remedial actions:*
1. All large motors will be scheduled out of service at least every five years for P.M. Where the service conditions warrant, this frequency will be higher. The mechanical department is responsible for scheduling this work.
2. Filter and fan operation should be monitored closely and any failure given high priority for repair.
3. Multiple-start criteria as laid down in the operating manual must be strictly adhered to. This varies with the design of the machine, but generally the machine must be allowed sufficient time between successive starts to return to normal operating temperature.

This failure could perhaps have been prevented by appropriate troubleshooting routines. Table 4-28 represents a typical troubleshooting guide for electrical motors.

Troubleshooting the Process

From time to time, the machinery troubleshooter will be called upon to solve problems in cases where machinery is a more integral part of the total process system. A refrigeration system utilizing compression machinery with different types of drivers is a good example. Figure 4-31 shows a simplified schematic of this system; Table 4-29 shows an appropriate troubleshooting guide.

Usually, the following problems may be encountered in a refrigeration system:

1. Compressor problems:
   a. Centrifugal compressor—See the troubleshooting guide in Table 4-14.
   b. Reciprocating compressor:
      1. Open or faulty clearance pockets.
      2. Leaking valves.
      3. See the troubleshooting guide in Table 4-18.
   c. Steam-turbine drivers:
      1. High exhaust-steam pressure.
      2. Governor not opening fully.
      3. Turbine hand valves not open.
      4. Fouling of turbine wheels.
      5. Overspeed trip set too low.
      6. See the troubleshooting guide in Table 4-25.
   d. Electric-motor drivers:
      1. Motor running too hot.
      2. Amperage trip set too low.
      3. See the troubleshooting guide in Table 4-28.

(text continued on page 346)

Table 4-28
Electric Motor Troubleshooting Guide

Symptoms

- A — Will Not Start
- B — Runs Backward
- C — Noise—Excessive
- D — Rapid Knocking Noise
- E — Fails to Pull in—Synchronous Motor
- F — Will Not Come Up To Speed—Synchronous Motor
- G, H — Overheating
- I — Hot Sleeve Bearing
- J — Hot Antifriction Bearing
- K — Hot Bearings (General)
- L — Vibration

	Probable Cause	#	A	B	C	D	E	F	G	H	I	J	K	L	#	Remedy
Electrical Wiring	Line Trouble—Overload Trip—Loose Connections	1	●					●							1	Check Out Power Supply
	Reverse Phase Sequence	2		●											2	Interchange Two Line Connections At Motor
	Open Circuit, Short Circuit, Or Winding Ground	3	●					●							3	Check For Open Circuits and Ground
	Motor Single Phased (3-Phase Machine)	4			●	●									4	Stop—Then Start Again—Motor Will Not Start
	Unbalanced Load Between Phases	5				●			●						5	Check For Single Phasing—All Three Leads
	Low Voltage/Improper Line Voltage	6	●						●						6	Check Motor Name Plate—Check Wire Size
	Short Circuit In Stator Winding	7	●						●						7	Check Insulation Resistance
	Ground In Stator Winding	8	●						●						8	Check For Ground

(table continued on next page)

Table 4-28 (cont.)

Symptoms

- A Will Not Start
- B Runs Backward
- C Noise—Excessive
- D Rapid Knocking Noise
- E Fails to Pull in—Synchronous Motor
- F Will Not Come Up To Speed—Synchronous Motor
- G Vibration
- H Hot Bearings (General)
- I Hot Antifriction Bearing
- J Hot Sleeve Bearing
- K Overheating
- L

System	#	Probable Cause	A	B	C	D	E	F	G	H	I	J	K	L	#	Remedy
Elec System	9	Field Excited													9	Check Field Control
	10	No Field Excitation					•								10	Check Field Contactor & Conn./Exc. Output
Mechanical System	11	Uneven Air Gap			•										11	Center Rotor. Replace Bearings If Needed
	12	Unbalanced Rotor			•				•					•	12	Check Mechanical & El. Balance
	13	Foreign Matter In Air Gap													13	Clean Motor
	14	Loose Parts Or Loose Rotor On Shaft			•				•					•	14	Repair
	15	Rotor Rubbing On Stator—Bent Shaft			•				•		•	•			15	Check Shaft Run-Out/Center Rotor/Repl. Bearing
	16	Ventilation Restricted							•						16	Inspect Motor Interior For Dirt
	17	Motor Tilted (Shaft Bumping)			•	•					•		•	•	17	Check Level & Realign If Needed
	18	Misalignment Or Excessive Thrust			•	•					•	•	•	•	18	Realign & Check Limited End Float Cplg.

Category	No.	Probable Cause	A	B	C	D	E	F	G	H	I	J	K	L	No.	Remedy
Mechanical System	19	Jammed	•							•					19	Disconnect Motor From Load, Check Solo Run
	20	Overloaded	•							•					20	Reduce Load Or Install Larger Motor
	21	Loose Coupling			•			•							21	Check Alignment & Tighten Coupling
Bearings	22	Motor Loose On Foundation Or Base Plate				•	•	•							22	Check Hold-down Bolts & Grout
	23	Noisy Ball Bearings				•									23	Check Lubrication & Replace Bearings
	24	Ball Bearings Installed Incorrectly			•						•		•	•	24	Replace
	25	Ball Bearings Damaged—Worn			•								•	•	25	Replace
Lubrication	26	Insufficient Oil					•				•		•	•	26	Add Proper Amount Of Oil
	27	Wrong Grade Of Oil									•		•	•	27	Use Recommended Lubricant
	28	Oil Contaminated									•		•	•	28	Drain Oil & Relubricate
	29	Oil Rings Not Rotating									•		•	•	29	Oil May Be Too Heavy, Check For Burrs
	30	Too Much Grease										•	•	•	30	Remove Excess Grease
	31	Insufficient Grease											•	•	31	Add Proper Amount Of Grease
	32	Wrong Grade Of Grease											•	•	32	Use Recommended Grease
	33	Grease Contaminated										•	•	•	33	Relubricate, Assure Grease Supply Is Clean

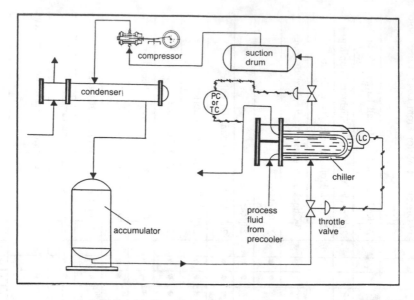

Figure 4-31. Simplified process refrigeration system.

(text continued from page 342)

2. Impurities in the refrigerant/composition:
   a. Formation of ice and hydrates.
   b. Insufficient circulation with speed-limited compressors.
   c. Violation of compressor minimum suction pressure.
3. Malfunction of throttle valve and accumulator:
   a. Throttle valve excessively open.
   b. Throttle valve excessively closed.
   c. Accumulator wrong liquid level.
4. Chiller:
   a. Low refrigerant level.
   b. Inadequate refrigerant feed.
   c. Low circulation on refrigerant side.
   d. Fouling on the process side.
   e. Seal-oil accumulation in refrigerant.
5. Condenser:
   a. Fouling due to low water velocities.
   b. Not enough air flow through air-cooled condensers.
   c. Recirculation of warm air through air fin coils.
   d. Poor refrigerant drainage from condenser.
   e. Excessive subcooling of refrigerant.

The foregoing concludes our discussion on the use of troubleshooting matrices.

Table 4-29
Troubleshooting Guide—Refrigeration System

	Symptoms					Symptoms			
	C Low Suction Pressure **B** High Discharge Pressure **A** Low Discharge Pressure					**High Suction Pressure** **D** **Process Side of Chiller** **E** **Temperature Too High**			
	Possible Causes	**#**	**A**	**B**	**C**	**D**	**E**	**#**	**Possible Remedies**
Compressor & Drivers	Turbine Driver Problem (HP Limiting)	1	1			4		1	* Consult steam turbine TSG
	Centrifugal Compressor Problem (Overload)	2	1			4		2	* See centrifugal compressor TSG
	Recip. Compressor Problem (Valves etc.)	3	1			4		3	* See TSG reciprocating compressor
	Wrong Composition For Speed Limited Compressors	4	2		3	5		4	* Adjust composition
	Wrong Composition For HP-Limited Compressors	5	2					5	* Adjust composition
Refrigerant	Wrong Composition	6	2	3	4		4	6	* Change composition
	Shortage Of Refrigerant	7	3				10	7	* Find & repair leak. * Recharge System
	Wrong Accumulator Level	8	4	4	6		9	8	* Adjust
EV†	Throttle Valve Plugged	9					8	9	* Change or repair on line
	Throttle Valve Out Of Adjustment	10	5	5	5	1	7	10	* Adjust superheat setting
Chiller	Insufficient Process Fluid In Chiller	11			1			11	* Check superheat at chiller outlet * Check for obstructions etc.
	Low Refrigerant Level	12				2	6	12	* Check LIC
	Inadequate Refrigerant Feed	13				3	5	13	* Check for restrictions in refrigerant flow
	Low Circulation Ratio On Refrigerant Side	14			2		2	14	* Investigate Cause
	Possible Causes	**#**	**A**	**B**	**C**	**D**	**E**	**#**	**Possible Remedies**

†EV = Throttle/Expansion valve

(table continued on next page)

Table 4-29 (cont.)

		Symptoms						Symptoms	
	C	Low Suction Pressure			High Suction Pressure				D
	B	High Discharge Pressure			Process Side of Chiller				E
	A	Low Discharge Pressure			Temperature Too High				
	Possible Causes	#	A	B	C	D	E	#	Possible Remedies
Chiller	Oil Accumulation	15					3	15	* Wash out with low temperature solvents
Chiller	Fouling On Process Side	16					1	16	* Clean
Chiller	Fouling On Water Side	17		6				17	* Check cooling water velocities * Backflush
Condenser	Vapors Leaking Out	18	6					18	* Find leak & recharge
Condenser	Not Enough Air Flow In Air Cooled Cond./Fouling	19	1					19	* Check for obstructions on condenser coils * Check for faulty fan operation * Ambient conditions too hot
Condenser	Improper Refrigerant Drainage From Cond.	20	2					20	* Check for obstructions
Condenser	Excessive Subcooling Of Refrigerant	21	7					21	* Check cooling water temperature
	Possible Causes	#	A	B	C	D	E	#	Possible Remedies

References

1. Bloch, A., *Murphy's Law and Other Reasons Why Things Go Wrong,* Price, Stern, Sloan, Los Angeles 1979, "Machinemanship" pp. 40-45, "Metalaws," p. 87.

2. De Bono, E., *Practical Thinking—4 ways to be right. 5 ways to be wrong. 5 ways to understand,* Penguin Books, 1981.

3. Sternberg, R.J. and Davidson, J.E., "The Mind of the Puzzler," *Psychology Today,* June 1982 pp. 37-44.

4. Hänny, J., "Interfaces—The Achievement of Engineering Excellence," *Sulzer Technical Review,* 3/1981, pp, 77-81.

5. Selye, H., *The Stress of Life*, McGraw-Hill Book Company, 1956, p. 308.

6. McKean, K., "Diagnosis by Computer—Caduceus is learning fast, has a prodigious memory and, some believe, a brilliant future as a diagnostician," *Discover*, September 1982, pp. 62-65.

7. Multilin Inc., "139 Series Motor Protection Relays," Multilin Inc., Markham, Ontario, Canada.

8. Vesser, W.C., "Predictive Maintenance of Heavy-Duty Engines Through Premalfunction Waveform Analysis," ASME Publication 74-Pet-34, p. 6.

9. Stengel, R., "Calling Dr. Sumex—The Diagnostician's New Colleague Needs No Coffee Breaks," *Time*, May 17, 1982, p. 73.

10. *Sawyer's Turbomachinery Maintenance Handbook*, First Edition, Turbomachinery International Publications, Norwalk, Conn., 1980, Volume I, p. 6/29.

11. Karassik, I.J., *Centrifugal Pump Clinic*, Marcel Dekker, Inc., New York, 1981, pp. 365-369.

12. Dufour, John W. and Nelson, William E., *Centrifugal Pump Sourcebook*, McGraw-Hill, New York, N.Y., 1993.

13. Worthington Pump Company, *Pump World*, Vol. 6., No. 2, 1980.

14. Bloch, H. P., "Root Cause Analysis of Seven Costly Pump Failures," Proceedings of 7th International Pump Users Symposium, Texas A&M University, 1990.

15. Makay, Elemer, and Barrett, James A., "Changes In Hydraulic Component Geometries Increase Power Plant Availability," Proceedings of 1st International Pump Users Symposium, Texas A&M University, 1984.

16. Goulds Pumps, Inc., "PRIME I Pump Reliability Improvement Seminar," 1987.

17. Exxon Chemical Americas, "Escoweld Epoxy Grouts," Sales Bulletin OFC-K81-1500, 1981.

18. Bloch, H. P. and Geitner, F. K., *Machinery Component Maintenance and Repair*, Second Edition, Gulf Publishing Company, Houston, Texas, 1990.

19. Piotrowski, John, "The Importance of Shaft Alignment—Answers to Pressing Questions," *P/PM Technology*, September/October, 1989.

20. Bloch, H. P., *Improving Machinery Reliability*, Second Edition, Gulf Publishing Company, Houston, Texas, 1988.

21. Cantley, R. E., "The Effect of Water in Lubricating Oil on Bearing Fatigue Life," ASLE Transactions, 20, (3), 1977, pp. 244–248.

22. Bloch, H. P., "Better Bearing Housing Seals Prevent Costly Machinery Failures," Proceedings of Industrial Energy Technology Conference, Houston, Texas, March 24 & 25, 1993.

23. CR Industries, Elgin, Illinois; Form 457466, CR Technical Report.

24. INPRO SEAL Company, Rock Island, Illinois; Technical Bulletin for RMS 700 Repulsion Magnetic Seal.

25. Bloch, H. P., "Reducing Pump Oil Degradation," *Hydrocarbon Processing*, December 1991, p. 25.

26. Bloch, H. P., *Oil Mist Lubrication Handbook*, Gulf Publishing Company, Houston, Texas, 1987.

27. Bloch, H. P., "Eliminating Cooling Water from General Purpose Pumps and Drivers," *Hydrocarbon Processing,* January 1977.
28. National Fire Protection Association, "Centrifugal Fire Pumps 1976," ANSI Z277.1—1976, pp. 20/108–109.
29. Sum, C., *A Practical Guide to Compressor Valves,* The IIC Group of Companies, 364 Supertest Rd., Downsview, (Toronto), Ontario, Canada, 1982, p. 8.
30. Blue, A. B., "Internal Combustion Engine Failures: A Statistical Analysis," ASME Publication 75-DGP-16, 1975, pp. 2–4.
31. *Handbook of Loss Prevention,* Allianz Versicherungs—AG, Springer Publishing, Berlin, 1978, pp. 137–140.

Chapter 5

Vibration Analysis

Machinery distress very often manifests itself in vibration or a change in vibration pattern. Vibration analysis is, therefore, a powerful diagnostic tool, and troubleshooting of major process machinery would be unthinkable without modern vibration analysis.

There are many ways that vibration data can be obtained and displayed for detecting and identifying specific problems in rotating machinery. Some of the more common techniques include:

1. Amplitude versus frequency.
2. Amplitude versus time.
3. Amplitude versus frequency versus time.
4. Time waveform.
5. Lissajous patterns (orbits).
6. Amplitude and phase versus rpm.
7. Phase (relative motion) analysis.
8. Mode shape determination.

By far the most important of these techniques involves the capture of amplitude versus frequency data. Dozens of different data collectors and/or acquisition modules are available for this task and manufacturers or vendors are almost always in a position to give advice on such issues as sensor type, placement, etc. The user's interest often gravitates toward data interpretation, a topic of prime importance to the failure analyst and serious troubleshooter.

While this is where we wish to place our main emphasis in this chapter, the reader should keep in mind that it is helpful and advisable to consider the machine and the circumstances which led to a requirement for vibration analysis.

Machine history. When the complaint is an excessive increase in vibration, much can be learned by reviewing the history of the machine. For example, if components such as couplings, sheaves, or bearings have been replaced, there is a good possibility that balance or alignment may be affected. Additional machines added to a structure can easily change the natural frequency of the structure, causing resonance. Changes in normal operating conditions of speed, load, temperature, or pressure may produce significant changes in machine vibration from unbalance, cavitation, aerodynamic/hydraulic forces, etc. Improper grounding when arc welding on machines can cause extensive damage to bearings and couplings. In any case, when the vibration of a machine increases, it is generally caused by wear or deterioration in the machine's mechanical condition or by changes which have been made to the machine or structure. In many cases, considerable analysis time and effort can be saved by reviewing the history of the machine to see if such changes might be associated with the vibration increase.

Machine characteristics. A review of machine characteristics such as rpm, type of bearings, gear frequencies, aerodynamic/hydraulic frequencies, etc. can help establish the expected vibration frequencies. These will help determine the type of instrumentation and vibration transducers needed for the analysis.

The operating characteristics of the machine may also dictate the type of analysis equipment required. For example, production machine tools with very short duty cycles may require a real-time spectrum analyzer because of the limited analysis time available. Machines which operate with continuously variable speed or load conditions may require real-time analysis or special instrumentation with tracking-filter capability. Machines with extremely complex, random, or transient vibration may require the spectrum averaging or "transient-capture" features of a real-time analyzer.

To summarize, a thorough evaluation of the machine and its history leading to the requirement for analysis can greatly simplify the analysis procedure by specifying the appropriate instrumentation and analysis data. Listing the faults typical of the machine, combined with past experience on this or similar installations, can suggest likely trouble areas. Discussion with the machinery manufacturer might also provide clues to common problems which have been encountered.

Making the Comparison

After the vibration data have been obtained, the next step is to compare the readings of significance with the characteristics of vibration typical of various types of trouble. The key to this comparison is frequency. A frequency comparison is made on the basis of the rotating speed or speeds of the machine parts. Figure 5-1 lists the frequencies of parts in terms of rpm and possible cause of the vibration. The trouble referred to will be associated with the part whose rpm is some multiple of the vibration frequency, and this comparison should identify the part causing the trouble.

The vibration identification chart in Figure 5-2 lists the most common causes of vibration likely to be encountered, together with the amplitude, frequency, and strobe-picture characteristics for each cause. The "remarks" column provides helpful information about any peculiar characteristics that will help pinpoint the trouble.

Since many machine troubles have similar characteristics, and since several troubles may be present in a machine simultaneously, it is often necessary to choose between several likely possibilities. One useful technique is to examine the relative probability of the occurrence of machine faults and start with the most probable fault first. The four-part vibration and noise identification chart in Figure 5-3 originally proposed by turbomachinery consultant John S. Sohre in 1968 at the ASME Petroleum Mechanical Engineering Conference in Dallas, Texas, will assist in this task. Subsequent refinements and updates have made this chart one of the most comprehensive listings of machinery problems commonly encountered. Chart segments 5-3 A and B provide a relative-probability-rating percentage system that indicates which trouble is most likely in a given set of circumstances. Where no probability number appears, this indicates that the specific fault will not exhibit the amplitude or frequency characteristics shown.

For example, looking at the vibration characteristics of unbalance, it can be seen that the number 90 appears in the predominant frequency column of 1 × rpm. This

VIBRATION FREQUENCIES AND THE LIKELY CAUSES

Frequency In Terms Of RPM	Most Likely Causes	Other Possible Causes & Remarks
1 x RPM	Unbalance	1) Eccentric journals, gears or pulleys 2) Misalignment or bent shaft — if high axial vibration 3) Bad belts if RPM of belt 4) Resonance 5) Reciprocating forces 6) Electrical problems
2 x RPM	Mechanical Looseness	1) Misalignment if high axial vibration 2) Reciprocating forces 3) Resonance 4) Bad belts if 2 x RPM of belt
3 x RPM	Misalignment	Usually a combination of misalignment and excessive axial clearances (looseness).
Less than 1 x RPM	Oil Whirl (Less than ½ RPM)	1) Bad drive belts 2) Background vibration 3) Sub-harmonic resonance 4) "Beat" Vibration
Synchronous (A.C. Line Frequency)	Electrical Problems	Common electrical problems include broken rotor bars, eccentric rotor, unbalanced phases in poly-phase systems, unequal air gap.
2 x Synch. Frequency	Torque Pulses	Rare as a problem unless resonance is excited
Many Times RPM (Harmonically Related Freq.)	Bad Gears Aerodynamic Forces Hydraulic Forces Mechanical Looseness Reciprocating Forces	Gear teeth times RPM of bad gear Number of fan blades times RPM Number of impeller vanes times RPM May occur at 2, 3, 4 and sometimes higher harmonics if severe looseness
High Frequency (Not Harmonically Related)	Bad Anti-Friction Bearings	1) Bearing vibration may be unsteady — amplitude and frequency 2) Cavitation, recirculation and flow turbulence cause random, high frequency vibration 3) Improper lubrication of journal bearings (Friction excited vibration) 4) Rubbing

Figure 5-1. This chart lists the vibration frequencies normally encountered and the most likely causes for each frequency.

indicates that the vibration frequency of unbalance will be 1 × rpm 90% of the time. In the predominant amplitude column, it can be seen that the predominant amplitude of unbalance will be in the horizontal direction 50% of the time, in the vertical direction 40% of the time, and in the axial direction 10% of the time. Further, the probable location of unbalance will be the shaft or rotor 90% of the time and caused by the bearings only 10% of the time, perhaps as the result of eccentricity of rolling-element bearings such as ball bearings or roller bearings.

Operation and installation-related vibration problems are tabulated in chart segments 5-3C and D. Here, one check mark indicates that the problem is likely to occur, and two check marks identify problems that occur frequently.

As can be noted from the vibration and noise identification charts, there are many mechanical and electrical problems which can generate very similar vibration characteristics, and a particular mechanical problem may develop different vibration characteristics under varying circumstances. The following subheadings discuss the common mechanical problems and will supplement the information available from the identification charts.

(text continued on page 363)

VIBRATION IDENTIFICATION

CAUSE	AMPLITUDE	FREQUENCY	PHASE	REMARKS
Unbalance	Proportional to unbalance. Largest in radial direction.	1 x RPM	Single reference mark.	Most common cause of vibration.
Misalignment couplings or bearings and bent shaft.	Large in axial direction 50% or more of radial vibration	1 x RPM usual 2 & 3 x RPM sometimes	Single double or triple	Best found by appearance of large axial vibration. Use dial indicators or other method for positive diagnosis. If sleeve bearing machine and no coupling misalignment balance the rotor.
Bad bearings anti-friction type	Unsteady - use velocity measurement if possible	Very high several times RPM	Erratic	Bearing responsible most likely the one nearest point of largest high-frequency vibration.
Eccentric journals	Usually not large	1 x RPM	Single mark	If on gears largest vibration in line with gear centers. If on motor or generator vibration disappears when power is turned off. If on pump or blower attempt to balance.
Bad gears or gear noise	Low - use velocity measure if possible	Very high gear teeth times RPM	Erratic	
Mechanical looseness		2 x RPM	Two reference marks. Slightly erratic.	Usually accompanied by unbalance and/or misalignment.
Bad drive belts	Erratic or pulsing	1, 2, 3 & 4 x RPM of belts	One or two depending on frequency. Usually unsteady.	Strob light best tool to freeze faulty belt.
Electrical	Disappears when power is turned off.	1 x RPM or 1 or 2 x synchronous frequency.	Single or rotating double mark.	If vibration amplitude drops off instantly when power is turned off cause is electrical.
Aerodynamic hydraulic forces		1 x RPM or number of blades on fan or impeller x RPM		Rare as a cause of trouble except in cases of resonance.
Reciprocating forces		1, 2 & higher orders x RPM		Inherent in reciprocating machines can only be reduced by design changes or isolation.

Figure 5-2. Vibration identification chart.

SYMPTOMS AND DISTRESS MANIFESTATIONS

CHART NO. 1 ← — VIBRATION ANALYSIS —

CAUSES OF VIBRATION	Rotor or Stator resonant frequencies	Still whirl frequency	40-50%	50-100%	1 x running frequ.	2 x RF	Higher multiples	1/2 RF	1/4 RF	Lower multiples	Odd frequ.	Very high frequ.	Line frequ.	Multiples of L.F.	Slip frequ.	Vert.	Horiz.	Axial	Shaft	Brgs.	Casing	Foundation	Piping	Coupl.
Predominant frequencies →													**Electrical**			**Direction and Location of predominant amplitude**								
Initial unbalance	5				90	5	5									40	50	10	90	10				
permanent bow or lost rotor parts (vanes)	30				90	5	5																	
Temporary Rotor bow	20				90	5	5																	
Casing distortion — Temp.			←10→		60	20	10																	
Casing distortion — Perm.			←10→		60	20	10																	
Foundation distortion			20		40	30					10					▼	▼	▼	30	40		→10		
Seal rub	10	10	10		20	10	10		10		10	10				30	40	30	60	20	10			10
Rotor rub, axial			←20→		30	10	10		10		10	10				30	40	30	70	10	20			
Misalignment	5	10			30	60	10					10				20	30	50	30	10	10		50	
Piping forces	5	10			30	60	10					10				20	30	50	30	10	10		50	
Journal & bearing eccentricity	60				40	60										40	50	10	90	10				
Bearing damage	20				40	20						20				30	40	30	70	20	10			
Bearg. & Support excited vibration (oil whirls, etc.)	20	80						10	10							40	50	10	50	20	20	10		
Unequal bearing stiffness, horizontal vs. vertical	80					80	20									40	50	10	40	30	30			
Thrust brg. damage	90←											10				20	30	50	30	20	20			30
Insufficient tightness in assy. of: *(PREDOMINANT FREQUENCY WILL SHOW AT LOWEST ORITIONAL OR REDONANT FREQUENCY)*																								
Rotor (shrink-fit)	40	40	10		10	10	10				10					40	50	10	60	20	10			10
Brg. liner	90→										10					40	50	10	70	10	10			10
Brg. case	90→				30						10					40	50	10	40	20	10			30
Casing & support	50→				30						50					40	50	10	30	20	30			20
Gear inaccuracy or damage						20				Gear-mesh freq. 20→60						20	20	60	20	20	30			30
Coupling inaccuracy or damage	10	20	10		20	30	10			Gear mesh freq. →80						30	40	30	70	20				30
Aerodynamic excitation	80	20			20			10	10							30	40	20	20	20	10	10		
Rotor & brg. system critical	100				100											40	50	10	70	30				
Coupling critical	100				100	ALSO MAKE SURE TOOTH FIT IS TIGHT!										20	40	40	10	10				80
Overhang critical	100				100											40	50	10	70	10				20
Structural resonance of:																								
Casing	100	10			70	10		10					30	30	30	40	50	10		40	40	10	10	
Supports	100	10			70	10		10					30	30	30	40	50	10		20	50	20	10	
Foundation	100	20			60	10		10					30	30	30	30	40	30		10	40	40	10	
Pressure pulsations	80	MOST TROUBLESOME IF COMBINED WITH RESONANCES														30	40	30	CAN EXCITE WHIRLS OR RESONANCE 30	30	40			
Electrically excited vibrations	80												←100→			30	40	30						
Vibration transmission	30	30										40				30	40	10	▼	▼	40	40	20	
Oil-Seal-induced vibr.	30	70							▼							50	50	100 Very severe						
NAME OF PROBLEM *(THE SECTION BELOW IS MEANT TO IDENTIFY BASIC MECHANISMS)*													POSSIBLY ELECTRIC ORIGIN											
Sub-harmonic resonance	100							←100→								30	30	40 IF BRG.-EXCIT 20 / 20		20	20	20		
Harmonic resonance	100						←100→							100		40	40	20	10	10	30	30		
Friction induced whirl	100	30	10													40	50	10	80 Very severe	20				
Critical speed	100				100											40	50	10	60	40				
Resonant vibration	100	20			60	20	5	10	5							40	40	20	20	10	20	30	20	
Oil whirl		80						10	5	WATCH FOR AERODYNAMIC ROTOR-LIFT (PARTIAL ADMISSION ETC.)						40	50	10	80	20				
Resonant whirl	100	100														40	50	10	20	20	20	20		
Dry whirl	100											100				30	40	30	40	20	10			10
Clearance induced vibrations	50	50			20	30	10									DEPENDS ON NATURE & LOCATION OF GAP								
Torsional resonance	100	10			40	20	20				Gear speeds Mesh-frequ.		←100→ IF ELECTRICALLY EXCITED ←100→			TORSIONAL			TORSIO NAL 40 / 40 LATERAL AMPLITUDE					10
Transient torsional	100				50							50				TORSIONAL MOTION 100	40	40						10

Numbers indicate % of cases showing above symptoms for causes listed in vertical column at left.

Figure 5-3. Vibration and noise identification chart.

(Sawyer's Turbomachinery Maintenance Handbook 1980 Edition)

◄─── OPERATIONAL EVIDENCE ───

Amplitude Response to Speed Variation during vibration-test runs											Effect of Operating Conditions										Effect of Oil p & t and Flow							
Coming up						Slowing down					No effect	Comes in or Goes out at:						Start up only	After load dump or surge	When other machine starts or shuts down	Increase p		Decrease p		Increase t		Decrease t	
Stays Same	Increases	Decreases	Peaks	Comes suddenly	Drops out suddenly	Stays same	In-creases	De-creases	Comes suddenly	Drops out suddenly		Full load in	out	No load in	out	Part load in	out				better	worse	better	worse	better	worse	better	worse
	100		PEAK AT					100			100 (LOAD-INDUCED WHIRL 100%)										90	10	10	90	10	90	90	10
	100							100			100										Slight changes only, due to variation of oil-film damping							
30	60	5		5		30	5	50	5	10		10	20	20	10			20	20									
30	50	5		5	10	30	5	50	5	10		10	20	20	10			20	20									
40	60							40			40	10	10	10	10			10	10									
20	80		(PEAKS AT CRITICALS)					20			40	10	10	10	10			10	10									
10	70			10	10	10		70	10	10	10	10	10	5	10			20	20	5								
10	40	10		20	20	10		50	20	20	5	30	10	5	10			10	30									
20	30	10		20	20	20		40	20	20	10	30	10	5	5	5	5	20	10									
20	40			20	20	20		40	20	20	10	20	10					20	20	20								
40	50	10				40	10	50			100																	
10	50	10	▼	20	10	10	10	50	10	20	90																	
	10			90				10		90	30	5	5	5	5	20	20	10 (MOSTLY MULTI-BEARING-SOLID COUPLED)	10						90	10	10	90
	40			50	10			40		10	100																	
20	50	10		10	10	20	10	50	10	10		30	5	5	30				30									
				90	10				10	90	30	30	5	5	10			30	10	20								
				90	10				10	90	90							10										
				90	10				10°	90	90							10										
				90	10				10	90	10	5	5	5	5	30	30	10										
20	20	20	20	10	10	20	20	20	10	10	30	10	10	20	10	10	10	(LOOSE SLEEVE, FRICTION, OR DIRT IN TEETH)										
10	20		20	40	10	10		20	10	40		10	10	30	10	10	10	10	20 (80% IF TOOTH FRICTION OR DIRT IN TEETH)									
20	20	20		30	10	20	20	20	10	30	0	60	40	20	80	80	20	50	30	20								
20		80						20			90																	
20		80	(LOOSENESS)					20	50% IF LOOSE		90	IF LOOSE	IF LOOSE					IF LOOSE	IF LOOSE									
30		70						30			90																	
20		80						20			90										NO EFFECT							
20		80						20			90																	
20		80						20			90																	
90		10%—DEPENDING ON ORIGIN OF DISTURBANCE				90	10	→			20	5	5	5	5	20	20			20								
90				▼		90	▼				50	40								10								
90						90	▼				50									50								
0	30	0	0	70			0	50		50	80	0	0	20					30		PRESSURE-DROP ACROSS SEAL HAS PRONOUNCED EFFECT							
20			20	30	30			20	30	30	40							20	20									20
20	20			60			20		20			80	10							10								
				90	10				10	90	40	20	40		60			10	30		90	10	10	90	90	10	10	90
20			80					20			100										10 (SMALL DAMPING EFFECTS ONLY)		10			10	10	
20			80					20			100																	
				100						100	60					20	20				90	10	10	90	90	10	10	90
				80	20				20	80	50	50	————→							50	90	10	10	90	90	10	10	90
				80	20	80				20	90	10					10				90						50	50
				80	20	20			20	60	40					20	20				SMALL DAMPING EFFECTS ONLY							
20		30	30	20	20			20	30	30	50				10	10												
		50	30	20				30	20									100										

Figure 5-3. Cont'd

SYMPTOMS AND DISTRESS MANIFESTATIONS CONT'D.

These charts are reproduced with the permission of Turbomachinery International Publications, Division of Business Journals, Inc., 22 South Smith Street, Norwalk, Connecticut, U.S.A. 06855, and may not be reproduced without its prior written permission.

CHART NO. 2 — CAUSES OF VIBRATION	Lo frequ. "Rumble"	Loud "Roar"	"Hum"	Comes and goes periodically ("beat")	High pitched "whine"	Very high loud "scream"	Very high "squeal"	Ultrasonic (measure w. sound analyzer)	Initial startup	1st year	1-10 years	Over 10 years
Predominant Sound (DURING VIBRATION TEST RUN) →									**History of Machine** →			
Initial unbalance		80	20						✓✓			
Permanent bow or lost rotor parts (vanes)		80	20						✓		✓	✓
Temporary Rotor bow	10	70	10	10					✓✓	✓		
Casing distortion	10	70	10	10							✓	✓
Foundation distortion	10	50	30	10					✓	✓	✓	✓✓
Seal rub	20	40		10		10	10	10	✓	✓	✓	✓
Rotor rub axial	30	40		20				10	✓✓	✓		
Misalignment		40	40	20					✓	✓		
Piping forces	30	40	30						✓✓	✓	✓	✓✓
Journal & bearing eccentricity		10	90						✓✓			
Bearing damage	20	40	10		10	10	10		✓	✓	✓	✓
Bearg. & Support excited vibration (oil whirls. etc.)	60	10		30					✓✓	✓		
Unequal bearing stiffness. horizontal--vertical		10	60	20	10				✓✓			
Thrust brg. damage	80	10	10						✓✓	✓	✓	✓
Insufficient tightness in assy. of:												
Rotor (shrink-fits)	60	20		20					✓	✓✓	✓	✓
Brg. liner	60	20		20					✓✓	✓		
Brg. case	60	10		30					✓✓	✓		
Casing support	60	20		20					✓✓	✓	✓	✓
Gear inaccuracy or damage	20	10	10	20	20	10		10	✓✓			
Coupling inaccuracy or damage	10		40	10	40	ALSO "METALLIC" SOUND			✓✓	✓		
Aerodynamic excitation	60			60					✓✓		✓	✓✓
Rotor & brg. system critical		50	30	20					✓✓	✓		
Coupling critical			20	40	20	20			✓✓	✓		
Overhang critical		50	40	10					✓✓	✓		
Structural resonance of:												
Casing	20		20	60					✓✓	✓		
Supports	20		20	60					✓✓	✓	✓	
Foundation		10	80	10					✓	✓	✓	
Pressure pulsations	40	20		40					✓	✓✓	✓	✓
Electrically excited vibrations		10	80	10					✓✓	✓		
Vibration transmission	20		20	60					✓			
Oil-Seal-induced vibr.	90 (il whirl)	40								✓	✓✓	✓✓
NAME OF PROBLEM												
Sub-harmonic resonance	80			20					✓✓	✓	✓	
Harmonic resonance			40	20	40				✓✓	✓		
Friction induced whirl	60	20		20					✓✓	✓✓	✓	
Critical speed		50	30	20					✓✓			
Resonant vibration	40		30	30					✓✓	✓		
Oil whirl	60	10		30					✓✓	✓	✓	
Resonant whirl	60	20		20					✓✓	✓	✓	
Dry whirl							20	80	✓	✓	✓	✓
Clearance induced vibrations	60	20		20					✓✓	✓		
Torsional resonance	10	20	20	30	10	10			✓✓	✓		
Transient torsional				60 (cough)		20	20		✓✓	✓		

Figures indicate the likelihood in % that the symptoms (top, horizontal) will occur where causes (left, vertical) are involved.

Figure 5-3. Cont'd

(Sawyer's Turbomachinery Maintenance Handbook 1980 Edition)

OPERATIONAL EVIDENCE, CONT'D

Damage or Distress Signal

LOOK FOR LOST BLADES & CRACKED DISKS, ETC. — DUE TO CORROSION FATIGUE OR STRESS CORROSION

Seals rubbed	Shaft bent	Thrust Bearing damage	Wiped	Fatigued	Babbitt squeezed out	Case distorted or cracked	Out of align-ment	Coupling burned or pitted	Gear teeth broken or pitted	Gear teeth marked on back side	Shaft cracked or broken	Galling or fretting marks under disks or hubs	Coupling bolts loose	Foundat. settled or cracked	Soleplates loose or rusted	Sliding surfaces binding	Thermal expansion restriced	Fluid marks on internals	Rotor components eroded	solids accumulated on vanes & rotor	Salt deposits on internals	Main flanges leaking	Seals leaking
10			10	10																			
50	+		30	10	10														20	20			10
90	10		50		20			20	20	20	5		5										20
90	10		50	10	10	+	10	20								30	30					50	20
90	10		50	30			50	30							30	30	10					20	10
+	15		15	10	10	10	20	10							10	10	10	10		10	10		50
90	30	30	50					20		10			10			20	40						50
50	30		10	20		20	+	40					10	50	50	10	50						10
50	10		10	20		40	40	40					10		10	20	50					50	10
			30	60																			
90	10		+	+		10	40	20	20	10	10	20	20	20	20	20	20	40	10	10	10	10	10
90	10		80	60	30			20	20	10	10	10											10
10			40	40																			
90	20	+	50		30						10		10					60	30	60	60		50
90	10		60	60	40			80	40	40	20	50	10					40	10				30
90	10		30	30	40			80	20	20	10	5	10										30
50	5		20	20	30			20	20	20	10	5	10										10
20			40	60	10			20															10
		40	20	20					40	20	10												
	60		20	20				40															
																							10
10		50	20	30				30	20	20	5	20	10	10	30				40	40	20		20
50	10		50	50				30	20	10	5												20
30	10		50	50				30															
50	20		60	60	10			40			5												25
20			20	20																			10
20			20	20				20						30	20	10							10
20			20	20			30	30						30	10	10	20						10
30	30		30	30			20	30						30	20								15
			20	20				30	10	10	5		5										
20			20	20			30	30															10
90	50		90		10	20		50			10		20	40	40								90
10			20	10				10	30	10	5		5	10	10'								
10			20	20				20	30	10	5		5	10	10								
90	10		60	60	40			80	40	40	20	50	10					40	10				
50	10		50	50				30	20	10	5												
20			20	20				20						30	10	10	20						
60	10		80	60	30			20	20	10	10	10											
80	15		90	80	40			40	30	20	20	20	10	10									
80	10		90	20	40			10	5	5	30	5	70										
90	10		30	30	20			80	20	20	10	5	10										
5			5	20				40	80	80	40	5	40										
5			5	5				30	80	80	50	10	50										

1968 John S. Sohre, Turbomachinery Consultant. Revised June 1980. One Lakeview Circle Ware, MA 01082 USA

Figure 5-3. Cont'd

POSSIBLE CAUSES OF DIFFICULTY

CONDITIONS LIKELY TO BE RESPONSIBLE FOR DIFFICULTIES

CHART NO. 3 — CAUSES OF VIBRATION	Started too quickly	Loaded too quickly	Sudden temp. variation (8 F/min. = max.)	Sudden press. variation	Fluid slugging	Carry over	Boiler priming	Chemicals in steam	Case filled with fluid	Fluid sucking up from drains	Trapped fluid flashing up	Dirt in oil	Drains not closed after warmup	Drains closed too early	Sealing steam left on standg. rotor	Purge gas left on standg. rotor	Hot leakage getting on standing rotor
Unbalance	✓	✓	✓	✓	✓	✓	✓	✓	✓								
Permanent bow or lost rotor parts (vanes) *(LOOK FOR LOST BLADES ETC. DUE TO CORROSION FATIGUE OR STRESS CORROSION)*	✓	✓	✓		✓	✓✓	✓✓	✓✓	✓	✓	✓				✓	✓	✓
Temporary Rotor bow	✓	✓✓	✓✓	✓	✓✓	✓	✓		✓✓	✓✓	✓✓	✓		✓	✓✓	✓✓	✓✓
Casing distortion	✓✓	✓✓	✓✓	✓	✓✓	✓	✓		✓✓	✓	✓		✓	✓			
Foundation distortion																	
Seal rub	✓	✓✓	✓✓	✓	✓✓	✓			✓	✓✓	✓	✓	✓	✓	✓	✓	✓
Rotor rub. axial	✓	✓✓	✓✓	✓	✓✓												
Misalignment	✓	✓	✓	✓	✓				✓✓								
Piping forces				✓													
Journal & bearing eccentricity																	
Bearing damage	✓✓	✓✓	✓✓	✓	✓✓							✓✓					
Bearg. & Support excited vibration (oil whirls. etc.)	✓	✓	✓	✓	✓												
Unequal bearing stiffness. horizontal—vertical																	
Thrust brg. damage	✓	✓		✓✓	✓✓	✓	✓					✓	✓✓				
Insufficient tightness in assy. of: Rotor (shrink-fits)	✓	✓	✓✓	✓✓	✓✓	✓✓	✓		✓	✓✓	✓✓			✓		✓	
Insufficient tightness in assy. of: Brg. liner																	
Insufficient tightness in assy. of: Brg. case																	
Insufficient tightness in assy. of: Casing & support	✓	✓	✓		✓	✓			✓								
Gear inaccuracy or damage					✓							✓✓					
Coupling inaccuracy or damage					✓✓	✓✓						✓✓		✓			
Aerodynamic excitation / Rotor & brg. system critical													✓	✓			
Coupling critical																	
Overhang critical																	
Structural resonance: Casing																	
Structural resonance: Supports																	
Structural resonance: Foundation																	
Pressure pulsations									✓				✓				
Electrically excited vibrations																	
Vibration transmission																	
Oil-Seal-induced vibr.												✓✓					
NAME OF PROBLEM																	
Sub-harmonic resonance																	
Harmonic resonance																	
Friction induced whirl	✓	✓	✓	✓								✓✓					
Critical speed																	
Resonant vibration																	
Oil whirl				✓													
Resonant whirl																	
Dry whirl	✓	✓	✓														
Clearance induced vibrations	✓	✓	✓	✓													
Torsional resonance																	
Transient torsional	✓✓	✓✓															

✓ LIKELY TO OCCUR   ✓✓ FREQUENT CAUSES OF DIFFICULTY

Figure 5-3. Cont'd

(Sawyer's Turbomachinery Maintenance Handbook 1980 Edition)

														INSTALLATION					
														FOUNDATION					
Cooling air on hot standing rotor	Exhaust was overheated	Excessive vacuum	Not operated at design speed	Over loaded	Off-design p & t	Surging	Shaft currents	In-correct leakoff pressure	Fluctuat. leakoff pressure	Hollow rotor filling w. fluid (Or bore-scope hole)	Syncronized out of phase	Had short circuit	Phase fault	Reso-nant	Settling or shrinking	Unevenly heated (hot lines too close)	Not separated from building	Grout swelling or shrinking rust or deteri-oration under soleplates	Cross-Beams Key-Bars. or horizontal stiffness insufficient
			√	√	√	√				√	√	√	√						
√	√	LOOK FOR LOST BLADES √√ √ √√ √√ → DUE TO EXCESSIVE BLADE LOAD								√	√	√					√	√
√√	√					√		√		√	√	√	√					√	√
	√√				√											√	√		√
										√√	√√	√		√	√	√	√√	√√
√	√					√√		√	√	√	√		√	√√	√√	√	√√	√√
						√√	√						√	√	√	√	√
	√√				√√	√				√√	√√	√	√√	√√	√	√√	√√
				√	√√								√√	√√	√	√√
	√		√		√	√	√√		√		√	√	√	√√	√√	√√	√
		√	√	√				√	√√	√	√	√	√√	√√	√√	√√
														√√	√√
√		√	√	√	√√	√√	√√	√	√						
√	√		√	√√	√√	√			√	√	√
					√					√	√
√√				√√	√				√√	√√	√	√	√√	√√
					√				√√	√√	√	
	√	√√		√√	√√				√√	√√	√	√	√	√	√√	√√
√	√√	√√	√√	√√		√√	√√						√√	√√	√√
												√	√
												√√		√	√
											√√	√√	√√
												√√	√√
			√	√		√	√				√	√	√	√√		√√
												√√		√√	√√
			√√√		√								
											√√		√√	√√	√√
											√√		√√	√√	√√
													√√	√√
													√√	√√
											√√		√√	√√	√√
													√√	√√
											√√		√√	√√	√√
											√		√ '	√√	√√
		√√	√	√	√					√√	√√	√√'			√	√
		√	√	√	√					√			

' 1968 John S. Sohre, Turbomachinery Consultant
Revised June 1980. One Lakeview Circle, Ware, MA 01082 USA

Figure 5-3. Cont'd

POSSIBLE CAUSES OF DIFFICULTY, CONTINUED

CHART NO. 4 — CAUSES OF VIBRATION	Soleplates loose	Rigidity insufficient	Excessive forces & moments	Expansion joints not properly installed	Piping not properly supported	Not properly sloped & drained	Resonant	Not taking off at top of headers	Water accumulates over valves	Casing drains run to common sewer or common header or into water	Traps not working	Drain pots under-size	Drain lines under-size	Pipe expansion restricted by contact with foundation or other pipes	Branch lines restricting expansion	Dead ends not drained
	FOUNDAT. CONT'D.		INSTALLATION — PIPING													
Unbalance																
Permanent bow or lost rotor parts (vanes)						✓✓		✓✓	✓✓	✓	✓	✓	✓	✓	✓	✓
Temporary Rotor bow						✓✓		✓✓	✓✓	✓✓	✓✓	✓✓	✓✓	✓	/	✓✓
Casing distortion		✓✓	✓✓	✓	✓✓	✓		✓	✓	✓	✓✓	✓	✓✓	✓✓	✓✓	✓
Foundation distortion		✓✓	✓	✓	✓									✓	✓	
Seal rub	✓	✓✓	✓✓	✓✓	✓✓	✓✓		✓✓	✓✓	✓✓	✓✓	✓✓	✓✓	✓✓	✓✓	✓✓
Rotor rub, axial		✓	✓	✓	✓	✓✓		✓✓	✓✓	✓	✓	✓✓	✓✓	✓		✓✓
Misalignment	✓		✓✓	✓✓	✓✓									✓✓	✓✓	
Piping forces		✓✓	+	✓✓	✓✓									✓✓	✓✓	
Journal & bearing eccentricity																
Bearing damage	✓	✓✓	✓	✓	✓	✓✓	✓	✓✓	✓✓	✓	✓	✓	✓	✓	/	✓
Bearg. & Support excited vibration (oil whirls, etc.)	✓✓	✓✓	✓	✓	✓		✓							✓✓	✓	
Unequal bearing stiffness. horizontal-vertical	✓	✓														
Thrust brg. damage						✓✓		✓✓	✓✓	✓	✓✓	✓✓	✓✓			✓✓
Insufficient tightness in assy. of																
Rotor (shrink-fits)						✓✓		✓✓	✓✓	✓✓	✓✓	✓✓	✓✓			✓✓
Brg. liner			✓	✓	✓									✓		
Brg. case			✓	✓	✓									✓	✓	
Casing support			✓✓	✓✓	✓✓									✓✓	/	
Gear inaccuracy or damage																
Coupling inaccuracy or damage		✓	✓✓													
Aerodynamic excitation				✓✓		✓✓	✓✓			✓✓	✓✓	✓✓	✓✓			✓✓
Rotor & brg. system critical	✓	✓✓														
Coupling critical																
Overhang critical																
Structural resonance of:																
Casing				✓	✓✓		✓✓							✓✓	✓✓	
Supports	✓✓			✓✓	✓✓		✓✓							✓✓	✓✓	
Foundation	✓✓	✓		✓	✓✓	✓✓	✓✓							✓✓	✓	
Pressure pulsations				✓✓			✓									
Electrically excited vibrations							✓✓							✓		
Vibration transmission		✓		✓✓	✓✓		/ /							✓✓	✓	
Oil-Seal-induced vibr.																
NAME OF PROBLEM																
Sub-harmonic resonance	✓✓	✓✓		✓	✓		✓							✓	✓	
Harmonic resonance		✓		✓	✓		✓ .							✓	✓	
Friction induced whirl		✓				✓✓		✓✓	✓✓	✓✓	✓✓	✓✓	✓✓			
Critical speed		✓	✓	✓	✓											
Resonant vibration	✓✓	✓✓		✓✓	✓✓		✓✓							✓✓		
Oil whirl																
Resonant whirl	✓✓	✓✓		✓✓	✓✓		✓✓							✓✓		
Dry whirl			✓													
Clearance induced vibrations	✓✓		✓✓	✓	✓✓		✓							✓	✓	
Torsional resonance																
Transient torsional																

Figure 5-3. Cont'd

(Sawyer's Turbomachinery Maintenance Handbook 1980 Edition)

ELECTRICAL				
Short-circuit or synchronized out of phase. Phase fault.	Reverse current relay failed	Synchronous motor starting pulsations excessive for system	Starting cycle improperly timed	COMMENTS
✓✓		CHECK FOR TORSION CRACKS		Long, high-speed rotors often require field-balancing at full speed to make adjustments for rotor deflection and final support conditions. Make corrections at balancing ring or coupling.
✓✓		✓✓	✓	Bent rotors can sometimes be straightened by "hot-spot" procedure, but this should only be regarded as a temporary solution, because bow will come back in time, and rotor failures have resulted from this practice. If buckets, vanes, or disks are lost or cracked, check for corrosion fatigue, stress-corrosion, resonance and off-design operation.
✓	✓	✓	✓	Straighten bow slowly, running on turning gear or at low speed. If rubbing occurs, trip unit immediately and keep rotor turning 90° by shaft-wrench every 5 min. until rub clears. Then resume slow run. This may take 24 hours.
✓	✓			Often require annealing of new case, but sometimes a mild distortion corrects itself with time (requires periodic internal and external re-alignment). Usually caused by excessive piping forces, thermal shock, unsuitable casing design, wrong material, welding, improper stress-relief.
✓✓				Caused by poor mat under foundation, or thermal stress (hot-spots), or unequal shrinkage. Usually requires extensive and costly repairs. Structural-steel baseplates warp from thermal strain and lack of stress relief.
✓	✓	✓	✓	Slight rubs may clear, but trip unit immediately if a high-speed rub gets worse. Turn until clear.
	✓	✓	✓	Unless thrust bearing has failed this is caused by rapid changes of load and temperature. Machine should be opened and inspected.
✓✓				Caused by excessive pipe strain and/or inadequate mounting and foundation, but sometimes also by local heat from pipes or sun on base and foundation. Some compressors and turbines have no provision to control thermal expansion (keys, flexplates, sliding feet).
				Most trouble is caused by poor pipe supports (should use spring hangers), improperly used expansion joints, and poor pipe line-up at casing connections. Foundation settling can cause severe strain.
				Bearings may get distorted from heat. Make hot-check. If possible, checking contact. Check bearing case for excessive heat and distortion. Provide for thermal expansion and install heat shields.
✓✓	✓	✓	✓	Watch for brown discoloration which often precedes recurring failures. This indicates very high local oil-film temperatures. Check rotor for vibration. Check bearing design and hot clearances, possible dilution of oil with process fluid (gasoline), gas (chlorine). Check viscosity and contamination. Mount pressure gage on bearing cap.
✓✓				Check clearances and roundness of journal, as well as contact and tight bearing fit in case. Watch out for vibration transmission from other sources and check frequency. May require tilt-shoe bearings. Check especially for piping or foundation resonance at whirl frequency.
✓				Can excite resonances and criticals and combinations thereof at 2x running frequency. Usually difficult to field-balance because as horizontal vibration gets better, vertical gets worse and vice versa. It may be necessary to increase horizontal bearing support stiffness (or mass) if the problem is severe.
✓				May result from slugging the machine with fluid, solids built up on rotor, or off-design operation, especially surging.
				Note: Looseness of bolting and insufficient interference fit of wheels, coupling hubs (f), and bearing assemblies is the most common cause of problems with turbomachinery.
✓✓✓	✓	✓	✓	The frequency at rotor/support critical is characteristic. Disks and sleeves may have lost their interference fit by rapid temperature change. Parts usually are not loose at standstill. Usually requires rotor disassembly.
✓✓	✓	✓	✓	Often confused with oil whirl because characteristics are essentially the same. Before suspecting any whirl make sure everything in bearing assembly is absolutely tight, with interference fit. Do not use silicone-rubber (RTV) to seal horizontal joint of bearing case!
✓✓				Make sure to check attachment-bolts and split-bolts with torque-wrench!
✓✓				Usually involves sliding pedestals and casing feet. Check for friction, proper clearance, and piping strains. Check all bolting using a torque wrench.
✓✓				To get frequencies, tape microphone to gear case and record noise on magnetic tape, play back through vibrograph. Strong axial vibration indicates pitch-line runout, bent shaft, gear wheel cracking.
✓✓		✓	✓	Loose coupling sleeves are notorious troublemakers! Especially in conjunction with long, heavy spacers. Check tooth fit by placing indicators on top, then lift with hand or jack and note looseness. Should not be more than 1-2 mils at standstill, at most. Use hollow coupling spacers. Make sure coupling hubs are tightly mounted (1 mil/inch).
				Check stage pressures and pressure fluctuations by installing pressure gages, transducers, and thermometers in all accessible openings (drains, pressure-balancing lines). Measure pressure drop across balance-line, and especially balance-flow temperature. High temperature - stage in surge or "stonewalling". Check Molweight.
				Try field balancing; more viscous oil (colder); larger, longer bearings with minimum clearance and tight fit; stiffen bearing supports and other structures between bearing and ground. Basically a design problem. May require additional stabilizing bearings or solid coupling. Difficult to correct in field. Sometimes adding mass at bearing case helps considerably, especially at speeds exceeding 8,000 RPM.
				These are criticals of the spacer-teeth-overhang sub-system. Often encountered with long spacers. Make sure of tight fitting teeth with slight interference at standstill, and make spacer as light and still as possible (tubular). Consider solid coupling if problem is severe. Check coupling balance!
				Can be exceedingly troublesome. Overhangs shift the nodal point of the rotor deflection line (free-free mode) toward the bearing, robbing the bearing of its damping capability. This can make criticals so rough that it is impossible to pass through the critical. Shorten overhang or put in an outboard bearing for stabilization. use lighter coupling and spacer (aluminum, titanium).
				Note: Resonances are very common with structural-steel baseplates, and with improperly supported piping. Oil piping resonance is very dangerous, especially in the vicinity of hot parts and steam piping.
				So called "case-drumming". Can be very persistent but is sometimes harmless. Danger is that parts may come loose and fall into the machine. Also, rotor/casing interaction may be involved. Diaphragm drumming is serious. It can cause catastrophic failure.
				Local drumming is usually harmless, but major resonances, resulting in vibration of entire case as a unit are potentially dangerous because of possible rubs and component failures, as well as possible excitation of other vibrations.
				As above, but with the added complications of settling, cracking, warping and misalignment. Also may get into piping troubles and possible case warpage. Foundation resonance is serious and greatly reduces unit reliability.
				Can excite other vibrations, possibly with serious consequences. Eliminate such vibrations by using restraints, flexible pipe supports, sway braces, shock absorbers. etc., plus isolation of foundation from piping, building, basement and operating floor (?). Unrestrained expansion joints can cause very large vibrations, especially during compressor surge.
✓✓		✓	✓	Mostly at 2 x line frequency (7,200 CPM), coming from motor and generator fields. Turn fields off to verify source. Usually harmless, but if foundation or other components (rotor!) are resonant, the vibrations may be severe. Shaft failures are likely if torsional critical coincides with line-frequency or 2 x line-frequency, and electrical fault or out-of-phase synchronization occurs.
				Can excite serious vibrations or cause bearing failures. Isolate piping and foundation and use shock absorbers and sway braces.
				Most likely caused by damage to seal faces (shaft-currents, dirt, nicks, surface roughness) and/or poor face lubrication. Also poor face contact. Lap seals with very fine compound. Use baked-on or bonded solid-film lubricant.

BASIC PROBLEM MECHANISMS LISTED FOR IDENTIFICATION

✓✓				Vibration at exactly 1x, ½x, ⅓x of exciting frequency. Can only be excited in non-linear systems, therefore, look for such things as looseness and aerodynamic or hydrodynamic excitations. It may involve rotor "shuttling", then check seal system, thrust clearances, couplings, and rotor-stator clearance effects.
✓✓				Vibration at 2x, 3x, 4x, exciting frequency. Fairly common and troublesome. Treatment is the same as for direct resonance: Change frequency, add damping, reduce excitation.
✓✓				Can be very serious. If intermittent, look into temperature variations. Usually rotor must be rebuilt, but should try to increase stator damping first; larger bearings (tilt-shoe), increased stator mass and stiffness, better foundation. This problem is usually caused by mal-operation such as quick temperature changes and fluid slugging. Use membrane-type coupling.
		✓	✓	Basically a design problem but often aggravated by poor balancing and poor foundation. Try to field balance rotor at operating speed, lower oil temperature, use larger and tighter bearings.
✓✓				Add mass or change stiffness to shift frequency. Add damping. Reduce excitation. Improve system isolation. Reducing mass or stiffness can leave the amplitude the same even if resonant frequency shifts. Check "mobility".
✓✓				Use tilt-shoe bearings. CHECK FOR LOOSE BEARINGS FIRST!
				As above, but with additional resonance of rotor, stator, foundation, piping; or external excitation. Find resonant members and sources of excitation. Tilt shoe bearings are best. CHECK FOR LOOSE BEARINGS!
✓✓				The "squeal" of a bearing or seal. May be ultrasonic and very destructive. Check for rotor vanes hitting stator, etc., especially if clearances are smaller than oil film thickness plus rotor deflection while passing through critical speed.
✓✓				Usually rocking motions and beating within clearances. Serious especially in bearing assembly. Frequencies are often below running frequency. Make sure everything is absolutely tight, with some interference. Line-in-line fits are not sufficient to positively prevent this type of problem.
✓✓		✓✓		Very destructive and hard to identify. Symptoms: Gear noise, wear on back side of teeth, strong electrical noise or vibration, loose coupling bolts, fretting corrosion under coupling hubs. Wear on both sides of coupling teeth, torsional fatigue cracks in keyway ends. Best solution: Install properly tuned torsional vibration dampers.
✓✓		✓✓		As above, but encountered only during start and shutdown due to very strong torsional pulsations. Occurs with: reciprocating machinery, synchronous motors (slip frequency excitation if torsional critical is below 2x line frequency). Check for torsional cracks.

©1968 John S. Sohre, Turbomachinery
Consultant. Revised June 1980.
One Lakeview Circle
Ware, MA 01082 USA

Figure 5-3. Cont'd

(*text continued from page 353*)

Interpretation of Collected Data

Learning how to interpret vibration spectra has been likened to learning a new language. The analyst must know which "expressions" are important, and what can be ignored. The significance of a vibration problem can be determined from a particular spectrum. Also, the analyst should keep in mind that:

1. The frequency displayed determines the type of fault or the source of the vibration.
2. The amplitude of the frequency determines the severity of the fault and increases the closer the sensor is to the fault source.

With this standard, we can now examine a series of actual vibration spectra and discuss the interpretation of each spectrum.

Imbalance, and Other Once-Per-Revolution Signals*

Although it would be reasonable to assume that vibrating machinery is out of balance, Figure 5-3 shows that there are many possibilities, imbalance being only one. Some of the characteristics of imbalance are:

1. It is a single frequency vibration whose amplitude is the same in all radial directions. In a pure imbalance, it will be a perfect sinusoidal vibration at the machine running speed, referred to as 1X.
2. The vibration is caused when the center of mass of the rotating element is not turning on the same axis as the rotating assembly.
3. The amplitude will increase with an increase in speed up to the first critical speed of the rotating element.
4. The spectrum generally does not contain harmonics of the 1X running speed.
5. In an overhung rotor there will be an axial component. (Wavetek 1984)

Experience shows that new pumps, fans, and motors are usually in good balance when purchased. Older pumps develop imbalance when the product being pumped causes uneven impeller wear or if foreign material passes through the pump causing damage to the impellers. One common problem with fans is uneven dirt buildup on the blades. This problem is sometimes compounded by the "helpful" person who, with time on his hands, either cleans off just *one* blade, or all *except* one blade. Balance problems with motors arise if too much grease is applied and the excess sticks to the rotor. Incidentally, more motors are ruined by excess grease than by lack of grease.

The first example spectrum, shown in Figure 5-4, is of a containment spray pump showing a vibration level of 1.5 mils. Figure 5-5 shows that when the pump was checked three months later, the amplitude had increased to 2.83 mils, a level that placed this particular pump in a safety alert range that called for action.

Notice that the only significant changes in the spectrum amplitudes occur at the running speed of the pump, 1800 CPM (1X). There are also very small harmonics of running speed at 3600 CPM (2X), 5400 CPM (3X) and 7200 CPM (4X). As will be

---

*Material compiled by Robert M. Jones, Ph.D., Principal Applications Engineer, SKF Condition Monitoring, San Diego, CA. Adapted by permission.

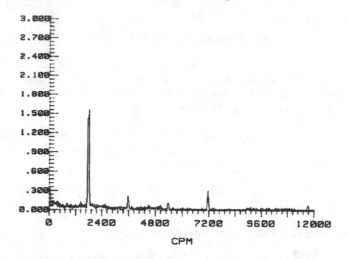

Figure 5-4. Initial vibration spectrum of a containment spray pump.

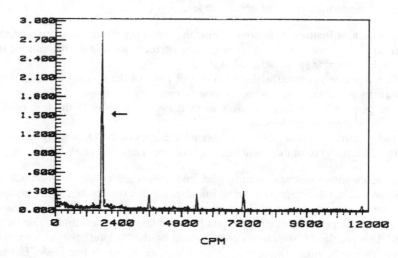

Figure 5-5. Vibration level of pump illustrated in Figure 5-4, three months later. Note multiples of running speed.

shown in later examples, these multiples of running speed have a specific meaning. These amplitudes are all under 0.3 mils and were considered insignificant for this case.

When the pump was taken out of service and dismantled, it was found that a foreign object had gone through the pump and knocked pieces out of the impeller blades. This damage changed the center of mass and led to the classic out-of-balance condition.

Another lesson was learned with the spare impeller. It had been purchased 12 years earlier and had been on the shelf the entire time since. When it was placed on a steady rest and spun, the impeller exhibited signs of static imbalance. Investigation revealed that the manufacturer had neglected to balance the rotor prior to shipment.

When pumps are subject to impeller wear, a computer method known as *enveloping* is available to warn of an increasing vibration level. In this technique, a baseline spectrum is taken of the pump with a new impeller. The decision is made to determine how much vibration will be allowed before action is taken. Based on this decision, the computer draws an envelope around the baseline spectrum. As time passes and the 1X amplitude increases, the 1X spike will penetrate the envelope triggering an appropriate alarm.

Figure 5-6 is a baseline spectrum taken soon after a new pump was placed in service. Figure 5-7 illustrates the computer-generated envelope placed around the existing spectrum. This envelope can be adjusted by the operator in both height and width to permit setting the desired amplitude and frequency (speed) shift limits.

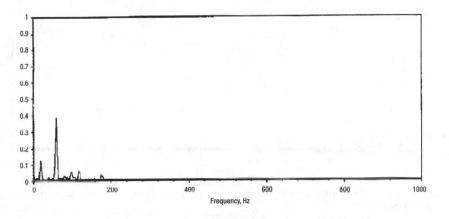

Figure 5-6. Baseline spectrum of a new pump.

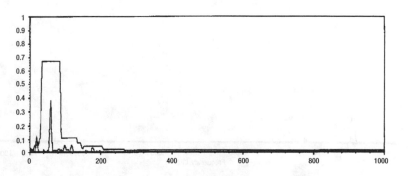

Figure 5-7. Computer-generated envelope delineates alarm levels.

Figure 5-8 is a spectrum taken nine months later after the impeller had developed sufficient imbalance to penetrate the envelope amplitude limit.

Another spectrum that will have a prominent 1X is taken from a large forced draft fan with a bent shaft. When the 1X first began to increase, a balance correction was applied, and the amplitudes were reduced from 6 mils to less than 0.5 mils. However, after a short period of time, the amplitude began to increase and again reached over 6 mils. Removal of the original balance weights returned the vibration level to less than 1 mil. However, after a few weeks, the amplitudes began to increase and again reached over 6 mils. Based on this history and thorough examination, it was concluded that the shaft was bent and possibly was flexing because the owner said the vibration now would come and go over a period of several weeks. Our two spectra show a high 1X in the radial position on the inboard bearing (Figure 5-9) and a high 1X in the axial position on the outboard bearing (Figure 5-10).

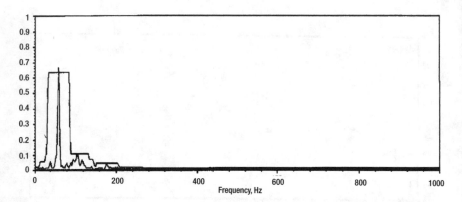

Figure 5-8. Impeller imbalance of pump shown in Figures 5-6 and 5-7 causes alarm envelope to be pierced.

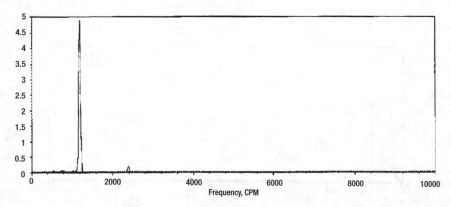

Figure 5-9. Vibration readings taken on a pump in radial direction.

It is the high reading in the axial position that points to a problem other than imbalance since, as mentioned earlier, imbalance causes vibration primarily in the radial direction. A bent shaft will cause a force in the axial direction. It was recommended that the shaft be changed at the next outage and, in the meantime, close attention be paid to the rate of change of the 1X amplitudes. A rapid increase in the rate would indicate that the shaft was cracking, in which case the fan would have to be shut down immediately.

Figure 5-10 is the spectrum taken on the inboard fan bearing. The prominent feature is the large spike at the prominent feature is the large spike at the rotating (1X) speed, 1203 RPM. The small amplitude excursion at 2X indicates a small amount of misalignment and is not significant. It is easy to understand why simple unbalance was suspected, because nearly the entire spectrum is at the rotating speed. However, when all the history was considered, primarily the variable amplitudes, a bent shaft was diagnosed. This assessment was proven to be correct at the next outage when the shaft was replaced.

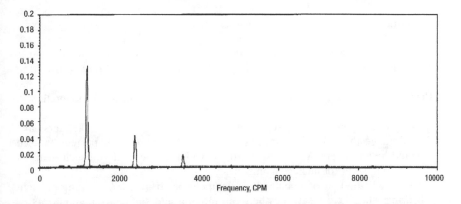

Figure 5-10. Vibration readings taken on pump in Figure 5-9 in axial direction. The spikes at speed multiples indicate a bent shaft condition.

Figure 5-10 is an axial reading taken on the outboard bearing. The units are in inches per second, and the amplitude at 1X is an indication of the push/pull on the bearing caused by the rotating bent shaft. The 2X and 3X perturbations with some very small additional harmonics are an indication of looseness. One hold-down bolt on the outboard bearing had become loose, probably from the continuous vibration.

Attaching a poor quality pulley or sheave that is not concentric with the shaft centerline will create a vibration problem on any electric motor rotor. Pulley runout values in excess of 0.035 inches are not at all unusual. The resulting once-per-revolution (1X) pulse as the belt passes over this high spot, is shown in Figure 5-11. The amplitude will be highest in the direction of the belt, but will also be noticed as amplitude excursions in the horizontal and vertical directions on the motor inboard bearing housing.

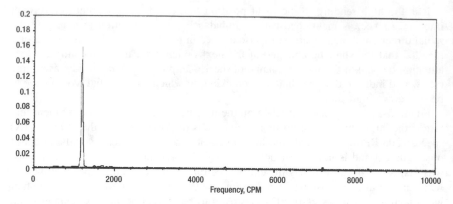

Figure 5-11. Simple imbalance will typically cause an at-speed (once-per-revolution) amplitude excursion.

Misalignment, a Twice-Per-Revolution-Signal

Although one might typically think of machinery vibration problems as "something being out of balance," the consensus of the vibration community is that 70 to 75% of vibration problems are caused by misalignment.

The following is a sequence of events leading to misalignment-induced vibration:

1. Every rotating element has some amount of imbalance. All manufacturers have a tolerance they allow for during assembly. To obtain near-perfect balance would be an extremely expensive process which is used only on such items as quiet equipment for nuclear submarines.
2. This small amount of imbalance is damped, or absorbed, by rolling element bearings that have clearances between moving and stationary components of perhaps 0.0005″ or, for all practical purposes, no clearance.
3. As the unit is operated in a misaligned condition, the push-pull of the rotating elements through the coupling causes unexpected wear on the bearings. Soon the wear leads to excessive clearances between the rolling elements and the races. The bearing now no longer provides the damping needed to restrain the inherent imbalance left in the rotating element.
4. The final step occurs when some well-meaning person notices the vibration and initiates a work order stating that the unit is out of balance. Without a correct diagnosis, unnecessary work is now performed. As an aside, statistics show that approximately 12% of any work done has to be repeated for some reason, and the costs keep going up.

Here, then, are some of the characteristics of a spectrum indicating misalignment:

1. There will be a display of energy at twice running speed (2X). Since connected machines are rarely perfectly aligned, it is not unusual to see some indications

at 2X. However, concern should be aroused if the 2X amplitude is more than 50% of the 1X amplitude when the 1X is greater than 0.15 inches per second.

2. As the misalignment increases, the axial readings will show an increase in amplitude. The trend chart of the axial readings will show a rise.
3. The phase relationship across the coupling will show that the motor and driven unit are 180° different, plus or minus about 30°. This difference is measured between the inboard bearing of the motor and the inboard bearing of the driven unit. The difference is caused by the push-pull action that for one-half of the rotation is pulling the two pieces of equipment together and then is pushing them apart on the second half of the rotation (Wavetek 1984 and Jackson 1979).

It should be kept in mind that there are two types of misalignment: parallel and angular. Parallel will be more prominent in the radial measurement and the angular more prominent in the axial direction. Both will have a 180° out-of-phase relationship across the coupling.

It is good practice to perform a vibration check on equipment when it is returned to service after repairs and then follow up in a short time to recheck the unit. The spectrum in Figure 5-12 was obtained after a seal water pump was returned to service. Although some misalignment was indicated, the 2X was less than 50% of the 1X and the unit was accepted for service. A week later, the 2X had increased to the point that it was in excess of the 1X and was clearly too high (Figure 5-13). Although the 2X increased greatly, the increase in the 1X was very small.

A point should be made here on the resolution used to collect spectra. This 400-line spectrum is giving us two important bits of information. Four hundred lines means that the 12000 CPM range is divided into 400 "bins" of 30 CPM each. If the range were 40000 CPM, each of these "bins" would be 100 CPM. Technicians occasionally take all their spectra at ranges of 120000 CPM, which equals a resolution of 300 CPM per line. Notice in Figure 5-13 that there are two spikes near 7200 CPM,

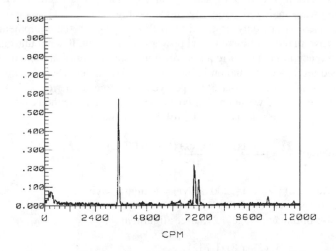

Figure 5-12. Pump vibration spectrum taken after a repair.

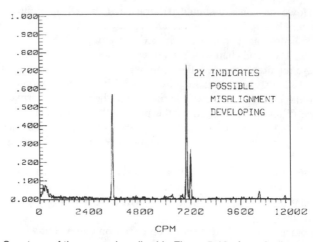

Figure 5-13. Spectrum of the pump described in Figure 5-12 after misalignment had developed.

the 2X at approximately 7100 CPM and another at exactly 7200 CPM, a difference of only 100 CPM. If each bin were 100 CPM or larger, the display could only show one spike, and an important bit of information would be lost. The 7200 CPM spike is generated by an electrical fault in the motor stator windings. This condition will be covered in more detail later. Whenever a unit exhibits signs of distress, it would be appropriate to first check the vibration levels and then take a spectrum to establish the cause of the vibration. When the injection water pump, Figure 5-13, was reported to be vibrating, the spectrum showed the high 2X signal. The unit was checked, and the face-to-face misalignment found to be 0.010″. This fault was corrected and the resulting spectrum, Figure 5-14, taken four days later. The operator then reported that the unit was running more quietly and smoothly.

Misalignment is not only destructive to the bearings in the equipment, but it is also expensive in terms of incremental power consumption. It is not unusual to find a 3–7% difference in the power required for a unit that is out of alignment vs. a properly aligned unit. One calculation involved a 100 HP motor where 2 amps of avoidable current draw cost the user $690 per year. In large refineries with hundreds of pumps and motors, the wasted power cost alone will run in the thousands of dollars.

This savings can be calculated in the following manner:

$$\text{Three phase KW} = \frac{(\text{volts}) \, (\text{amperes}) \, (\text{P.F.}) \, (1.732)}{1000}$$

(KW difference) ($/KW) (7200 hrs/yr) = $ annual savings
7200 hrs/yr is assuming 6 day-week – 50 wks/yr
Example: Initial amperage = 27, after alignment = 25

$$\text{Initial KW} = \frac{(575) \, (27) \, (.8) \, (1.732)}{1000} = 21.5$$

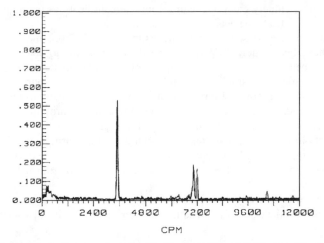

Figure 5-14. Spectrum of pump described in Figure 5-12 and 5-13 after misalignment condition was corrected.

$$\text{After alignment} = \frac{(575)\,(25)\,(.8)\,(1.732)}{1000} = 19.9 \text{ KW}$$

21.5-19.9 = 1.6 KW difference
(1.6KW) ($.06/KWH) (7200 Hrs/yr) = $691.20/yr

One final note on misalignment: many coupling manufacturers claim that their couplings can absorb the stress of misalignment. This may be true, but the bearings cannot. The energy created by misalignment will eventually destroy the bearings, no matter what coupling is installed.

Mechanical Looseness

Even the most carefully assembled machines can develop looseness over time. Nuts work themselves loose, pillow block bolts break, shafts become worn and undersized, and, occasionally, an incorrectly dimensioned component is installed, allowing excessive clearances.

The predominant spectrum characteristic of equipment that has loose components is a display of multiple harmonics of running speed. The running speed of the equipment is referred to as 1X, and, for an example, assumed to be 1800 CPM. The second harmonic (2X) would then be at 3600 CPM, the third (3X) at 5400 CPM, etc. Spacing between each harmonic equal to the running speed would indicate mechanical looseness. It is not unusual to see harmonics extending to 10X. Some texts will refer to 1X, 2X, etc., as "orders" in which the nomenclature 5X would be referred to as five orders of running speed. Although the amplitudes of the harmonics vary because of the phase relationship between the harmonics, there is no particular interpretation that can be applied to this feature. Some researchers have noted an exception to this statement, stating that on machinery equipped with rolling element bearings, an amplified third harmonic would indicate the bearing is loose on the shaft.

When the bearing is loose in the housing, as in the end bell of a motor, high fourth harmonics have been reported (Wavetek 1984). It is possible to determine to some extent the source of the looseness by taking amplitude measurements at several locations on the machine. The amplitudes will be higher closer to the source of the looseness. For example in Figure 5-15, the amplitude on the loose foot was higher than on the other three, indicating this location was the source of the harmonic signals.

Figure 5-15 is a spectrum of the outboard bearing of a motor that has a loose foot because the hold-down bolt had broken, allowing the outboard end of the motor to move. Replacement of the bolt removed the looseness, and Figure 5-16 is a normal spectrum. Note the number of energy spikes, each separated from the other by the 1X running speed.

Figure 5-17 was an interesting case involving a vacuum pump that was used to pull air and lint off the floor of a thread manufacturing plant. The air was then routed

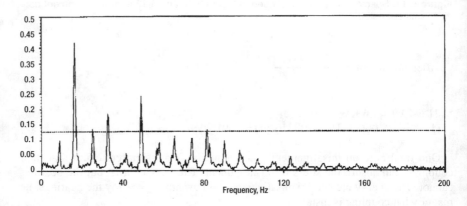

Figure 5-15. Vibration behavior of a motor with one unsupported foot.

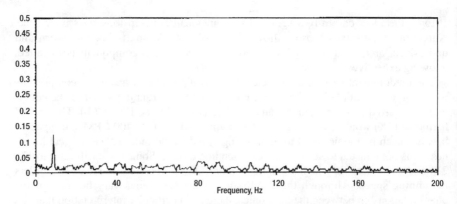

Figure 5-16. Motor of Figure 5-15 after proper support was re-established.

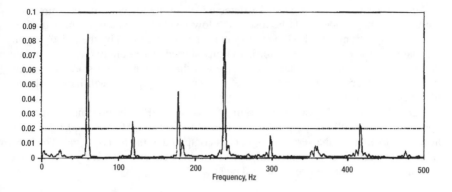

Figure 5-17. Vacuum pump with indications of internal looseness.

through a bag house filter, to the pump, and then returned to the plant environment. The unit had been rebuilt several months prior and had shown no indication of any problems. However, as can be seen in the spectrum, there was, on this day, a display with multiple harmonics with an overall amplitude, in some positions, of 0.7 IPS. Because the pump had been rebuilt within the previous three months, the interpretation was that something had come loose inside the pump. Since the rotor was a built-up, multi-stage unit, it was suspected that one stage was loose on the shaft.

When the unit was taken apart, the inspection revealed looseness of an unexpected nature. A bag in the filter house had detached itself and had been sucked into the pump. It had wrapped itself around part of an impeller, and this entanglement forced the impeller out of balance. When the unit was overhauled, the outboard sleeve bearings had been bored oversize, and with the out-of-balance rotor performing as the forcing function, the rotor was bounding around inside the bearing, creating the harmonics of running speed and inducing the looseness indications. This example was a case of a good interpretation of the signals—looseness, but for the wrong reason; the impellers were tight on the shaft.

However, whatever the cause, looseness is not a condition that can be tolerated for an extended period of time. In this case, the sleeve bearing might have quickly worn to the point that an impeller would have contacted the stator and a catastrophic failure taken place.

Evaluating Electric Motor Conditions, AC and DC

Some rather expensive methods for examining the condition of an electric motor are available to the sophisticated user. However, it is possible to adapt the standard battery-powered, portable, electronic data collector so that an inexpensive, on-the-spot evaluation is quite feasible.

Early attempts using a clamp-on ammeter attached to one lead of a three-phase motor were successful in detecting internal high resistance, whether this was caused by broken rotor bars, broken end rings or bad connections anywhere in the circuit. The limitation was that the fault resistance had to equal the resistance of at least four broken rotor bars before it could be detected in the motor current spectrum. Any fault of less magnitude would be lost in the low level electronic noise near the bottom of the spectrum. This breakage of motor internals is caused by the twisting from the magnetic forces when the motor is initially started. There are two types of motor design: those designed for continuous usage and those designed for start-stop operation. If the motors are interchanged, the continuous duty motor will have a shorter life (Kryter and Haynes 1989).

Figure 5-18 is the log scale spectrum of a good 5 HP induction motor that was operating at near full load, a condition required to evaluate AC motors. Log scale is necessary to enhance the very low signals as compared to the 60 Hz signal. The large spike in the center of the spectrum is from the 60 Hz current used in the United States. In Europe, this frequency would be at 50 Hz.

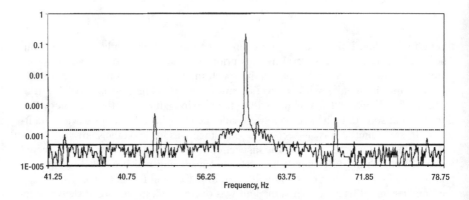

Figure 5-18. Log scale spectrum of 5hp induction motor in good condition.

Figure 5-19 is the log scale spectrum of another 5 HP induction motor, also operating at near full load, but this motor has been damaged by cutting four rotor bars and putting a cut in the end ring. What is seen is the same 60 Hz line frequency signal and additional signals equidistant from those shown in Figure 5-18. These smaller signals are known as indicators of multiple slip frequency.

Slip is the term applied to the amount of lag in the rotor speed vs. the theoretical speed. For example, a four pole AC induction motor operating on 60 Hz power theoretically will rotate at 1800 RPM. Due to friction, load, and inertia, the rotor cannot attain this speed and will usually rotate under full load at about 1730 RPM. The difference between 1800 RPM and 1730 RPM is the normal 70 RPM slip of the motor. One must remember this is not a constant, but will vary primarily with the load applied to the motor, friction and inertia remaining the same.

Changes in the amplitude of these slip frequencies as compared to the amplitude of the 60 Hz signal, when measured in decibels (db) indicate a problem. In Figure 5-19, the signal amplitude in volts reads 8.7772 at 60 Hz and 0.0908 at 51.4 Hz. Note that 51.4 Hz is the slip frequency. The other energy excursions represent the harmonics of this signal and are not used in the calculations.

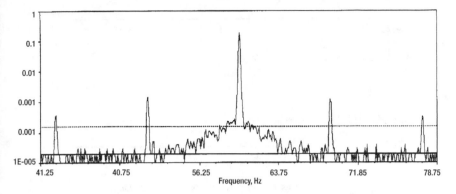

Figure 5-19. Log scale spectrum of 5hp induction motor with cut rotor bars. Note multiple slip frequencies.

In order to convert these voltage measurements to db, the following computations are accomplished. The 60 Hz amplitude is divided by the 51.4 Hz amplitude and the log of the result is multiplied by 20.

20 Log 8.7772/0.0908 = 39.71 db

Experience has shown that if this db resultant is less than 45 db, then there is a fault in the motor. It does not locate the fault; it only indicates that there is a high resistance that equals the resistance of at least four broken rotor bars (Jones 1990).

These initial attempts to evaluate the mechanical condition of electric motors have been adapted by the marketplace and are now an accepted diagnostic tool. More recent research has carried this principle even further as motor operators demanded evaluation tools that could detect flaws much earlier in the life of the motor. Users of DC motors were also seeking the same information for the same reasons. One of the largest users of DC motors is the railroad industry. In 1991, the industry costs for overhaul of locomotive DC traction motors was in excess of $100 million dollars.

Research at the Oak Ridge National Laboratories on motors used as valve operators in nuclear power plants has resulted in a patented electronic device that filters the 60 Hz line frequency signal and enhances the remaining signals. By this process, it is possible to see the actual slip frequency signal. Figure 5-20 is a spectrum of a good 100 HP AC motor under full load with no induced faults. Figure 5-21 is the spectrum of the same motor with one rotor bar cut 25% of the way through. Figure 5-22 is the same motor with the bar now cut 75% of the way through.

The method of evaluation is to calculate the amplitude of the slip frequency or its harmonic, if present, in relationship to the overall amplitude. For example, in the good motor, the slip frequency amplitude is 0.0038 volts at 1.5 Hz, and the overall amplitude is 0.01599 volts.

$$\frac{0.00380 \text{ v}}{0.01599 \text{ v}} = 24.3\%$$

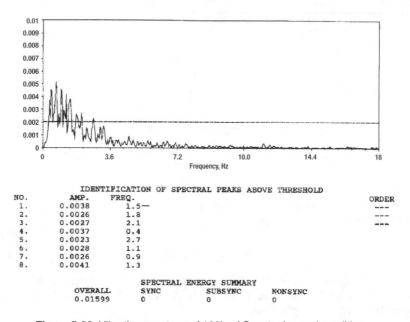

IDENTIFICATION OF SPECTRAL PEAKS ABOVE THRESHOLD

NO.	AMP.	FREQ.	ORDER
1.	0.0038	1.5—	---
2.	0.0026	1.8	---
3.	0.0027	2.1	---
4.	0.0037	0.4	
5.	0.0023	2.7	
6.	0.0028	1.1	
7.	0.0026	0.9	
8.	0.0041	1.3	

SPECTRAL ENERGY SUMMARY

OVERALL	SYNC	SUBSYNC	NONSYNC
0.01599	0	0	0

Figure 5-20. Vibration spectrum of 100hp AC motor in good condition.

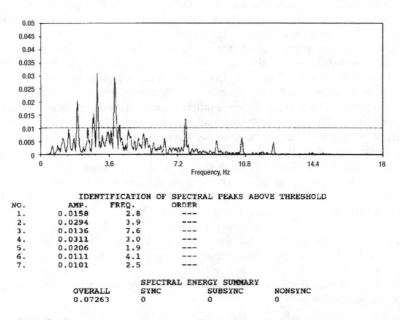

IDENTIFICATION OF SPECTRAL PEAKS ABOVE THRESHOLD

NO.	AMP.	FREQ.	ORDER
1.	0.0158	2.8	---
2.	0.0294	3.9	---
3.	0.0136	7.6	---
4.	0.0311	3.0	---
5.	0.0206	1.9	---
6.	0.0111	4.1	---
7.	0.0101	2.5	---

SPECTRAL ENERGY SUMMARY

OVERALL	SYNC	SUBSYNC	NONSYNC
0.07263	0	0	0

Figure 5-21. Spectrum of motor described in Figure 5-20 after cutting one rotor bar 25% of the way through.

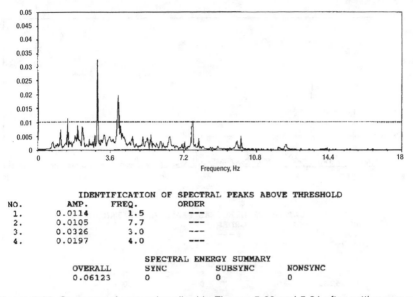

```
            IDENTIFICATION OF SPECTRAL PEAKS ABOVE THRESHOLD
NO.        AMP.     FREQ.         ORDER
 1.       0.0114    1.5           ---
 2.       0.0105    7.7           ---
 3.       0.0326    3.0           ---
 4.       0.0197    4.0           ---

                          SPECTRAL ENERGY SUMMARY
            OVERALL       SYNC          SUBSYNC        NONSYNC
            0.06123        0             0              0
```

Figure 5-22. Spectrum of motor described in Figures 5-20 and 5-21 after cutting one rotor bar 75% of the way through.

Testing of known good motors has shown that a good motor will have a ratio of about 25%, which this motor has, as shown above.

Next, we do the calculations for the motor that has been damaged. The slip frequency is at 1.9 Hz, the amplitude is 0.0206 and the overall amplitude is 0.07263. This ratio calculates to be 28.3%, not a large increase, but there is a change. Later, it was found that the current in the laboratory was noisy and was raising the overall amplitude. When this interference was corrected, another reading was taken with the 25% damage; Figure 5-23 is the spectrum. As can easily be seen, the spectrum is much cleaner; the slip amplitude changed to 0.0196; the overall became 0.05381. Under these conditions, the ratio became 36.4%, a clear increase and clear indication that damage had occurred.

Doing the same calculations with the rotor bar cut 75% (Figure 5-22) and using the second harmonic at 3 Hz results in a ratio of 53.2%, with the higher number indicating the additional damage. Although this current conditioning device is commercially available at the present time, additional empirical data needs to be studied to gain confidence that the results are repeatable.

Another AC motor fault that can be detected in the field, using just the data collector, is an electrical breakdown in the motor stator. Independent of the motor speed and size, if a fault develops in the stator, a signal at twice line frequency will be generated. Line frequency is 60 Hz (3600 CPM); so, twice line frequency will place the signal at exactly 120 Hz (7200 CPM), as illustrated in Figure 5-24. This exact measurement is important because motors that are rated at 3600 RPM will generate a misalignment signal *near* 7200 CPM. As discussed earlier, no induction motor can operate under load at 3600 RPM; slip is always present, so the actual rotational speed will be about 3540 RPM. A misalignment signal would then be generated at 2X or 7080 CPM.

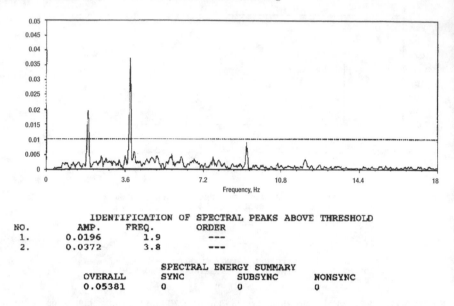

```
          IDENTIFICATION OF SPECTRAL PEAKS ABOVE THRESHOLD
NO.       AMP.      FREQ.      ORDER
 1.       0.0196     1.9        ---
 2.       0.0372     3.8        ---

                     SPECTRAL ENERGY SUMMARY
          OVERALL    SYNC         SUBSYNC        NONSYNC
          0.05381     0            0              0
```

Figure 5-23. Motor from Figure 5-20 after "cleaning up" the electric power supply.

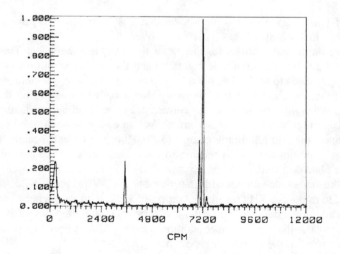

Figure 5-24. Electric motor with breakdown in stator insulation.

Another fault that can occur in AC induction motors is a breakdown of the insulation in the stator. The unusual aspect of this fault is that it will show up at a frequency equal to the RPM times the number of rotor bars and is referred to as the slot or rotor bar pass frequency. For example, on an 1800 RPM motor with 36 rotor bars, the frequency would be 64,800 CPM. However, current research has shown that the frequency is actually at 32 times rotation speed or 57,400. In the test case, the 1800

RPM motor has a four-pole rotor. The multiplier of 32 is obtained by subtracting the quantity of poles, 4, from the number of rotor bars, 36, resulting in the multiplier, 32. The reason for this is not understood at the present time. Figure 5-25 is a spectrum of a motor with degraded insulation and shorting in the stator.

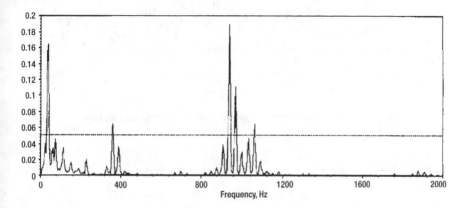

Figure 5-25. Spectrum of electric motor with degraded insulation and shorting in the stator.

If the frequency span of the digital data collector is not set this high, and it usually isn't for route data collection, this energy will not be observed in the spectrum. However, an observant operator will notice that the motor is vibrating at a higher amplitude than is observed on the data collector. The solution, of course, is to increase the range of the spectrum to cover this area. If removal of electrical power immediately stops the vibration, an electrical fault would be indicated.

Modern data collectors can often be used to evaluate DC motors as well. A rather common fault in DC motors is the breakdown of motor winding insulation due to overvoltage. In view of this common failure mode, a DC motor was first run in a no-fault condition, and the spectrum in Figure 5-26 was the result; Figure 5-27 is the time domain signature for the same operating conditions.

The rotor was then removed and the insulation was nicked between one pair of wires on the rotor winding. The wires were not forced together; only the insulation was scratched down to bare wire.

Figure 5-28 is the resultant spectrum of the motor with one fault. The interpretation of the spectrum is not complete, but there is one outstanding fact: this small flaw resulted in an increase of the overall RMS amplitude in excess of 1300%.

As mentioned before, it is good practice to observe the motor at the instant power is turned off. If the vibration ceases the instant power is removed, the vibration is electrically induced. If the vibration slowly decreases as the motor slows down, it can be deduced that the rotor is out of balance or has a bowed shaft.

The rotor shaft can become bowed when an electrical short develops in the rotor, creating heat on one side of the shaft. This uneven heating will cause the shaft to bow and induce a once-per-revolution vibration. The difficulty in detecting this fault

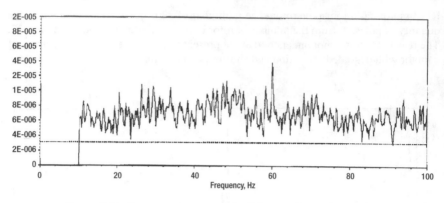

Figure 5-26. Frequency spectrum of a DC motor in good condition.

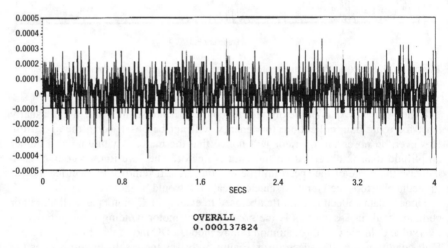

OVERALL
0.000137824

Figure 5-27. Time domain spectrum of DC motor in Figure 5-26.

is that it takes time for the heat to develop and bow the shaft. If the motor is checked soon after starting, no fault is likely to be found. Temperature checks of equipment are most informative when performed after operating temperatures and loads have been reached.

Aerodynamic Flow-Induced Vibrations

When a water leak was reported in an internal room cooler, it was generally assumed that the fan was out of balance and that the resulting vibrations caused the brazed joint on the water pipe to begin leaking. This assumption was borne out when

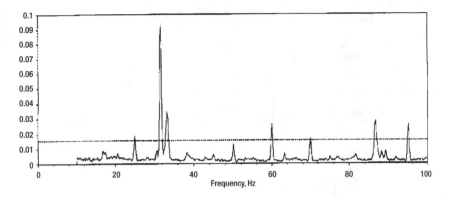

Figure 5-28. Motor from Figure 5-26 after implanting one fault.

the spectrum in Figure 5-29 revealed a large vibration, 19.92 mils, occurring at a frequency equal to the rotating speed of the fan, 1182 CPM (1X).

Examination of the fan housing revealed turning vanes that had broken loose and were hanging down in the air flow. These were repaired and the fan restarted. Much to everyone's surprise, the original vibration was still present. Another spectrum, Figure 5-30, was taken with an increased frequency range, 40 Hz increased to 500 Hz, and the real source of the vibration was found. Note the large vibration signal at 7 times the fan rotation speed, 0.6 IPS at 8250 CPM (7X). This signal was generated by the seven blades of the fan as each blade went past some interference (Wavetek 1984).

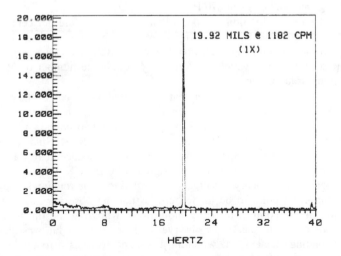

Figure 5-29. Vibration spectrum of a defective cooling fan.

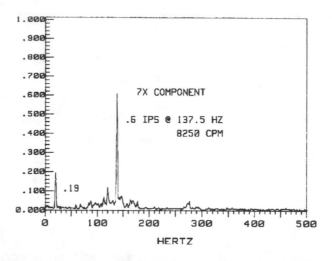

Figure 5-30. Vibration spectrum of Figure 5-30 expanded to show amplitude excursion at seven times running speed.

Closer examination of the fan housing revealed that a straightening vane, located 2 inches above the plane of the fan blades, had a broken weld at one end. This faulty weld allowed the vane to vibrate as each fan blade forced air over it. The event occurred seven times per revolution of the fan, thereby creating the vibration frequency at 7 times rotating speed. After the broken vane was repaired, the vibration level on the water line decreased from 19.1 mils to an acceptable 2.4 mils.

Figure 5-31 is a spectrum of another fan where the duct work had dropped down and was interfering with the air flow. In this case the fan speed, 30 Hz, times the nine blades generated energy at 270 Hz.

This same type of vibration can occur with pumps that are not centered in the pump housing. As the impeller forces liquid from pumps with incorrect clearances, it is possible to detect surges as each impeller vane discharges. This surge sequence will create a vibration in which the frequency equals the number of pump vanes times the pump rotation speed.

Figure 5-32 is the spectrum of a fan that is mounted on the end of the motor shaft. It was reported to be making a large amount of noise, and, when examined, the reason for this was easy to understand. This spectrum has two important bits of information. Note that the fan speed is seen at 60 Hz or 3600 RPM. Since the energy is displayed at 1X, the rotating speed of the fan, one could deduce there was an out-of-balance condition. However, also note the signal at 30 Hz, ½X. When a rotating element rubs against a stationary object, a signal equal to ½ the rotation speed is generated. An important point to be made here is that the observer has to know the construction of the equipment being monitored. Knowing the fan was attached directly to the end of the shaft, the observer deduced that the fan was loose on the shaft. The rotating element was wobbling back and forth as it rotated and rubbed against the side of the fan housing. There was no out-of-balance condition. When the fan was taken out of service and examined, this diagnosis proved to be correct. Both

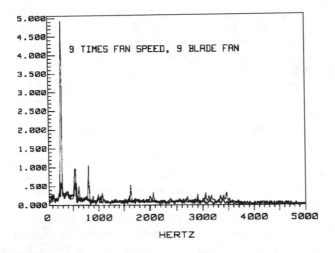

Figure 5-31. Blade passing frequency predominates in this fan spectrum.

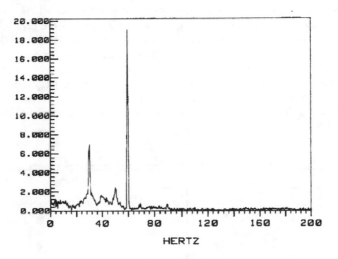

Figure 5-32. Internal rub and looseness caused one-half running speed frequency to be rather pronounced.

fan and motor were replaced due to damage to the original, and the replacement ran with less than 0.5 mils vibration.

Gear Defects

Just as the number of fan blades will create a vibration signal at the number of blades times the rotation speed, a gear set will generate a vibration signal at the number of teeth on the gear times the shaft rotation speed. It is important to note that the gear count being used is multiplied by the shaft speed that the gear is mounted on. For example, the input shaft speed is 1800, and the bull gear has 130 teeth. The vibration signal generated will be $130 \times 1800 = 234{,}000$ CPM. If the output shaft has a speed of 3600 RPM, there are 65 teeth on the pinion, and the resultant multiplication will again equal 234,000 CPM. If the two shafts don't give equal gear mesh frequencies, the tooth count and RPM should be rechecked; they must be equal for the gear box to operate (Wavetek 1984).

When measuring vibration on a gear box, it is necessary to measure the amplitudes in acceleration "Gs." The close fit required in a gear box does not allow sufficient space for any significant velocity to develop, but acceleration forces will be generated as the gear teeth mesh. Velocities on a good gear box measuring in the order of 0.02 IPS and readings of 2–3 Gs are not unusual. As the gear box develops such problems as misalignment, for example, the IPS reading may increase by a very small amount, while the acceleration amplitudes can double and triple.

Figure 5-33 is the spectrum of a gear box that was checked after a maintenance shutdown. One audible characteristic of a misaligned gear box is the modulating sound. When this characteristic was observed, the information in Figure 5-33 was collected. It is often difficult to obtain the number of gear teeth from drawings; for some reason they are rarely included. In this case, it was necessary to go to the storeroom and count the teeth on the spare gear. Shaft speed times the gear tooth count gave the frequency that is circled on the spectrum, approximately 3800 Hz, indicating a possible gear problem. When it was learned that the motor had been shifted 0.004″ during the outage, misalignment was suspected. The motor was reset to the original location , and Figure 5-34 shows how the gear mesh frequencies were no longer being generated. It has been hypothesized that the energy represented in the area of ½ gear mesh frequency is generated by the rubbing action of the gear teeth as

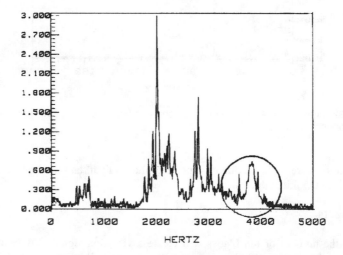

Figure 5-33. Gear misalignment caused a modulating sound and prompted capture of this spectrum, with gear mesh frequency circled.

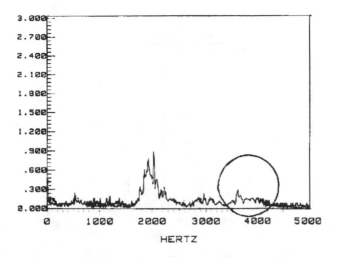

Figure 5-34. The gear of Figure 5-34 after realignment. Note amplitude reduction throughout entire spectrum.

they mesh. Also note on the two spectra that there is a large amount of energy displayed at a number of frequencies when the gear mesh frequency is observed. If asked what they represent, the answer is "We don't know." These are three words every technician needs to learn. Sometimes there is no answer and if one attempts to offer a guess that turns out be wrong, credibility may be sacrificed. Credibility is the backbone of an accepted program.

Figure 5-35 is the spectrum taken on a gear oil pump. During operation, the gears are immersed in oil. Some oil is trapped between the teeth and is forced out the discharge in a pumping action. While not always efficient, it is usually inexpensive and functional. One characteristic of machinery degradation is that as the fault progresses, sideband frequencies are generated around the primary gear mesh harmonics. In this case, note the small energy spikes on both sides of the harmonics. The spacing of these sidebands is equal to the rotating speed of the gear shaft, in this case, 900 RPM. Whenever the gear mesh frequency is seen to increase, there is a problem. The problem is more severe when harmonics of the gear mesh frequency are observed, and when the stage is reached where sidebands are present. The unit is then near failure.

That condition proved to be the case when the pump was taken apart; the bearings had completely failed, and the unit fell apart when the cover was removed. The overlay on this plot is a reading taken on the new gear pump, a bit difficult to see here, but a notable improvement nevertheless.

Figure 5-36 is the spectrum taken on a large, slow, ball mill gear box operating in a Western gold mine. The owner stated that he was aware of a problem. Past history had shown that about once per year a tooth would break off the bull gear. Since it was a recurring event, he had on hand a spare bull gear, which he planned to install during the next shutdown.

Figure 5-37 was generated, using the time domain feature of the data collector. Instead of showing different frequencies, this display indicated events that occur

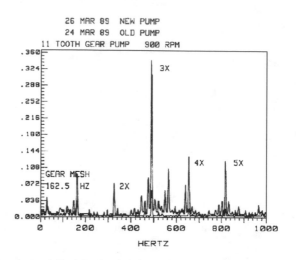

Figure 5-35. Gear pump with defective bearings produced the high-amplitude spectral line. The new spectrum, just barely above the lower boundary of our plot, was obtained after bearing replacement.

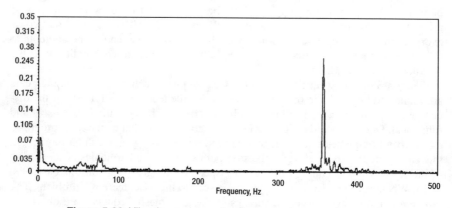

Figure 5-36. Vibration spectrum from a large, slow-speed gear unit.

over a set time span. As an event occurs, a signal is generated and displayed as a "spike" of energy. Since the owner suspected a broken tooth on the bull gear, the time between events should equal the time of one rotation of the bull gear shaft (600 RPM), or in this case, one pulse every 0.100 seconds. When the observed pulse rate was measured, it was found to be 0.272 seconds, which equalled the pinion shaft speed of 220 RPM. There was a tooth problem, but not on the gear that was suspected. As it turned out, there was no spare pinion on site, and one had to be flown in, resulting in a three-day delay. If this problem had been discovered during the shutdown, the shutdown would have been extended three additional days, incurring

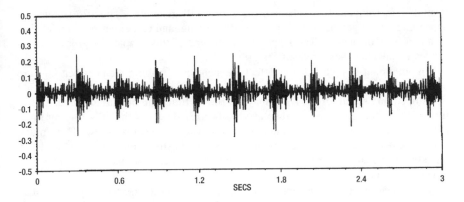

Figure 5-37. Time domain signal taken on the unit in Figure 5-36 shows pulse rate equal to pinion shaft RPM—a clear indication of tooth distress.

heavy loss of income, since the value of daily production at this plant is approximately $750,000.

The lesson here is never to assume you know what the problem is; verify by analysis and inspection.

Rolling Element Bearing Defects

One of the most useful inventions at the beginning of the machine age was rolling element bearings. For example, one of the early limitations on the speed of railroads was the oil-soaked rags stuffed into the wheel axle journal boxes, appropriately called stuffing boxes. Either speed or heavy loads would overheat the boxes, igniting the oil-soaked rags, thus stopping the train. Now, with tapered roller bearings, the speeds and loads have increased dramatically and wheel fires are rare.

Nearly all pumps and motors of small-to-medium size operate with ball or roller bearings. The largest bearing manufacturer, with 26% of the market, sells approximately three billion dollars worth of bearings each year. These bearings are designed to provide up to one million hours of running time. However, from the day the bearings leave the factory, events take place that reduce this number to as low as tens of hours. A short list of problems would include mishandling of the bearings, including improper storage; improper installation; poor lubrication; harsh operating environment; overloads; overspeeds; and the biggest cause of bearing failure—estimated at 70%—misalignment of the driver and driven unit. Chapter 3 discussed the telltale results of abnormal operation.

One of the reasons that vibration data are collected on the bearing housing is that all the forcing frequencies are transmitted through the bearings from the rotating element. These detectable forces are also at work degrading the life of the bearing.

Degradation will manifest itself in any of four ways: as damage to the outer race referred to as BPFO (ball pass frequency outer), damage to the inner race, referred to as BPFI (ball pass frequency inner), damage to the rolling elements, referred to as BSF (ball spin frequency), or damage to the bearing cage, referred to as FTF (fundamental train frequency). Each of these flaws will generate a specific frequency, depending on the geometry and speed of the bearing, and can either be calculated or obtained from databanks. The Appendix at the end of this chapter lists the formulas needed to calculate these frequencies.

It is important to note that the defect frequencies of rolling element bearings are non-integer multiples of machine operating speeds. For example, the BPFO is 8.342 times operating speed, or the BPFI may be 10.365 times the operating speed. Therefore, if the defect frequency is displayed, a bearing defect is the only forcing function that will generate that frequency. The question then becomes, "How much amplitude is too much?", and at this point research is still being conducted to determine the best answer. For the foreseeable future, Figures 5-38 and 5-39 can be used as general guidelines.

A new bearing analysis technique that is just entering the market in portable data collectors is known as *enveloping* or *demodulating*. The technique allows the user to examine the condition of the bearing by observing the high-frequency harmonics generated by bearing flaws that are not evident in the low-frequency range, and displaying the information in the low-frequency band available on current data collectors (Jones 1992).

Figure 5-40 is an example of a normal spectrum of a ball bearing. Velocity processing shows no evidence of any defects in this spectrum. When the data were collected at the same point and the information enveloped, Figure 5-41 was the result. In this bearing, the BSF (ball spin frequency) is calculated to be 95 Hz for the speed at which the shaft was turning. Not only is the 95 Hz signal present, there are also multiple harmonics, a further indication of damage.

Figure 5-42 is another bearing where the defect was not apparent in a velocity spectrum. However, Figure 5-43 displays a frequency of 225.0 Hz that is calculated to be generated by a defect in the outer race, BPFO. This new technique provides the user with very early warning that there is a defect in the bearing and that closer, or more frequent, observation may be required to prevent unexpected failure.

Figure 5-44 is the spectrum of a newly installed fan, which exceeded acceptable start-up vibration levels measured on the bearing cap. Using the enveloping techniques, the spectrum display indicated a cage problem, FTF at 11.3 Hz, a possible cage rub at ½ the FTF, 6.3 Hz, and some light damage in the outer race.

Laboratory analysis of the bearing found a large gouge in the bearing cage that was rubbing on the housing as the cage rotated. Three small defects in the outer race were also found. A check of the bearing history revealed that it was more than 20 years old and had been purchased from a nonauthorized dealer. The buyer probably got the bearing for a low price, but in the long run, it cost him much more than the retail price of a good bearing.

One interesting sidelight on this spectrum is the interaction between forcing frequencies, which by themselves are not a problem. The rotating frequency is 30 Hz. The second harmonic of the running frequency is 60 Hz (2×30 Hz). The third harmonic of the running frequency is 90 Hz (3×30Hz). The FTF is 11.25 Hz, and the eighth harmonic of this is 90.0 Hz. What occurred in the field was a reinforcing of the running speed 3X by the FTF 8X resulting in an unacceptable vibration. This is the type of interaction that often occurs when several pieces of equipment are mounted on the same frame or base. Each individual piece of equipment can operate

**Vibration Velocity
(Inches/Second Peak)**

0 - .005 in./sec	Extremely Smooth
.005 - .01 in./sec	Very Smooth
.01 - .02 in./sec	Smooth
.02 - .04 in./sec	Very Good
.04 - .08 in./sec	Good
.08 - .16 in./sec	Fair
.16 - .32 in./sec	Slightly Rough
.32 - .64 in./sec	Rough
Above .64 in./sec	Very Rough

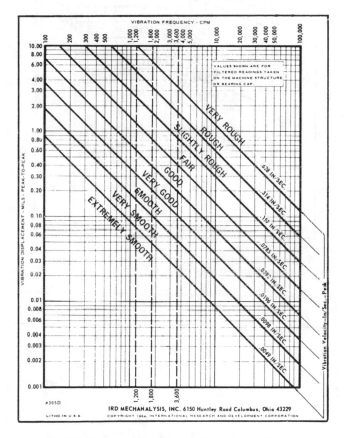

Figure 5-38. General machinery vibration severity chart.

satisfactorily by itself, but the combination of units results in an unacceptable vibration of the total.

Another area in which the demodulating circuits show spectacular results is in the analyzing of low-speed bearings. In the past, it was necessary to use accelerometers

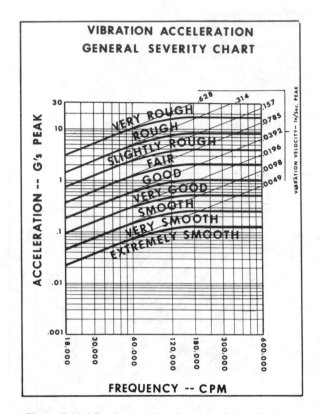

Figure 5-39. Vibration acceleration general severity chart.

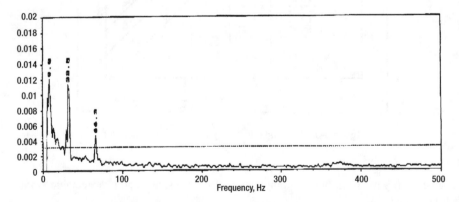

Figure 5-40. Normal velocity spectrum obtained from a ball bearing.

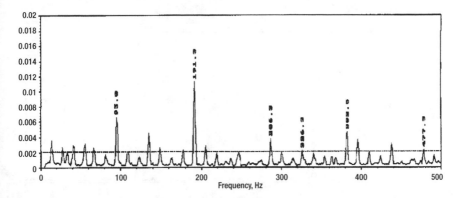

Figure 5-41. "Envelope processing" of the data obtained from the bearing in Figure 5-40 shows multiples of ball spin frequency, a sign of defects.

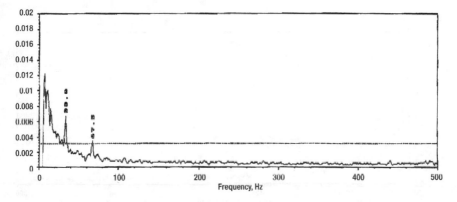

Figure 5-42. No defect is apparent in this vibration velocity spectrum of a typical ball bearing.

with high sensitivity, that is, 500 mv/EU, and fight to get through the integration noise common to all data collectors. Now, using a standard 100 mv/EU sensor, it is possible to obtain clear frequency spectra with units operating as slow as 8 RPM. It has even been reported that good results have been obtained down to 3.5 RPM. One special case involves large radio and radar antennas, turning as slow as one revolution in nine minutes. By incorporating the enveloping circuits and presenting the data in the time domain, it is possible to get a direct measurement of the energy being generated in the bearing as it turns. When the bearings are subsequently ranked by overall amplitude, the unit most in need of inspection is revealed. From the study of physics, we know that F = ma. Since the mass of each antenna is the

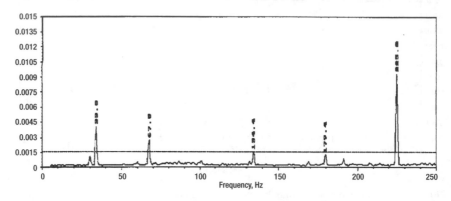

Figure 5-43. The data for Figure 5-42 "envelope processed" and highlighting an outer ring defect.

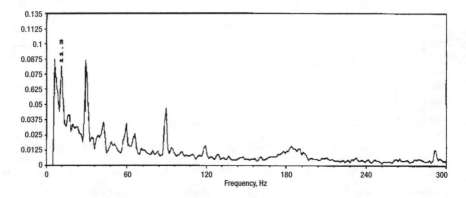

Figure 5-44. Excessive vibration of a newly installed fan.

same, we can equate m = 1; the forces generated in the bearing are therefore equal to the acceleration that we can directly measure. The amount of force generated is then proportional to the amount of damage or contamination in the bearing.

As an example of a very slow bearing, the following spectra were obtained on a conveyor chain drive rotating at 8.5 RPM. The concern is the bottom thrust bearing. The last time this bearing failed, the line was down for two-and-a-half days, causing the complete loss of production.

Figure 5-45 is a velocity spectrum taken on the bearing housing. If there were bearing damage, we would expect activity at the calculated BPFO of approximately 103 CPM. As can be seen, there was no energy displayed at the calculated BPFO, and it

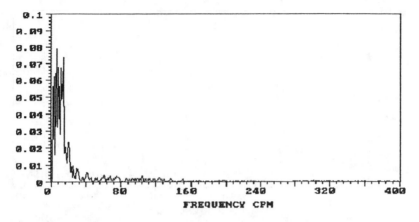

Figure 5-45. Velocity spectrum taken on the bearing housing of a very slow conveyor chain drive.

was initially assumed that there was no damage. Next, the bearing was examined at the same location using plain acceleration; the spectrum in Figure 5-46 is the result.

Although there is energy displayed in the area of 110 to 120 CPM, there is no distinct frequency, just the "haystack" that might alert the viewer, but with such low amplitudes, it is likely to be missed. Figure 5-47 is a spectrum using enveloped acceleration. Not only does the specific frequency clearly show, it also shows its harmonics.

A word about harmonics in an enveloped spectrum: if a bearing is heavily loaded, it is possible to see the BPFO generated as the rolling elements enter and exit the

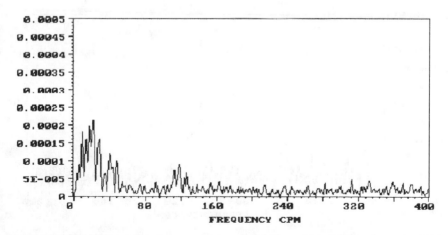

Figure 5-46. Acceleration spectrum of the bearing housing in Figure 5-45. No distinct amplitude excursions are seen here.

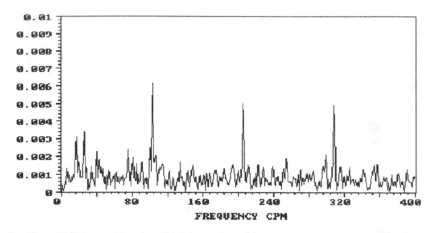

Figure 5-47. "Envelope processing" of the acceleration signal in Figure 5-46 shows a predominant frequency and its harmonics.

load zone. However, the presence of harmonics of the defect frequency would be a good indication that some damage has occurred. The next level of damage can be seen when sidebands of rotating speed begin to appear around the fundamental frequency and its harmonics. In essence, then, the presence of harmonics can be used as a indicator of moderate damage.

Finally to show just the enveloped BPFO, the Fmax is reduced to 200 CPM and the frequency is displayed (Figure 5-48).

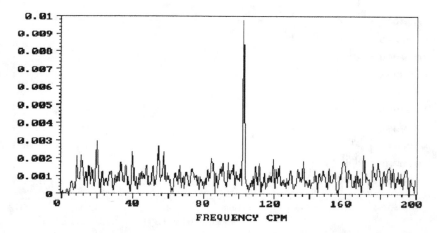

Figure 5-48. Enveloped ball passing frequency of the outer ring (BPFO) of the bearing described in Figure 5-47.

Acoustical Evaluations

Although we often think of vibrations as something we can physically feel, such as a tire out of balance or a pump shaking on its base, sound waves are also vibrations. Instead of the equipment vibrating, the air molecules vibrate, transmitting sound. A state-of-art data collector may be able to "listen" to the sound of a rotating leaf chopper. As long as product is passing through the chopper, the sound is a low-pitched growl. When the product hangs up in the chute, the chopper speeds up and the aerodynamic noise changes pitch or frequency. By setting frequency shift alarm limits, the equipment can be made either to shut down, or to sound an alarm when the product hangs up.

A steam trap is a non-rotating piece of equipment that can be monitored using the capabilities of current data collectors. A baseline spectrum is taken when the trap is in operation and is operating satisfactorily. If the trap sticks open and begins to bypass steam, the high-frequency noise level will increase in amplitude and alert the operator.

Similarly, the condition of safety valves can be monitored by sensing acoustic signals. Figure 5-49 is a spectrum taken with the sensor placed on the valve body with the valve known to be closed (Jones 1989).

Figure 5-50 is a spectrum taken in the same location immediately after the valve was opened. The increase in the energy detected indicates that flow is taking place in the valve. Figure 5-51 is the third spectrum taken after the valve was returned to the closed position. Note the similarity to the spectrum in Figure 5-49.

All three of these spectra can be plotted on the same graph with three colors, giving a clear picture of the opening and closing of the valve. The United States Nuclear Regulatory Commission has accepted this method as confirming operation of valves that cannot be physically observed or are in hazardous locations.

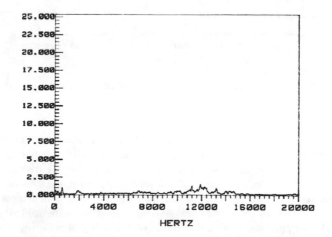

Figure 5-49. Spectrum obtained from sensor mounted on a fluid valve body, in Gs. The valve is in its closed position.

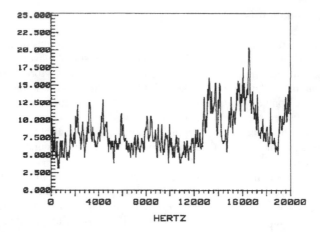

Figure 5-50. Data obtained from the sensor in Figure 5-49; the valve is now open.

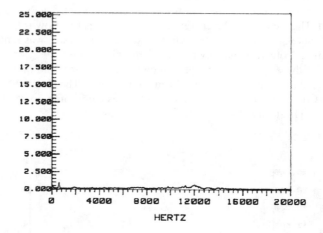

Figure 5-51. The same valve (Figure 5-49) being closed again. Note the good repeatability.

Without extensive empirical data, it is not possible to estimate flow quantity using sound measurements. However, it is possible to rank a set of valves or steam traps attached to a manifold to determine if any are leaking. Using the overall amplitude measurement, data are collected from all the valves or traps in the system. If, in fact, all the valves or traps are operating under the same conditions of temperature and pressure, the units with the largest leaks will be those with the highest indicated amplitude. Experience shows that this mode of acoustical monitoring has allowed significant savings to be realized in a power plant using this technique. The process allows the inspection and repair of only those valves that indicate a problem.

For an illustration of this technique, the lower trace in Figure 5-52 is an acoustical measurement on a water valve that was closed. When the flow was increased to 400 ml/min, the resulting upper trace shows the increase in overall energy.

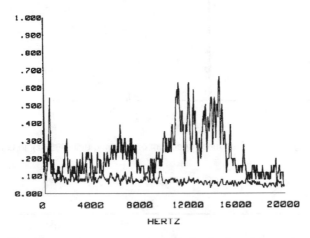

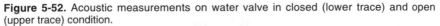

Figure 5-52. Acoustic measurements on water valve in closed (lower trace) and open (upper trace) condition.

Resonance

Resonance is often a little-understood phenomenon that can cause extensive damage while the operator is trying to figure out what is going on and what to do. The two most dramatic examples of the destructive forces that can occur when a piece of equipment is operating near or at a resonance frequency involved the case of the Tacoma Bay Bridge in Washington State and a turbo-prop passenger aircraft that suffered several mysterious crashes. In the case of the bridge, wind blowing through the support cables caused them to be excited, and they began to vibrate. This vibration was transferred to the support structure and the roadbed began to sway and buckle. The entire structure collapsed and fell into the water.

In the case of the aircraft, it was determined that the speed, or frequency, of the engines, was a forcing function that fed into the wing root and caused a very small, high-rate bending action. Just as a paper clip breaks by bending it back and forth, the same action of this forcing function caused the wing to break off.

Another good example deals with bells. A bell is a resonant device which, when struck, rings at a frequency determined by material and shape. This frequency is described as its natural frequency. Bells are tuned by altering the material and shape parameters so as to obtain the desired sound. Figure 5-53 is the frequency, 685 Hz, and Figure 5-54 the time response, 0.24 seconds, of a large dinner bell. Since it is desirable to hear the bell "ring," the decay rate, as seen in the time domain, occurs over a relatively long period of time.

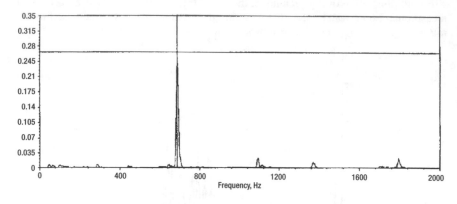

Figure 5-53. Frequency spectrum of a ringing bell (685Hz), in Gs.

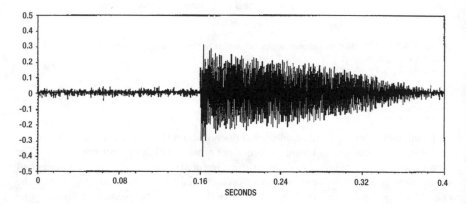

Figure 5-54. Time domain response plot of ringing bell described in Figure 5-53 illustrates long decay time.

Every piece of equipment has a natural frequency that, depending on the mounting of the equipment, should have a very fast decay rate. Figure 5-55 and 5-56 are examples of a low frequency and a fast decay. The equipment being tested is a cold-rolled steel mill and the "hammer" used was the roll of steel as it rolled into the spool holder. Note that the time display decays very rapidly, indicating a stiff piece of equipment, something one would expect in a steel rolling mill. The first-time response at 0.16 seconds occurs when the roll dropped onto the platform and the second response at approximately 0.24 seconds occurred when the roll tail piece hit the roll nip.

Trouble usually arises when the operating speed of the equipment is close to the natural frequency. For example, a fan suspended in the air with a sensor attached is

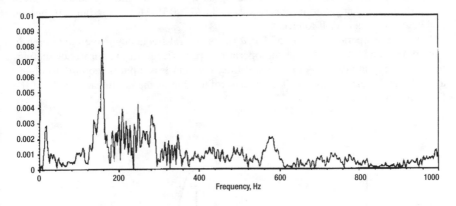

Figure 5-55. Low frequency response obtained at a steel mill.

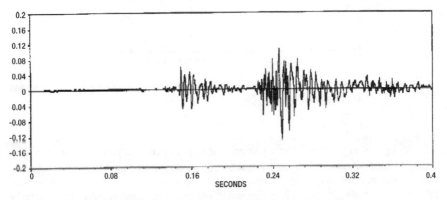

Figure 5-56. Rapid decay time of steel mill roll vibration.

struck. Although the ear may not detect the sound, it will "ring" at its natural frequency, and a frequency will be displayed as in Figures 5-53 and 5-55. The determination of the fan's natural frequency should be made when it is mounted in its operating position in the same manner: mount a sensor and strike the frame of the fan. It will respond and display the frequency, and, for purposes of this discussion, we will assume it to be at 37.5 Hz.

A good design will ensure that the equipment does not operate within 20% of a natural frequency. These frequencies can also be estimated by mathematical means. If a fan is to operate at 1800 RPM (30 Hz), there should be no natural frequencies from 24 Hz to 36 Hz. Problems arise when the operator decides that this fan, which is moving product, needs to be speeded up so it can move more product. Sheaves and pulleys are changed to increase the speed to 2200 RPM, and the fan is turned on and begins to shake itself to pieces.

Say a fan is operating within 50 CPM of its natural frequency. Note the tracing in Figure 5-57 and observe that as the operating speed (frequency) approaches the natural frequency indicated by the vertical line, less and less input is required to obtain a greater and greater response. This is also referred to as the nonlinear portion of the frequency response curve.

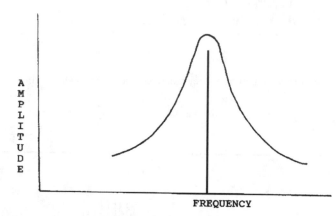

Figure 5-57. A typical frequency response curve. Frequencies producing high amplitudes of vibration must be avoided.

Many operators have had the unsuccessful experience of trying to balance a fan, as given in this example. One solution is to add braces to stiffen the fan structure and increase the natural frequency so that the new operating speed is again outside the 20% range. Another solution is to increase the mass of the fan by placing a large mass either directly on top of the equipment, or by mounting a "torsional pendulum" on the machine whose natural frequency is to be modified. Such a pendulum may simply consist of a vertical rod attached to the equipment casing. A mass of suitable magnitude is then clamped to the rod, and the height of this mass successively adjusted relative to the machine centerline. It is thus possible to experimentally arrive at a point of least vibration for the entire assembly. This approach is often chosen in cases where it is impractical to lower the natural frequency by attaching bracing to the surrounding structure. When either of these solutions is accomplished, the fan can be balanced and operated without resonance problems.

When the testing lab of a major oil company was operating a diesel engine, the entire structure vibrated extensively. The vibration was so severe, it was not possible to perform the tests. The engine was programmed to operate at 2000 RPM during the test, and it was suspected that a resonant condition might exist. Figure 5-58 is the time domain of the bump test showing a fairly fast decay and Figure 5-59 is the frequency

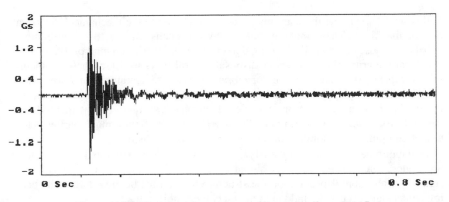

Figure 5-58. Time domain response of an engine "bump test."

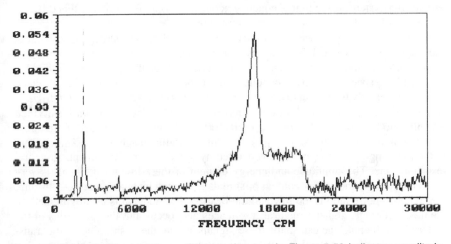

Figure 5-59. Frequency spectrum of the engine test in Figure 5-58 indicates amplitude excursion at 2025 CPM—undoubtedly one of the running speeds of the engine.

domain. Note the resonant frequency marked at 2025 CPM!!! The only solution was to stiffen the structure and raise this natural frequency to 2400 CPM or higher.

It should be recalled that there is absolutely no way to operate equipment that is in itself resonant at its operating speed or mounted to a structure that is resonant at the equipment operating speed. And, it is impossible to balance rotating equipment that is operating near a resonant frequency.

SEE Technology and Enveloped Acceleration

It is well-known that there are many factors that lead to the mechanical deterioration of machines. These include, but are by no means limited to, misalignment, unbalance, loose components, bearing defects, lack of lubrication, and problems with structural integrity. We have at our disposal several ways to detect these problems. For example, it is possible to monitor vibration levels, temperatures, and power levels. The customary approach is to log one or more of these indications and to develop trends of the amplitudes or severities. However, two more recent methods are available for monitoring bearing conditions. These will allow a much earlier detection of equipment deterioration or machinery component distress.

To capture bearing defects by analyzing vibration signals generated from velocity or acceleration sensors, extensive signal processing is needed. Since the vibration signal is often more dependent on the structure than on the bearing defect, the bearing defect signal is usually hidden in the machinery noise.

Techniques such as tuned low-frequency filtering and envelope detection can help to reveal the bearing defect signal, but these techniques have to be adapted to the application and depend on machine dimensions and bearing sizes. A different system implementation is therefore needed for every different application being monitored. Monitoring systems using high-frequency acoustic emissions overcome this difficulty; they work well over a wide range of applications.

Whenever a rolling element runs over a crack or spall mark, a stress wave signal of short duration is generated. This means that the signal contains a very wide range of frequencies. The range of frequencies generated depends on the shape and width of the crack and the speed of the balls or rollers. It has been demonstrated in the lab that a spall defect of only 0.5 mm in diameter will generate detectable signals up to 6 MHz.

Vibration monitoring techniques for bearing defect detection can be divided roughly into low (0–20 kHz), medium (20–100 kHz), and high (>100 kHz) frequency techniques. Most of the signals detected during routine machinery monitoring are in the low range. These signals are generated by unbalance, misalignment, structural resonance, etc. The amplitude and energy content of these low-level signals is very high compared to the energy content of signals generated by bearing defects. In sum, the signal-to-noise ratio is too low and the bearing defect signal is usually hidden. In the medium range, higher structural resonances also occur, but the properties of the materials in the machine can shape these signals so that the positioning of the transducer often becomes critical. Transducer location can cause the signals to be either amplified or attenuated, which results in a variable signal-to-noise ratio. In other words, sometimes you have something to work with and sometimes you don't.

In the high frequency range above 100 kHz, the machinery material problems are eliminated since they usually do not show up in the higher frequencies. Therefore, by using proper care, these high-frequency signals can be examined for their energy content, a measurement that will provide us with the desired bearing fault information. And, we can avoid the high cost of high-frequency analysis by using an enveloping technique to process the signals generated by bearing defects.

The high frequency acoustic emission transducer is a very wide band piezo-electric accelerometer that responds to frequencies in excess of 1 MHz. The signal is filtered, rectified, and processed to produce a low-frequency signal that includes the fundamental repetition frequencies of any pulses from the original data and their higher harmonics. These pulses are generated by the overrolling of the defect by the balls or rollers in the bearing. As the metal alloy grains are deformed by the bearing forces, the acoustic signal is emitted as a pulse of energy and is referred to as a spectral emitted energy (SEE) signal. This pulse is repeated at a fixed frequency determined by the geometry of the bearing and the speed of rotation. The signal produced by the defect is a train of pulses, which has as its Fourier transform another train of pulses. When used in conjunction with computer-generated bearing frequency programs, these pulses can be easily assigned to a particular type of bearing fault by its specific frequency. As in the case of any fault analysis using the frequency spectrum, the frequency identifies the source of the fault and the amplitude identifies the severity.

The manner in which these signals are processed is often referred to as enveloping, a technique which has been used since the 1960s. The primary difference is the frequency range that is being monitored: 250 to 350 kHz. This frequency band is well above any machinery signals such as imbalance, misalignment, or structural resonances and therefore contains primarily the acoustic pulses generated by the over-rolling of bearing faults. It is only as the severity increases to near-failure that mechanical vibrations are detected in the acceleration or velocity mode.

Experience has thus shown that relevant acoustic signals appear much earlier than the usual vibration signals, in both velocity and acceleration. It is very important to note that *no* acoustic signal is generated when the fault is not severe enough to break through the oil film and no metal-to-metal contact takes place. Tests have shown in these cases that the acceleration or velocity signals can be processed through the same enveloping circuits to enhance the repetitive vibration signals and present them to be examined for bearing fault frequencies. One note of caution is that the enveloped signal amplitudes are in fact a mathematical summation of the harmonic series from the bearing defect. Therefore, the amplitude cannot be compared to an amplitude obtained without enveloping. For example, a normal acceleration signal may read 0.10 Gs and with the sensor in the same position on the same bearing, the enveloped signal amplitude may be 0.30 Gs. However, the enveloped amplitudes can be trended to detect increasing degradation in the bearing just as the non-enveloped amplitudes are trended for the same purpose.

The following spectra, taken in the lab and at several work sites illustrate the differences between the various techniques. It should be noted, however, that deliberately induced bearing flaws are usually of such magnitude that they will also be detected by the more traditional monitoring techniques.

Figure 5-60 is the spectrum of a compressor bearing in which a fault was placed in the outer race. The ball pass frequency outer race (BPFO) was calculated to be 285 Hz. This was marked and the harmonics marked. This spectrum was taken using normal acceleration processing. Figure 5-61 is obtained by looking at the same bearing and taking the measurement in velocity. Again the BPFO is seen at 285 Hz with some harmonics marked. The spectrum shown in Figure 5-62 is generated using the high-frequency sound generated as the ball rolls over a defect in the outer race. The BPFO of 285 Hz is marked with harmonics.

Referring to Figure 5-63, this spectrum is again in Gs, but this time it is enveloped Gs. The signal is again at 285 Hz with additional side bands with spacing at running speed. In comparing this spectrum with Figure 5-60, note that the amplitudes are dif-

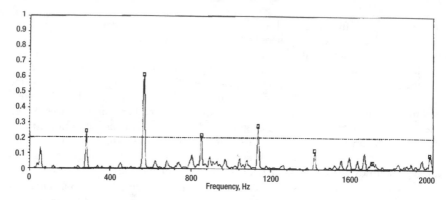

Figure 5-60. Acceleration spectrum (Gs) of rolling element bearing with outer ring defect. The calculated BPFO was 285 Hz. Note harmonics.

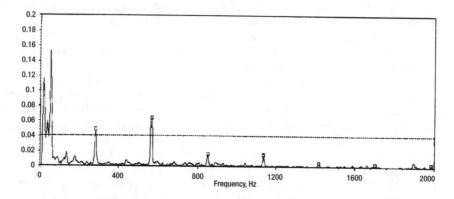

Figure 5-61. Velocity spectrum (ips) of the bearing described in Figure 5-60. BPFO and harmonics are marked.

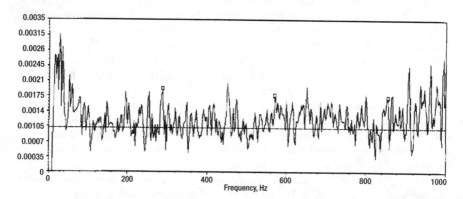

Figure 5-62. Acoustic spectrum of bearing from Figures 5-60 and 5-61. Note again BPFO at 285 Hz and its harmonics.

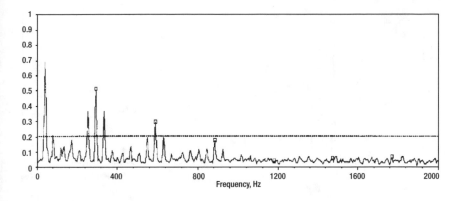

Figure 5-63. "Enveloped acceleration" of Figure 5-60 shows BPFO and harmonics, all with side bands spaced at running speed.

ferent. In Figure 5-60, the amplitude is 0.23 Gs and, with the enveloping, the summed amplitude is 0.51 Gs. Over time the two amplitudes could be independently trended, but as previously mentioned, never mixed.

Figure 5-64 is a field test of an installed bearing. The customer complained of a noisy running bearing, and the cause was investigated. A small amount of energy was seen at 224 Hz, which is the BPFO for this bearing, although the amplitude is only 0.0025 IPS. In order to enhance the spectrum, the signal was then processed as an enveloped acceleration signal for Figure 5-65.

With the machinery noise degraded by signal processing and the repetitive BPFO signal at 223 Hz enhanced, it was apparent that there was a flaw in the outer race of the bearing. Again note that the enhanced amplitude is larger than actual, in this case, 0.009 Gs. The bearing was removed and the inspection revealed scratch marks on the raceway. It should be noted that the number of flaws or, in this instance

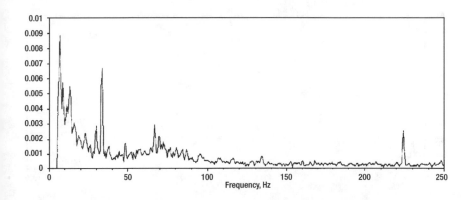

Figure 5-64. Vibration velocity detected during field test of ball-bearing-equipped machinery.

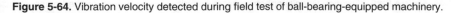

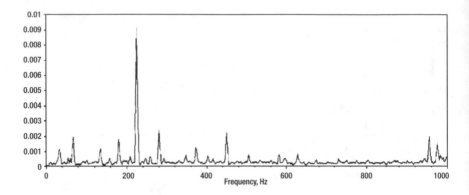

Figure 5-65. The signals obtained from the machine in Figure 5-64 after being processed as "enveloped acceleration." Note indication of flaw at a BPFO frequency of 223 hertz.

scratches, will not change the frequency. It doesn't matter if there is one flaw, two flaws, or more, the BPFO will remain the same. However, the more flaws, the greater the amplitude.

Noisy bearings seem to be the most common complaint among bearing customers. Several techniques were thus used to investigate the cause of a high-pitched whine in a motor generator. Figure 5-66, taken in IPS, did not reveal any apparent problems. Next, the bearing was checked using plain acceleration and Figure 5-67 was obtained.

With so much energy displayed, it is difficult to determine what is important and what is not. The information was downloaded to the computer and, while displaying the spectrum, the bearing information from the frequency analysis module (FAM) in Figure 5-68 was overlaid. With this information available, there appears to be a fault in either the outer race (BPFO), or on the balls (BSF), although the BSF is only showing in the second harmonic.

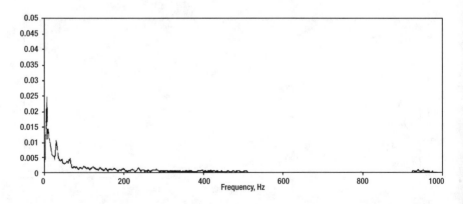

Figure 5-66. Vibration velocity spectrum from electric motor with high-pitched noise.

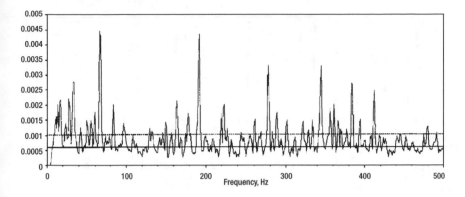

Figure 5-67. Plain acceleration signal obtained from electric motor in Figure 5-66.

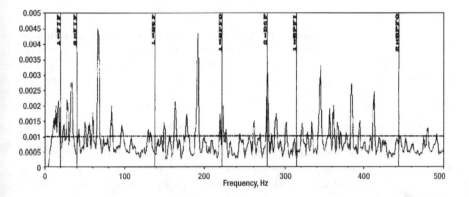

Figure 5-68. Vibration velocity signals overlaid with frequency analysis module (FAM)—electric motor from Figures 5-66 and 5-67.

The next spectrum was taken, using enveloped acceleration. The enveloping circuits degraded all the machinery noise and their harmonics, leaving just the repetitive signals generated by the bearing.

As can be seen in Figure 5-69, the BPFO and the BSF are clearly enhanced and appear to be the problem. Note: the slight offset from the FAM markers and the actual signal can be attributed to bearing geometry or a slight change in actual speed vs. what was used in the FAM calculations.

When the bearing was torn down, there were corrosion marks on the bearing race and on the balls. The history of the bearing was traced, and it was found to be more than nine years old. After the original purchase, it had changed hands three times within the same company and had not been properly maintained in storage. The shipping lubrication had dried, allowing atmospheric moisture to get inside the bearing. This was apparently acidic and had etched the surfaces during the nine years. Result: a noisy bearing.

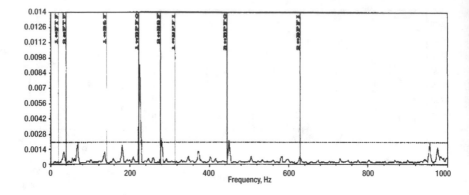

Figure 5-69. "Envelope acceleration" of motor described in Figures 5-66 through 5-68 finally points to BPFO and BSF as defect generators.

Another example of the same type of problem is shown in the next three figures. When the noisy motor in Figure 5-70 was examined using only velocity, there were no apparent problems. Another spectrum, Figure 5-71 was taken using acceleration and the displayed energy was found to be very low. Nothing of significance was apparent. To complete the examination another spectrum, Figure 5-72 was taken, this time using enveloped acceleration. Having consulted FAM, it was evident that the BSF was 95 Hz and it was very apparent in this spectrum.

In summary, acoustic and enveloping techniques have been combined in another portable battery powered analyzing tool for the plant vibration analyst. As with any tool, it has its limitations that must be understood and allowed for by the operator. With that understanding, a good vibration monitoring program will be further enhanced.

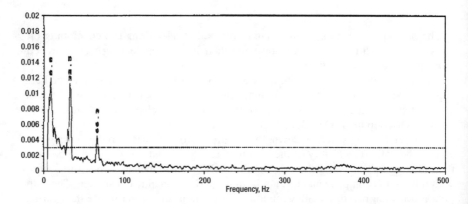

Figure 5-70. No apparent problems are found in "noisy" motor bearings analyzed in this velocity spectrum.

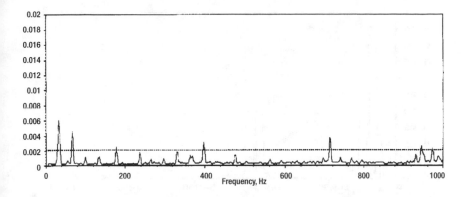

Figure 5-71. Acceleration spectrum of noisy motor bearing from Figure 5-70 shows no apparent deficiency.

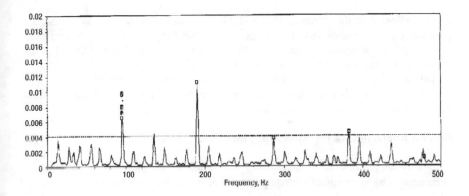

Figure 5-72. "Envelope acceleration" of the signal in Figure 5-71 enhances amplitude at BSF (95 Hz) and its harmonics.

Quality Control

A manufacturer of large compressors was confronted with a unique problem. This company had a standard compressor casing that was designed to accept several different components. This "modularization" was to provide their customers with the internals that matched their needs. Problems arose when it was necessary to confirm that the correct internals were provided as called for in the purchase order. Performance testing was too expensive; the units were thus given only a mechanical run test on air.

The solution to the problem was found using frequency spectra of known correct compressors. When the spectrum was taken, an alarm envelope was created around the existing display. Figure 5-73 is the plot of a compressor with the proper internals.

This acceptance envelope was generated for each compressor and stored in the computer, readily available whenever needed. To use the automatic program, the

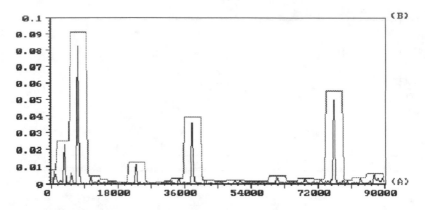

Figure 5-73. Acceptance envelope drawn around acceptable velocity peaks in a compressor spectrum.

technician enters in the computer the model being tested. This loads a data collection program that, in turn, collects the data, displays the frequency spectrum, and overlays the proper acceptance envelope of the unit under test. To illustrate what is displayed when the wrong internals are installed, see Figure 5-74.

Although the acceptance envelope is developed to check for the proper internals, it also provides another check on proper gear mesh, verifies gear condition and the fact that the bearings are acceptable and installed properly.

Engine Analyzers for Reciprocating Machinery

Performance analyzers contribute to the economical operation of engines and compressors by showing which equipment is operating efficiently, which is not, and

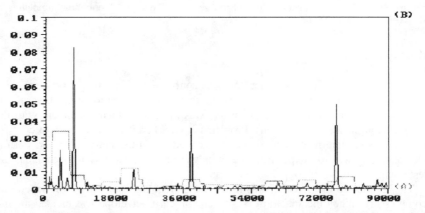

Figure 5-74. When a particular compressor spectrum showed velocity peaks only under the envelope drawn here, comparison with the stored peaks identified that the wrong compressor internals were installed here.

the reason why it is not. Performance analysis obviously cannot eliminate all engine problems. What it can do is alert reciprocating engine and compressor operators to problems and potential problems before serious equipment damage occurs.

The condition of reciprocating machinery is revealed by efficiency calculations involving horsepower, pressure vs. time relationships, pressure vs. swept volume, ignition and vibration waveforms, or acquisition of data containing clues as to ignition malfunction, valve leakage, worn bearings, and other mechanical problems.

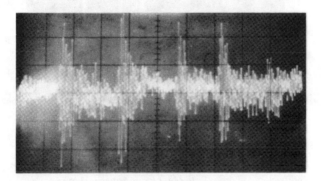

Figure 5-75. Time wave-form analysis of reciprocating compressor shows vibration impulses due to opening and closing of inlet and discharge valves.

In the mid-1960s, electronic performance analyzers became commercially available that can analyze higher-speed equipment better than can be analyzed with older instrument types. Not only are pressure/time and pressure/volume displays available on its oscilloscope readout, but vibration, ignition, and ultrasonic traces can be displayed also. Various connecting rod length-to-stroke ratios can be more easily accommodated, and changes in pressure scaling are feasible. Because of this relative versatility, many of these electronic performance analyzers are used in the petro-chemical and gas pipeline industry.

Typical problems diagnosed by one such analyzer* are shown in Figures 5-76 through 5-83.

Figure 5-76 shows both the vibration (low frequency) and ultrasonic (high frequency) signatures of a gas engine with piston ring "rattle" due to a worn cylinder liner.

A more severe problem is illustrated in Figures 5-77 through 5-79. Progressing from normal electronic signal traces in Figure 5-77 to incipient defect indications in Figure 5-78, we finally observe severely notched cylinder port bridges in Figure 5-79.

Low-frequency vibration and high-frequency ultrasonic signals are compared for a small gas engine operating normally in Figure 5-80 and experiencing cylinder scuffing in Figure 5-81.

* Beta Machinery Analysis, Ltd., Calgary, Alberta, Canada.

(*text continued on page 414*)

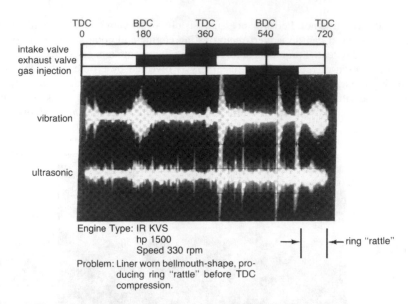

Figure 5-76. Analyzer signatures of a gas engine with worn cylinder liner. Note ring "rattle" before TDC compression. (*Source: Beta Machinery Analysis, Ltd.*)

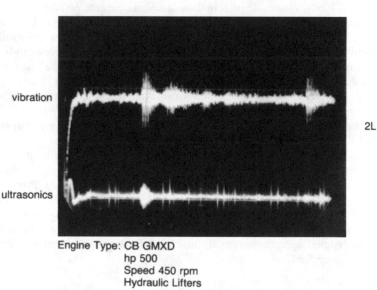

Figure 5-77. Gas engine analyzer signature with normal port bridges.

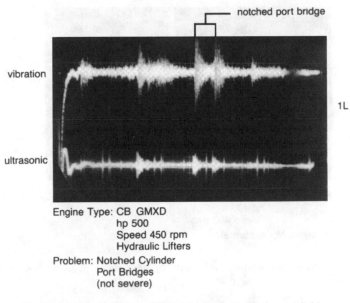

notched port bridge

vibration

1L

ultrasonic

Engine Type: CB GMXD
hp 500
Speed 450 rpm
Hydraulic Lifters
Problem: Notched Cylinder
Port Bridges
(not severe)

Figure 5-78. Gas engine with slightly notched cylinder port bridges.

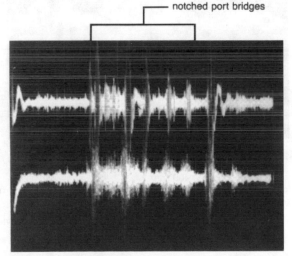

notched port bridges

ultrasonic

vibration

Engine Type: CB GMXD
hp 500
Speed 450 rpm
Hydraulic Lifters
Problem: Notched Cylinder
Port Bridges
(severe)

Figure 5-79. Gas engine with severely notched cylinder port bridges.

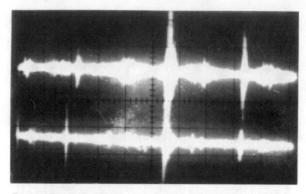

Engine Type: IR SVG
hp 400
Speed 400 rpm
Problem: None, normal

Figure 5-80. Gas engine with normally worn cylinder walls. (*Source: Beta Machinery Analysis, Ltd.*)

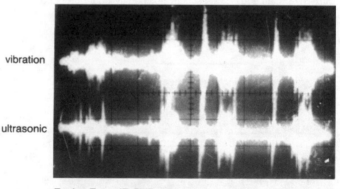

vibration

ultrasonic

Engine Type: IR SVG
hp 400
Speed 400 rpm
Problem: Cylinder scuffing

Figure 5-81. Gas engine with scuffed cylinder walls. (*Source: Beta Machinery Analysis, Ltd.*)

(*text continued from page 411*)

Figure 5-82 depicts the oscilloscope traces for cylinder 1R with worn rocker arm bushings, and for cylinder 2R whose piston and cylinder liner are badly scuffed.

Power cylinder pressure vs. time displays are shown in Figures 5-83 through 5-86. Figure 5-83 indicates power cylinder unbalance. The underfueled No. 1 cylinder

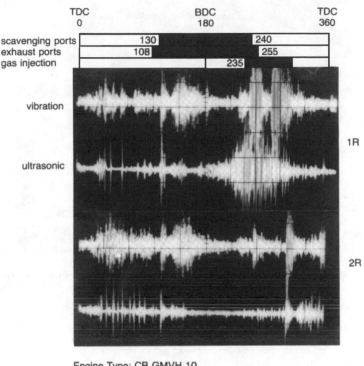

Engine Type: CB GMVH 10
hp 700
Speed 330 rpm
Problems: No. 1 R, worn rocker arm bushing
No. 2 R, piston and liner badly
scuffed

Figure 5-82. Analyzer traces for gas engine with cylinder 1R exhibiting worn rocker arm bushings, and cylinder 2R exhibiting badly scuffed piston and cylinder liner. (*Source: Beta Machinery Analysis, Ltd.*)

shows wide fluctuation of peak pressures, typical of a lean condition. Adjusting the fuel balancing valve raised peak firing pressures to average levels and reduced peak pressure deviations, as shown in Figure 5-84.

Misfiring and "soft" firing by excessive scavenging air pressure led to significant deviations in peak cylinder pressures, Figure 5-85. Reducing the scavenging air pressure resulted in steadier peak pressures, as illustrated in Figure 5-86.

Pressure-time and pressure-volume curves of a 14-inch diameter reciprocating compressor cylinder are shown in Figures 5-87 and 5-88, respectively. The discharge valve on the head end of this cylinder had failed. With engine load remaining normal, there was no significant change in discharge temperature. However, a significant reduction in interstage pressure was noted.

(*text continued on page 419*)

Power Cylinder Pressure Time Display

Right Bank

Cylinder No.   1   4   3   2   5

13° BTDC

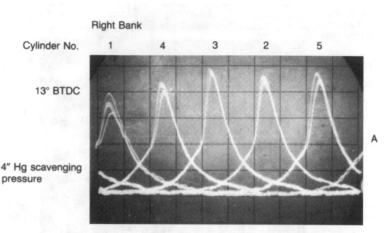

A

4″ Hg scavenging
pressure

Figure 5-83. Gas engine with cylinder no. 1 underfueled. (*Source: Beta Machinery Analysis, Ltd.*)

After balancing

13° BTDC

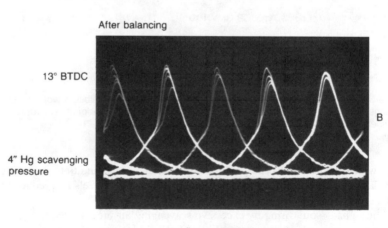

B

4″ Hg scavenging
pressure

line separation = 150 psi

Engine Type: CB GMVH 10
           hp 700
           Speed 330 rpm

Problem: Power Cylinder unbalance. Underfueled No. 1 cylinder shows wide fluctuation of
peak pressures typical of a lean condition.
Adjusting fuel balancing valve raised peak firing pressure to average levels and
produced reduced deviation in peak pressures.

Figure 5-84. Gas engine properly balanced. (*Source: Beta Machinery Analysis, Ltd.*)

Power Cylinder Pressure Time Display

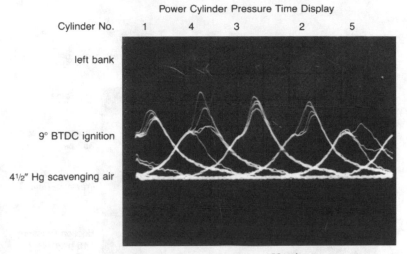

1 cm = 150 psi

Figure 5-85. Misfiring in gas engine caused by excessive scavenging air pressure. (*Source: Beta Machinery Analysis, Ltd.*)

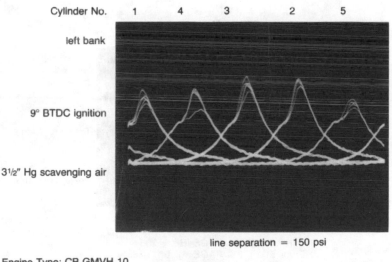

line separation = 150 psi

Engine Type: CB GMVH 10
        hp 700
        Speed 330 rpm

Problem: Lean mixture causing excessive misfiring and soft firing. Note steadier peak pressures when scavenging air pressure is reduced.

Figure 5-86. Gas engine exhibiting much steadier pressure after scavenging air pressure was reduced. (*Source: Beta Machinery Analysis, Ltd.*)

Compressor Data

Note: The pressure shown adjacent to the Pressure-Volume curves are based on the deadweight markers.

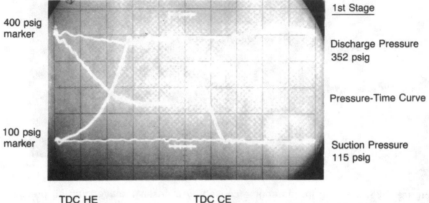

1st Stage

400 psig marker

Discharge Pressure 352 psig

Pressure-Time Curve

100 psig marker

Suction Pressure 115 psig

TDC HE   TDC CE

Figure 5-87. Pressure-time curve for reciprocating compressor cylinder with failed discharge valve. (*Source: Beta Machinery Analysis, Ltd.*)

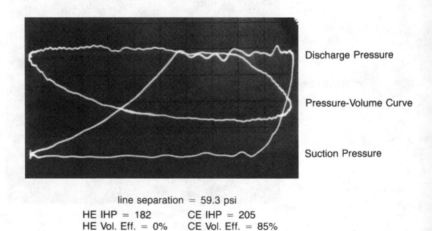

Discharge Pressure

Pressure-Volume Curve

Suction Pressure

line separation = 59.3 psi

HE IHP = 182   CE IHP = 205
HE Vol. Eff. = 0%   CE Vol. Eff. = 85%

Compressor: IR KVG 14″ cylinder diameter
Speed 330 rpm

Problem: Failed discharge valve on head end. No significant change in discharge temperature but a significant reduction in interstage pressure. Engine load remained normal.

Figure 5-88. Pressure-volume curve for reciprocating compressor cylinder showing significant reduction in interstage pressure. (*Source: Beta Machinery Analysis, Ltd.*)

(text continued from page 415)

In summary, vibration and pressure-volume analysis techniques can be applied effectively to reciprocating machines to detect and identify mechanical *and* operational problems. However, since many problems encountered on reciprocating machines have similar vibration characteristics, further investigation may be necessary in order to pinpoint the exact cause. Analysis and identification of defects can be greatly enhanced if prior vibration history or baseline signatures can be obtained on units which are known to be operating in satisfactory condition. This will help identify the normal or inherent vibration frequencies and their corresponding amplitudes. Then, when vibration increases do occur, a new signature can be obtained and compared with the original baseline signature to quickly reveal which frequency components have changed. This approach can greatly simplify the analysis of reciprocating machinery vibration, which in many cases is inherently complex.

Ultrasonic Analyzers

There are two ways in which ultrasonic detection can be used for condition monitoring of compressors: the walk-around method, by which an inspector literally walks around and takes readings with an ultrasonic scanner, and the use of an ultrasonic scanner with a compressor analyzer, an oscilloscope that shows the effects of vibration, pressure, and other events inside the engine by traces on its screen. Scanner and analyzer used together allow the engineer to examine, review, and record the mechanical condition of the compressor.

Similarly, ultrasonic inspection and monitoring of bearings is considered an acceptable method of detecting incipient bearing failure. Experiments performed by NASA revealed that changes in frequency from 24kH through 50kH in a bearing gave warning long before other indicators, such as heat and vibration.

The structural resonances of one faulty component, such as the repetitive impact of a ball passing over the pit of a fault in the race surface, produces ultrasounds. Faults are made evident by increases in amplitude in the monitored frequencies. Brinelling of bearing surfaces produces similar increases in amplitude as balls get out of round. The repetitive ringing of the flat spots is also detected as an increase in amplitude of the monitored frequencies.

As the ultrasound detector reproduces the ultrasonic frequencies it detects as audible sounds, technicians must become familiar with the sounds of good bearings so that they can detect the sounds of faulty bearings. Because operators must know the sounds of bearing wear, some user companies confine the use of their ultrasonic detector to one or two members of the maintenance crew, who become expert in recognizing the different sounds of good and bad bearings.

Generally, a good bearing makes a rushing or hissing sound. Loud rushing sounds similar to those of a good bearing, only slightly rougher, indicate lack of lubrication. Crackling or rough sounds indicate a bearing is at its point of failure. In some cases a damaged ball makes a clicking sound. These sounds are detected by touching one of the detector probes to the bearing housing, and observing the changes in intensity displayed on the meter.

There are two basic procedures in testing bearings: comparative and historical. The first involves testing two similar bearings and "comparing" their differences. Historical testing means monitoring a given bearing over a period to establish its history. By analyzing a bearing's history, patterns of wear at particular frequencies become obvious, and allow for the early detection and repair of problems.

Severity of Vibration Reduced with Synthetic Lubricants

Compressor and gear operation with synthetic lubricants has been steadily increasing since the early 1960s, making this application one of the oldest and most widely recognized. Phosphate esters, silicones, synthetic hydrocarbons, polyol esters, and diesters, as well as various combinations of these base stocks, have been used in systems ranging in size from a few gallons to several thousand gallons per year.

Synthetic compressor oils can provide a number of important performance improvements, including: reduced sludge and deposit formation, extended oil drain intervals, reduced energy consumption, lower make-up oil requirements, and year-round operation in areas subject to wide temperature extremes.

In the final analysis, however, the value of a synthetic lubricant can only be measured by savings in day-to-day maintenance and operating costs. Any major supplier of synthetic lubricants will be able to produce well-founded cost justifications and reference lists of satisfied clients. Few references, however, are more striking than the oscilloscope traces obtained at a gas-engine compressor water-pump drive before and after adding a high-grade synthetic lubricant (such as supplied by the Royal Purple Company, Humble, Texas 77396) to the drive train (Figures 5-89A through 5-89C). There is clear evidence that the greater film strength and film thickness of these oils can arrest wear and dampen the impact loading of contacting machine-internal components.

Establishing Safe Operating Limits for Machinery

The establishment of safe operating limits for major machinery requires a knowledge of machine sensitivity and failure risk. Single-train, critical machines are usually in the spotlight because their shutdown—scheduled or unscheduled—can cause costly interruptions in production and unit output. Moreover, serious failures of major machinery may incur burdensome repair expenditures even for a relatively wealthy owner company.

Some safe operating limits are firmly set and require neither debate nor discussion. Regardless of the critical nature of a potential shutdown or its impact on plant output, when the suction drum level rises to the point where liquid is about to enter into a centrifugal compressor, or when the lube oil supply pressure at the bearings dips to zero, the machine *must* be shut down. But what about high vibration? How high is still acceptable?

In 1977, S. M. Zierau of Essochem Europe ventured to put years of troubleshooting experience into a set of "limit guidelines." He prefaced these guidelines by stating that the decision to shut down vulnerable turbomachinery trains calls for risk evaluation by a team of process-operations, maintenance, and machinery-oriented

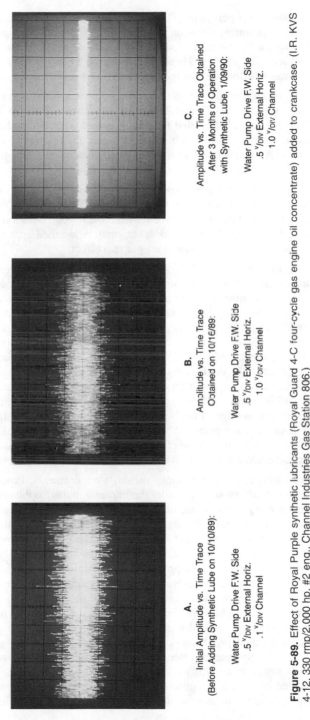

A.

Initial Amplitude vs. Time Trace
(Before Adding Synthetic Lube on 10/10/89):

Water Pump Drive F.W. Side
.5 ᵛ/ᴅɪᵥ External Horiz.
.1 ᵛ/ᴅɪᵥ Channel

B.

Amplitude vs. Time Trace
Obtained on 10/18/89:

Water Pump Drive F.W. Side
.5 ᵛ/ᴅɪᵥ External Horiz.
1.0 ᵛ/ᴅɪᵥ Channel

C.

Amplitude vs. Time Trace Obtained
After 3 Months of Operation
with Synthetic Lube, 1/09/90:

Water Pump Drive F.W. Side
.5 ᵛ/ᴅɪᵥ External Horiz.
1.0 ᵛ/ᴅɪᵥ Channel

Figure 5-89. Effect of Royal Purple synthetic lubricants (Royal Guard 4-C four-cycle gas engine oil concentrate) added to crankcase. (I.R. KVS 4-12, 330 rmp/2,000 hp, #2 eng., Channel Industries Gas Station 806.)

technical staff personnel. The machinery engineer must bring to this risk-assessment meeting pertinent vibration baseline and available trend data. He should be prepared to explain the machine's sensitivity, possible problem causes, and available repair and remedial options. He should also give his expert view on the probable consequences of allowing anticipated defects to progress further.

Knowing the machine's sensitivity is of paramount importance. All machines are certainly not created equal. Experiencing six mils of shaft vibration at 5000 rpm may be high, but still tolerable for one machine, whereas another machine experiencing the same vibration amplitude at 5000 rpm may literally start to come apart at the joints. A qualified machinery engineer must have a feel for the probability of his machine belonging to one category or the other.

Similarly, we would expect the process operations person to know the value of unit downtime to the plant. For the low-loss case, it would not be appropriate to take high risks by operating a defective machine beyond conservative limits, while for the high-loss case, additional risk-taking may be deemed appropriate as long as supplementary protective measures are put into effect.

In Figures 5-38 and 5-39, we had given general vibration guidelines for typical machinery. These guidelines are approximate at best and can be exceeded when the specific machine behavior in its physical environment of structural supports and its response to load, speed, and temperature changes are known. It is important to understand that experiences of successful operation at a higher than maximum guideline level are not transferable to other machines, since such a level usually reflects a specific installation peculiarity. However, recording such incidents is valuable because it provides needed experience background. Again, we need to say that there is no easy method for establishing such "custom-tailored" maximum allowable levels, independent of the sophistication of the analysis equipment used. Such limits require knowledge of one's machine, confirmative field testing, experience, and risk judgment. In any event, S. M. Zierau's experience led to the development of Figures 5-90 and 5-91, which are offered as a supplement to the many charts available from machinery vendors.

Finally, a few observations on skills requirements. Process plant engineers and managers are sometimes under the impression that sophisticated analysis equipment and/or computerized machinery monitoring alleviates the need for experienced analysts. This impression may well stem from the need to justify the expensive and often elaborate equipment, and one way chosen was to overemphasize actual and potential benefits. However, while signal or spectrum analyzers and trend/trip data from computer systems provide valuable information which eases the job of the technical support engineer, the engineer's skill is still needed to make use of this information. Manufacturing excellence will not be achieved through black boxes, but will continue to be the result of skill applied by all levels of plant personnel. Computer prediction of faults and action recommendations are only possible on clearly defined repeat problems or by remaining at conservative guideline levels. Repeat problems are generally rare, and conservative guideline levels are unacceptable for high-outage-loss units. In other words, we need the sophisticated tools to *assist* the skilled personnel in optimizing plant operations, but skilled personnel are needed more than ever to take advantage of the additional information now available.

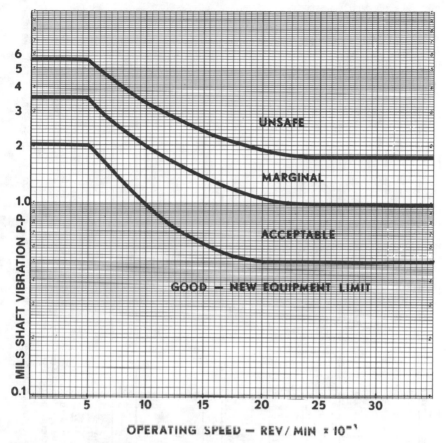

OPERATING SPEED — REV / MIN × 10⁻¹

Notes:
1. Operation in the "unsafe" region may lead to near-term failure of the machinery.
2. When operating in the "marginal" region, it is advisable to implement continuous monitoring and to make plans for early problem correction.
3. Periodic monitoring is recommended when operating in the "acceptable" range. Observe trends for amplitude increases at relevant frequencies.
4. The above limits are based on Mr. Zierau's experience. They refer to the typical proximity probe installation close to and supported by the bearing housing and assume that the main vibration component is *1 × rpm frequency*. The seemingly high allowable vibration levels above 20,000 rpm reflect the experience of high-speed air compressors (up to 50,000 rpm) and jet-engine-type gas turbines, with their light rotors and light bearing loads.
5. Readings must be taken on machined surfaces, with runout less than 0.5 mil up to 12,000 rpm, and less than 0.25 mil above 12,000 rpm.
6. Judgment must be used, especially when experiencing frequencies in multiples of operating rpm on machines with standard bearing loads. Such machines cannot operate at the indicated limits for frequencies higher than 1 × rpm. In such cases, enter onto the graph the predominant frequency of vibration instead of the operating speed.

Figure 5-90. Turbomachinery vibration limits proposed by experienced machinery engineers.

SPEED HARMONICS

MACHINE TYPE	1xRPM	2xRPM	3xRPM	4xRPM	1xVane Pass	2xVane Pass	1xGear Mesh Freq.	2xGear Mesh Freq.	1xBlade Pass	2xBlade Pass
Blowers, Up to 6,000 RPM Maximum	.50	.40	.25	.25	.10	.050				
Centrifugal Compressors Horizontal	.25	.20	.15	.15	.10	.050				
Barrel	.15	.10	.10	.10	.05	.025				
Steam Turbines Special Purpose	.30	.25	.15	.15					.10	.05
General Purpose	.50	.40	.25	.25					.10	.05
Gas Turbines and Axial Compressors	.50	.40	.25	.25					.10	.05
Centrifugal Pumps Between Brgs.	.25	.20	.15	.15	.10	.05				
Overhung Type	.50	.40	.25	.25	.10	.05				
Electric Motors	.25	.20	.15	.15						
Gear Units Parallel, Spec.Purp.	.25	.20	.15	.15	.15 (Intermed. Freq. 1,000–3,000 Hz)		.10	.05		
Parallel, Gen. Purp.	.50	.40	.25	.25	.15		.10	.05		
Epicyclic	.15	.10	.10	.10			.10	.05		

	1xRPM	2xRPM	1xLOBE Pass	2xLOBE Pass	3xLOBE Pass	4xLOBE Pass	5xLOBE Pass
Screw Compressors	.25	.20	.20	.20	.20	.20	.20

NOTE: (1) The significance of vane, blade and lobe pass frequencies is not yet fully understood. More field data must be evaluated to arrive at universally meaningful maximum levels.

(2) Vane, blade, lobe pass and gear mesh frequency amplitudes vary with load and/or speed changes. The actual sensitivity should be part of the data base.

Figure 5-91. Maximum allowable vibration limits as a function of speed harmonics (all measurements are in inches-per-second and are taken on bearing cap).

Appendix

Glossary of Vibration Terms

Acceleration. The time rate of change of velocity. Units are measured in Gs, ft/sec squared.

Accelerometer. Transducer whose output is directly proportional to acceleration. The output is generated by deforming either a natural crystal or a man-made piezo-electric crystal.

Acoustic emissions. Sound emissions that are emitted when an object or material vibrates. These emissions may or may not be heard but can be detected with the proper equipment. For example, flow noise through a valve can be heard with the ear and with equipment. Early bearing raceway damage may or may not be heard by the ear but can be detected with the proper acoustic accelerometers.

Aerodynamic and flow-induced vibration. Air flow from fans and fluid flow from pumps induce vibrations each time the fan or pump impeller discharges air or fluid. These pulsing discharges can be detected at a frequency equal to the shaft speed times the number of fan blades or pump impellers.

Amplitude. A measure of the severity of the vibration. For acceleration measured in Gs. For velocity, measured in inches-per-second. For displacement, measured in mils.

Anti-aliasing filter. A low-pass filter designed to filter out frequencies higher than ½ the sample rate in order to prevent aliasing.

Axial measurement. Measurements taken with the sensor held to the object, parallel to the axis of the rotating shaft and parallel to the ground.

Average. The sum of the values of the measurements taken, divided by the number of measurements taken.

Balance. When the centrifugal forces are equal as an object rotates, it is referred to as being "in balance." If the center of mass of the rotating object does not coincide with the center of rotation, unequal centrifugal forces will be generated and the rotating element will be "out of balance." There are two balance conditions: static and coupled. Static balance refers to a single-plane mass in which a single weight opposite the unbalanced mass will balance the unit. Coupled unbalance refers to a two-plane unit in which an unbalance exists on each end and the unbalance masses can be up to 180° out of phase with each other.

Ball pass frequency. The frequency generated when a rolling element passes over a flaw in the inner race, BPFI, or over the outer race, BPFO. See calculation formulas, later.

Baseline measurement. Usually the first performance measurements taken on a piece of equipment when it is new or rebuilt. Later performance parameters are then compared to this baseline to determine if any changes have taken place.

Bin. One spectral line in the frequency display of an FFT analyzer.

Blade or vane pass frequency. The number of fan blades or pump vanes times the rotational speed equals the specific frequency.

Calibration. A test to verify the accuracy and repeatability of measurement instruments.

Condition maintenance. Maintenance that is performed on equipment only when it is required to maintain desired performance. In contrast to periodic and preventative maintenance.

CPM. Cycles per minute. A cycle is one complete turn of a shaft or one complete display of a sine wave. In rotating equipment, the term is interchangeable with RPM. To convert to Hz, divide by 60, i.e., 1800 CPM is equal to 30 Hz.

Critical equipment. Usually considered non-spared equipment that, if it is out of service, will shut down production. Can also refer to safety equipment, such as fire pumps, that are not part of the production loop but are required for plant operation.

Critical speed. The rotational speed that coincides with the resonant frequency of the machine or the structure on which it is mounted.

Decay rate. The rate at which an object stops vibrating after being struck. A bell will have a long decay rate and a pump will have a fast decay rate. A slow decay rate is applicable to an object that is more resonant than an object that is less resonant.

Decibel (dB). A logarithmic representation of amplitude ratio defined as 20 times the base ten logarithm of the ratio of the measured amplitude to a reference amplitude.

Displacement. The distance traveled by the vibrating part, measured peak-to-peak in mils.

Dynamic range. The difference between the highest voltage level that will overload the instrument and the lowest voltage level that is detectable. Dynamic range is usually expressed in decibels.

Electrical runout. An error signal that occurs in eddy current displacement measurements when shaft surface conductivity varies.

Envelope processing, acceleration, and velocity. The signal processing technique in which the higher frequency harmonic signals are electronically processed to provide a mathematical sum of these harmonics over a selected range. The purpose of selecting a higher frequency range is to process signals that are well above the normal machinery signals of balance, misalignment, rubs, etc.

FFT. The accepted abbreviation for Fast Fourier Transform. Fourier was a French mathematician who devised the mathematical formula to derive the frequency domain from the time domain. See illustration following.

Forcing function. The force that causes the equipment to vibrate. For example, when the rotating mass is out of balance, it is the force that creates the 1X frequency. Misalignment is that force that creates the 2X frequency. See Chapter 5 for forcing functions and examples.

Frequency. The repetition rate of a periodic event, expressed in cycles per second (Hertz), cycles per minute (CPM), or multiples of rotational speed (Orders).

Frequency domain. The display of energy generated and displayed in the frequency spectrum in contrast to the time domain display. See FFT.

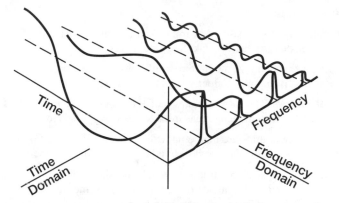

Time vs. frequency domain.

Gear mesh frequency (GMF). The frequency generated by improperly meshing gears. It is calculated by multiplying the speed of the shaft time the number of gear teeth on that shaft. Example: 1800 RPM × 30 teeth equals 54,000 CPM or 900 Hz.

Hanning window. The selectable window function that provides the best frequency resolution.

Harmonic. The repetitive signals generated by looseness in the equipment being measured. The frequency between signals will equal the 1X frequency. It is not unusual to see harmonics out to 15X. Sideband harmonics in gear mesh spectra are smaller energy spikes equally spaced around the primary harmonic of gear mesh in which the spacing will equal the gear shaft speed.

Hertz, Hz. A frequency measurement of the number of cycles of the sine wave per second. To convert Hz to CPM, multiply by 60, i.e., 30 Hz × 60 equals 1800 CPM.

Horizonal measurement. A measurement in which the sensor axis is placed 90° to the axis of the rotating shaft and parallel to the ground.

Imbalance. A condition in which the center of the rotating mass does not coincide with the center of rotation. In a spectrum, the imbalance signal will appear at the running speed of the unit being examined; also stated as 1X.

Lines. A measurement of the resolution of the spectrum. For example, in a 400-line spectrum measuring from 0 to 800 Hz, each line would equal 2 Hz. Any energy that occurred in that 2 Hz band would be displayed as one spike on that line. If the same spectrum was taken at 800 lines, the resolution would be 1 Hz per line, or twice as good.

Linear, nonlinear. When the vibration levels are trended over time and the trend is a straight line, either rising or falling, the trend is referred to as linear because the amount of increase is the same for each equal increase in time. A nonlinear increase would be the case in which, as time progresses, the amplitude increases or decreases, at a larger and larger amount, each time frame. Projections can be made from linear trends; they cannot be made from nonlinear trends.

Measurement units:

Mils. Displacement is measured in mils. A mil is one thousandth of an inch. Displacement is stated in peak-to-peak. See sine wave.

IPS. Inches per second. A measurement of velocity; the speed at which the item being measured is moving. Velocity is stated in peak.

Gs. acceleration. The rate of change of the velocity. A measure of the force being applied to the item being measured. Acceleration is stated in peak values.

These measurement units are mathematically related. IPS can be derived from the integration of Gs and displacement derived by integration of velocity.

Misalignment. A physical condition in which the shafts of two coupled units are not parallel (angular misalignment) or are not in the same vertical and horizontal planes (offset). Misalignment will generate a spike on the frequency spectrum at twice the operating speed of the units.

Modulating. When the vibration signal amplitude rises and falls over time. For example, a flaw on the inner race of a bearing will rotate in and out of the load zone. When in the zone, the amplitude will be high, and when it rotates out of the zone, the amplitude will fall. In the frequency spectrum, modulating signals will generate sideband harmonics; the spacing of the harmonics will equal the speed (CPM) of the shaft.

Natural frequency. Also referred to as resonant frequency. The frequency of free vibration of a system when excited with an impact force, as in a bump test.

Nonintrusive examination. The technique of determining the mechanical condition of equipment without stopping, opening, or modifying the equipment.

Non-synchronous. The amplitude sum of all frequencies that are not below 1X or multiples of 1X. See synchronous and subsynchronous.

Oil analysis. A laboratory technique to analyze the composition of lubricating oil to determine if any bearing material is present. Presence of bearing material would indicate wearing of the bearing and the amount would indicate the amount of wear.

Outage. The two types of outages are planned or forced. A planned outage occurs when the unit is shut down and work is performed as planned. A forced outage occurs when the unit fails and work is performed, usually on an emergency basis.

Overall amplitude. Total amount of vibration occurring in the frequency range selected.

Peak. The maximum positive amplitude shown on a sine curve. See sine wave.

Peak-to-peak. The sum of the maximum and minimum amplitudes shown on a sine curve. See sine wave.

Peak hold. A menu choice on data collectors. The data collector will continuously collect data and, as the amplitude varies, will capture and hold the latest peak amplitude. This will continue until the data collection is halted.

Period. The time required for a complete oscillation or for a single cycle of events. The reciprocal of frequency, $F = 1/T$.

Periodic maintenance. Maintenance that is performed on a calendar or some measure of time basis, i.e., every 12 or 18 months, every so many RPM, or every so many hours.

Phase. The angular method of determining how one object is moving in relation to another. Phase information is required to perform balance operations and requires a once-per-revolution trigger signal from the object being balanced. This trigger can be supplied by a fiber optic system, strobe light, or eddy current probes, and establishes the 0° starting point for balance weight placement.

Predictive maintenance. Usually maintenance that is performed based on a measurement.

Radial measurement. Measurements taken perpendicular to the axis of rotation to measure shaft dynamic motion or casing vibration.

Resonance. See natural frequency.

RMS. Root mean square. The measure of energy displayed in a frequency spectrum. It is derived by squaring each spectrum line, summing the results, and taking the square root of the sum. It also equals (peak) × 0.707. See sine wave.

Rolling element bearing. Bearings whose low friction qualities derive from lubricated rolling elements (balls or rollers).

Rotor. The rotating portion of a pump, fan, or motor.

Runout. The amount of wobble at the end of a rotating shaft.

SEE technology (Spectral Emitted Energy). The analysis process in which the high-frequency acoustic signals generated when the rolling element in a bearing passes over a flaw in the bearing surface. The signals are emitted by the microscopic movement of the metal grains as they rub against each other. These signals are then enveloped and presented in the low-frequency spectrum. The display signal will be at the characteristic bearing frequencies, BPFO, BPFI, etc.

Sensor (vibration). The device used to measure vibrations. It can refer to non-contact eddy probes (displacement), velocity pickups (velocity), or accelerometers.

Sine wave. A waveform with deviation that can be expressed as the sine or cosine of a linear function of time or space or both.

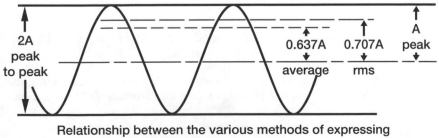

Relationship between the various methods of expressing
the amplitude of a sine wave.

Spectrum. In a frequency display, the energy generated by rotating equipment in which the specific frequency generated will identify the fault creating the energy. In a time spectrum, the energy will display based on the time interval between events.

Stator. The nonrotating portion of a pump or motor.

Sub-synchronous. The amplitude sum of all vibration frequencies below 1X.

Synchronous vibration. The amplitude sum of all vibration frequencies that are multiples of the 1X frequency, i.e., 1X + 2X + 3X + . . . nX.

Time domain. A spectrum that displays events as they occur over a set period of time with amplitude on the "Y" axis and time on the "X" axis.

Vertical measurement. A measurement taken in which the sensor is placed 90° to the rotating shaft and 90° to the plane of the earth.

Vibration. Oscillation of an object about a point where the point may or may not also be moving.

Vibration severity. The chart on the following page illustrates generally accepted bearing cap vibration limits. The operator must decide what levels he can accept for operation.

Formulas

Rolling Element Bearing Defect Frequencies

These formulas assume the inner race is rotating with the shaft while the outer race is fixed (stationary). See also Note 1.

$$\text{(1) Inner Race} = \text{BPFI} = \frac{N_b}{2}\left(1 + \frac{B_d}{P_d}\cos\theta\right) \times \text{RPM}$$

$$\text{(2) Outer Race} = \text{BPFO} = \frac{N_b}{2}\left(1 - \frac{B_d}{P_d}\cos\theta\right) \times \text{RPM} = N_b \times \text{FTF}$$

$$\text{(3) Ball (or Roller)} = \text{BSF} = P_{d.}\left[1 - \left(\frac{B_d}{P_d}\cos\theta\right)^2\right] \times \text{RPM}$$

$$\text{(4) Cage} = \text{FTF} = \frac{1}{2}\left(1 - \frac{B_d}{P_d}\cos\theta\right) \times \text{RPM} \approx .35 - .45 \times \text{RPM}$$

NOTE: If the inner ring is fixed (stationary) while the outer race rotates, the minus sign must be changed to a plus sign within the parentheses of the Cage Frequency Equation (4). In this case, $N_b \times$ FTF will not equal BPFI rather than BPFO; and FTF $\approx .55 - .65 \times$ RPM.

where:

N_b = number of balls or rollers
B_d = ball or roller diameter (in. or mm)
P_d = bearing pitch diameter (in. or mm)
θ = contact angle degrees

For a simple rule of thumb when the bearing geometry is not known:

BPFO = RPM $\times$ (number of balls) $\times$ 0.4
BPFI = RPM $\times$ (number of balls) $\times$ 0.6

Notes 1 and 2 give additional explanations.

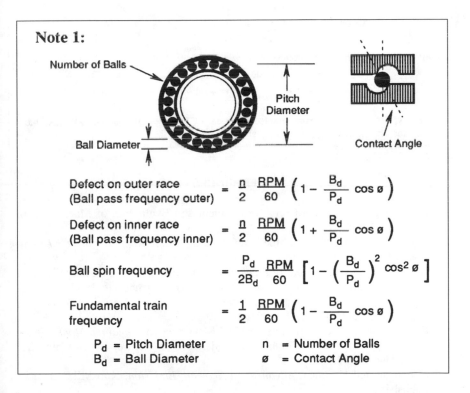

Note 1:

Number of Balls

Pitch Diameter

Ball Diameter

Contact Angle

Defect on outer race
(Ball pass frequency outer) $= \dfrac{n}{2} \dfrac{RPM}{60} \left(1 - \dfrac{B_d}{P_d} \cos \varnothing \right)$

Defect on inner race
(Ball pass frequency inner) $= \dfrac{n}{2} \dfrac{RPM}{60} \left(1 + \dfrac{B_d}{P_d} \cos \varnothing \right)$

Ball spin frequency $= \dfrac{P_d}{2B_d} \dfrac{RPM}{60} \left[1 - \left(\dfrac{B_d}{P_d} \right)^2 \cos^2 \varnothing \right]$

Fundamental train
frequency $= \dfrac{1}{2} \dfrac{RPM}{60} \left(1 - \dfrac{B_d}{P_d} \cos \varnothing \right)$

P_d = Pitch Diameter   n = Number of Balls
B_d = Ball Diameter $\varnothing$ = Contact Angle

Note 2:

Because: *(sin α)(sin β)* = $\frac{1}{2}$ cos *(α – β)* – $\frac{1}{2}$ cos *(α + β),*

Assume an acceleration signal is filtered to pass only the higher orders of a bearing defect frequency greater than the 50th harmonic. When a harmonic series is multiplied by itself, the resultant series is a summation of all the sum and difference components that are developed during the multiplication process.

f (A) × *f (A)* =

   f(A) = sin (51*A*) + sin (52*A*) + sin (53*A*) ... + sin (99*A*) + sin (100*A*)

   ×

   f(A) = sin (51*A*) + sin (52*A*) + sin (53*A*) ... + sin (99*A*) + sin (100*A*)

   = sin (51*A*) sin (51*A*) + 2 sin (51*A*) sin (52*A*) + 2 sin (51*A*) sin (53*A*) ...
      + 2 sin (51*A*) sin (100*A*) + 2 sin (52*A*) sin (51*A*) + sin (100*A*) sin (100*A*)

Since the sum components *(α + β)* are generally outside the analysis measurement range, we are interested only in the difference components.

The products which are one unit apart (such as 52A and 51A) produce a 1X component according to:

$$1X\ component = \sum_{n=51}^{n=100} \sin[(n+1)-n]A = \sum_{n=51}^{n=100} \sin A$$

Similarly, any mX component produces:

$$mX\ component = \sum_{n=51}^{n=100} \sin[(n+m)-n]A = \sum_{n=51}^{n=100} \sin mA$$

These 1X, 2X, 3X, etc., components, produce FFT peaks at the 1X, 2X, 3X, etc., frequencies, thus permitting normal FFT analysis.

REFERENCES

1. ASA S2/WG88 Standard for Measurement and Evaluation of Machine Tool Vibration. Draft 04/27/93.
2. Barclay, J., "Keep Your Uptime Up: Condition Monitoring," *Engineer's Digest,* March 1991.
3. Berry, J., *Tracking of Rolling Element Bearing Failure Stages Using Vibration Signature Analysis.* Technical Associates of Charlotte: Charlotte, 1992.
4. Buscarello, R. and Kruger, Wm., *Machinery Vibration Control: A Manager's Approach.* Atlanta: Update International, 1990.
5. F. Berryman, P. Michie, A. Smulders and K. Vermeiren, *Condition Monitoring—A New Approach,* SKF Engineering and Research Centre, Nieuwegein, The Netherlands, 1989.
6. Dunnegan and Hartman, *Advances in Acoustic Emission,* Knoxville: Dunhart Publishers, 1981.
7. *Manager's Approach—Going Beyond the Limits,* Predictive Maintenance Technology National Conference, Atlanta.

8. Jackson, C., *The Practical Vibration Primer*. Houston: Gulf Publishing Company, 1979.

9. James, R., *Directions in Predictive Maintenance*. Cincinnati, 1991.

10. Jones, R., *Bearing Analysis Using Enveloping*. SKF Report, February 1992.

11. Jones, R., *Monitoring Electric Motors Using Electronic Data Collectors*, Vibrations 6, December 1990.

12. Jones, R., *Acoustical Determination of Nuclear Valve Operation*. TVA report, July 1989.

13. Jones, R., *TVA Annual Maintenance Report*. Chattanooga, TN, 1988.

14. Kryter, R. and Haynes, H., "How to Monitor Motor-Driven Machinery by Analyzing Motor Current," *Power Engineering,* October 1989.

15. Mitchell, J., *Machinery Analysis and Monitoring*. Tulsa: PennWell Publishing Company, 1981.

16. Wavetek Corporation, *Machinery Vibration Diagnostic Guide,* Application Note 22, 1984.

17. Zierau, S., "Machinery Vibration Spectrum Analysis: A Blend of Problem Solving and Advancement of Diagnostic Know-How," Petrotech Maintenance Symposium, Amsterdam (Netherlands), April 1976.

Chapter 6

Generalized Machinery Problem-Solving Sequence

In our introductory chapters we developed the need for a uniform approach to problem solving, progressing from the specifically machinery-related "trouble-shooting" to a more generally applicable "Professional Problem Solver's Approach (PPS)." This direction was prompted by the fact that:

1. Machinery failure analysis and troubleshooting do not only entail cause identification.
2. Follow-up action is almost always necessary.
3. Our goal is future failure and trouble prevention.
4. We must act.

Consequently, we may occasionally be looking at a chain of one problem-solving event after another as part of a single problem situation.

We also alluded to so-called executive processes in thinking and problem solving as part of the Professional Problem Solver's Approach. They are the systematic techniques common to all problem-solving activities. Five suitable techniques will be discussed*:

1. Situation analysis
2. Cause analysis
3. Action generation
4. Decision making
5. Planning for change

These techniques are parts of what we earlier identified as problem-solving sequences. Table 6-1 is an overview of what will be presented in this chapter.

---

*From information furnished by T. J. Hansen Company, 6711 Meadowcreek Drive, Dallas, Texas, and Alcoa, "Eight-Step Quality Improvement Process," copyright © 1989. Adapted by permission of Aluminum Company of America.

434

Table 6-1
Problem-Solving Sequence

LOGICAL QUESTION	What's the problem?	Why did it happen?	What are some answers?	Which one is best?	How can I avoid trouble?
TECHNIQUE OR PROCESS	Situation Analysis	Cause Analysis	Action Generation	Decision Making	Planning for change and avoiding surprises
END RESULT	Identify areas requiring my action	Discover the critical cause/ effect relation- ship	Create some new solutions	Select the action which best meets our objectives	Set up actions to: • avoid trouble • reduce damage • capitalize on key opportu- nities

Situation Analysis

Situation analysis considers the following three steps:

1. Separating the elements of a problem.
2. Assigning priorities.
3. Action planning.

Separating the elements. The most difficult part of the separation step is accepting that there may be more than one situation to deal with and more than one action required. After all, who wants to solve three problems if you only have to solve one?

Big problems can rarely be solved easily or quickly, because they are truly multiple problems. So when partial answers are proposed, they get rejected because they don't solve the whole problem. For example, the statements "This large motor suffers from too many starts, the operators need more training, and nobody knows from one moment to the next whether or not the machine is needed" are probably too general to be handled in one step. They are all candidates for situation analysis.

Steps in separating. Take for example "the operators need more training":

1. Define the key words and ask, for example: What kind of training? What is happening (or not happening) that tells you they need more training?

2. List the effects of the problem. Are these some troubles caused by specific lack of training or instruction?
3. Make each separation a complete sentence with modifiers, if possible. An example might be: "Several operators have complained that they cannot find proper instructions on how to start the machine."
4. Write everything down where you can see it. An easel or an overhead projector is good for this.

Figure 6-1 shows two different examples of the separating process.

EXAMPLE I

High early failure rates	Lack of pump maintainability review	Lack of operator/machinist training

Cost overruns on
pump maintenance budget

In this case there are three problems which can now be attacked separately. Also the "high early failure rates" seem to be most important.

EXAMPLE II

No supervision on major machine overhaul/repair	High tower feed pump has major problem	Two trainees need to be indoctrinated

Joe will be off
3 weeks with a
sudden illness

This example shows separating by effects. Unless the question, "What effects will Joe's absence have?" is asked, these potential problems may become unwelcome surprises.

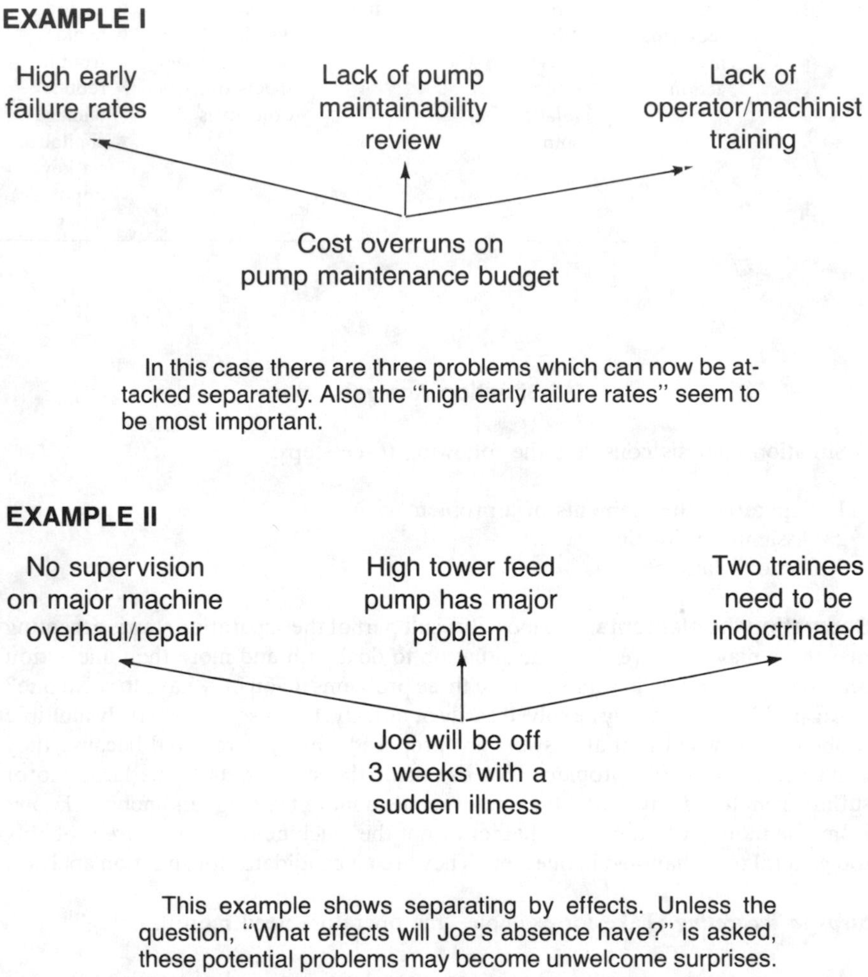

Figure 6-1. The separating process.

Assigning priorities. If we have trouble agreeing on priorities, our objectives may be unclear. When we are in accord about our destination, how we get there is an easier decision. Objectives and goals ought to be clearly stated and understood.

Here are some priority-setting methods:

1. Do the quickest or easiest first.
2. Pick the items that have been waiting the longest.
3. Do some you know you can complete.
4. Do the most urgent ones.
5. Let the boss do the priority setting.
6. Make a list and begin at the top.
7. Take each item in order from the pile or in-basket.

Most of us have at least a subconscious method for deciding what is worth doing and when to do it. A good technique should save time, be conscious, and be repeatable. Three questions will help set priorities in most situations:

1. What is the objective here?
2. What is it worth to fix the problem?
3. When do I have to act to be effective?

For potential or future problems, there is a fourth question:

4. How likely is it to happen?

The issue is really to decide which problem will be addressed first—or next—as the case may be. Several techniques will prove useful:

Check sheets. These are very simple and easy-to-understand forms used to answer the question: "How often are certain events happening?" It starts the process of translating opinions into "facts." Making a check sheet entails the following steps:

1. Agree what event or failure is being observed. All participants have to be looking at the same thing.
2. Determine the time period during which data will be collected. This could range from hours to weeks. See Figure 6-2.
3. Design a format that is clear and easy to use, making sure that all columns are clearly labeled and that there is enough space to enter the data.
4. Collect the data consistently and honestly. Make sure there is time allowed for the data-gathering task.

The check sheet is a simple, clear record of problems or failures that occur during each time period. Using the check sheet, workers can easily keep track of problems.

Pareto charts. Pareto charts help us quickly see the order of many different factors contributing to a problem. A form of vertical bar charts, Pareto charts (Figure 6-3) can be used to identify a major problem or a root cause, or to monitor success. The bars are arranged from largest to the left to smallest on the right. This helps focus on the most important factors of an issue.

From the top of the tallest bar and moving upward from the left to right, a line can be drawn to show the cumulative measure of the categories. This answers a question like: "How much of the total is accounted for by the first three categories?" Once the most significant problem or category is singled out, it can be given the special attention it deserves.

Part #: XL-3894-2 Description: Product Final Flaws   Date: 9-1

Source: MILLING MACHINE # 3 Name: B. SMITH   Time Frame: 9/1 - 9/29

Flaws	Tally	Total
Sharp Edge	7HH /7HH /7HH	15
Burrs	7HH /7HH /7HH /7HH /7HH /7HH //	32
Scratches	/7HH /7HH /7HH /7HH 7HH /7HH /7HH /7HH /7HH ///	48
Chips	7HH /7HH 7HH /7HH 7HH 7HH 7HH /7HH //	42
Total		137

Figure 6-2. Check sheet.

                        SOURCE OF
Part #:   Description: COMPUTER DOWNTIME   Date: 1/1

Source: TIME SHARING SYSTEM Name: R. SWANSON   Time Frame: 9/1 - 12/31

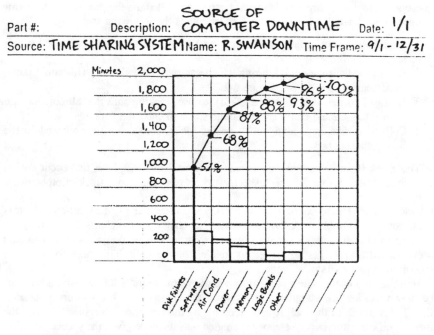

Figure 6-3. Pareto chart.

Brainstorming. Charting techniques are aids to thinking. They focus the attention of the troubleshooter on the truly important dimension of a problem. It is equally important, however, to expand one's thinking to include *all* the dimensions of a problem or its solution. Brainstorming is used to help a group create as many ideas as possible in as short a time as possible.

In a structured brainstorming session, every person is asked to present an idea. This can force shy people to speak up, but it can also create tension. In an unstructured brainstorming session, group members give ideas as they come to mind. It's more relaxed, but the most vocal people may dominate the session. Good guidelines for any brainstorming session are:

- Write down and keep visible the central topic.
- Don't criticize ideas or suggestions.
- Write down all ideas as spoken, and keep them visible to the group.
- Keep the session short, but ask people to keep a list of any additional ideas they might have after the session is over.

Flow charts. If the problem or trouble is tied to a process, it would be well to construct a flow chart in order to examine how various steps in a process are related to each other. After you document the actual flow of a process, you can compare it to the flow the process should follow. This will often reveal problem areas. Figure 6-4 shows a simple flow chart and its symbols.

Part #:   Description: WATCHING T.V.
Date: 3/1   Source: T.V. SET
Name: B. LITTLE   Time Frame:

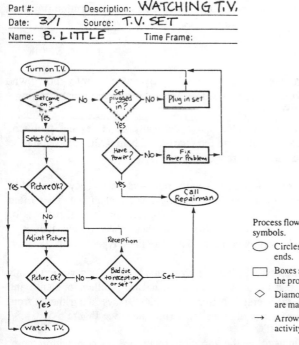

Process flow charts use a set of consistent graphic symbols.

⬭ Circles indicate where a process begins and ends.

▭ Boxes show the steps or the activities within the process.

◇ Diamonds indicate points where decisions are made.

→ Arrows show the direction of flow from one activity to another.

Figure 6-4. Flow chart.

Cause Analysis

We need cause analysis if:

1. Something is wrong and we are not sure why.
2. It is a high priority.
3. We intend to take action—if not, we need not bother.
4. Knowing the cause will help us come up with a better solution.

Process steps. These steps require you to:

1. State the problem (headline or problem statement).
2. Describe the problem (What is wrong and what is not wrong with it?).

Specifications	Problems −	Non-Problems +
Identify		
Locate		
Timing		
Dimensions		

3. State differences which seem significant. Also mention relevant changes.
4. Develop possible causes or hypotheses from Step 3.
5. Test causes against the known facts about the problem to determine the most probable cause. Try to shoot it down. (Test destructively.)
6. Devise a next step to assure you are correct.
7. Return to Action Plan.

Seeking the cause of a problem is like being a detective: you want to know "who did it." The cause of most trouble is evident either by observation or by asking a simple question: "Why did this happen?"

In order to help arrive at a statement that specifically describes the problem, establishing check sheets, Pareto charts, and pie charts will help. Other tools could be run charts, histograms, and cause-and-effect diagrams.

Run charts. These charts are employed to visually represent data. They are used to monitor a system to see whether or not the long-range average is changing. They often communicate far better than words and show sets of numbers and reveal patterns or relationships more effectively than tables or lists of numbers. Accordingly, they help you and others understand and interpret complex data. Common types are line, scatter, and stratification graphs.

Line graphs give you the simplest possible display of trends within observation points over a period of time. They depict a sequence of the data, and are excellent for identifying meaningful trends or shifts in the average. A line graph that displays data gathered from sequential observations of a process is often called a run chart. See Figure 6-5.

Histograms. Histograms help us to see the variations that processes have within them. This is a type of bar chart that shows the spread of a set of data. They also highlight the distribution pattern of the variation and help frame an issue. In a histogram, the categories are arranged in numerical order. Histograms are often particularly useful in preliminary data analysis. Figure 6-6 shows a histogram of a production process.

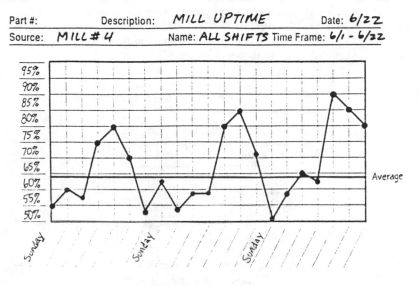

Figure 6-5. Run chart.

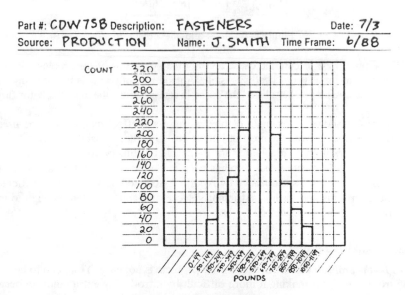

Figure 6-6. Histogram.

Cause-and-effect diagrams. These represent another important technique in developing a complete picture of all possible causes of a problem. Such a diagram helps us explore, identify, and display the possible causes of a specific problem or condition.

A cause-and-effect diagram—also called "fish bone" or "Ishikawa" diagram—has a primary horizontal line that lists the effect or symptom on the right. The major categories of possible causes are arranged as branches slanting off above and below this line to the left. Typical branches include the four "M's"—man, methods, materials, and machines. Or, policies, procedures, people, plant, and time. Specific possible causes become the detailed twigs, as shown in Figure 6-7.

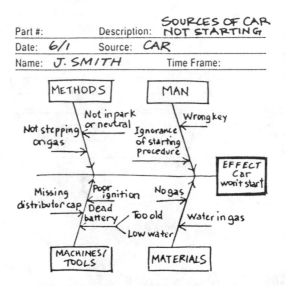

Figure 6-7. Cause-and-effect diagram.

Remember, look to cure the cause and not the symptoms of the problem. Push the causes back as far as is practically possible. A good rule of thumb is to ask "why" four or five times. After you have listed the possible causes, select the most likely for further analysis.

Figure 6-8 shows an example of how a cause-and-effect diagram can be used to identify the cause(s) with the greatest impact on "L-pump" reliability. Figure 6-9 may be used to work up similar solutions for unreliable machinery systems.

The cause-and-effect relationship has been made the cornerstore of a modern failure analysis technique spearheaded by Apollo Associated Services, Inc., of Richland, Washington. This approach is sufficiently successful and important enough to merit detailed discussion in Chapter 11.

Cause Analysis Steps

Step 1—the problem statement. Stating the problem is not easy. You need to headline the trouble without making it more difficult by introducing extra emotion because when things go wrong, people are emotional enough.

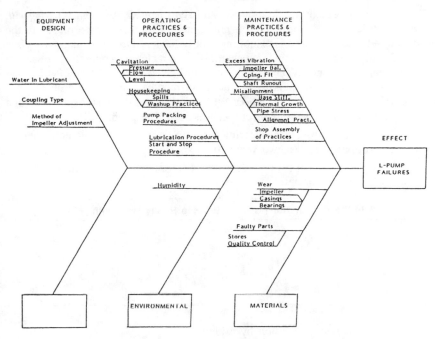

Figure 6-8. Cause-and-effect diagram for "L-pump."

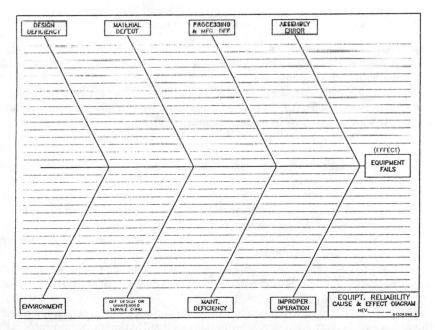

Figure 6-9. Cause-and-effect diagram that can be used for unreliable machinery systems.

The problem statement should show *what is having trouble, and what is wrong with it.* For example, "drive motors are burning out." In the case of people, state the unwanted behavior and who is exhibiting that behavior. Do not evaluate or place blame. Usually, you know what is expected. If it is not clear or generally known, put it in the problem statement. Then you have a third piece—*how it should be.* For example, "drive motors are wearing out after six months" *may* be a problem, but there should be some more information. When we discover the design specs call for a two-year life, we *know* we have trouble. Restated, the problem is:

> Drive motors (the object) are wearing out after six months (the defect) when they are supposed to last two years (the expected performance).

If you had some typical problems each week, you might list them as problem statements in tabular format:

	What Is the Expectation?	What Is Really Happening?
1.		
2.		
3.		

Test the value of your tabulation by asking:
1. Does the way you stated each problem tell both what (or who) you are having trouble with and what the trouble is?
2. If something is wrong, do you know specifically how it should be?
3. Will it be more clear if you reword the statement?

Step 2—problem description. This is the most critical step—information gathering. And you need to gather information not only about the problem that *is*, but you must also describe the problem that *is not*.

A good set of questions and a format are helpful. You will not get answers to all the questions, and they will not all be relevant. But you can often save considerable time and effort with a good problem specification.

Let us take the easy part first—describing the problem you have:

Identity. Ask, "What's the thing that's affected? What's wrong with it? Important: make certain you separate these two. For example, if the problem *is* cold flow (rolling and peening) of #1 gear set, the *is not* or non-problem portion could be listed as Broken Teeth or Interference Wear. Even better would be: *is* Cold Flow, *is not* Broken Teeth, and *is* #1 gear set, *is not* #2 and #3 gear sets.

Location. Ask, "Where is the trouble found? Any special part? Where geographically? Where is the process?"

Time. Try to keep this to time and dates. Ask, "When did you first notice this? Did it happen before? What has been the pattern since then? Is there any cycle or period?"

Size. Ask, "When it occurs, is it progressive? Are all objects affected? Is the whole thing affected? Is there additional information observed at the site of the problem?"

Now we want to turn each of these questions around and note similar problems which you might have but do not:

Identity. Ask, "What similar objects, products, symptoms, etc., are there where there is no problem like this? Are there other things we might have expected to go wrong, but have not? Is there one that works?"

Location. Ask, "What other similar items, operating units, machines, etc., have we? Any trouble there?"

Time. On the surface, this looks simple. Do *not* write down "no other time" or similar statements. If the problem occurs Monday a.m. and Wednesday p.m., say so, but additionally, state that it does not happen Monday p.m., Tuesday, Wednesday a.m., Thursday, or Friday. Also, ask, "When would we have expected such trouble?"

Size. If the *whole* item is affected, then it is not *partially* damaged. Also, if *all* items in a defective package are bad, it is not just the top or bottom items.

A pitfall to avoid. Do not build information into your description or specification for the purpose of supporting your pet cause. Remain objective and *ask questions.* Otherwise, you will have a piece of information looking for a home.

Step 3—analyze differences which seem significant; also relevant changes. The first question to ask is, "What is different about where we have a problem from where we do not have a problem?" (Now you see one of the reasons for describing the problem you *do not have.*) Next you ask, "Has anything changed with this element? When?"

Step 4—possible causes (hypotheses). For *each* significant difference under Item 3 ask, "What about this difference or change could have caused the trouble?"

Step 5—test causes against both (−) and (+) (minus & plus) problems and non-problems. In testing, you want to shoot down possible causes that do not fit.

A pitfall to avoid. It is tempting here to "prop up" a possible cause that you fancy. *Do not do it.* If it really did cause the trouble, it will fit the facts.

Questions to ask. Does this cause fully explain both the (−) and (+) of the problem description? What additional data do I need? What conditions do I have to check out?

Step 6—devise a next step to assure you are correct. Asking a few questions will usually do this, although a field test or thorough lab test may be called for.

Step 7—return to action plan. We now return to the action plan under cause. The Fix or Action should be better for having discovered the cause.

About Day-One Deviations

Definition: A deviation from planned, designed, or expected standard. It has been present since startup. It never has worked out.

Actions:

a. Use the minus (problem)/plus (non-problem)/difference approach:

Reality	Expected (+)	Difference (−)
Actual construction	Design specs	
Actual use	Designed use	
Actual operating conditions	Operation assumptions in design	
Actual location	Location initially assumed	
Actual implementation (timing, staffing, budget, etc.)	Implementation (startup assumptions)	

b. Find one somewhere that is working or that has worked properly. What is different? Seek out similar situations, equipment, objectives, elsewhere. Any trouble? What is different?

c. Ask these questions:
   Has anyone else attempted to solve this problem? What approach did they take? Did it work? What did they do differently?
   Has this system, equipment, technique, material, etc., been used elsewhere with success? What is different?

When looking for the cause of the troubles relating to people, ask:

1. What *behavior* am I seeing that troubles me?
2. Who is exhibiting that behavior?

1. What things are going well? (What things are not problems?)
2. Who in a similar situation or position is *not* showing the troublesome behavior?

Your goal: try to find some factors *under your control* that contribute to the problem. Look for environmental rather than personal factors, since you can manage environment or arrangements. People manage themselves.

Another valuable use of the expected/reality pairing of information is with problems or conflicts with people. If you carefully state what you expect and what the other individual expects, often the source of the trouble is evident.

The same is true for describing reality, or what is really happening. Two people observing the same event or set of facts may have totally different feelings about it. They may even see two different realities.

Communication Techniques

In other parts of this book, we are making the point that our success as machinery failure analysts and troubleshooters is highly dependent on how well we communicate with others. Examples of successful problem solving show time and again how we depend on the art of communication because frequently we are just not "there" when things go wrong: other people have to supply the background information, the subtleties and details of a trouble-causing event. In the following, we look at some of the more important aspects of communication.

Roles. Successful communication about a problem involves three roles and any number of individuals. The roles or positions are:

1. *Data source*—The person, people, or records that contain data about the problem.
2. *Analyst*—The individual or group of individuals who will gather the information and draw conclusions or recommendations.
3. *Decision maker*—The person or group who will take action.

These roles may be overlapped. The person who knows the details may analyze and take action. Then again, the data source may answer questions and let others analyze and recommend. For any productive communication to occur, there must be willingness and ability. It is critical to recognize these as a pair. One alone will not do.

Willing. If individuals who know the facts about a trouble are threatened or put off by your manner or your technique, they may not give you all the information or they may give you false information. Either way, communication breaks down because they were unwilling.

Able. There are some things you can do to increase the willingness of others to supply the information you need. Your best chances of success will come if you first assume people are doing the best they can. In other words, they are not out to specifically undermine your efforts. Therefore, you can assist by:

1. Finding the person who has the information.
2. Tell them how you will use the information.
3. Ask specific, direct questions to get the selected information you need.

The first is self-explanatory. Someone who does not know the facts or cannot get them is clearly unable to help you. The second will help defuse some emotional hesitancy (defensiveness). If you can show you are objective and avoid jumping to quick conclusions, you will get more cooperation. There is a big difference in seeking the true cause(s) of trouble and seeking merely to place blame.

Ideally the data source, analyst, and decision maker share the same method and concern for the problem. Not always so. The data source almost always provides screened information (that is what good process questions are designed to do). If you are not asking people precise, specific questions, they will give the information *they* consider important. Also some data may be withheld to avoid blame.

An analyst will often be hampered by a lack of time (demand for a quick solution) and bombardment by speculations. Therefore, an analyst will give you only the information *he* thinks is important or necessary. In other words, an analyst will select or screen information based on: (1) what he thinks you want, (2) what he knows, (3) what he thinks, and (4) what he feels. Let us look at these one at a time:

1. *What he thinks you want.* The questions you ask have a lot to do with this. Also, the use you make of the information tells the other person something. You have a lot of control. *You* decide what questions to ask, and you can show the other individual what you are going to do with the data. Crisp, fact-oriented questions in a logical order are necessary. You also need structure to place the answers in.

2. *What he knows.* People usually know more about a situation than they are aware of. Here again, crisp directed questions are necessary. Further, a critical consideration is reaching the *correct people.* Individuals who know or can find out the critical facts.

3. *What he thinks.* People will often have speculations or hypotheses about a problem. Premature judgment turns people off. Write it *all* down.

4. *What he feels.* If the others involved know your method, they will be more eager to work with you. Questions to discover cause can sound quite aggressive (they are). And they can turn off people's feelings if they do not know where you are going.

Pattern Matching

When there is a major disaster, for instance a machinery wreck, it is critical to discover as soon as possible what happened. Often, as with many problem situations, there is no clue that allows us to understand immediately what went on before the wreck occurred—all that remains is the physical evidence that something has gone wrong. We cannot get accurate accounts of what was happening at the time. All we see are the results.

Pattern matching is a useful tool in such instances. It calls for openness or divergence of thought when speculating or hypothesizing about the cause. And it calls for skillful logical processing of the data in testing.

Here is how it works:

1. Examine the evidence (firsthand if possible).
2. Do some "Hypothetical What-Iffing" (HWI) Do not judge or evaluate at that point!
3. Beginning with the top of the list, or anywhere you like, built your test model. You do this on the cause analysis worksheet (Figure 6-12) by asking: "If _____ caused the trouble/accident/etc., what item(s) would we see evidence on? What items would we see *none* on?" And so on down the list of questions from the specification sheet. To do this you will need some experience or expertise. If you do not have it, seek someone who does. In the absence of expertise (perhaps the problem has not happened before), you can picture what you would expect to find.
4. Return to the actual evidence. Again, on-site is best.
5. Look for what you would expect if your hypothesis were correct.
6. When you have misfit, go to another "Hypothetical What-If."

This should help solve any mystery you encounter. Whether it involves machinery, people, process, natural occurrences, or a combination, some evidence either physical or behavioral will remain. We conclude our discussion of cause analysis by presenting Figures 6-10, 6-11, and 6-12 as working aids.

INC

Question:

<u>I</u>s

<u>Not</u>

<u>C</u>hanged

Figure 6-10. Problem specification worksheet—short form.

PROBLEM SPECIFICATION WORKSHEET

Instructions: Fill in the answers to the specifying questions as precisely as you can.

DIMENSION	PROBLEMS −	NON-PROBLEMS +
ID	1a. What item specifically has or shows trouble?	1b. Are there other similar items? What are they?
		Are they affected?
	2a. What is wrong with it?	2b. What similar or related trouble could we be seeing?
		Are we?
TIME	3a. When did it happen? (as precisely as possible)	3b. What were other possible or likely times?
PLACE	4a. Where exactly did it happen on the item?	4b. What parts of the item are unaffected?
	5a. Where was/is the item with the problem located?	5b. Are there other locations? Are they having trouble?
	6a. Did the problem happen in a certain geographical location?	6b. Where could it have happened but didn't?
SIZE	7a. When the trouble happens how much is affected?	7b. Is some portion consistently not involved?
	8a. Is there a pattern or a cycle to the trouble? What is it?	8b. Is this usual? Is there another pattern, cycle, or lack of one you would expect?

Figure 6-11. Problem specification worksheet—long form.

I. Problem Statement or Headline

II. DESCRIBE		III. ANALYZE	IV. POSSIBLE CAUSES	V. QUESTIONS
PROBLEM −	NON-PROBLEM +			
Existing conditions that are off standard or unwanted	Related existing conditions, things, people that aren't off standard	What's different or changed with the—? (Problem Areas)	What about this could have caused the trouble? W H E N ?	Questions for missing information or confirmation of possible cause
ID				
TIME				
PLACE				
SIZE				

Based on this analysis, what caused this trouble?

Figure 6-12. Cause analysis worksheet.

Action Planning and Generation

Too often problems are looked at or defined only one way. We are in a hurry to do something, so we grab the first statement of the problem and run. This limits the number and the range of possible solutions. A complaint frequently heard from top management is, "My folks aren't considering enough different answers to problems."

There are some helpful techniques. Understanding them is not difficult. But using them effectively requires some familiarity and practice. Thinking of new ideas is fun. And it can be channeled in a productive way. To do it you need:

1. Some people willing to work with you as resources. Three or four people would typically be involved here.
2. Visibility—either a flipchart, blackboard, or someplace to write things down in view of the others.
3. Adequate time to work on the problem. If it's worth fixing, it should be worth taking time to define the best possible action.
4. An agreed-to method, including role assignments. One or two roles could be assigned to a participant.

The essential roles are:

1. *Client*—the person whose problem we wish to solve.
2. *Recorder*—The individual who writes the solutions, questions, etc. This should not be the client (too much power of omission in the pen).
3. *Chairperson*—This person keeps the discussion on track, watches the time, and especially guards against premature evaluation of any sort.
4. *Data source*—The person most familiar with what happened.
5. *Clout*—The individual who can authorize the expenditure of money, authorize bringing in outside work forces, extend the downtime, allocate overtime, etc.
6. *Worker*—The one who is expected to implement the agreed-upon action. In this context, "worker" may pertain to a responsible engineer, supervisor, technician, operator, etc. Having this person present at the meeting will go a long way towards him "buying into" the agreed-upon action. This will also avoid misunderstandings and foster a spirit of cooperation and motivation.

Action Planning

The heart of action planning is active personal intent. *You* are going to *do something*. (Either act, recommend, or drop it because you are wasting time.) If you are not going to act on the results, analyzing is frustrating and nonproductive.

Recommending, for example, is taking action. Not everyone gets to be the final decision maker. Ask: "Am I the recommender, the decision maker, the implementor, or a combination? What is my role in the situation?" Ask these questions: "If I am the recommender, should the person implementing my recommendation/decision be consulted or involved?" If the answer is yes, add some structure and visibility to your process. Communications will be better, and you will have a better shot at success.

A list of actions is shown in Table 6-2. The kind of action taken depends on the situation. But you should be conscious of your action, the purpose of your action, and what triggers the action.

What kinds of action are you most often involved in?

1. Corrective
2. Innovative
3. Opportunistic
4. Interim
5. Adaptive
6. Replacement
7. Selective
8. Preventive
9. Contingent

If you chose the first four, you are concerned with problems and opportunities. Situation analysis and cause analysis should be helpful tools. Adaptive, Replacement, and Selective actions involve deciding. If you chose those, you'll find decision making useful. Preventive and Contingent actions are associated with planning for change. Table 6-2 shows appropriate examples.

An action plan shows what *you* intend to do. Here are the elements:

1. The priority concern—what you are taking action against. This could be a piece of the problem, its cause, or one of its effects.
2. The cause of that concern, if known.
3. The "fix," or action, you plan. List several—be imaginative.
4. Who is responsible—names, which should include the Clout, the Data Source, the Worker, the Reporter (one of these people should be *you*).
5. Your specific actions.
6. The timing. This should include start, report, and completion times. Dates and times only—not "immediately" or "right away."

In other words, what action, against which area, with whom, and by when.

An action plan often includes both short-term and long-term actions. Troubleshooters frequently need to take interim or short-term action against symptoms or effects, then they can take long-term actions to seek and cure the cause. For instance, sometimes machinery distress symptoms must be temporarily treated—like fractional-frequency oil whirl in a centrifugal compressor—before attacking the root of the problem.

Visibility techniques. Frequently, the troubleshooter has to explain a machinery malfunction, distress situation, or component failure to an audience not familiar with the technical details in question. Visibility techniques will help in this case.

Components of a situation can be displayed or made visible several ways. The "cloud technique" is commonly used. At least three others are worth discussing. They may even be more useful:

1. *Force Field**
   - The force field is useful in showing the positive and negative elements affecting a situation.

Table 6-2
Types of Action

Triggering Situation	Action Types	Purpose
Something goes wrong. (Inlet Guide Vanes—IGVs—in forced-draft fan jammed because operating temperature exceeds design temperature of IGV shaft bearings.)	Corrective	Removes the cause and returns to standard performance. A fix, (i.e., install bearings suitable for temperature).
Something goes wrong. (IGV in forced-draft fan jammed)	Innovative or Opportunistic	Extends the cause of improved performance in one area and spreads it around. Raises the standard (i.e., review if IGVs needed at all).
Something goes wrong. (IGV in forced-draft fan jammed)	Interim	Attacks symptoms or effects to buy time. A *patch* (i.e., fix in wide-open position & use by-pass damper for modulation. Fix it for now!)
Something goes wrong. (IGV in forced-draft fan jammed)	Adaptive	Attacks the effects to allow normal or near normal results. Allows you to live with the cause. A more permanent patch (i.e. keep repairing, learn to live with it)!
A *need*. (Sudden coupling Failure)	Replacement or Selective	Fills the void while meeting a set of objectives (i.e., find custom shop with quick turnaround).
Recognition that something in a planned action may go wrong.	Preventive	Removes future causes of trouble. Reduces risk. It is set up and activated ahead of time.
Recognition that something in a planned action may go wrong.	Contingent	Reduces or eliminates bad effects of future problems. Allows continued progress toward goal, even though trouble occurs. Set up ahead of time, but not activated until the problem happens or is imminent.

- Either the driving forces can be strengthened or the restraining forces reduced to improve things.
- Again the main purpose is to support action to make decisions more rational.

Action usually comes from identifying which elements you wish to attack so the current situation line can be moved to the right. Your question should be "What items here can I act against?" and "How?"

2. *Pictorial*

   Often it is better, cleaner, and faster to show a diagram or a picture of how something is (or is not) working (see Figure 6-13). This is especially true if:

   - You are explaining a situation to people who do not understand the technical terms or system.
   - You are showing an expert how the situation exists. The expert can then compare the present state with how it should be and attack the problem at its causes.

3. *Stair-Step*

   Problem

       (cause)
     Problem

         (cause)
       Problem

           Root Cause

It is helpful most of the time to keep digging for the root cause of the trouble and solve it there. The stair-step method is excellent for showing a long cause-effect relationship. It is also good for showing where you are attacking the problem. The thinking that goes on behind this is that distinguishing between cause and effect is no more than chopping a chain of happenings at some convenient point and calling what

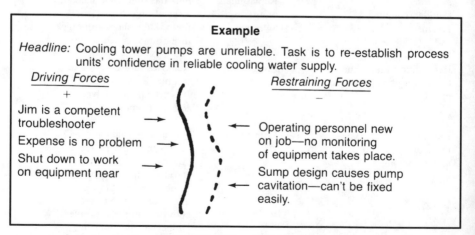

Figure 6-13. Example of the force-field method.

goes before the chop "cause," and what comes after the chop "effect." If there are numerous "causes," you ask "which, if any of these can I reduce or eliminate?"

This is an important ingredient of at least some successful failure analysis approaches and will be further elaborated on in Chapter 11. Figures 6-14 and 6-15 are convenient worksheets for using any of the previous techniques.

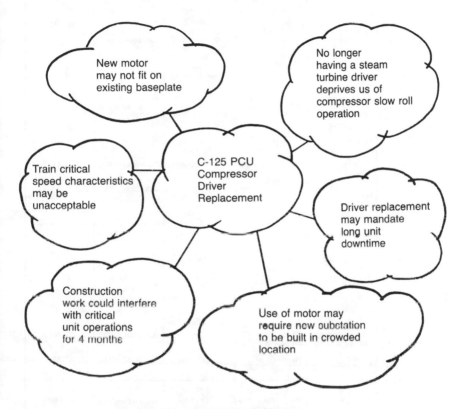

ACTION PLAN

Priority concern	Cause	Fix or other action	Responsible persons and roles	Your actions	When
Interruption of unit operations	Work activity on C-125 slab	Surround with fence	H. G. Smith to coordinate and execute	Sign work order	3/22/83
Driver replacement may mandate long downtime	Unforeseen difficulties with mounting dimensions	Obtain certified drawings for all equipment	P. C. Monroe	Reduce his remaining workload	3/29/83

Figure 6-14. Action plan worksheet—preliminary.

ACTION PLAN

Priority concern	Cause	Fix or other action	Responsible persons and roles	Your actions	When
Interruption of unit operations	Work activity on C-125 compressor floor slab	Erect 8 ft high wooden board fence	H. G. Smith	Sign work order. Notify unit supervisors	3/22/83
Driver replacement may mandate excessive downtime	Unforeseen technical and dimensional interference problems	Arrange with HQ staff for alternate source of unit product (contingency plan)	Primary: P. C. Monroe Secondary: E. D. Visser	Communicate with HQ product coordination personnel	3/23/83
Train critical speed problems	Torsional critical speed coincident with train speed	Locate Holset-type couplings	Lou Rizzo to tabulate all available options	Authorize advance funding for computer study	4/4/83
New substation located in congested part of plant	Equipment density too high	Install special fire monitors. Obtain variance from safety committee	T. L. Hernandez	Confirm in writing to TLH. Set up meeting with safety coordinator	4/4/83 4/4/83
Lack of slow roll capability may cause rotor to bow	High temperature process gas surrounds compressor rotor	Change operating procedures. Develop clutch system.	G. F. Moeller P. C. Monroe	Request opinion Request feasibility study	3/29/83 3/23/83
Long lead time for motor delivery	Manufacturers too busy. Strikes.	Develop 5 possible sources. Negotiate penalties with Union.	F. D. Corpute J. D. Crumb	Authorize contact Initiate via letter	3/22/83 5/17/83
Project gets cancelled	Economic downturn	Negotiate cancellation charges	J. D. Crumb	Request work to commence	5/17/83

Figure 6-15. Action plan worksheet—final.

Action Generation

A method for action generation proven successful has been popularized by the Synectics Corporation of Cambridge, Massachusetts. They suggest eight steps:

Step 1—briefly define the task. Tell the group what is wrong. Give them a brief history, and tell them what you expect in terms of group output. If you know cause/effect relationships, show them. Try to present the problem in some non-verbal way as well. Pictures, diagrams, clouds, stair-steps, etc., are ways to do this.

Do not over-define the problem. Remain non-judgmental. Avoid stifling the group with a list of "we've already trieds."

It is helpful here to restate the problem in several ways. For example, a problem stated as: "The inlet guide vane bearings of the forced draft fan are seized" could be restated:

"How to get operations to live with it."
or
"How to enlist operations cooperation in the effort to try to run with inlet guide vanes positioned permanently at one setting."

The client picks one problem to work on. The rest are saved for possible action later.

Step 2—goal wishing. This is a free-thinking approach to solutions. It is totally without regard to feasibility.
Opening the doors—First, let us discuss getting the ideas started. Some of the more important things to do and avoid:

 a. *Do* go first for quantity.
 b. *Avoid* evaluating (looking for quality)—that comes later.
 c. *Do* include some other people.
 d. *Avoid* assumptions about limits, restrictions, salability, etc.
 e. *Do* write everything down.
 f. *Do* have fun.
 g. *Do* loosen up your brain a little before starting.

A couple of other aids are beginning statements with "I wish . . ." and building on other's comments or ideas.

Step 3—narrowing. The client picks one or more directions or wish statements that hold some intrigue.

Step 4—excursion and force fit. There are several ways to do this:
 a. *Key Words.* You select unrelated key words like "circus," "landscaping," "art," "sports," etc., and again let your mind open up with examples that relate to the selected direction (from Step 3). Then you relate these to problem solutions.
 b. *Role Playing.* You might ask them to assume a different identity. For instance, if the problem has to do with motivating first-line supervisors, you might ask your associates to assume they are first-line supervisors, half of whom are "turned on" the other half of whom are disgruntled. Now they are in a frame of mind to invent ways to stay motivated or to get that way. These are then worked into possible solutions.
 c. *Imaging.* This is only slightly different from role playing. The difference is the use of objects along with or instead of people. For example, if expensive tools were being stolen and solutions were needed, you might have part of the people imagine they were prisoners (the tools) and the prison was the machine shop.

The other people would be the guards. Half would generate ideas on how to escape. The guards, of course, would devise all possible ways to keep them confined. From this you look for new security measures.

Step 5—potential solution. From the wishes and the excursions, the client again narrows to a possible solution. It need not be feasible or viable yet.

Step 6—Pro/con or itemized response. This is the initial evaluation step. *First* you list the good points, the pros, about the idea. Stretch for as many as you can. *Second,* the client states the cons or concerns. These must be proved and worded as "how to" statements.

For example, "Costs too much," might be reworded, "How to do the job with less-expensive materials" or "How to recover some of the $." These "how to" statements are more action-oriented rather than purely negative shoot downs.

Steps 7—actions to reduce concerns. This again is creative and open. However, you are really trying to come up with ways to make the idea fly now. You do this by removing obstacles or concerns.

New ideas often start out as totally ridiculous. As concerns are addressed, they get a little more realistic until finally they reach a critical point: the threshold of "tryability."

Step 8—next steps. If the concerns have been adequately addressed, the client should be ready to say what is next.

An Actual Example to Illustrate the Process

When four pusher centrifuges were started up at an overseas petrochemical plant in 1969, it became immediately evident that for the particular medium being centrifuged (paraxylene slurry) at 2,000 rpm, the gap between feed pipe and centrifuge inlet (Figure 6-16) was big enough to permit unacceptably large amounts of flammable liquor to spill on the operating floor. The equipment owners called for an "action generation" meeting and all involved parties, from mill-wrights to operators to start-up engineers and managers assembled in a conference room. The group designated a "chairman;" the chairman designated a "recorder." Someone in authority identified the "worker," an individual who was told before the meeting even had its official kickoff that it would be *his* responsibility to implement the decision that was to result from an open listing and discussion of the various remedial action steps the meeting participants were to call out just as quickly as they came to mind.

The recorder went up to a flip chart and quickly wrote down the rapid-fire suggestions tossed up by the eager audience:

1. Mechanical seal
2. Packing
3. Baffle plates
4. Labyrinth seal

5. Elastomeric tire
6. Plumber's putty
7. etc., etc., etc.

Because the chairman had specifically and emphatically instructed the meeting attendees to go for quantity first, and *not* to discuss the merits of any suggestion until there was no longer any out-pouring of thoughts, not even a second was wasted in discussing or ridiculing what was obviously an awful suggestion: plumber's putty! In some settings, the poor individual who dared to blurt out plumber's putty would have been ridiculed. The result would have been a hearty laugh, the "putty man" would have felt hurt, and the others would have bitten their tongues rather than make any additional suggestions.

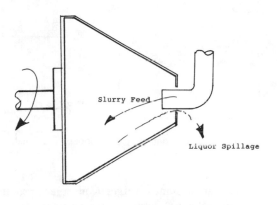

Slurry Feed

Liquor Spillage

Figure 6-16. Spillage problem on pusher centrifuge inlet.

But these plant owners were smarter. They encouraged the participants to make other suggestions, and only after everyone had truly run out of ideas, the chairman instructed the persons who had made suggestions to *withdraw* them *if* they saw a more appropriate idea on the flip chart. "Putty man" was among the first to ask that his suggestion be stricken; others followed. The group agreed that an attempt should be made to design and fabricate a mechanical seal. The "worker" took the design sketches to the shop, and 24 hours later, a mechanical seal was installed in the first of the four centrifuges.

Unfortunately, the seal proved rather unsatisfactory. The "action generation" meeting was reconvened in the same conference room where the earlier course of action had been decided upon. The various flip charts were still posted on the four walls and the attendee/participants had little difficulty in refocusing their attention on another promising remedial step. It combined the ideas "baffle plate" and "labyrinth" into the suggestion to fabricate a bronze bushing with an internally machined square thread—a visco-dynamic seal, so to speak. This device was sketched out, fabricated, and installed. It worked so well that a patent was applied for in a commercial effort to deprive the competition from using this particular fix (Figure 6-17).

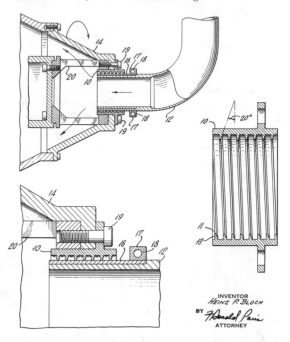

Figure 6-17. Non-contacting seal for centrifuge inlet.

Summary

As can be seen, this is a constant divergent/convergent process. We begin by expanding the number of problem statements, then narrow to one. We proceed with multiple wishes or approaches, then narrow to one, and so on.

There are some other keys about the group and the technique:

1. No more than seven people.
2. Include people who have knowledge of the problem.
3. Someone must be willing to take action.
4. Visibility is critical.
5. Method must be agreed to.
6. Six roles assigned.
7. Strive for freshness. (Creativity is not necessarily a new idea, but one that is new to you.)

The core step in solving problems is developing appropriate action. Indeed, supporting the action step is the primary reason for any other analysis, and perhaps the only sound reason.

A novel, new, or creative solution may also be worth a lot in terms of:

1. Support by fellow employees and management.
2. Interest.
3. Motivation of workers.

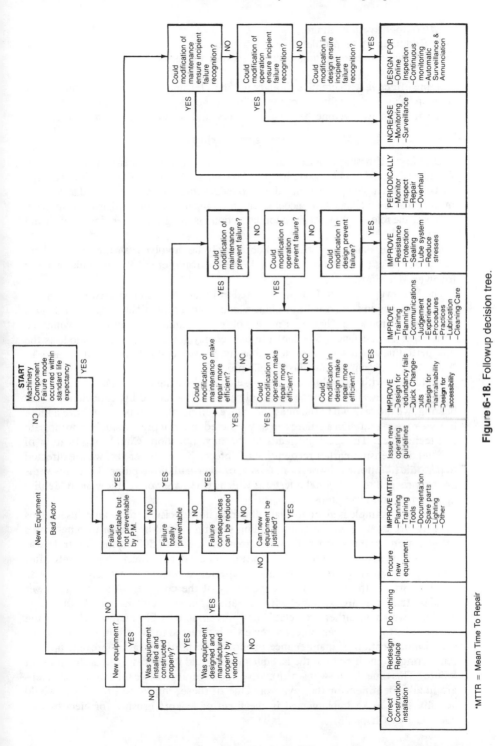

Figure 6-18. Followup decision tree.

*MTTR = Mean Time To Repair

4. Competitive advantage.
5. Reducing other problems.
6. Allowing previously unusable techniques to become feasible.

Being effective with creative thought involves letting go—not an easy thing to do. It involves using the right side of the brain—something we are not used to. Only by doing this can we become balanced thinkers and balanced problem solvers.

Decision Making

Deciding is choosing a course of action. Frequently, it is in this area that the machinery failure analyst and troubleshooter has to "show his stuff." Since all action involves risk, we believe that a good process should help to reduce this risk. Keep in mind that decision making may be facilitated by encouraging group input or group participation. In many instances the role assignments given for the Action Planning and Generation phase should be continued.

Usually the machinery troubleshooter will be involved at the point where everybody asks the question "Where do we go from here?" We are talking about followup decisions.

The Followup Decision Routine depicted in Figure 6-18 can be used to determine appropriate action steps after a machinery component failure has occurred. It begins with the decision as to whether or not the observed failure mode turned up within its standard life expectancy. Standard life expectancy for a given failure mode is the time period in which we, quite often subjectively, expect this failure mode to appear. A few simple examples may illustrate the process.

Let us start with the failure of a rolling-element bearing in a refinery pump. After verifying that this failure mode occurred within the standard life expectancy of the bearing, we move down to the question. Since rolling-element bearings have a finite life, we reason that such a failure can be predicted, but not prevented, by appropriate PM techniques. This decision leads to the next question: Could modification of maintenance ensure earlier recognition of defect? If we answer yes, we are directed to periodic monitoring, inspection, repair, or overhaul of the pump. If the answer is no, we proceed to additional questions which, when answered, allow us to identify appropriate followup steps.

A second example is given to again illustrate how decision trees can be used. This time we are dealing with a gear coupling serving a 3000-hp steam turbine-driven refrigeration compressor rotating at a nominal speed of 5000 rpm. The train was taken out of service when shaft vibration reached excessively high levels after approximately two years of operation. Experience tells us that this failure mode occurred within the standard life expectancy of the coupling. Also, the reviewer decides that we are dealing with a failure mode—electric spark discharge erosion—which is neither predictable nor preventable by PM methods. Judging the failure to be totally preventable, the reviewer is directed to two additional questions. Could modification of maintenance or modification of operation prevent failure? Answering no, he arrives at the last question: Could modification in design prevent failure? This time the answer is affirmative and several possible followup action steps are finally identified for the reviewer. One of these, improved protection, would actually have to be implemented in the form of leakoff brushes for electrostatic currents (see Chapter 3).

The process just developed will suffice for most of our day-to-day decision making. In the following, we would like to deal with the more critical and unusual decisions, such as "select best way to get back on line after a machinery wreck," or "decide whether or not to continue running after a major vibration problem on a critical machine has appeared." The systematic part of decision making included both the correct steps (doing the right things) and following them in the correct order (doing things right). Decision making is a step in the total problem-solving continuum, and the fourth item in it coming up for discussion. As we remember:

1. What is the problem?
2. Why did it happen?
3. What are some solutions?
4. What is the best answer, and how can I present it?

Often people in decision or action situations do not realize they are trying to *solve* a *problem*. Others involved in the decision or who need to approve the action may simply be expecting something to be fixed. As a result, the chosen action is rejected or fails to meet expectations. To help avoid this, you should ask, "what *problem* are we trying to solve?" Here is an example:

	First	Second
Choose best action to fix inlet guide vanes in F.D. fan	Ask: Why do they have to work? State objectives: 1. Combustion air control for furnace 2. Energy savings 3. Startup needs	Seek and analyze solutions that meet needs and constraints

A good decision-making process must be usable individually or in a group. It is almost always desirable to have some visibility—to write down some of your logic.

There are seven steps to the process. Depending on the situation, some steps will receive more of your attention than others:

Step 1—The decision statement (your mission).
Step 2—Objectives or criteria—what you want to accomplish or avoid.
Step 3—Ranking of objectives. Which are mandatory? Rank order the others, and give them a relative weight.
Step 4—Possible actions—different people, methods, machines, etc., which might meet your objectives.
Step 5—Find the best fit (usually done with a matrix).
Step 6—Assess risk (can be incorporated into Step 2).
Step 7—Take best choice back to the Action Plan.

Three Pitfalls to Avoid

1. Sometimes you will hear, "I don't have time to think things through or talk them over. I've got to make split-second, on-the-spot decisions." Even in tight situations, it *never* hurts to examine quickly objectives, alternatives, and risk.

2. Do not confuse decision making with following a plan where the decisions have been made. The follower of a program is not really deciding, the person who wrote it made the decisions.

3. We are talking about *choosing,* making a choice from several possible actions. Probably you are going to change something—a method, a machine, a person in a new job. As in any change, you would like to reduce uncertainty. Also, you would like your ideas to be willingly accepted or "bought." Notice, it was not said you want to *sell* them. Acceptance is vital when others will be implementing or using your ideas. Sometimes those who are "sold" have not really accepted and may fight the implementing.

Step 1—the decision statement. A decision statement should allow a good selection of alternatives. And, it should reflect decisions already made. For example, "Select new centrifugal pump" is too narrow if what you wanted is to increase pressure from one pressure level to a higher one and deliver a certain quantity of fluid. Another example, "Select new gear coupling" eliminates all other types of couplings. The point is you may be limiting your choices to one kind of coupling when other ideas could better meet your objectives. Also you should ask, "What problem are we trying to solve?" During World War II, the military strategists were stymied trying to decide "How can we avoid the disastrous losses from German U-boats?" The convoy system was helping, but the situation in England was critical. Finally, some bright person suggested the most urgent problem was, "How to get top-priority goods to Great Britain." Flying them in became an obvious alternative. Unusual? Far fetched? Not at all. Failing to seek out the problem to be solved, the reason the action must be taken, can easily result in totally misdirected solutions.

Step 2—objectives. This is a critical step. It is also difficult. Many decision criteria are not conscious. "I don't know what I want, but I'll know it when I see it," is not an irrational statement. It is the legitimate remark of someone who has unconscious objectives difficult to describe, write down, or communicate. Even with oneself! Recognizing some gut-feel objectives probably enter everyone's process, there is considerable value in making your objectives visible and conscious (not necessarily *public*—but conscious). If you went through the action generation step, the "pro" and "con" approach provides some input. You already have an idea of what you like and dislike in the solutions offered.

These can be written in selection criteria form. For example, "a reliable component" (a pro) becomes: "Make best contribution to system reliability." This expression can be used to measure several possibilities.

"How to avoid operator errors" (a con) becomes: "Cause least (or no) machinery related incidents," which can again help in selection.

A simple and always useful question is: "What do we want to consider?" For example, reliability, operating and maintenance staff, time, energy, space, efficiency, etc. Develop your objectives from such a list. The worksheet "Do This Before Deciding" (Figure 6-10) is a series of questions to help you state your objectives. Let us consider an example and follow it through all seven steps. Here is the situation: you are involved in a task group assigned to recommend the best course of action to reduce early failures of pumps in a petrochemical complex. After some discussion, the decision statement or mission is determined:

Decision Statement—Select best method to reduce early failures of repaired pumps.

To set our objectives, we consider the needs of the organization, specific problems our recommendation is to address, and any constraints or directives that have been imposed on us. Using the worksheet, we came up with this list of objectives (see Figure 6-19):

a. Achieve minimum average runtime of one year after repair.
b. Minimize Mean-Time-To-Repair (MTTR).

DO THIS BEFORE DECIDING

This will be your decision statement

What is the headline of the decision you have to make?

Does it reflect the reason for this decision or the problem you're trying to solve?

These will be objectives. They will be used as selection criteria.

List at least three things you especially want your choice to accomplish—benefits you expect.

List at least one thing you wish to avoid.

Who will be affected by or will implement this decision?

What are at least three things *they* would like for this to do?

What resources are to be spent?

What are the limits on these, if known?

Figure 6-19. "What to do" before deciding.

    c. Minimize impact on operations.
    d. Minimize cost of repair (overtime).
    e. Minimize additional supervision.
    f. Maximize personnel safety.

We are now ready for Step 3.

Step 3—Ranking of objectives. First determine which objectives are absolutely mandatory (which means if the solution does not meet them, you throw out the solution). These objectives are used as a screen and, therefore, need recognizable limits. In the example, the achievement of a minimum average run time of one year after repair is probably in the mandatory category. The other objectives have relative importance. Numerical weights are helpful and almost any set of numbers will do. A one, two, three scale (low, medium, high) is easy to use. A 10-point scale is good for complex decisions with a larger number of objectives.

For our example, we have used a three-point scale. We went through the list assigning each objective high, medium, or low importance. Then we converted the highs to three, mediums to two, and lows to one.

Now we have obtained:

    A. Decision Statement—Select best method to reduce early failures of repaired pumps.

    B. Objectives

Achieve Minimum Average Run Time of One Year	Mandatory
Maximize personnel safety	3
Minimize impact on operations	3
Minimize additional supervision	2
Minimize MTTR	2
Minimize repair cost (overtime)	1

A pitfall to avoid. It is important to agree on objectives before selecting your final action. There is a good reason to identify objectives first. Possible actions are possible choices, possible solutions. Like political candidates, they often have backers. Naturally, these backers are potential enemies of all ideas except theirs. If you have good agreement on the objectives *prior* to considering these actions, the chances of win/lose confrontations are reduced.

Step 4—possible actions. Possible actions may come from:

    a. A meeting called for that purpose: What factors affect pump repair quality?
    b. Vendors asked to propose solutions to a mechanical problem that is recurring and affecting repair quality.
    c. Solicited or unsolicited solutions to a problem—five or six suggestions from mechanics and/or operating technicians.

A pitfall to avoid. "There's really only one way to do this," should merit our immediate attention. If objectives are examined, there is almost always at least one other approach.

Possible actions should be stated as completely as is necessary to clearly understand the various choices. If you fix one up or improve it, you need to write in beside the action how you improved it. For example, if additional mechanical-technical help is being considered, you might add: "Find way to release mechanical-technical service personnel from other duties."

Getting back to our example, let us imagine we settled on many possible actions or recommendations.

a. Incorporate formal failure analysis step for every repair.
b. Introduce training sessions for mechanics.
c. Select "teams" of top mechanics to work on pumps only.
d. Advise operations of any repair compromises made due to lack of sufficient time to repair.
e. Introduce repair quality checklists.
f. Define what is expected of mechanics and communicate this expectation.
g. Critique bad workmanship on a positive note.
h. Plan to have mechanics present at first startup after repair and develop consistent startup procedures.
i. Review spare parts policy to avoid stock-outs and "make do" compromises.
j. Subcontract some or all pump repairs—with one-year guarantees on parts and workmanship.

The final matrix would look something like the one in Figure 6-20. The content of this matrix is further explained on page xxx.

Step 5—find the best fit. Now we decide—at least tentatively:

a. Each possible action is measured against the mandatory objectives on a "go or no-go" basis.
b. Then measure the possible actions against each other and rank them on how well they meet each objective. One way is to give each a number showing how it ranked. If you have three, then the best gets three; if there are six possible actions, then give the best one six, etc. Ties are permitted.

A choice rarely fits all conditions perfectly. There will be some trade-off. However, the mandatory conditions *must* be met, or the action is dropped from consideration.

Step 6—assess risk. Risk can be considered after the choice is made. Asking the question, "What'll happen if we go with choice A?" will sometimes highlight potential trouble. This step is emotionally difficult. We have just built up a choice, a direction. We have *decided*. Now we are asking ourselves to reconsider. Not easy. There is an emotional release even when the answer is adverse—we are no longer under the strain of indecision. Still, risk must be considered. Anything else is avoiding reality. There are two ways to do this:

HEADLINE OF DECISION:

OBJECTIVES	WEIGHT	POSSIBLE ACTIONS			

$$\Sigma W \cdot \Sigma \overset{I}{\underset{PA}{N}} = \Sigma T$$

Figure 6-20. Decision-making worksheet.

   a. Look at potential risks of the actions and build them into your objectives with a weight. For example, "Subcontracting makes us dependent upon too many extraneous circumstances beyond our control." It becomes "Subcontract must maximize insurance against extraneous changes—bond, etc."

   b. The other way is to ask, "What about this choice is risky?" Then determine if it can be overcome, or if it is serious enough to warrant dropping the choice in favor of the next best alternative.

Step 7—take best choice of the action plan. The selected alternative comes back in under Fix or Action.

Fine tuning and details

   a. *The irrational part of deciding.* There is also an irrational portion to most decisions. All of us have intuitive feelings about what to do. Some we listen to and some we do not. Whether we act on these "gut feelings" may depend on how well they have worked in the past. If a new solution is especially appealing, but does not seem to meet your objectives, perhaps your objectives need looking at. There generally are reasons why the creative new idea looks good, and these reasons can be put into words. Your "public" objectives may not be the real ones. Ask, "What do I really like about this new idea?" Then list the reasons as objectives. Show *why* this answer is the best. This will help sell your idea, and everyone will feel good knowing why they decided the way they did.

Or, in the case of a lone decider, you will know why you picked this choice. There is no guarantee of success, but you will be more confident.

b. *Making recommendations.* All recommendations are risky. The only way to be sure you have made the correct choice is to have a controlled laboratory situation. Even then, you cannot be sure you have tested all options. The best solution may never be found. When you are recommending or selling you can bet your listener is comparing your solution with one or more others. But how are they being measured? This is the major rule in recommending: *discover the presentee's objectives.* Remember, little is sold in this world. Successful transactions involve people *buying* things, and they buy for their reasons, not yours. Therefore, you must find out what those reasons are. Some options are to ask "what do you want?" show a possible list of selection criteria and have them respond or rank order, ask what they especially like in your idea, ask what would make it better, ask what they like about other ideas. The point is, you probe. You help them think it through. Many people do not actually know what they want on a conscious level, and if they do not know, they cannot tell you. Everyone wins if you help them think this through. Once you have the criteria or list of objectives, you need to rank them, possibly assigning relative weights. You need to agree on how these will be measured. Then it is only a matter of comparing the options or actions to see which best fits the criteria.

c. *Recommendation variation.* There is a variation which works a little better when you need to sell your idea, when you are not sure you are deciding among the best choices, or when you would like to highlight areas which are especially good or bad about the possible actions.

Now listen carefully. The difference is subtle. Here is how it works:

a. Arrange the Possible Actions across the top like before.
b. Add the criteria or objectives down the left side.
c. Give each criterion a weight (at least high/medium/low).
d. Now check the Possible Actions against each of the criteria. How well do they do? Good? So-so? Poorly? Give them some points. The simplest is three for good, two for so-so, one for poor.
   If you assigned numerical weights to your criteria, you can multiply to get the final weighted score.
e. *Improving the actions.* Wrong idea won? Either there are some important criteria you have not stated or you need to strengthen your favorite. Look at the areas where your pet idea fared poorly. How can you improve it? When you are in a group situation, the hardest part is to stick with the objectives and their weights. For example, you are deciding which person to promote. There are several mandatory criteria, one of which is: "must have technical degree." Jackie's name comes up. Jackie does not have the necessary degree, but is a favorite of one of the key executives. If you decide to keep Jackie in the running, you *must* remove the mandatory nature of that criterion for *all* candidates, or else Jackie is dropped from consideration. A positive way to use this is to figure out a way around the requirement such as making it a condition that Jackie enroll in a course of study to get the degree and in the meantime

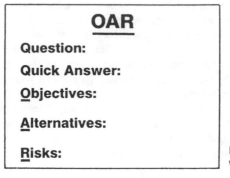

Figure 6-21. Decision-making worksheet—short form.

receive the necessary technical support. To conclude this discussion of the decision process, we are including Figures 6-21 and 6-22 as working aids.

Planning for Change

A good plan should contain problem-preventing steps. If risks are high, you also need protection as part of your plan. The process to do this *must* be quick and simple. It can be done by asking a couple of mental questions:

1. What could go wrong?
2. What can we do about that?

In cases where the exposure of money, people, time, or other resources is high and critical, the plan should be written.

Where Problems Come From

When a certain concentration or combination of factors occurs, we have trouble. Avoiding unwelcome surprises means discovering these factors or conditions. That is not enough, though. The person who can observe the conditions needs to know what to look for—both warnings and symptoms. In addition, someone must be willing and able to take the right action.

If you are to avoid or minimize trouble effectively, you need a good process. And, the simpler it is, the more likely it will be effective. Unused safety devices prevent few accidents.

The process is simple. There are six parts:

1. Statement of plan (what, where, when, and how much?)
2. Chronological steps.
3. Separate potential problems (think of everything that will change and ask "What could go wrong?")
4. Pick two or three with highest priority. Assign priority with these questions:
   a. What is it worth to solve?
   b. When must I act to be effective?
   c. How likely is it to happen?
5. Make an Action Plan for the highest priority items.
6. Update your plan with the Action Items.

HEADLINE OF DECISION: SELECT BEST METHOD TO REDUCE EARLY FAILURES OF REPAIRED PUMPS

OBJECTIVES	WEIGHT Must	POSSIBLE ACTIONS			
		Formal FA Contributing	Repair QC Contributing	No Stock-Outs Contributing	Subcontract Unknown
Achieve Minimum Average Run Time of One Year					
Maximize Personnel Safety	3	Perhaps: End Result 3 / 9	Improvement 4 / 12	No Impact 1 / 3	Perhaps 2 / 6
Minimize Impact on Operations	3	Follow-up Required 2 / 6	No Impact 3 / 9	Positive 4 / 12	Subcontractor Needs to be called 1 / 3
Minimize Additional Supervision	2	No Impact 3 / 6	Need Supv. 1 / 2	No Impact 4 / 8	Needs follow-up & co-ordination 2 / 4
Minimize MTTR	2	Extra Time Required 1 / 2	Need extra time 2 / 4	Eliminate waiting time 4 / 8	Could take longer (week-ends) 3 / 6
Minimize Repair Cost	2	No additional cost 4 / 8	Extra Time=$$ 3 / 6	Increase in cost 2 / 4	Higher cost 1 / 2
ΣW		T 31	T 33	T 35*	T 21

$$\Sigma W \cdot \Sigma \underset{PA}{\overset{1}{N}} = \Sigma T \qquad 12 \cdot 10 = 120$$

$$\left.\begin{array}{r} 31 \\ 33 \\ 35 \\ 21 \end{array}\right\} \; 120$$

*Highest total score = Preferred Action

Figure 6-22. Decision-making worksheet—example of long form. Each of the possible actions is given a value ranging from 4 (high) to 1 (low). This value is multiplied by the weight of each given objective and the resulting products added up to give the total (T) for each possible action.

Gathering, organizing the information, and setting up the system is not easy. The concept is simple, but it is emotionally difficult. Common sense tells you that removing likely causes of problems reduces the probability of their occurring. Since most problems have interactive causes, you can cut down trouble by attacking the various conditions which act together.

A working system of problem prevention has immense value. Operating technicians, mechanical-technical service personnel, and the machinery troubleshooter and failure analyst usually gain increasing confidence when they apply this system.

Details of the System

1. *The Plan*
   - What is going to change?
   - Where will it be done?
   - When will it begin and end?
   - How much is involved?
   - What is your part—are you in charge, an observer,
     or a participant?

2. *Steps*
   List them in order of occurrence. What has to be done first, second, etc. Put in times and dates. This is not much detail. If you are going to plan at all, you should do at least this much.

3. *Problems or Opportunities*
   List things that might go wrong *without regard to whose fault or control.* Also any positive surprises that may occur. Here are the questions to ask:
   - What could go wrong?
   - What trouble have we had before?
   - Are we dealing with new people, equipment, technology, vendors, supplies, etc.?
   - Anything where we have no experience?
   - Where is timing or scheduling tight?
   - Is this event dependent on another?
   - Are we counting on people not under our control? Other departments? Other companies?
   - Would trouble for us benefit anyone else? Are they in a position to hinder us? To withhold or be less than enthusiastic with support?

List them in the same way you separated items under Situation Analysis.

4. *Assign Priorities*
   - Value—what is it worth to avoid or capitalize on this situation?
   - Timing—when must you act (or when must action be taken) to be effective. If you want to order a standby item in case your schedule is delayed, and the lead time is 30 days, you obviously must act at least 30 days ahead. This may seem obvious, but why are there so many crash programs and emergency

expedite orders in our industry? The key here is "when timing is really tight, plan backward."

- How likely is this to happen? Has it happened before? What incentive do others have to make sure it does not happen? The more likely it is to occur, the more you need to consider actions to prevent (if a trouble) or to take advantage (if an opportunity)

5. *Action Planning*

Use the same Action Planning format as before. The priority concerns are clearly just that—items which are critical in importance, timing, and likelihood of occurrence.

Under causes, the concept of contributions is helpful.

Contributors

List the conditions or factors which make this problem happen. Some of them merely increase the chances of trouble. Be *sure* to list those that increase probability. List the factors, events, or conditions that could cause it, indicate its approaching, or raise the probability of its happening. Questions to ask:

- What usually happens just before this problem?
- What conditions make this problem likely? Possible?
- What key material, people, timing, etc., *must* be correct for all to go well? Is it all in order?
- What tells us trouble is imminent? Is happening? Has happened?
- Are those indicators functioning? Are they monitored?
- What caused trouble last time? Any other time?

Fix or Other Actions

a. *Before Actions.* These are actions to prevent trouble and are specifically designed for each condition factor. The idea is to reduce the probability of its happening, or to prevent it entirely if possible. Frankly, there are few perfect preventives. Inoculations, total physical isolation from risk, total removal of a vital link (for example, no fire in the absence of oxygen) are some examples. Total removal of risk is often prohibitively expensive. So, we reduce the probability to an acceptable risk level and monitor the key indicators of trouble. Examples of Before Actions are:

Problems	Contributors	Before Actions
Large motor is damaged when transported to electrical shop	Suffers traffic accident	Provide front & back escort
	Falls off float	Check & doublecheck tie-down
Large motor fails	Dirt in windings	Clean and inspect regularly.

b. *When actions.* These are arranged ahead of time to activate when the indicators show the problem has happened or is imminent. They are your contingency plan. They are like insurance. They do not keep the trouble from happening, but they reduce the damage. Effects of the problem are their target. Examples of When Actions are:

Problem	Contributors	When Actions
Large motor suffers winding failure	Dirt & overload	Enact contingency plan, i.e., use reserve copper bar stock for windings.

Note that When Actions include the specific symptoms or warning which tells you to begin the action. Also, both Before and When Actions are appropriate in connection with serious problems.

6. *Update your plan with the Action Items*

You have accomplished nothing until the plan itself is altered. If you pick out some actions designed to prevent problems as you decide to change the timing or sequence in your plan, *go back and physically make the change.*

7. *Assign responsibilities and see that the plan gets executed*

Figures 6-23 and 6-24 show appropriate examples of a "planning for change" sequence. Figure 6-25 may be used as a convenient working aid for future planning work.

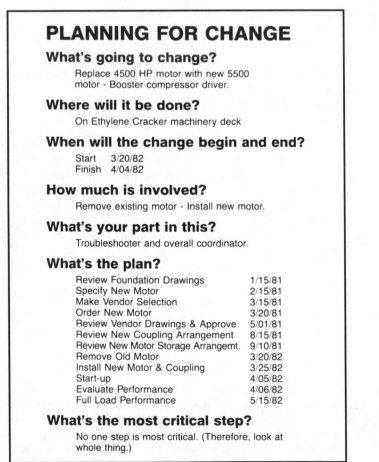

PLANNING FOR CHANGE

What's going to change?
Replace 4500 HP motor with new 5500 motor - Booster compressor driver.

Where will it be done?
On Ethylene Cracker machinery deck

When will the change begin and end?
Start   3/20/82
Finish   4/04/82

How much is involved?
Remove existing motor - Install new motor.

What's your part in this?
Troubleshooter and overall coordinator.

What's the plan?

Review Foundation Drawings	1/15/81
Specify New Motor	2/15/81
Make Vendor Selection	3/15/81
Order New Motor	3/20/81
Review Vendor Drawings & Approve	5/01/81
Review New Coupling Arrangement	8/15/81
Review New Motor Storage Arrangemt.	9/10/81
Remove Old Motor	3/20/82
Install New Motor & Coupling	3/25/82
Start-up	4/05/82
Evaluate Performance	4/06/82
Full Load Performance	5/15/82

What's the most critical step?
No one step is most critical. (Therefore, look at whole thing.)

Figure 6-23. Planning for change example.

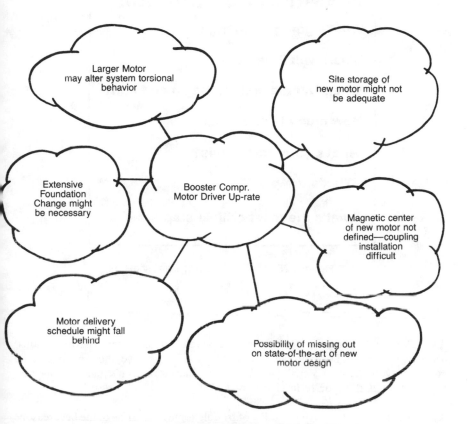

ACTION PLAN

Priority concern	Cause	Fix or other action	Responsible persons and roles	Your actions	When
Larger motor chg. system torsional	1. Different mass-elastic character	Define torsional criticals early	DFM/vendor	Assist	3/15/81
	2. Wrong couplg. selection	Specify	Vendor	Review	4/01/81

Figure 6-24. Action plan example.

Figure 6-25. Planning for change worksheet.

A tip. The people who suffer from the problem are ideal to enlist for Before Actions. They are motivated. For example, prior to a long sailing voyage, the captain asks the crew, "Anyone here who can't swim?" When Charlie raises his hand, the captain says, "Ok, Charlie, you're in charge of leaks!"

A final word. Reducing the probability of trouble tomorrow is one of the best reasons to spend part of today seeking the cause of yesterday's problems.

References

1. Edwards, B., *Drawing on the Right Side of the Brain,* Houghton-Mifflin Co., Boston, 1979.
2. Stein, M.I., *Stimulating Creativity, Volume 1: Individual Procedures, Volume 2: Group Procedures,* Academic Press, New York, 1974-5.
3. Gallwey, W.T., *The Inner Game of Tennis,* Random House, New York, 1974.
4. *The Journal of Creative Behavior,* The Creative Education Foundation, Buffalo, New York.
5. Parnes, S.J., *Guide to Creative Action,* Charles Scribner's Sons, New York, 1977.
6. Parnes, S.J., *This Magic of Your Mind,* The Creative Education Foundation, Inc., Buffalo, NY 1981.

Chapter 7

Statistical Approaches in Machinery Problem Solving

In the preceding chapters it was shown how machinery failure and distress behavior can be influenced by appropriate design techniques. Ideally, we would like to verify low failure risk during the specification, design, installation, and start-up phases. However, experience shows that good, basic design alone may not prevent failures. Failure statistics in earlier chapters demonstrated clearly that many machinery component failures result from equipment operation at off-design or unintended conditions. Our failure analysis and troubleshooting activities will thus often be post-mortem and commence after installation and start-up. The maintenance phase has then begun, and failure analysis and trouble-shooting will become an integral part of that phase.

Machinery maintenance activities, together with failure analysis and trouble-shooting, can be regarded as failure fighting processes. Failure fighting calls for appropriate strategies in order to be successful. However, before we can apply any strategies, we must gather as much information as possible about failure modes and failure frequencies. Suitable statistical approaches will accomplish this task and therefore, have to be part of the failure fighting arsenal.

This chapter discusses statistical approaches and other important issues in machinery maintenance as they relate to failure analysis and troubleshooting.

Machinery Failure Modes and Maintenance Strategies

Most petrochemical plants have large maintenance departments whose sole reason for being is to take care of various and numerous forms of plant deterioration. Typical forms of plant deterioration are machinery component failure events, malfunctions, and system troubles.

As in most areas of human endeavor, we would like to perform process machinery maintenance in a rational and planned manner. Planning in turn could imply that the object of our efforts also behaves in a predictable way, i.e., it responds favorably to planned maintenance actions. Is that always true in machinery?

To answer this question satisfactorily, we must examine how machinery component failure modes appear and behave as a function of time. This discussion concerns the concept of machinery component life.

Life Concepts

Failure is one of several ways in which engineered devices attain the end of their useful life. Table 7-1 shows the possibilities.

Since most machinery failure analysis takes place at the component level, one must define what constitutes life attainment in connection with component failure modes. The term "defect limit"[1] is used in this context and needs to be explained first.

Defect limit describes a parameter which we are all familiar with. For instance, original automotive equipment tires often have built-in tread wear indicators. These are molded into the bottom of the tread grooves and appear as approximately ½-inch wide bands when the tire tread depth is down to $1/16$ inch. The wear indicators are "defect limits."

Other examples of defect limits are clearance tolerances of turbomachinery assemblies which are observed and documented during machinery overhauls. But not all defect limits are obvious. Mechanical shaft couplings are an example of a difficult-to-define-limit where we wish we had something like a tread wear indicator to determine serviceability. Another example, but connected with a random failure mode, is the decision point at which a leak from a mechanical seal or a vibration severity value on a pump is considered excessive and deserving of maintenance attention. Here, the defect limit defies ready quantification and, more often than not, depends on subjective judgment.

One noted machinery troubleshooter[2] refers to this phenomenon as "user tolerance" to the distress symptoms exhibited by machinery equipment. In his discussion of flow recirculation in centrifugal pumps, he describes a pump designer's view as to how users determine acceptable minimum flow for centrifugal pumps. He

Table 7-1
Reaching End of Useful Machinery Life

Way in Which the End of Useful Life Is Reached		Example
Failure	Slow	Mechanical seal leakage
	Sudden	Motor winding failure
Obsolescence		Recip. steam pump/engine
Completion of mission		
		Oil mist
		Packaging
		Wear pads
Depletion		Electric battery

states while some users will accept that an impeller subject to cavitation may have to be replaced every year, others will not tolerate impeller replacements in the same service more frequently than every three or four years. Similarly, noise and pulsation levels perfectly acceptable to one user are totally unacceptable to another.

In any case, defect limit defines the degree of failure mode progress that would initiate such maintenance activities as replacement, repair, more thorough or more frequent checks, or perhaps a calculated risk decision to continue without resorting to these conventional maintenance activities.

"Life" of a machinery component can now be defined as the time span between putting a machinery component into service and the point in time when the component attains its defect limit. "Standard Life" is the average lifetime acceptable to any particular plant failure analyst or troubleshooter. Have you arrived at defect limits and therefore "standard life" of *all* the failure modes encountered on machinery components in your plant? Are you willing to live with them? Or are you looking for improvement?

These questions remind us that we need to set defect limits objectively, albeit just for our own plant. Two implications seem obvious: (1) if we do not, we have no baseline from which to improve, and (2) if we allow subjective judgment to prevail, we will most likely experience unpleasant surprises, because those judgment calls will range from the conservative to the optimistic. The conservative judgment will reduce component life arbitrarily and the optimistic judgment call will expose us to unjustifiably high risks. A team approach to documented defect limit setting seems therefore appropriate.

From experience we know that some failure modes occur rather often, for example, "wear of coke crusher rotor fingers" in tar service. Other failure modes are hardly ever encountered; for instance, electrical pitting (FM: erosion) on surfaces of spherical rollers caused by current passage through the bearing. Obviously, different failure modes have different frequencies of occurrence. In the past, we have generally used mean-time-between-failure (MTBF) to express these differences. What we are describing is the probability of machinery failure and breakdown events as a function of operating time. While this is of some interest to all of us, it is of particular concern to the maintenance failure analyst and troubleshooter. No doubt at one time or another we all have to grapple with the realization that some machinery failure modes appear slowly and predictably while others occur randomly and unpredictably. In any event we have probably encountered a mix of both types of failures.

The Swedish mathematician W. Weibull developed mathematical models to describe characteristics of component failures. In "Weibull functions" we find a constant β which characterizes the life cycle of a given group of components. Figure 7-1 shows these Weibull functions. In its simplest form, the Weibull relationship looks like this:

$$\text{failure rate} = at^{(\beta-1)} \qquad (7\text{-}1)$$

where:

a = scale factor.

$1/\beta$ = Weibull slope.

t = time in service.

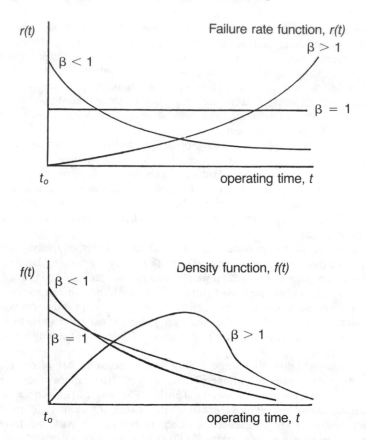

Failure rate function, $r(t)$

Density function, $f(t)$

Failure Rate, $r(t)$

Consider a test where a large number of identical components are put into operation and the time to failure of each component is noted. An estimate of the failure rate of a component at any point in time may be thought of as the ratio of a number of items which fail in an interval of time, say, a week to the number of the original population which were operational at the start of the interval. Thus the failure rate of a component at time t is the probability that the component will fail in the next interval of time given that it is good at the start of the interval, i.e., it is a conditional probability.[3] A noteworthy relationship is

$$r(t) = \frac{1}{\text{MTBF}}$$

Other terms for failure rate are hazard rate, force of mortality, and failure intensity.

Density Function or Frequency Function, $f(t)$

The density function, similar to a histogram, shows the percentage of failing components during a time interval relative to the total population operating at time t_o.

Figure 7-1. Weibull functions.

Where β is equal to 1.0, the analysis describes random failures such as experienced in ball bearings and many other mass-produced machinery components before the effects of wear-out set in. Those failures occur with constant probability during component operating time.

A very useful example in this context is the failure experience of mechanical shaft seals. A classic study[4] shows how to use the Weibull function to describe the nature of seal failures in a petrochemical plant. The Weibull indices (β) published in the study suggested that seal failures were largely a mixture of "infant mortality," i.e., seal defects, fitting errors, and random failures caused by the inability to maintain design assumptions (dry running and excessive vibration).

Other examples are the failure experience on diesel engine control units shown in Figure 7-2 and diesel engine pistons, Figure 7-3.[5]

Another important contribution to the proper understanding of random failure experience of machinery components was made by Bergling.[6] He applied a β of 1.1 for rolling bearings based on numerous laboratory tests and demonstrated that, if a bearing is removed and inspected after operation and no signs of fatigue as manifested by

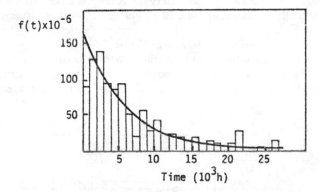

Figure 7-2. Density function $f(t)$ of failures on diesel engine control units (Ref. 5).

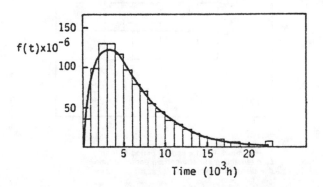

Figure 7-3. Density function $f(t)$ of failures on diesel engine pistons (Ref. 5).

spalling, are detected, then this bearing is practically just as serviceable from a fatigue point of view as a brand new bearing. Figure 7-4 shows probable remaining life for used, undamaged rolling bearings. We see from this that if L_{10} is 28 years, the bearing after 40 years of operation can be expected to function for a further period corresponding to approximately 83% of the basic rating life. There is no reason to discard a bearing because of age.

The condition $\beta > 1.0$ describes machinery component wear-out modes which occur with increasing probability over operating time. Carbon brushes on electric motors are one example. Most wear- and aging-related failure modes fall into this category.

Since our discussions have mainly been at the machinery component level, we should at this point ask how life performance of machinery assemblies, units, and systems could be evaluated. What life should be expected for our machinery systems? Figure A-1 in Appendix A shows typical life spans for machinery equipment and related systems. Figure 7-5 shows life dispersions for different industrial products compared with those of human beings in industrial countries. This information provides a greater understanding of life concepts.

To conclude our discussion of machinery life concepts, a popular view of failure rate dependence from operating time of machinery and other systems should be considered. If the failure rate is plotted as a function of time, a curve, known as the "bathtub curve" is generated as shown in Figure 7-6.

In this curve three conditions can be distinguished: (1) early failures, (2) random failures, and (3) wear-out failures. One would be hard pressed to construct an actual curve like this for a given machinery system. However, it has been used to give an

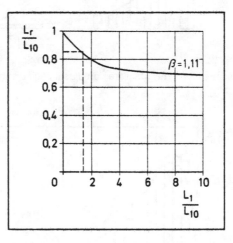

L_{10} = basic rating life
L_1 = attained life at a certain period of time
L_r = remaining life at a certain period of time
β = dispersion exponent

Figure 7-4. Probable remaining life for used, undamaged anti-friction bearings (Ref. 6).

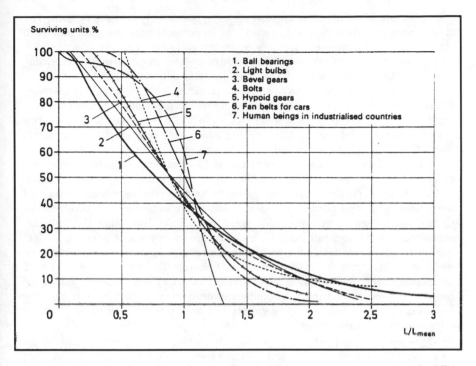

Figure 7-5. Life distribution for various industrial products and for human beings (L = Life) (Ref. 6).

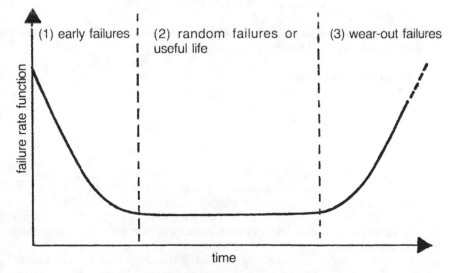

Figure 7-6. Bathtub curve concept.

overall picture of the life cycle of many systems, particularly complex systems. The closest approximation is the life cycle of a petrochemical process unit with compression and pumping machinery as an integral part of it.

Condition 1 describes the early time period of the system by showing a decreasing failure rate over time. It is usually assumed that this period of "infant mortality" is caused by the existence of material and manufacturing flaws together with assembly errors. Condition 2, the area of constant failure rate, is the region of normal running. Here, failure occurs as a consequence of statistically independent factors. This period is also termed "useful life," during which time only random failures will occur. Condition 3 is characterized by an increase of failure rate with time. As described previously, here failures may be due to aging effects. Also, for a good percentage of lubrication-related failures, the failure rate increases with time period.[5]

The broad applicability of the bathtub concept at the system level has unfortunately been postulated as an ideal-failure or life-cycle model all the way down to individual component levels. We agree with J. R. King[7] that the too easy representation of failure rate behavior has often been a deterrent to effective product improvement. Quick debugging procedures to remove infant mortality have been overemphasized, and the random *times to failure* in the useful life period have erroneously been understood to mean that the *causes of failure* are random. Real progress in process machinery reliability can only be made by diligently isolating individual failure modes and their causes and then finding ways to eliminate or minimize them permanently.

The Weibull concept, when applied properly, can become a powerful tool for the machinery failure analyst who is trying to make sense of a large amount of field data that will typically be incomplete. Later in this chapter we will make our readers familiar with such a tool, which is known as hazard analysis, a special case of the Weibull concept.

Machinery Maintenance Strategies

Different failure modes and their behavior in time are met with different maintenance strategies. Machinery maintenance strategies are:

1. Preventive maintenance
2. Predictive maintenance
3. Breakdown maintenance
4. Bad actor management

Preventive Maintenance

All over the world the term "preventive maintenance" means "periodic" maintenance.[1] Preventive maintenance in this context should therefore be understood as a periodic or scheduled activity which has as its objective the direct prevention of failure modes or defects. In its simplest form this activity entails periodic lubrication, coating, impregnating, or cleaning of machinery components to increase life spans.

Some failure modes respond very well to periodic preventive maintenance, whereas others do not. Moreover, quite often deterioration cannot be completely eliminated even in those failure modes that do respond to periodic preventive maintenance. For instance, by periodically renewing the lube oil in a pump bearing (if one has not decided to use superior oil mist lubrication) one will effectively reduce wear (FM) of the bearings by postponing or even preventing the failure modes "age" and "contamination" of the lube oil itself. Another example is the periodic cleaning of large pipe-ventilated motor drivers of process gas compressor trains where excessive winding dirt contamination (FM) and, more important, insulation failures (FM) can be effectively prevented by this maintenance activity. We presented an example related to this maintenance action in Chapter 4.

Predictive Maintenance

The other maintenance activity is periodic inspection, followed by replacement or overhaul if incipient defects are detected. These actions do not directly reduce the deterioration rate but indirectly control the consequences of accidents, breakdowns, malfunctions, and general trouble. We refer to this maintenance activity as *predictive maintenance.*

While preventive maintenance should look after the predominantly time-dependent failure modes, predictive maintenance addresses the randomly and suddenly occurring failure modes as far as possible by searching for them and by effecting timely repairs. Through the use of this maintenance technique, we expect not to prevent the failure mode, but to influence the consequences of the unexpectedly occurring defect. An appropriate example would be the inspection and change of a major compressor face-type oil seal where random heat checking (FM) has been observed over the years.

Ideally, predictive maintenance strategy should dictate a continuous search for defects, i.e., continuous monitoring of machinery condition and performance coupled with continuous feedback. Usually, however, depending upon a chemical plant's management philosophy and cost-benefit considerations based on failure severity and risk evaluations, the monitoring task will be shared to a varying degree among vigilant operators or maintenance people and sophisticated onstream monitoring equipment. In reality, most predictive maintenance activities are somewhat periodic and involve the observer's senses of seeing, hearing, feeling, and smell.

Breakdown Maintenance

Many people have the tendency to be overly optimistic about the possibilities of machinery preventive maintenance. We believe that there are several limitations to machinery preventive maintenance (PM) concepts.

First, the limitations set by random failure events. Random machinery failure events, according to our definition, could occur with equal probability in time. They do not appear periodically and we conclude from this that PM, or periodic maintenance, will be ineffective in that case. Continuous monitoring is the only resort.

Second, the life dispersion of machinery components. Even time-dependent failures are not all that predictable. They do not appear after absolutely equal operating intervals, but after very dissimiliar time periods. Consider the life dispersion of mechanical gear couplings on process compressors. Obviously, they are components subject to wear. If we conclude that their MTBF (mean-time between failure), or mean-time-between-reaching-of-detect-limit is 7.5 years, it is possible to have an early failure after 3 years and another after 15 years. The assumptions are that there are no undue extraneous influences, such as excessive misalignment, lubricant loss, etc. The longest life span in this case will be five times as long as the shortest life span. Figure 7-7 describes this relationship.

According to H. Grothus,[8] most time-dependent failures show even larger life dispersions. Accordingly, we must consider ratios of maximum to minimum of 4:1 and even 40:1. Appendix A, Figure A-1 provides additional insights into this phenomenon. As a rule of thumb, it can be said that relative life dispersion increases with the absolute value of MTBF. This means that wear items with relatively short life expectancy such as rider rings on reciprocating compressors, will have a comparatively smaller dispersion than components such as gear tooth flanks, which can be expected to remain serviceable for long periods of time.

The third limitation of preventive maintenance in process machinery is that in order to inspect we have to shut down and open up. Who is not familiar with the risk of this procedure? The Canadian Navy uses the phrase "Leave well enough alone." There have been numerous studies that show how machinery troubles started to mount right after preventive maintenance inspections had made a disassembly of the equipment necessary. Figure 7-8 shows how trouble incident frequencies can increase contrary to expectations after inspection and overhaul.

We conclude from this that a certain amount of breakdown maintenance of our machinery will always be necessary, even if we succeeded to strike that fine balance

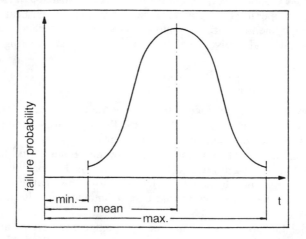

Figure 7-7. Failure probability (density) and minimum–maximum life of machinery components (Ref. 8).

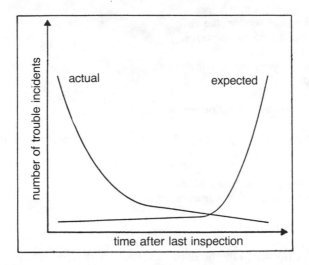

Figure 7-8. Number of trouble incidents per specified operating hours as a function of time after last inspection and overhaul (Ref. 8).

between too much predictive/preventive maintenance and too little. There will be failure modes that will not respond to periodic servicing, such as lube-oil changes, nor can they be detected by some form of inspection or monitoring. For instance, we have hardly any possibility of predicting a motor winding failure, the insulation breakdown of a feeder cable for a large compressor motor driver, or pump shaft fracture caused by fatigue.

Breakdown maintenance is here to stay! It is the maintenance activity necessary to restore machinery equipment back to service after failure modes developed that were:

1. Preventable but not prevented.
2. Predictable but not predicted.
3. Predicted but not acted upon.
4. Not preventable or predictable.

Bad Actor Management

Bad actor management addresses that highly variable portion of our process machinery population that wants to behave as though it did not belong.

Bad actors among our process pump population, for instance, sometimes prove Pareto's law, i.e., 20% of all pumps cause 80% of the trouble. All or some of the causes previously identified in connection with machinery component failure modes can give rise to bad actor behavior. Some of the predominant causes are:

1. Not suited for service due to:
   a. Wrong design assumptions.
   b. Incorrect material.
   c. Change in operating conditions.
2. Maintenance:
   a. Improper repair.
   b. Design and/or selection errors.

Bad actor management is preoccupied with the following necessary problem-solving steps:

1. Bad actor identification and tracking.
2. Failure analysis and documentation.
3. Follow-up.
4. Organizational aspects.[9]

Since a large portion of this book concerns itself with steps 2, 3, and 4, we would like to limit the discussion at this point to step 1, namely, bad actor identification and tracking.

Bad Actor Identification and Tracking

As in reliability work, a prerequisite for bad actor identification is the knowledge of machinery MTBF by type, unit, and service. The next step is to decide what time to failure (TTF) is acceptable to us vis-à-vis a machinery break-down, keeping in mind that our resources will be limited to allow us to investigate *all* items that deviate from an acceptable or expected TTF. So let us make sure we do not set our sights too high at the beginning. Even though this sounds complicated, do we not all know our bad actors? It is necessary to institute this kind of decision step covering *all* failures. Later in our program we will not be able to tell our bad actors from acceptable ones quite that easily. Figure 7-9 shows the methodology of this process.

The third step is to proceed to the component level because "the devil is hiding in the detail."* At this point, Table 7-2 will be a good aid in decision making. It shows machinery component failure modes commonly encountered in machinery failure analysis together with suggested standard life values, Weibull indices (β), and responsiveness to preventable or predictable maintenance strategies. Finally, Figure 7-10 is a tracking sheet that will allow the analyst to again document bad actor failure modes on the component level.

In conclusion, bad actor identification and tracking is the first step in the painstaking effort to eliminate recurring and costly machinery failures. Later we will deal with all other necessary steps that will allow us to reach our objective.

In the two following sections we are going to take a look at some useful statistical tools that will help us to define failure experience trends in a given machinery population.

(text continued on page 495)

---

*Old German proverb.

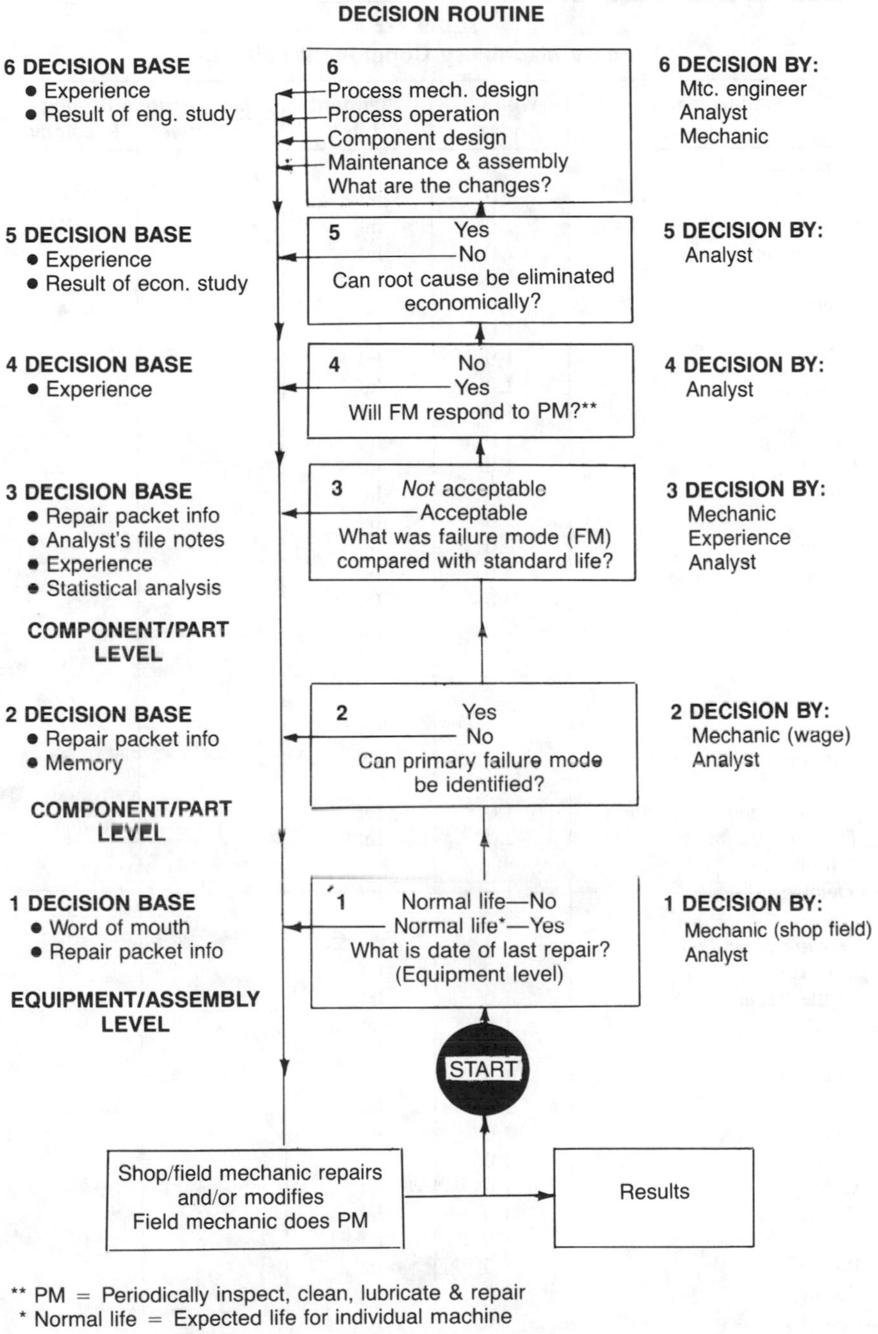

DECISION ROUTINE

6 DECISION BASE	6	6 DECISION BY:
• Experience	Process mech. design	Mtc. engineer
• Result of eng. study	Process operation	Analyst
	Component design	Mechanic
	Maintenance & assembly	
	What are the changes?	

5 DECISION BASE	5 Yes	5 DECISION BY:
• Experience	No	Analyst
• Result of econ. study	Can root cause be eliminated	
	economically?	

4 DECISION BASE	4 No	4 DECISION BY:
• Experience	Yes	Analyst
	Will FM respond to PM?**	

3 DECISION BASE	3 *Not* acceptable	3 DECISION BY:
• Repair packet info	Acceptable	Mechanic
• Analyst's file notes	What was failure mode (FM)	Experience
• Experience	compared with standard life?	Analyst
• Statistical analysis		
COMPONENT/PART LEVEL		

2 DECISION BASE	2 Yes	2 DECISION BY:
• Repair packet info	No	Mechanic (wage)
• Memory	Can primary failure mode	Analyst
COMPONENT/PART LEVEL	be identified?	

1 DECISION BASE	1 Normal life—No	1 DECISION BY:
• Word of mouth	Normal life*—Yes	Mechanic (shop field)
• Repair packet info	What is date of last repair?	Analyst
EQUIPMENT/ASSEMBLY LEVEL	(Equipment level)	

START

Shop/field mechanic repairs and/or modifies Field mechanic does PM → Results

** PM = Periodically inspect, clean, lubricate & repair
* Normal life = Expected life for individual machine

Figure 7-9. Machinery failure analysis methodology.

Table 7-2
Primary Machinery Component Failures

Failure Mode	Weibull Index*	Standard Life*	Maintenance Strategy	
			Preventive	Predictive
Deformation				
Brinelling	1.0	Inf		●
Cold flow	1.0	Inf		●
Contracting	2.0	Inf	●	
Creeping	2.0	Inf		
Bending	1.0	Inf		●
Bowing	1.0	Inf		
Buckling	1.0	Inf		
Bulging	1.0	Inf		
Deformation	1.0	Inf		
Expanding	1.0	Inf		
Extruding	1.0	Inf		
Growth	1.0	Inf		
Necking	1.0	Inf		
Setting	2.0	Inf		
Shrinking	2.0	Inf		
Swelling	3.0	Inf		
Warping	1.0	Inf		
Yielding	1.0	Inf		
Examples:				
Deformation of springs	1.0	Inf		●
Extruding of elastomeric seals	1.0	4.0Y		●
Force-induced deformation	1.0	Inf		
Temperature-induced deformation	2.0	Inf		●
Yielding	1.0	Inf		
Fracture/Separation				
Blistering	1.0	Inf		
Brittle fracture	1.0	Inf		
Checking	1.0	Inf		
Chipping	1.0	Inf		
Cracking	1.0	Inf		
Caustic cracking	1.0	Inf		
Ductile rupture	1.0	Inf		
Fatigue fracture	1.0	Inf		
Flaking	1.0	Inf		
Fretting fatigue cracking	1.0	Inf		
Heat checking	1.0	Inf		
Pitting	1.0	Inf		
Spalling	1.0	Inf		
Splitting	1.0	Inf		

*From References 3-5, 8 and author's estimates.   *(table continued on next page)*

Table 7-2 (cont.)

Failure Mode	Weibull Index*	Standard Life*	Maintenance Strategy	
			Preventive	Predictive
Examples:				
Overload fracture	1.0	Inf		
Impact fracture	1.0	Inf		
Fatigue fracture	1.1	Inf		●
Most fractures	1.0	Inf		
Change of Material Quality				
Aging	3.0	5.0Y	●	
Burning	1.0	Inf		
Degradation	2.0	3.0Y	●	
Deterioration	1.0	Inf		
Discoloration	1.0	Inf		
Disintegration	1.0	Inf		
Embrittlement	1.0	Inf		
Hardening	1.0	Inf		
Odor	1.0	Inf		
Overheating	1.0	Inf		
Softening	1.0	Inf		
Examples:				
Degradation of mineral oil-based lubricant	3.0	1.5Y	●	●
Degradation of coolants	3.0	1.0Y	●	●
Elastomer aging	1.0	4.0-16Y		●
'O'-Ring deterioration	1.0	2.0-5Y		●
Aging of metals under thermal stress	3.0	4.0Y		●
Corrosion				
Exfoliation	3.0	2.0–4.0Y	●	
Fretting Corrosion	2.0	3.0Y	●	
General Corrosion	2.0	1.0–3.0Y	●	
Intergranular Corrosion	2.0	1.0–3.0Y	●	
Pitting Corrosion	2.0	1.0–3.0Y	●	
Rusting	2.0	0.5–3.0Y	●	
Staining	2.0	0.5–3.0Y	●	
Examples:				
Accessible Components	2.0	2.0-4.0Y	●	●
Inaccessible Comp.	2.0	2.0-4.0Y		
Wear				
Abrasion	3.0	0.5–3.0Y		●
Cavitation	3.0	0.5–3.0Y		●
Corrosive Wear	3.0	0.5–3.0Y		●
Cutting	3.0	0.5–3.0Y		●
Embedding	3.0	0.5–3.0Y		●

*From References 3-5, 8 and author's estimates.

Table 7-2 (cont.)

Failure Mode	Weibull Index*	Standard Life*	Maintenance Strategy	
			Preventive	Predictive
Erosion	3.0	3.0Y		●
Fretting	3.0	2.0Y		●
Galling	3.0	2.0Y		●
Grooving	3.0	2.0Y		●
Gouging	3.0	2.0Y		●
Pitting	3.0	1.0Y		●
Ploughing	3.0	1.0Y		●
Rubbing	3.0	3.0Y		●
Scoring	3.0	3.0Y		●
Scraping	3.0	0.5–3.0Y		●
Scratching	3.0	3.0Y		●
Scuffing	3.0	1.0Y		●
Smearing	3.0	1.0Y		●
Spalling	3.0	0.5–16Y		●
Welding	3.0	0.5–3.0Y		●
Examples:				
Non-lubed relative movement	3	1.0Y		●
Contaminated by	3	3.0M		●
lubed sleeve bearings	3	4.0-16Y	●	●
Spalling of antifriction bearings	1.1	16.0Y		●
Displacement/seizing/adhesion				
Adhesion	1.0	Inf		
Clinging	1.0	Inf		
Binding	1.0	Inf		
Blocking	1.0	Inf		
Cocking	1.0	Inf		
Displacement	1.0	Inf		
Freezing	1.0	Inf		
Jamming	1.0	Inf		
Locking	1.0	Inf		
Loosening	1.0	Inf		
Misalignment	1.0	Inf		
Seizing	1.0	Inf		
Setting	1.0	Inf		
Sticking	1.0	Inf		
Shifting	1.0	Inf		
Turning	1.0	Inf		
Examples:				
Loosening (locking fasteners)	1	Inf		
Loosening (bolts)	1	Inf		●

*From References 3-5, 8 and author's estimates.   *(table continued on next page)*

Table 7-2 (cont.)

Failure Mode	Weibull Index*	Standard Life*	Maintenance Strategy	
			Preventive	Predictive
Loosening (nuts)	1	Inf		•
Misalignment (process pump set)	2	1.5-3.0Y		•
Seizing (linkages)	1	Inf	•	•
Seizing (components subject to contamination or corrosion)	1	Inf		•
Shifting (unstable design)	1	Inf		
Leakage				
Joints with relative movement	1.5	3.0M-4.0Y		
Joints without relative movement	1	16.0Y		•
Mechanical seal faces	0.7-1.1	0.5-1.5Y		
Contamination				
Clogging	1.0	Inf		
Coking	2.0	0.5–3.0Y		•
Dirt accumulation	2.0	0.5M–3.0Y		•
Fouling	1.0	Inf		•
Plugging	1.0	Inf		•
Examples				
Fouling gas compressor	3	1.5-5.0Y	•	•
Plugging of passages with moving medium	1	Inf		•
Plugging of passages with non-moving medium	1	Inf		
Conductor Interruption				
Flexible cable	1	Inf		
Solid cable	1	Inf		
Burning through insulation				
Motor windings	1	16Y	•	
Transformer windings	1	16Y	•	

Legend: Inf = Infinite
      M = Month(s)
      Y = Year(s)
      • = Yes. Absence of (•) indicates that break-down maintenance is appropriate strategy.

*From References 3-5, 8 and author's estimates.

BAD ACTOR TRACKING			FAILURE ANALYSIS	NO: 023
DESCRIPTION: P-101 , Hightower Feedpump				
INFORMATION	FAILURE	FAILURE	FAILURE	FAILURE
DATE	80/11/05	81/01/15	81/05/10	
COMPONENT / PART	SEAL	SEAL	SEAL	
VENDOR DRWG.NO.				
ELEMENT/PART	Secondary sealing O-ring Part # 13	Same Part # 13	(3)Springs Part # 5	
FAILURE MODE	Swelling	Swelling	Breakage	
TTF *	?	1.5M	5.0M	
CORRECTIVE ACTION	Replaced	Changed to Viton	Corrected internal alignment	
BAD ACTOR PART (Check)	?	✓		
FAILURE CAUSE	Changed pump service	Changed pump service	Internal Mis-alignemnt due to flange loading	
FAILURE RESISTANCE	O-ring material	O-ring material	Pump casting strength	
ANALYSED BY:	J.Pickel		DATE: 81/05/30	
REVISON NO.	REMARKS		DATE	
1				
2				
3				
4				
* Time-To-Failure				

Figure 7-10. Bad actor tracking.

(text continued from page 488)

Hazard Plotting for Incomplete Failure Data*

Plotting and analysis of data must take into account the form of the data. Failure data can be complete or incomplete. If failure data contain the failure times of all units in the sample, the data are complete. If failure data consist of failure times of failed units and running times of unfailed units, the data are incomplete, are called "censored," and the running times are called "censoring times." If the unfailed units all have the same censoring time (which is greater than the failure times), the data are singly censored. If unfailed units have different censoring times, the data are multiply censored.[10]

Complete data result when all units have failed. Singly-censored data result in life testing when testing is terminated before all units fail. Multiply-censored data result from:

1. Removal of units from use before failure.
2. Loss or failure of units due to extraneous causes.
3. Collection of data while units are still operating.

The following are examples of situations with these reasons for multiply-censored data:
1. In some life-testing work, units are put on test at various times but are all taken off together; this is done to terminate testing by some chosen time to meet a schedule.
2. In medical followup studies on the length of survival of patients after treatment, contact with some patients may be lost or some may die from causes unrelated to the disease treated. Similarly, in engineering studies, units may be lost or fail for some reason unrelated to the failure mode of interest; for example, a unit on test may be removed from a test or destroyed accidentally.
3. In many engineering studies, field data are used which consist of failure times for failed units and differing current running times for unfailed units. Such data arise because units are put into service at different times and receive different amounts of use.

Probability plotting is commonly used for graphical analysis of failure data. Knowledge of probability plotting is not essential to understanding the hazard plotting method presented here. Probability plotting can easily be used for complete data and for singly-censored data but is difficult to use for multiply-censored data. One simple expedient that is used to plot multiply-censored data is to ignore all failures that occurred after the earliest censoring time and plot only those that occurred before the first censoring time. This procedure may be acceptable, if there are relatively few failures after the first censoring time and thus little information is lost by excluding them from the plot. However, if there are relatively many failures after the earliest censoring time, considerable information is lost. Moreover, if the earliest censoring time is in the lower tail of the distribution and only the failure times

---

*Adapted by permission of Dr. Wayne Nelson, G.E. Research and Development, and of the American Society for Quality Control.

before that time are used, the distribution function must be estimated by considerable extrapolation beyond the earliest censoring time. Then the estimate of the distribution function may be considerably in error; that is, a theoretical distribution that agrees with the data below the earliest censoring time may differ considerably from the true distribution function above the earliest censoring time. This is a risk of extrapolating beyond the data that are plotted. For these reasons, it is preferable to plot all failure times occurring both before and after the earliest censoring time to make full use of the information in the data.

The method presented here of plotting data on hazard paper achieves full use of all failures and gives the same information but with less effort. Hazard papers are shown here for the exponential, Weibull, normal, log normal, and extreme value distributions.*

In the foregoing, failure is described in terms of time to failure. In various situations other measures of the amount of exposure before failure may be relevant, for example, cycles to failure, miles to failure, etc. Also, failure is used here in a general sense to denote any specific deterioration in performance of a unit. If units are put into service at different times, the time in use for each unit is figured from its starting time.

The following sections contain a detailed presentation of the hazard plotting method. The next section contains step-by-step instructions on how to make a hazard plot of multiply-censored data. The instructions are illustrated with an example based on actual failure data. The section after that shows how to interpret and use a hazard plot to obtain engineering information on the distribution of time to failure. This section also contains plots and analyses of simulated data from different known theoretical distributions; this permits us to assess the performance of the hazard plotting method. The last section contains additional practical advice on how to use hazard paper. Appendix B contains a presentation of theory underlying hazard plotting and descriptions of the theoretical distributions for which there are hazard papers.

How to Make a Hazard Plot

Step-by-step instructions are given here for making a hazard plot of multiply-censored failure data. Data used to illustrate the hazard-plotting method are shown in Table 7-3. These data for 70 diesel generators consist of hours of use on 12 generators at the time of fan failure, and hours on 58 generators without a failure. The ordered times to failure for the failed fans and the running times for the unfailed fans are shown in Figure 7-11. The engineering problem was to determine if the failure rate of the fans was decreasing with time. Information on the failure rate was to be used to aid in arriving at an engineering and management decision on whether or not to replace the unfailed fans with a better fan to prevent future failures. There was the possibility that the failures had removed the weak fans, and the failure rate on the remaining fans would be tolerably low. Steps in the hazard-plotting method follow with the fan failure data used as an example:

---

*These hazard plotting papers can be obtained from TEAM, Box 25, Tamworth, NH 03886.

Table 7-3
Generator Fan Failure Data and Hazard Calculations

n(K)	Hours	Hazard	Cumulative Hazard	n(K)	Hours	Hazard	Cumulative Hazard
1 (70)	4,500*	1.43	1.43	36 (35)	43,000		
2 (69)	4,600			37 (34)	46,000*	2.94	18.78
3 (68)	11,500*	1.47	2.90	38 (33)	48,500		
4 (67)	11,500*	1.49	4.39	39 (32)	48,500		
5 (66)	15,600			40 (31)	48,500		
6 (65)	16,000*	1.54	5.03	41 (30)	48,500		
7 (64)	16,600			42 (29)	50,000		
8 (63)	18,500			43 (28)	50,000		
9 (62)	18,500			44 (27)	50,000		
10 (61)	18,500			45 (26)	61,000		
11 (60)	18,500			46 (25)	61,000*	4.00	22.78
12 (59)	18,500			47 (24)	61,000		
13 (58)	20,300			48 (23)	61,000		
14 (57)	20,300			49 (22)	63,000		
15 (56)	20,300			50 (21)	64,500		
16 (55)	20,700*	1.82	7.75	51 (20)	64,500		
17 (54)	20,700*	1.85	9.60	52 (19)	67,000		
18 (53)	20,800*	1.89	11.49	53 (18)	74,500		
19 (52)	22,000			54 (17)	78,000		
20 (51)	30,000			55 (16)	78,000		
21 (50)	30,000			56 (15)	81,000		
22 (49)	30,000			57 (14)	81,000		
23 (48)	30,000			58 (13)	82,000		
24 (47)	31,000*	2.13	13.62	59 (12)	85,000		
25 (46)	32,000			60 (11)	85,000		
26 (45)	34,500*	2.22	15.84	61 (10)	85,000		
27 (44)	37,500			62 (9)	87,500		
28 (43)	37,500			63 (8)	87,500*	12.50	35.28
29 (42)	41,500			64 (7)	87,500		
30 (41)	41,500			65 (6)	94,000		
31 (40)	41,500			66 (5)	99,000		
32 (39)	41,500			67 (4)	101,000		
33 (38)	43,000			68 (3)	101,000		
34 (37)	43,000			69 (2)	101,000		
35 (36)	43,000			70 (1)	115,000		

*Denotes failure.

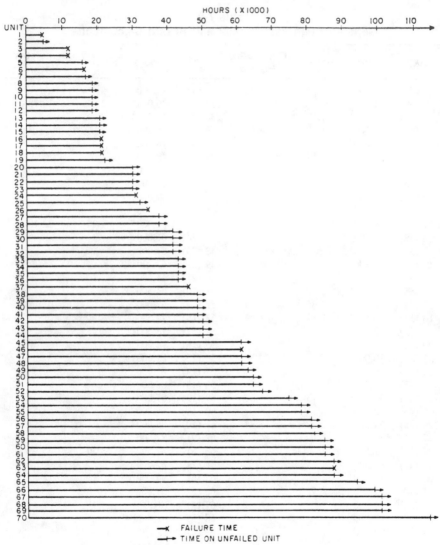

Figure 7-11. Graph of fan failure data.

1. Suppose that the failure data on n units consist of the failure times for the failed units and the running (censoring) times for the unfailed units. Order the n times in the sample from smallest to largest without regard to whether they are censoring or failure times. In the list of ordered times the failures are each marked with an asterisk to distinguish them from the censoring times. This marking is done in Table 7-3 for the generator fan failure data. If some censoring and failure times have the same value, they should be put into the list in a well-mixed order.

2. Obtain the corresponding hazard value for each failure and record it as shown in Table 7-3. The hazard value for a failure time is 100 divided by the number K of units with a failure or censoring time greater than (or equal to) that failure time. The hazard value is the observed conditional probability of failure at a failure time, that is, the percentage $100\,(1/K)$ of the K units that ran that length of time and failed then. This observed conditional failure probability for a failure time is readily apparent in Figure 7-11, where the number of units in use that length of time and surviving is easily seen. If the times in the ordered list are numbered backward with the first time labeled n, the second labeled $n - 1, \ldots$ and the nth labeled 1, then the K value is given by the label. For the example, these K values are shown in parentheses in Table 7-3. Three figure accuracy for K is sufficient. Note that the hazard value of a failure is determined by the number of failure and censoring times following it; in this way, Hazard plotting takes into account the censoring times of unfailed units in determining the plotting position of a failure time.

3. For each failure time, calculate the corresponding cumulative hazard value which is the sum of its hazard value and the hazard values of all preceding failure times. This calculation is done recursively by simple addition. For example, for the generator fan failure at 16,000 hours, the cumulative hazard value 5.93 is the hazard value 1.54 plus the cumulative hazard value 4.39 of the preceding failure. The cumulative hazard values are shown in Table 7-3 for the generator fan failures. Cumulative hazard values can be larger than 100%.

4. Choose a theoretical distribution of time to failure and use the corresponding hazard paper to plot the data on. Various hazard papers are shown in the figures. The theoretical distribution should be chosen on the basis of engineering knowledge of the units and their failure modes. On the vertical axis of the hazard paper, mark a time scale that includes the range of the sample failure times. For the generator fan failure data, Weibull hazard paper was chosen, and the vertical scale was marked off from 1000 to 100,000 hours, since the range of failure times for the data is from 4500 to 87,500 hours. This is shown in Figure 7-12.

5. Plot each failure time vertically against its corresponding cumulative hazard value on the horizontal axis of the hazard paper. This was done for the 12 generator fan failures and is shown in Figure 7-12.

6. Check whether the failure data on the hazard plot follow a reasonably straight line. This can be done by holding the paper up to the eye and sighting along the data points. If the plot is reasonably straight, draw a best-fit straight line through the data points. This line is an estimate of the cumulative hazard function of the distribution that the data are from. This was done in Figure 7-12

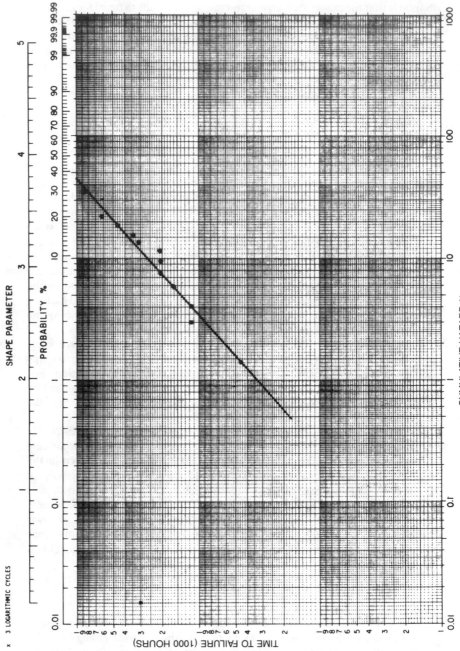

for the generator fan failure data. The straight-line estimate of the cumulative hazard function is used to obtain information on the distribution of time to failure. It provides estimates of the cumulative distribution function and parameters of the distribution of time to failure. If the data do not follow a reasonably straight line, then plot the data on hazard paper for some other theoretical distribution to see if a better fit can be obtained. Curvature of a data plot such as shown in Figure 7-13 indicates poor fit.

It may happen that data plotted on two different hazard papers give a reasonably straight plot on each. For practical purposes, it usually does not matter much which plot is used to interpolate within the range of the data. If extrapolation beyond the range of the data is necessary, the choice of which plot to use must be determined by engineering judgment.

The generator fan failure data were also plotted on exponential, normal, and log normal hazard paper (see Figures 7-13, 7-14, and 7-15). The exponential plot is a reasonably straight line; this indicates that the failure rate is relatively constant over the range of the data. The probability scale on this exponential hazard plot is crossed out, since it is not appropriate to the way the data are plotted; this is explained later. The normal plot is curved concave upward; this indicates that a normal distribution is not appropriate to the data. The log normal plot is a reasonably straight line. The log normal σ parameter of a line fitted to the data is approximately 0.7. For σ values near 0.7, a log normal distribution has a relatively constant failure rate over a large percentage of the distribution, particularly in the lower tail. For prediction purposes, the exponential, Weibull, and log normal plots are all comparable over the range of the data. For prediction to times beyond the data, it is necessary to decide which distribution is best on the basis of engineering knowledge.

The hazard plotting method, like all other methods for analyzing censored failure data, is based on a key assumption that must be satisfied in practice if it is to yield reliable results. It is assumed that if the unfailed units were run to failure, their failure times would be statistically independent of their censoring times. That is, there is no relationship or correlation between the censoring time of a unit and the failure time. For example, suppose that a certain type of unit has two competing modes of failure which can be identified after failure, but there is interest in only one of them. Then failure of a unit by the other mode might be regarded as a censoring of the time to failure for the mode of interest. The two times to failure that a unit potentially has may not be independent, but correlated, for physical reasons. If this is so, the hazard-plotting method fails to give valid results on the distribution of time to failure for the mode of interest.

The theoretical basis for the hazard plotting method and paper is given in Appendix B. Some readers may wish to refer to Appendix B at this point before proceeding to the next section which covers how to use and interpret hazard plots.

How to Use Hazard Plots

The line that is fitted to data on a hazard plot is used to obtain engineering information on the distribution of time to failure. Methods of obtaining such

(text continued on page 505)

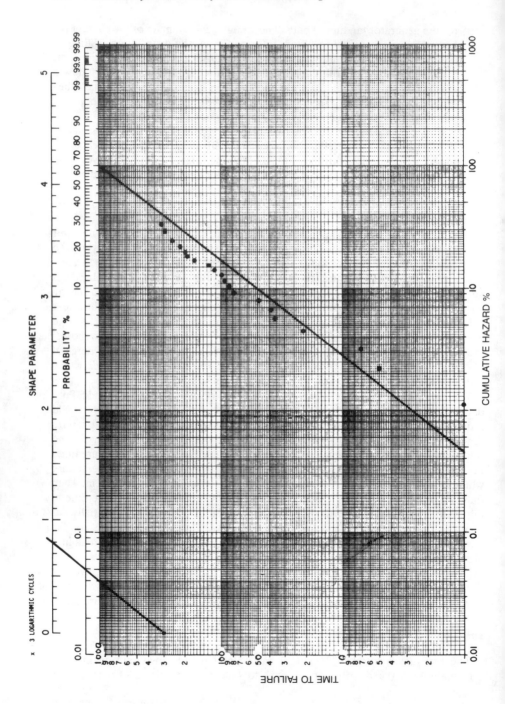

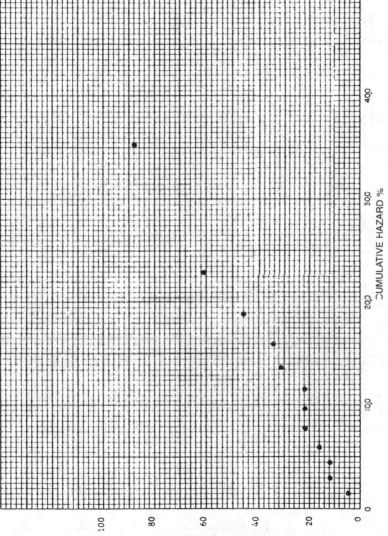

Figure 7-14. Normal hazard plot of generator fan failure data.

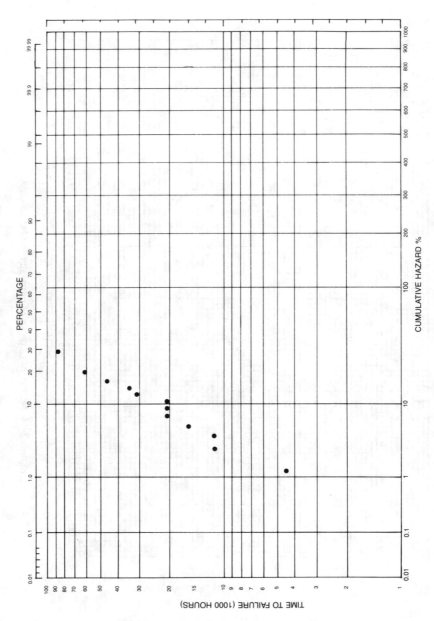

Figure 7-15. Generator fan failure data plotted on log normal hazard paper.

(text continued from page 501)

information are given here and are illustrated by the fan failure data and simulated data. The methods provide estimates of distribution parameters, percentiles, and probabilities of failure. The methods that give estimates of distribution parameters differ slightly from the hazard paper of one theoretical distribution to another and are given separately for each distribution. The methods that give estimates of distribution percentiles and failure probabilities are the same for all papers and are given first.

Estimates of Distribution Percentiles and Probabilities of Failure

It is shown in Appendix B that for any distribution the cumulative hazard function and the cumulative distribution function are connected by a simple relationship. The probability scale for the cumulative distribution function appears on the horizontal axis at the top of hazard paper and is determined from that relationship. Thus, the line fitted to data on hazard paper provides an estimate of the cumulative distribution function. The fitted line and the probability scale are used to obtain estimates of distribution percentiles and probabilities of failure. This procedure is explained next.

Suppose, for example, that an estimate based on a Weibull fit to the fan data is desired of the fifth percentile of the distribution of time to fan failure. Enter the Weibull plot, Figure 7-12, on the probability scale at the chosen percentage point, 5%. Go vertically down to the fitted line and then horizontally to the time scale where the estimate of the percentile is read and is 14,000 hours.

An estimate of the probability of failure before some chosen specific time is obtained by the following. Suppose that an estimate is desired of the probability of fan failure before 100,000 hours, based on a Weibull fit to the fan data. Enter the Weibull plot on the vertical time scale at the chosen time, 100,000 hours. Go horizontally to the fitted line and then up to the probability scale where the estimate of the probability of failure is read and is 38%. In other words, an estimated 38% of the fans will fail before they run for 100,000 hours.

Given next is an estimate of the conditional probability that an unfailed unit of a given age will fail by a certain time. Such estimates for units operating in the field are useful for planning. An example of this is in the next paragraph. Suppose that an estimate based on the Weibull plot is desired for the probability that the unfailed fan with 32,000 hours will fail by 40,000 hours. Enter the Weibull plot on the time scale at 32,000 hours. Go horizontally to the fitted line and then down to the cumulative hazard scale where the value is read as a percentage and is 13%. Similiarly, obtain the corresponding cumulative hazard value for 40,000 hours, which is 16.8%. Take the difference between the two values, that is, $16.8 - 13 = 3.8\%$. Enter the plot on the cumulative hazard scale at the value of the difference, and go directly up to the probability scale to read the estimate of the conditional probability of failure, which is 3.7%. For small percentages, cumulative probabilities and corresponding hazard values are approximately equal; this can be seen on the hazard papers. Thus, one could use the difference, if it is small, of the two cumulative hazard values as an approximation to the conditional probability of failure; this is 3.8% here.

Conditional probabilities of failure can be used to predict the number of unfailed units that will fail within a specified period of additional running time on each of the units. For each unit, the estimate of the conditional probability of failure within a specified period of time (8000 hours here) must be calculated. If there is a large number of units and the conditional probabilities are small, then the number of failures in that period will be approximately Poisson distributed* with mean equal to the sum of the conditional probabilities, which must be expressed as decimals rather than percentages. The Poisson distribution allows us to make probability statements about the number of failures that will occur within a given period of time.

It should be emphasized that the estimation methods presented previously apply to any hazard paper and, in addition, to a nonparametric fit to data obtained by drawing a smooth curve through data on any hazard paper.

Given next are the different methods for estimating distribution parameters on exponential, Weibull, normal, log normal, and extreme-value hazard papers. Descriptions of these distributions and their parameters are given in Appendix B. The methods are explained with the aid of simulated data from known distributions. Thus, we can judge from the hazard plots how well the hazard-plotting method does.

Exponential Parameter Estimate

Simulated data are shown in Table 7-4 and are a sample from an exponential distribution with a mean 1000 hours. The data are failure censored, which is a special case of multiply-censored data. That is, the data simulate a situation where 50 units were put on test at the same time and, according to a prespecified plan, five unfailed units were removed from test when the fifth failure occurred, five more unfailed units were removed when the tenth failure occurred, . . . and the last five unfailed units were removed when the twenty-fifth failure occurred. Unfailed units would be removed for inspection for deterioration and to hasten the life test. A plot of these data on exponential hazard paper is shown in Figure 7-16. In Figure 7-16 the theoretical cumulative hazard function for the distribution is the straight line through the origin. On exponential hazard paper only, the time scale must start with time zero, and the straight line fitted to the data must pass through the origin.

For the purpose of showing how to obtain from an exponential hazard plot an estimate of the exponential mean time to failure, assume that the straight line on Figure 7-16 is the one fitted to the data. Enter the plot at the 100% point on the horizontal cumulative hazard scale at the bottom of the paper. Go up to the fitted line and then across horizontally to the vertical time scale where the estimate of the mean time to failure is read and is 1000 hours. The corresponding estimate of the failure rate is the reciprocal of the mean time to failure and is $1/100 = 0.001$ failures per hour.

---

*The Poisson distribution is a special form of the normal distribution as described in Appendix B, Figure B-2.

Table 7-4
Simulated Exponential Data and Hazard Calculations

n(K)	Hours	Hazard	Cumulative Hazard	n(K)	Hours	Hazard	Cumulative Hazard
1 (50)	3*	2.00	2.00	26 (25)	385		
2 (49)	52*	2.04	4.04	27 (24)	385		
3 (48)	58*	2.08	6.12	28 (23)	385		
4 (47)	71*	2.13	8.25	29 (22)	385		
5 (46)	77*	2.17	10.42	30 (21)	385		
6 (45)	77			31 (20)	391*	5.00	46.49
7 (44)	77			32 (19)	410*	5.26	51.75
8 (43)	77			33 (18)	562*	5.56	57.31
9 (42)	77			34 (17)	611*	5.88	63.19
10 (41)	77			35 (16)	621*	6.25	69.44
11 (40)	101*	2.50	12.92	36 (15)	621		
12 (39)	130*	2.56	15.48	37 (14)	621		
13 (38)	161*	2.63	18.11	38 (13)	621		
14 (37)	180*	2.70	20.81	39 (12)	621		
15 (36)	235*	2.78	23.59	40 (11)	621		
16 (35)	235			41 (10)	662*	10.00	79.44
17 (34)	235			42 (9)	884*	11.11	90.55
18 (33)	235			43 (8)	1101*	12.50	103.05
19 (32)	235			44 (7)	1110*	14.29	117.34
20 (31)	235			45 (6)	1232*	16.67	134.01
21 (30)	301*	3.33	26.92	46 (5)	1232		
22 (29)	306*	3.45	30.37	47 (4)	1232		
23 (28)	319*	3.57	33.94	48 (3)	1232		
24 (27)	334*	3.70	37.64	49 (2)	1232		
25 (26)	385*	3.85	41.49	50 (1)	1232		

*Denotes failure.

Weibull Parameter Estimates

Table 7-5 contains a randomly censored sample of 90 times from a Weibull distribution with shape parameter equal to 0.8 and scale parameter equal to 1000 hours. The data are randomly censored with independent censoring times from a Weibull distribution with shape parameter equal to 2.0 and scale parameter equal to 300 hours. The data were obtained by generating an observation from the failure distribution and an independent observation from the censoring distribution. If the censoring time was smaller than the failure time, the censoring time was used and vice versa. Such data arise if there are independent, competing modes of failure for a unit and the mode of failure can be determined on a failed unit. The 20 failure times are plotted in Figure 7-17 on Weibull hazard paper. The data follow the cumulative hazard function of the distribution of time to failure but do not fall on a particularly straight line.

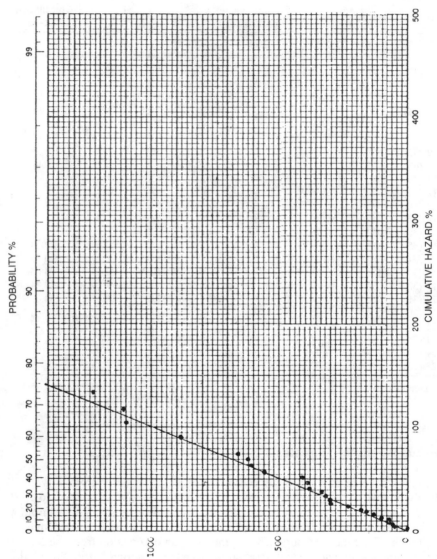

Figure 7-16. Exponential hazard plot of simulated exponential data.

Table 7-5
Randomly Censored Simulated Weibull Data

Number	Time	Cumulative Hazard	Number	Time	Cumulative Hazard	Number	Time	Cumulative Hazard
1	1*	1.11	31	199*	20.26	61	264	
2	5*	2.23	32	199		62	276	
3	7*	3.37	33	200		63	279	
4	21*	4.52	34	201		64	279	
5	37*	5.68	35	205		65	281	
6	38*	6.86	36	206		66	283	
7	49*	8.05	37	207		67	283	
8	78*	9.25	38	212		68	293*	29.59
9	85*	10.47	39	213		69	298	
10	91*	11.70	40	214		70	301*	34.35
11	96		41	217		71	301	
12	98*	12.97	42	217*	22.30	72	309	
13	111*	14.25	43	217		73	315	
14	125*	15.55	44	218		74	318	
15	141		45	220		75	322	
16	145		46	222		76	329	
17	156		47	223		77	330	
18	158		48	224		78	343	
19	160		49	226		79	344	
20	162		50	231		80	345	
21	163		51	232		81	349	
22	165*	17.00	52	239		82	354	
23	168		53	244		83	355	
24	176		54	244		84	363	
25	181		55	245		84	363	
26	182		56	251		86	370	
27	187		57	254*	25.24	87	380	
28	188*	18.59	58	256		88	385	
29	191		59	261		89	407	
30	198		60	262		90	412	

*Denotes failure time.

For the purpose of showing how to obtain from a Weibull hazard plot estimates of distribution parameters, assume that the straight line on Figure 7-17 was fitted to the data points. The estimate of the Weibull scale parameter is given first. Enter the plot at the 100% hazard point on the horizontal cumulative hazard scale on the bottom of the paper. Go up to the fitted line and then horizontally across to the vertical time scale where the scale parameter estimate is read and is 1000 hours. The estimate of the shape parameter is given next. Draw a line that is parallel to the fitted line and passes through the heavy dot in the upper left-hand corner of the grid. This line is shown in Figure 7-17. At the point where the line intersects the shape parameter scale at the top of the paper, read the estimate of the shape parameter, which is 0.80.

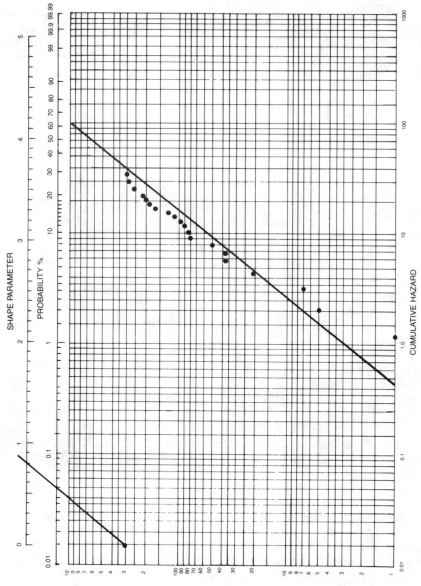

Figure 7-17. Weibull hazard plot of simulated Weibull data.

Similarly, we could investigate the shape parameter of Figure 7-12, the generator fan failures. However, we leave this up to the reader: were those failures the "infant mortality" kind, the random (antifriction bearings?) failure experience, or was it wear-out mode? Also, it would seem logical—after finding an answer to those questions—to start our investigation on the part/component level. Hazard plotting on this level might possibly lead us to new insights.

Some Practical Aspects of Data Plotting

Several aspects of data plotting that a user of hazard and probability plotting should be aware of are discussed here. The discussions cover potential difficulties that may be encountered in using plotting papers. Also, tips are given on how to use plotting papers effectively. Most of the discussions apply both to hazard and probability papers in the sense that good plotting techniques for probability plotting are also appropriate to hazard plotting.

If it is not known through engineering experience which hazard paper is appropriate to use for a set of multiply censored data, it may be useful then to try different papers to determine which does the best job. This may be done most efficiently by first plotting the sample cumulative hazard function on exponential hazard paper, which is just square grid. Then compare the plot with the theoretical cumulative hazard functions in Figures B-1, B-2, B-3, and B-4 of Appendix B and choose the distribution with the cumulative hazard function that agrees most closely with the shape of the sample cumulative hazard function over the range of cumulative hazard values in the sample. Plot the data on the corresponding hazard paper. The cumulative hazard scale on the horizontal axis of exponential hazard paper shown runs from 0% to 500%, which may be too big a range, particularly if the data consist of some early failure times and many censoring times and the sample cumulative hazard values are small. Then the scale should be changed to run from 0% to 50%, say, by adding decimal points. The probability scale then no longer corresponds to the hazard scale and must not be used. This was done in Figure 7-14 where the probability scale has been deleted.

If a data plot on a chosen hazard paper departs significantly from a straight line by being bowed up or down, it is an indication that the data should be replotted on a different hazard paper. Whether a plot of points is curved can best be judged by laying a transparent straight edge along the points. If a plot has a definite double curvature on exponential paper, log normal hazard paper should be tried. If the plot is still not straight, then this may be an indication that the units have two or more failure modes with different distributions of time to failure. Then a mixture of such data from different distributions usually will not plot as a straight line on hazard or probability paper but will be a line with two or more curves to it. Discussions on how to interpret curved probability plots are given elsewhere.[11, 12]

When a set of data does not plot as a straight line on any of the available papers, then one may wish to draw a smooth curve through the data points on one of the plotting papers, and use the curve to obtain estimates of distribution percentiles and probabilites of failure for various given times. With such a nonparametric fit to the data, it is usually unsatisfactory to extrapolate beyond the data because it is difficult to determine how to extend the curve. Nonparametric fitting is best used only if the data contain a reasonably large number of failures.

The behavior of the failure rate as a function of time can be gauged from a hazard plot.[13] If data are plotted on exponential hazard paper, the derivative of the cumulative hazard function at some time is the instantaneous failure rate at that time. Since time to failure is plotted as a function of the cumulative hazard, the instantaneous failure rate is actually the reciprocal of the slope of the plotted data, and the slope of the plotted data corresponds to the instantaneous mean time to failure. For data that are plotted on one of the other hazard papers and that give a curved plot, one can determine from examining the changing slope of the plot whether the true failure rate is increasing or decreasing relative to the failure rate of the theoretical distribution for the paper. Such information on the behavior of the failure rate cannot be obtained from probability plots.

On exponential hazard paper, if a smooth curve through the data points clearly passes through the time axis above the origin instead of through the origin, it is an indication that the true distribution may have no failures before some fixed time. This may arise, for example, if time to failure is figured from date of manufacture, and there is a minimum possible time it takes to get a unit into service. The minimum time is estimated as the time on the axis that the smooth curve passes through. This minimum time is then treated as time zero, and all times to failure are calculated from it for purposes of plotting the data on another hazard paper.

In a data plot on exponential hazard paper, the early failure times are crowded together by the origin in the lower left-hand corner of the plot and the later failure times are spaced farther apart (see Figure 7-14, for example). Thus, on exponential hazard paper, the later failures influence the eye and placement of the best-fit line more than the early failures. If one is interested in the early failure part of the distribution, which is typical of reliability work, then the early failures should be emphasized more in the analysis. This can be done by plotting data on Weibull hazard paper. The data points are then spaced out more equally throughout the range of the data, and the early and later failures are weighted about equally by eye (see Figure 7-12). If an exponential fit to the data is still desired, then a best-fit line with 45° slope should be drawn through the data points. This gives an exponential fit to the data, since the exponential is a special case of the Weibull with the shape parameter equal to one. If a nonparametric fit to the data such as described previously is desired that emphasizes early times to failure, then the data should be plotted on Weibull hazard paper and a smooth curve drawn through the data to provide a nonparametric estimate of the cumulative hazard function of the distribution. Also, percentiles and probabilities of failure for various given times can then be estimated from the smooth curve by the methods given earlier.

If estimates of distribution parameters are desired from data plotted on a hazard paper, then the straight line drawn through the data should be based primarily on a fit to the data points near the center of the distribution the sample is from and not be influenced overly by data points in the tails of the distribution. This is suggested because the smallest and largest times to failure in a sample tend to vary considerably from the true cumulative hazard function, and the middle times tend to lie close to it. Similiar comments apply to probability plotting.

Failure times are typically recorded to the nearest hour, day, month, hundred miles, etc. This is so because the method of measurement has limited accuracy or because the data are rounded. For example, if units are inspected periodically for failure, the exact time of failure is not known but only the period in which the failure occurred. For data plotting, the amount that failure times are rounded off should be considerably smaller than the spread in the distribution of time to failure. A rough rule of thumb for tolerable rounding for data plotting is that the amount of rounding should be no greater than one-fifth of the standard deviation of the distribution. Another way of saying this is that data are coarse if a large number of the failures are recorded for just a few different times. If data are too coarse, a plot on hazard or probability paper will have a staircase appearance with flat spots at those times with a number of failures. Plots of coarse data can be misleading unless the coarseness is taken into account; for example, estimates of failure probabilities and distribution parameters and percentiles can be somewhat in error if coarseness of the data is not taken into account.

Methods for analyzing coarse data can be found in the statistical literature[14, 15] where it is called grouped data. One method of taking into account the coarseness of data in plotting is to plot the failures with a common failure time at equal spaced times over the time interval associated with that time. For example, if failure times are recorded to the nearest 100 hours and 5 failures are at 1000 hours, then the corresponding interval is 950 to 1050, and the corresponding plotting times assigned to the 5 failures are 960, 980, 1000, 1020, and 1040 hours. This smooths the steps out of the plot and makes the plot more accurate.

When a set of data consists of a large number of failure times, one may not wish to plot all failures but just selected ones. This is done to reduce the labor in making the data plot or to avoid a clutter of overlapping data points on the plot. Any selected set of points might be plotted. For example, one might plot every kth point, all points in a tail of the distribution, or only some of the points near the center of the distribution. Care must be taken in plotting just a subset of the data to ensure that the plot of the subset leads to the same conclusions as the plot of the entire set. In practice, the choice of the selected points to be plotted depends on how the plot is going to be used. For example, if one is interested in the lower tail of the distribution, that is, early failures, then all of the early failures should be plotted. In general, one plots the data points that are most relevant to the information wanted from the data. Ideally, no fewer than 20 failure times, if available, should be plotted from a set of data. Often, in engineering practice there are so few failures that all should be plotted. When only a small number of failures is available, it should be kept in mind that conclusions drawn from a plot are based on a limited amount of information. Note that, if only selected failures from a sample are to be plotted on hazard paper, it is necessary to use all of the failures in the sample to calculate the appropriate cumulative hazard values for the plotting positions. Wrong plotting positions will result if some failures in the data are not included in the cumulative hazard calculations. A similar comment applies to the calculation of plotting positions for probability plotting.

For plots of some data, the ranges of the scales of the available hazard and probability papers shown throughout this chapter may not be large enough. To enlarge the time scale on any of the papers, join two or more pages together. To enlarge the hazard scale on Weibull or extreme-value hazard paper to get smaller hazard and probability values, add on extra log cycles for the lower values, and label them in the obvious manner; the cumulative hazard and cumulative probability scales are equivalent for the small probabilities in the added cycles. Weibull hazard papers with two different ranges are available to allow a user to choose the ranges that best suit his data (see Figures 7-12 and 7-17).

Concluding Remarks

The cumulative hazard plotting method and papers presented here provide simple means for statistical analyses of multiply-censored failure data to obtain engineering information. The hazard-plotting method is simpler to use for multiply-censored data than other potting methods given in the literature and directly gives failure-rate information not provided by others.

For further in-depth studies of statistical approaches to machinery failure analysis, please refer to Wayne Nelson's book on applied life data analysis.[11] Dr. Nelson is a consulting statistician with the General Electric Company.

Method to Identify Bad Repairs from Bad Designs*

Earlier we dealt with Weibull analysis and noticed how data-demanding that method can become. This is especially true when one wants to examine a sample of data from a time continuum such as in an established process plant. The sample of data may be a one-year period where the failure pattern can be considered similar to that in the previous year and in the succeeding year. However, this analysis method involves only a knowledge of the total sample population, total number of failures within the time period, the number of units on which these failures occurred, and time intervals between failures on units which failed more than once within the sample period. A good analogy would be a hospital operation with a yearly number of patients, of whom some are returning within that period for repeated treatment. Some of the returning patients only return once, others more than once during that year, and so forth.

In an earlier section, we acquainted ourselves with commonly used reliability terms: mean time between repairs (MTBR), mean time between failures (MTBF), etc. We saw that these terms have similar meanings but often include minor deviations. To avoid confusion, mean time to failure (MTTF) will be used in the following discussion. It can be expressed in terms of any time periods, i.e., days, months, years, etc.:

$$MTTF = \frac{P \times t}{F_\tau} \qquad (7\text{-}2)$$

---

*By permission from "Learn From Equipment Failures," by B. Turner, *Hydrocarbon Processing,* Nov. 1977, pp. 317-320.

where:
MTTF = mean time to failure.
P = population.
t = time periods in sample time.
F_τ = failures in sample time.

If t is unity, F_τ is failures

$$MTTF = \frac{P}{F_\tau}, \text{years or other period} \qquad (7\text{-}3)$$

While such a number is very useful as an index of performance, it gives no indication of what needs to be done to correct the situation if the index is low.

Inherent defects in either the system or the equipment, or infant mortality cause a low MTTF. Inherent defects will result in units failing more frequently than is desirable. These defects can be corrected by locating and correcting them or by taking action to reduce their effects, thus improving machine life.

Infant mortality results in repaired units failing shortly after their return to service. It can be corrected by simplifying repair techniques, quality control of repairs and repair parts, improved starting techniques, etc. Units which have persistent abnormally short lives may have an inherent defect which cannot be corrected by the previous methods. However, the usual characteristic of infant mortality is its variability. Very good pumps of last year become the very bad ones this year and vice versa. Any pump is a potential bad actor.

Two problems have different solutions.

Consider the beginning of any time period, i.e., this instant. Almost all the units are either running or available to run. Further, the lives of these units will be distributed, from the unit which was repaired yesterday to the unit which has run many years without failure.

If the life at which these units fail is a fixed interval, the failures expected during the period will also occur evenly throughout the year. Alternatively, if the unit life is random, the failures will also be evenly distributed, and so will any combination in between. The result is that the number of units being repaired remains fairly constant on a monthly basis. Such a system can be described as an exponential decay function, i.e., the number of failures occurring in a short period is proportional to the number available to decay (fail) at the start of the period. This is similar to atomic nuclei half life, i.e.,

$$R = Pe^{-Yt} \qquad (7\text{-}4)$$

where:
R = number of unfailed units at the end of sample time.
P = original population.
e = base of natural logarithms.
Y = decay constant or fraction which will fail in a unit time period.
t = time periods, in sample time.

Failures can be expressed as $P - R$, or

$$F = P(1 - e^{-Yt}) \qquad (7\text{-}5)$$

where F is the number of units from the original population which fail within R. If a unit fails and is repaired during the sample period, any subsequent failures are not included in F.

Consider a one-year sample period. All of the units which did not fail at all during the period are obviously R. The decay constant Y can be found from

$$-Y = \log_n \frac{R}{P} \qquad (7\text{-}6)$$

since R is 1, but $R = P - F$ where F is the number of units which failed at least once

$$\therefore -Y = \log_n \frac{P - F}{P} = \log_n \left(1 - \frac{F}{P}\right) \qquad (7\text{-}7)$$

Since $-Y$ is based on R, it cannot include any infant mortality effects from the repairs during the sample period.

By definition, F is the total units which fail at least once. However, this is not the total number of failures that can be expected. The total number of failures expected, F_{TE}, is the total of F plus the failures expected on these repaired units, F_1, plus the failures expected on units repaired twice, F_2, and so on. The repaired units have a similar expectation of failure to the original population, i.e.,

$$F_1 = F(1 - e^{-Y}) \qquad (7\text{-}8)$$

These will also have further failures, F_2, where

$$F_2 = F_1(1 - e^{-Y}) \qquad (7\text{-}9)$$

However, there is a difference. All of P are available to fail from the start of the period. If the failures, F are distributed evenly through the sample period, the average length of time that the repaired units are available to fail is only half of the period. Therefore, t is 0.5 for F_1 failures and the exponent to be used to calculate F_1 is $0.5Y$. The use of $t = 0.5$ for F_2, F_3, F_4, etc. is an exaggeration, since the average time available for the third failure will be less than half a year and the average time for the fourth and fifth failures will be even less. However, the assumption that $t = 0.5$ does not significantly distort the total number of failures expected and does simplify the calculations. This simplification is conservative, i.e., it increases the expected failures and slightly reduces the apparent infant mortality.

Total expected failures in a sample period are:

$$\text{Total failures} = F_{TE} = F + F_1 + F_2 + F_3 + F_4 \qquad (7\text{-}10)$$
$$= F + F(1 - e^{-Y/2}) + F_1(1 - e^{-Y/2})$$
$$+ F_2(1 - e^{-Y/2}) + F_3(1 - e^{-Y/2}) \qquad (7\text{-}11)$$

F_{TE} can be easily calculated using a hand calculator.

If F_{TE} is significantly below the total failures observed, there is a significant infant mortality problem. Further, the effect of the mortality is expressed in numerical terms:

$$F_{TA} - F_{TE} = \text{number of repairs which could be eliminated by solving infant mortality problems.} \qquad (7\text{-}12)$$

where:
F_{TA} = actual failures.
F_{TE} = expected failures.

Also

$$\frac{F_{TA} - F_{TE}}{F_{TA}}\% = \text{percent of actual failures due to infant mortality.} \qquad (7\text{-}13)$$

This conclusion about significant infant mortality can be confirmed in the following manner. Group the repaired machines which had two or more failures within the year into specified intervals of life; we have used 0.1 years and also months, for our graphs. These periods will be some fraction of a year, t. Total the repeated failures in each sample period, i.e., so many failing within 1 month or 0.1 years of being repaired, another number failing between 1 month and 2 months of being repaired, and some failing with a life of more than 0.2 of a year but less than 0.3 years. Now determine sequential failures.

The failures to be expected during the first sample period are

$$\frac{2 - t}{2} (1 - e^{-tY})F_{TE} \qquad (7\text{-}14)$$

The reason for the first multiplier is to correct for the failures which will take place outside the period under study. For example let $t = 0.1$ years. For the sixth period, i.e., units expected to fail in the period between 0.5 and 0.6 of a year, all such units repaired in the first 146 days of the year will have the second failure within the year. All such units repaired later than 183 days or July 2 will not have the second failure within the sample period. For units repaired between 146 and 183 days, there is an equal chance that the second failure will or will not be observed. The multiplier is to correct this effect. The total probability that a unit with life t will fail again within the year is

$$\text{probability} = 1 - \frac{t}{2} = \frac{2 - t}{2} \qquad (7\text{-}15)$$

During the second period, the expected failures will be

$$\frac{2 - 3t}{2}\left[(1 - e^{-2tY}) - (1 - e^{-tY})\right]F_{TE} \qquad (7\text{-}16)$$

For each subsequent period, the multiplier for t in the first expression increases by 2 and, for t in each exponent, it increases by 1.

For example, in the sixth period the expression will be

$$\frac{2 - 11t}{2}\left[(1 - e^{-5tY}) - (1 - e^{-5tY})\right]F_{TE} \qquad (7\text{-}17)$$

While the previous method is rigorous, it is unnecessarily complicated where F_{TE}/P is less than 0.5.

The expression

$$\left[(1 - e^{-ntY}) - (1 - e^{-(n-1)tY})\right] \simeq (1 - e^{-tY}) \qquad (7\text{-}18)$$

will give adequate accuracy. $F_{TE}(1 - e^{-tY})$ can be calculated and then multiplied by the initial multiplier from the rigorous equation for each period. The simplification becomes increasingly inaccurate for the latter periods, but the multiplier becomes increasingly smaller and reduces the significance of the error.

Calculation of the expected failures appears complicated, but it is quite easy with an electronic calculator (Table 7-6).

Significant differences between the actual failures and those expected during the first three periods (and especially in the first) are telltales of infant mortality. However, there are other methods of indicating this. The advantage of the method described is that it indicates what progress is possible if the problem is corrected. This justifies measures to correct the problem.

The expected failure rate F_{TE}/P can be inverted and used in a similar way to MTTF.

P/F_{TE} is expressed in time units and indicates the expected life of the units without infant mortality. Low P/F_{TE} ratio population obviously contain units which are unsuitable for the conditions of service, i.e., have inherent defects. Such cases can only be improved by changes to the machines or the conditions of service. Alternatively, analysis may show that samples with an unacceptable MTTF have an acceptable life when expressed as P/F_{TE}.

This indicates that the problem is due to infant mortality and not to inherent defects. However, units with a high MTTF will probably not justify extensive work to correct a high failure rate due to infant mortality because even the unnecessarily high repair rate is still very low.

Figure 7-18 shows the variation in F_{TA} and F_{TE} by 0.1 years for pumps in a refinery. Figure 7-19 shows the same variation by months for pumps in a chemical process plant. Table 7-7 indicates the range of observed values of MTTF, P/F_{TE} and $(F_{TA} - F_{TE})/F_{TA}$. Table 7-8 shows parameter variations.

Table 7-8 seems to indicate that the in-line pump has a higher MTTF and P/F_{TE} than the horizontal pump. The in-line also appears to be harder to assemble, since its infant mortality rate is higher. It is not necessarily correct that the in-line pump has a better inherent life since, in the sample they may have been installed in less severe services. Such data on the relative effect process services have on pump life is unknown. Therefore, Table 7-8 should not be used as a sole means of selecting a pump for a specified service. However, this approach may offer a way of evaluating pump bids, provided data on the effect of the pumped fluid on service life is established and a larger sample population is collected.

Table 7-6
Calculation of F_{TE} Using Electronic Calculator

	Assume P-500 F-200		
	Enter	**Display**	
1	200	200	(F)
2	+	200	
3	500	500	(P)
4	−	0.4	
5	CS	−0.4	
6	+	−0.4	
7	1	0.6	
8	F	0.6	
9	$\log_n X$	−0.510825	(Y)
10	+	−0.510825	
11	2	2	
12	−	−0.2554125	($Y/2$)
13	F	−0.2554125	
14	e^x	0.774598	
15	CS	−0.774598	
16	+	−0.774598	
17	1	1	
18	×	0.225402	($1 - e^{-t/y}$)
19	MC	0.225402	
20	200	200	(F)
21	M+	200	
22	−	−45	
23	M+	−45	
24	−	10.16	
25	M+	10.16	
26	−	2.29	
27	M+	2.29	
28	−	0.516	
29	M+	0.516	
30	−	0.116	
31	M+	0.116	
32	MR	258.16	(F_{TE})

Note: Most calculators have a single keyboard, and it is necessary to press the F button to activate the functions log, etc.

Other calculators have a double keyboard with separate buttons for the functions. Obviously, there is no need to press the F button on the latter models. The above procedure is based on the single keyboard model.

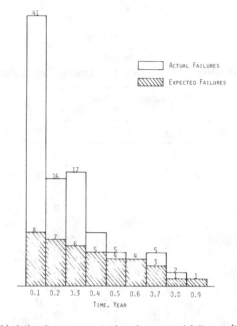

Figure 7-18. Variation between actual and expected failures of pumps (refinery).

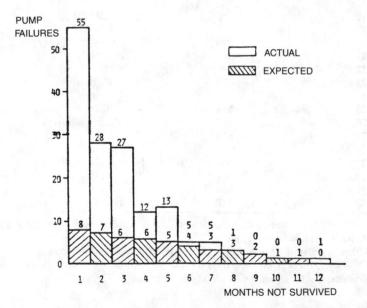

Figure 7-19. Variation between actual and expected failures of pumps (petrochemical process plant).

Table 7-7
Centrifugal Pumps Failure Analysis

Refinery	MTTF Years	$\dfrac{P/F_{TE}}{\text{Years}}$	$\dfrac{F_{TA}\text{-}F_{TE}}{F_{TA}}$ %
A	2.0	2.5	20
B	2.45	3.3	19
C	2.33	2.6	6
D	5.13	5.2	1

Table 7-8
Centrifugal Pumps Failure
Analysis (Type Pump)

Pump Type	MTTF Years	$\dfrac{P/F_{TE}}{\text{Years}}$	$\dfrac{F_{TA}\text{-}F_{TE}}{F_{TA}}$, %
Inlines	2.54	3.45	26.4
Horizontals	2.13	2.54	15.3
Deepwells	2.00	2.09	4.4

Nomenclature

MTTF Mean Time To Failure.

F Number of units from original population that fail in sample time.

F_T Total failures in sample time.

F_1 Expected failures after first repair.

F_2 Expected failures after second repair.

F_3 Expected failures after third repair.

F_{TE} Total failures expected.

F_{TA} Total failures in time period.

P Population.

t Time periods in sample time.

Y Decay constant or fraction that will fail in a unit time period.

References

1. Grothus, H., "Total Preventive Maintenance," Course Notes by F.K. Geitner, Toronto, Ontario, 1975.

2. Karassik, I.J., "Flow Recirculation in Centrifugal Pumps: From Theory to Practice," *Power and Fluids,* 1982, Vol. 8, No. 2.

3. Summers-Smith, D., "Performance of Mechanical Seals in Centrifugal Process Pumps," *pumps-pompes-pumpen,* 1981-181, pp. 464-467.

4. Czichos, H., "Tribology—A Systems Approach to the Science of Technology of Friction, Lubrication and Wear," Elsevier Scientific Publishing Co., 1978, pp. 237-239.

5. Bergling, G., "The Operational Reliability of Rolling Bearings," *SKF Ball Bearing Journal,* 188, pp. 1-10.

6. Czichos, H., "Tribology—A Systems Approach to the Science of Technology of Friction, Lubrication and Wear," Elsevier Scientific Publishing Co., 1978, pp. 240.

7. TEAM, "TEAM Easy Analysis Methods," Vol. 3, No. 4., Issued by TEAM, Box 25, Tamworth, NH 03886

8. Grothus, H., "Die Total Vorbeugende Instandhaltung", Grothus Verlag, Dorsten, W. Germany, 1974, pp. 63-66.

9. Jardine, A.K.S., *Maintenance, Replacement, and Reliability,* Pitman Publishing, Bath, U.K., 1973, pp. 13-24.

10. Nelson, W., "Hazard Plotting for Incomplete Failure Data," *Journal of Quality Technology,* Vol. 1, No. 1, Jan. 1969, pp. 27-52.

11. Nelson, W., *Applied Life Data Analysis,* J. Wiley and Sons, N.Y., 1982, pp. 634.

12. Hahn, G.J. and Shapiro, S.S., *Statistical Models in Engineering,* Wiley, New York, N.Y., 1967.

13. King, J.R., "Graphical Data Analysis with Probability Papers," TEAM Report, Box 25, Tamworth, NH 03886

14. Buckland, W.R., "Statistical Assessment of the Life Characteristic: A Bibliographic Guide," Hafner, New York, N.Y., 1964.

15. Govindarajulu, Z., "A Supplement to Mendenhall's Bibliography on Life Testing and Related Topics," *Journal of the American Statistical Association,* Vol. 59, No. 308, Dec. 1964, pp. 1231-1291.

Chapter 8

Sneak Analysis

Major machinery systems are inconceivable without hydraulics, electronic interfaces, control circuitry, shutdown instrumentation, and the like. The increasing complexity of these devices makes them prime initiators of unexpected malfunctions and unscheduled equipment downtime. Advanced troubleshooting methods must be employed to resolve existing problems or, better yet, find and eliminate potential problems before they can lead to expensive outages of plant and equipment. One such method is called sneak analysis, an advanced pre-event troubleshooting and failure prevention tool.

Within the context of overall machinery systems troubleshooting and general reliability reviews, it is deemed appropriate to present a description of the Sneak Analysis (SA) with primary emphasis on its purpose, potential benefits, and limitations, as well as cost factors and tradeoffs. It is not the intent of this section to define the exact methods or techniques for the performance of the analysis. However, it is hoped that our treatment of this topic will serve to introduce the concept of SA to management and engineering personnel responsible for the design and development of systems and equipment for process plants. Being familiar with SA principles will enable them to make appropriate decisions regarding its applicability to a given project. Guidelines are provided for the management and implementation of the SA.

Sneak Analysis Use[1]

SA is a powerful tool to identify system conditions that could degrade or adversely affect plant safety or basic equipment uptime and reliability. It is a technique that can

523

be applied to both hardware and software. For hardware, SA is generally based on documentation used in engineering and manufacturing. Its purpose is to identify conditions which cause the occurrence of unwanted functions or inhibit desired functions, assuming all components are functioning properly.

SA examines system operations during normal conditions for design oversights. It consists of two subanalyses: (1) Sneak Circuit Analysis (SCA) for electrical-electronic systems, and (2) Software Sneak Analysis (SSA) for computer programs. SCA represents a mature and useful technique that can be performed on both analog and digital circuitry. The purpose of SSA is to define logic control paths which cause unwanted operations to occur or which bypass desired operations without regard to failures of the hardware system to respond as programmed. After an SCA and an SSA have been performed on a system, the interactions of the hardware with the system software can readily be determined. The effect of a control operation that is initiated by some hardware element can be traced through the hardware until it enters the system software. The logic flow can then be traced through the software to determine its ultimate impact on the system. Similarly, the sequence of action initiated by software can be followed through the software and electrical circuits until its eventual total impact can be assessed. Finally, the analysis should be considered for critical systems and functions where other techniques are not effective but should not be applied to off-the-shelf computer hardware such as memory or data processing equipment.

The SA concept has potential applications in any design or development project for systems, equipment, or software, and its merits should be evaluated as part of the development planning effort. In the past, SA was generally considered applicable only to electrical and electronic circuits, but the concept has evolved to where (in theory at least) it can be used to evaluate other systems, including mechanical, pneumatic, hydraulic, power generation, etc. The complexity and criticality of such systems or devices, the extent of other analyses and test programs, and the associated cost factors are the primary determinants in the decision to require (or not to require) SA.

Definitions

As with any new or emerging technology, there soon develops a jargon which needs to be translated or defined for the uninitiated. Specialized terms unique to the SA technique are defined as follows:

Clue. A known relationship between a historically observed sneak-circuit condition and the underlying causes which created it. Clues are used to evaluate systems for possible sneak circuit conditions.

Glitch. An anomalous circuit response which occurs without apparent reason, generally due to signal overlap or timing problems.

Highlight. The process of isolating a particular design feature which is likely to generate a system malfunction so that it can be subjected to intensive analysis.

Manual SA procedures. SA procedures that are performed without computer assistance.

Network. A group of interconnected elements intended to perform some system function. Elements may be electrical components, electromechanical components, computer program instructions, operator actions, procedures, mechanical or pneumatic functions, or any other portion of a system that can be considered as an independent entity. Although the networks most frequently considered in SA consist of electrical and electromechanical components, the definition is not restricted to networks of this type.

Path search. A process of searching out all possible paths through a network.

Software. A computer program which may be either an independent (stand-alone) program or an embedded (within other system hardware) program.

Tailoring. The process of adapting a specification to fit the constraints of a particular program.

Topological methods. This term simply describes the application of clues at each mode found in the networks when networks are depicted "topologically"; that is, with flow initiating at the top of the diagram and terminating at the bottom.

Trade-off. The process of comparing the relative advantages and disadvantages of different courses of action.

Verification and validation. Terms generally referring to those activities in a software development program that serve to assure the adequacy and the accuracy of the software.

Sneak Circuits and Their Analysis

A sneak circuit is an unexpected path or logic flow within a system which, under certain conditions, can initiate an undesired function or inhibit a desired function. The path may consist of hardware, software, operator actions, or combinations of these elements. Sneak circuits are not the result of hardware failure but are latent conditions, inadvertently designed into the system or coded into the software program, which can cause it to malfunction under certain conditions. Categories of sneak circuits are:

1. *Sneak paths* which cause current, energy, or logical sequence to flow along an unexpected path or in an unintended direction.
2. *Sneak timing* in which events occur in an unexpected or conflicting sequence.
3. *Sneak indications* which cause an ambiguous or false display of system operating conditions and thus may result in an undesired action taken by an operator.
4. *Sneak labels* which incorrectly or imprecisely label system functions, e.g., system inputs, controls, displays, buses, etc., and thus may mislead an operator into applying an incorrect stimulus to the system.

Figure 8-1 depicts a simple sneak circuit example. With the ignition off, the radio turned to the on position, the brake pedal depressed, and the hazard switch engaged, the radio will power on with the flash of the brake lights. Although we can be reasonably certain that this automotive sneak circuit does not exist in many automobiles today, it was fairly common in certain types of late-1960s models. The radio should only be powered by current flowing through the ignition switch, but it may also be powered by current that flows along the indicated path, totally bypassing the ignition switch. The flasher module drives the radio in sync with the tail lights.

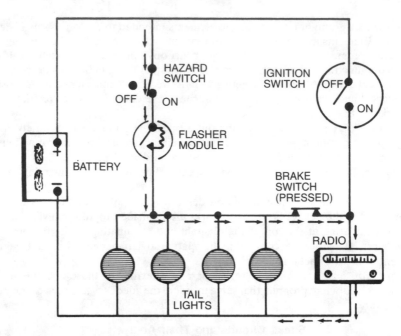

Figure 8-1. The automotive sneak circuit (sneak path), common in the 1960s. The radio should be powered by current flowing through the ignition switch; however, the electric current flows along an unintended sneak path (→), causing the radio to "blink" with the tail lights.

This type of sneak circuit is basically a stray current path. The path was designed into the system, and no hardware failures are required to activate it. Rather, a peculiar circumstance of combined operating conditions is the trigger. Until these circumstances are met, the sneak will lie latently dormant through testing and operation. Formal sneak circuit analysis can now uncover such potential "glitches" before they occur.

SCA is the term that has been applied to a group of analytical techniques that are intended to identify methodically sneak circuits in systems. SCA techniques may be either manual or computer-assisted, depending on system complexity. Current SCA techniques which have proven useful in identifying sneak circuits in systems include:

1. *Sneak path analysis.* This method of SCA attempts to discover sneak-circuit conditions by means of methodical analysis of all possible electrical paths through a network. Because of the large volume of data involved, sneak path analysis normally mandates computer data processing. It has been found that sneak-circuit conditions generally have certain common characteristics which are directly related to topological patterns within the network. Sneak path analysis uses these common characteristics as "clues" to look for sneak circuits in the system being analyzed.

2. *Digital sneak circuit analysis.* This SCA method is intended to discover sneak-circuit conditions in digital systems. Digital SCA may involve some

features of sneak path analysis, but it may also involve additional computer-assisted techniques such as computerized logic simulation, timing analysis, etc., to handle the multiplicity of system states encountered in modern digital designs. In general, digital SCA will identify the following types of anomalies:

a. Logic inconsistencies and errors.
b. Sneak timing, that is, a convergence of signals which causes an erroneous output due to differing time delays along different signal paths through a digital network.
c. Excessive signal loading or fan-out.
d. Power supply cross-ties, grounding, or other misconnections of signal pins.

3. *Software sneak path analysis.* Software sneak path analysis was adapted from hardware sneak path analysis. It was found that computer program flow diagrams which contained known sneak paths were most often associated with certain common flow diagram topologies and had other common characteristics. These common characteristics served as a basis for establishing "clues" which could be used to analyze new computer program flow diagrams. Computerized path search programs developed to do SCA on hardware were adapted or rewritten to accept software logical flows, and new clues were developed to analyze them. Software SCA can be done either manually or with computer assistance, depending primarily on the size and complexity of the software. It may be combined with the hardware SCA and is most often used on embedded software in a complete minicomputer or microcomputer-controlled hardware system. It has been used on both assembly language and higher-order language programs.

4. *Other sneak circuit analysis techniques.* Variations of SCA have been developed to analyze particular types or combinations of systems, such as hardware/software interfaces. The application of any new SCA procedure to a particular system or situation must be judged on its demonstrated effectiveness in detecting sneak circuits in similar cases or on its anticipated benefit in the specific situation being considered. It is difficult to predefine a course of action for handling new SCA types which may emerge, but the general ground rules for evaluating applicability hold true also.

All of the SCA types given above, i.e., sneak path analysis, digital SCA, and software sneak path analysis may be performed manually under some limited conditions. Computer-assisted SCA data processing is also possible with each of these types and is absolutely essential for more complex systems. The availability and thoroughness of computer aids strongly influences the selection of a contractor to perform SCA on a complex system.

A few additional words relating to complexity are in order. Sneak conditions result from the following three primary sources: (1) system complexity, (2) system changes, and (3) user operations. Hardware or software system complexity results in numerous subsystem interfaces that may obscure the intended functions or produce unintended functions. The effects of even minor wiring or software changes to specific subsystems may result in undesired system operations. A system that is relatively sneak free can be made to circumvent desired functions or generate undesired functions as a consequence of improper user operations or procedures. As systems become more complex, the number of human interfaces multiplies because of the involvement of more design groups, subcontractors, and suppliers. Hence, the probability of overlooking potentially undesirable conditions is increased proportionately.

Historical Development of SCA

Systems analysis techniques, in one form or another, have been employed throughout the evolution of the various technologies. As systems become more and more complex, and as greater emphasis is placed on safety and reliability, the need for more sophisticated techniques evolves.

Formal sneak circuit analysis was begun in late 1967 by the Boeing company as part of the Apollo Technical Integration and Evaluation contract with the Manned Spacecraft Center, National Aeronautics and Space Administration (NASA), in Houston, Texas.[2]

The initial effort evolved from critical concern for Apollo crew safety and involved detailed review of historical incidents of sneak circuits in various electrical systems. For the study a sneak circuit was defined as ". . . a designed-in signal or current path which causes an unwanted function to occur or which inhibits a wanted function." The definition was meant to exclude component failures and electrostatic, electromagnetic, or leakage paths as causative factors. The definition also excluded improper system performance due to marginal parametric factors or slightly out-of-tolerance conditions.

The review involved a number of unexpected occurrences suspected of resulting from sneak circuits. Notable incidents were:

1. Missiles accidentally launched.
2. Bombs accidentally armed and dropped.
3. "Unannounced failures" of aircraft electrical systems that led to crashes.
4. Electrocution of an electric-utility lineman.
5. Hydraulic analog, wherein fire alarms were falsely initiated at a public school by the automatic-sprinkler monitoring system.

One conclusion from the incident investigations was that sneaks are universal in complex systems and their analogs. Traditionally, they happened and then were corrected. The successful development of a means of detecting susceptibility to sneak conditions was triggered by study of an early space project test mission, described below.

Mercury Redstone Incident

On November 21, 1961, the Redstone booster's engine roared to life, and the vehicle began to lift off the pad. Then, after a "flight" of a few inches, the engine inexplicably cut off. The booster settled back onto the pad, while the payload, a Mercury capsule, jettisoned and came to rest about 1,200 feet away. Damage was slight, and both the booster and the Mercury capsule were subsequently used in other tests. However, a potentially hazardous condition existed on the pad, and the area had to remain clear for 28 hours for the Redstone batteries to drain down and the liquid oxygen to evaporate.

Investigation showed that no components had failed. A sneak circuit had occurred to shut off the engine, and this in turn initiated the capsule jettison. When the booster left the pad, a ground tail plug pulled out about 29 msec prior to control umbilical disconnect. The disconnect timing interrupted the normal return path to missile skin for the launch-pad equipment. Thus for several milliseconds the current flowed through a sneak return path, as shown in Figure 8-2. The sneak was of sufficient duration to operate the engine cutoff coil and create a surprise abort.

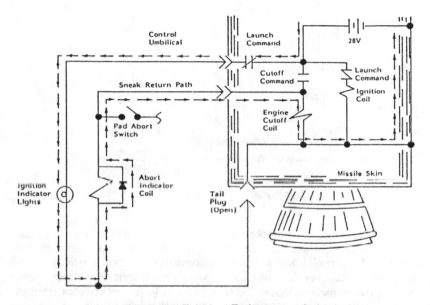

Figure 8-2. The Reluctant Redstone sneak circuit.

Study of the "Reluctant Redstone Circuit" in Figure 8-2 shows that characteristics of the sneak were an interrupted ground (return) path and current reversal conducted through a suppressor diode. Additionally, the ground path interruption can be considered a timing problem similar to a relay race. However, these characteristics are merely specific clues relating to the incident. Recognition of the sneak potential of the circuit as a general case is attained by drawing the circuit as shown in Figure 8-3.

The approach is "topological" in the sense that unswitchable (unremovable) power is always assumed to be at the top of the "tree" and unswitchable ground at the bottom. Figure 8-3 shows two trees because, barring short-circuit modes, the ignition coil circuit cannot affect the cutoff coil circuitry; that is, it is hard-wired to power above the highest level of switching in the cutoff circuit.

Inspection shows that there is no way for a sneak to occur in the ignition coil circuit. The circuit can only fail, or it may not be enabled when required, which is typical of single-line trees. However, the cutoff coil tree is somewhat more involved. Its "H" pattern should be immediately noted. Further inspection shows that current can reverse through the horizontal branch, depending on switch configurations. (In this context,

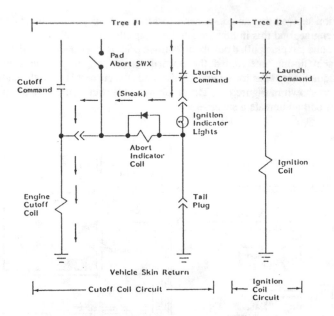

Figure 8-3. Topological-tree drawings of Redstone circuit.

anything that is designed to interrupt path continuity is considered to be a switch. This includes manual switches, umbilical disconnects, relay contacts, fuses, circuit breakers, transistors, and other solid-state switches.) Thus current reversal is typical of "H" patterns in topological trees. However, the probability and magnitude of reverse current of course depends on many factors, including switches in the vertical branches, interlocks on switch operations, relative impedances in the branches, and existence of diodes or other current limiters.

The principal lesson to be learned from this example is that topological-tree orientation provides a pattern-recognition basis for application of specific sneak clues. Clues can be stated many ways and are quite numerous, as derived from literature searches plus Apollo experience. In fact, the number of clues and many ways of stating them can become confusing for most analysts. Fortunately topological-pattern recognition provides an ordering of clue application that keeps the analysis simple. For example, few people immediately see the sneak possibility in Figure 8-2 when the sneak path is not highlighted with arrows. However, if an analyst is forewarned that "H" patterns lead to reverse current possibilities (depending on timing, etc., per a set of specific clues), then the sneak is readily apparent in the topological tree of Figure 8-3, even without arrows. Thus the Reluctant Redstone provided the breakthrough of sneak circuit analysis development around topological recognition criteria.

Topological Techniques

The Redstone example illustrates only one of four possible basic patterns to be considered in the topological approach to sneak circuit analysis. Before the technique is further detailed, however, it is necessary to redefine the scope of the analysis to its current use. The historical definition remains valid in that a sneak circuit is still taken to be a latent path that causes an unwanted function or inhibits a desired function without regard to component failure. However, it should be recalled that the path may consist of wires, components, software logic, fluid, people (as switch manipulators), and energy (as links between coils and their contacts, for example). The sneak path will cause current or energy to flow along an unexpected route created by power-supply cross ties, unanticipated ground-switch operations, etc. Sneak timing may be involved to cause current or energy flow or to inhibit a function at an unexpected time, as in the case of simultaneous opposing commands. Sneak indications may give ambiguous or false display of system operation and cause confusion or fail to give warnings. Finally, sneak labels on control and display panels may cause operator confusion to result in application of improper stimuli or inhibits.

As previously stated, failed components are not specifically included, nor are electromagnetic interferences or parametric concerns of marginal voltages, impedances, frequencies, etc. This is not to imply that the topological technique will not be helpful in dealing with such matters; it just is not the primary emphasis and objective in sneak circuit analysis. Such concerns in general address component selection in design implementations, whereas sneak circuit analysis has traditionally delved into identification of all possible modes of operation or inhibits of a system.

Sneak circuit analysis is based on the postulate that all topological trees consist of one or more of the four possible topographs, which can be assessed for sneak potential at each node, as illustrated in the general examples given in Figure 8-4: (1) the single line, (2) the ground dome, (3) the power dome, and (4) the H-pattern. The topological patterns

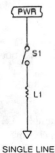
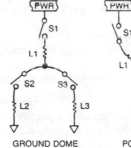
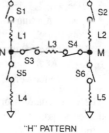

SINGLE LINE   GROUND DOME   POWER DOME   "H" PATTERN

S = GENERAL SWITCH; MAY BE A FUSE,
    CIRCUIT BREAKER SWITCH, ETC.

L = GENERAL LOAD; MAY BE LOGIC
    GATE INPUT, RELAY COIL,
    MOTOR, ETC.

N = GROUND NODE

M = POWER NODE

Figure 8-4. Basic topological patterns appearing in network trees.

containing branch or jump instructions have the highest incidence of problems. This includes the ground, power, and combination domes. The crossing of module or function interfaces as a result of the branch instruction is a prime problem cause.

Although at first glance a given software tree may appear to be more complex than these basic patterns, closer inspection will reveal that the code is actually composed of these basic structures in combination. As each node in the tree is examined, the analyst must identify which pattern or patterns include that node.

Single-Line (No-Node) Topograph

Some sneak possibilities (clues) for the single-line topograph (Figure 8-4) are: (1) switch S1 will be open when load L1 is desired, (2) S1 will be closed when L1 is not desired, and (3) the label of S1 may not reflect the true function of L1. Obviously such sneaks are rarely encountered owing to their simplicity. Of course, this is the elementary case and is given primarily as the default case, which covers circuitry not included by the other topographs.

Ground Dome or Double-Ground Node Topograph

One sneak possibility of the pattern of this topograph (also shown in Figure 8-4) becomes obvious when constructing the tree. If neither ground is the true return for the power supply, then there is a distinct possibility that the grounds will be at different absolute potentials. Under such conditions current can flow through L2 and L3, assuming S2 and S3 are closed, even if S1 is open to remove power. A few of the more straightforward concerns (clues) are given below.

1. S1 open: L1, L2, and/or L3 desired.
2. S2 open: L2 desired (and inverse of S1, S2 closed; L2 not desired).
3. S3 open: L3 desired (and inverse).
4. Circuit loading through L2 bypasses L3 (or inverse).
5. Label of S2 may reflect the function of only part of the circuit, say L2 and not L3, or vice versa.
6. Label of S1 may reflect the function of only part of the circuit, say L2 and not L3, or vice versa.

There are many more complicated clues for this node topograph. For example, L2 could be a indicator required to monitor operation of L3. Yet, if S3 is open (L3 off), the indicator (L2) can falsely state that L3 is on. The purpose here is not to list all possible clues for each pattern, but rather to illustrate the systematic approach for analytical investigations of design possibilities.

Double-Power Node (Power Dome) Topograph

In the double-power node topograph (Figure 8-4), again, one sneak possibility should be immediately evident. That is, if the power supplies are at different voltages, current can flow from one to the other and affect L1 and L2 even with S3 open. Such a power-to-power path is also potentially hazardous for other reasons. For example, with L1 and L2 near zero impedance, equipment damage can result for even small voltage differentials, and the connection negates redundancy provisions if the power points represent a primary bus and its backup. Otherwise, the applicable clues are similar to those for the double-ground node topograph with slight reorientation. One example is that opening S1 may be intended to disable L3, but S2 can still enable the load. Such networks, after tagging out the breakers (S1), electrocute people working in the load area (L3) if they have not tagged out the breakers represented by S2. Of course, the entire problem appears so simple as to be almost absurd at the node-topograph level, but unfortunately it is all too often hidden in volumes of drawing sheets in industry.

"H" Node or "H" Pattern Topograph

The Redstone historical example has already illustrated the unique "reversing-current" characteristic of the "H" topograph (Figure 8-4, lower right). During sneak circuit analysis in all types of industry, the reverse-current clue has consistently identified high potential for sneak paths and timing. In fact, all categories of sneaks are likely to occur within the "H" topograph. Clues similar to those given for the other topographs apply but in a greater number of interrelations due to the compound nature of the "H." That is, the "H" is equivalent to both double-ground and double-power topographs at nodes N and M. The "H" is required, however, when a ground-seeking branch of the topograph for any node contains another node, which, in turn, can lead to power via a different route. A majority of the critical sneak circuits found by Boeing Aerospace Company's analyses in all industries has been attributed to the "H" pattern. Such a design configuration should be avoided whenever possible.

Cost, Schedule, and Security Factors[3]

Estimating the cost of an SA program can be complicated because of a number of factors. There is a variation in SA techniques among contractors which can lead to significant differences in their cost estimates for the analysis. The process of tailoring or partitioning the system and using different techniques and different levels of scrutiny in the analysis of certain subsystems may also cause a wide variation in cost estimates for what is apparently the same job. Other factors, such as the degree of accessibility to the SA data base, the level of owner involvement, amount of liaison required, frequency of reports and specific data submission requirements, will all have an effect on the price quoted for the SA.

Cost effectiveness. The application of SA on a project must always be done with a serious consideration for cost effectiveness. There are obvious situations in which SA is clearly mandatory, such as a manned space flight, a nuclear power plant control system, or a nuclear missile system. In such cases the consequences of sneak conditions are potentially disastrous; the costs of a serious error are virtually incalculable. The decision in most cases involving industrial machinery systems or complex offshore systems is not so clear-cut. The owner's engineer must weigh the potential benefits of SA against the cost and decide either to do it or not.

Tailoring the SA to reduce costs while maintaining the essential benefits of the analysis is strongly encouraged. A vigorous assessment of the areas in the design most likely to generate serious sneak conditions can do a great deal to improve the cost effectiveness of SA on a given system. Thus, the built-in test equipment circuitry, or the interconnection to a test set, or the control and indicator subsystems could be singled out as the primary focus for SA. Experience has shown that power distribution circuits are especially susceptible to sneak conditions and are thus potential candidates for SA. Of course, this top-level tailoring of the effort should remain flexible until SA contractor inputs regarding proper tailoring are heard and considered.

Schedule impact. Both the number and the types of problems found in an SA can have an impact on project schedules. One of the more desirable situations is to run a controlled test to demonstrate the problem and its consequent system effects. Some problems are of this category and are relatively easy to demonstrate. Other problems require repeated cycling in order for the condition to occur. Normally these equipment or software problems are associated with race conditions or equipment sizing and may only occur after some component degradation. These apparently non-repeatable or non-recurring problems give rise to the likelihood of occurrence argument. The reported condition may then be dimissed solely on the basis of a value judgment. Extreme care should be exercised in dismissing the report conditions and the possible system effects.

Analysis scope and depth. Scoping factors which should be considered in an SA application include the following:

1. The number and type of components and/or computer instructions.
2. Schedule requirements.
3. Data availability.
4. Change analysis.
5. Estimated project cost.

The most important scoping factor in an SA application is the number and type of components involved in the system or systems to be analyzed. The greater the number of components, the higher the cost and the longer the required project schedule. The type of equipment may be digital, analog, or relay for hardware systems, and high-order or assembly language for software systems. No particularly significant scoping requirements are associated with any of these system types. Digital systems incorporate much more circuitry in small physical dimensions, but the analysis is basically performed using the same approach, changing only some of the SA clues. Relay and digital systems are the two most prevalent hardware categories selected for SA.

Security considerations. When proprietary data are involved in a SA, it may be necessary to obtain a signed agreement from the SA contractor not to release such data to a third party and to use such data only in the performance of the SA. The SA data base including input data, computer data base, output reports, worksheets, etc. would normally be retained by the performing contractor for a specified period to ensure its availability should the need arise to perform SA on system modifications. Maintenance by the performing contractor will be more cost effective and the risk of loss is less.

Summary

Establishing Need for SA

1. Reliability improvements in the overall project result from the identification and resolution of system problems. SA is very effective in identifying problems which may be missed by other analyses and testing.
2. Independent analysis is currently the only established approach for the analysis. The analysis must be performed by a contractor independent of the design group to preserve the integrity of the effort. It is also an excellent analysis tool which can be used to verify or cross-correlate the results or findings of other analyses.
3. Problem detection to eliminate the need for costly retrofits or redesigns in mass-produced systems and possible loss of critically important one-of-a-kind systems such as offshore platforms are immediate considerations for performing the analysis.
4. High criticality of the systems to be analyzed also warrants the analysis. Critical machinery protection, safety, and shutdown systems are the most likely candidates. Low criticality systems may be elminiated from consideration as long as no active control functions are performed in these systems.
5. Unresolved system problems or "recurring glitches" that have not been found by other analyses or tests are also good candidates for SA. If the analysis is undertaken to identify or isolate these system problems (typically during late-development phases), allow the contractor some additional leeway in cost and interpretation of instructions included in his work scope. Frequently, the unidentified problem causes are located in "unrelated" equipment/software areas. Analysis of only problem-prone functions, or areas where the problem is manifest, such as an instrument panel or test equipment, may be insufficient to locate the cause of the system problems.
6. A high change rate in the baseline design can also be used to justify the analysis. Loss of the design configuration baseline resulting from greater-than-expected change rates can be rectified by the detailed analysis of each change before the change is implemented in the hardware or software system.
7. SA is a cost-effective tool in all phases of project development, but the analysis results exhibit a pronounced effectiveness in early development phases, and particularly in the full-scale engineering development phase.

Determining Application Systems

1. Systems which perform active functions are the primary candidates for SA. Electrical power, distribution, and controls have traditionally been the main areas for hardware analysis. Computer programs which actively control and sequence system functions are good software candidates. In general, those systems which occur in the safety, command, and control areas are the primary candidates. Non-repairable systems are especially good candidates.
2. Passive systems that do not affect the overall plant operation can be omitted from analysis consideration. Systems which do affect plant operation but are

highly redundant may also be excluded. Redundancy in control areas, however, is not a ground for eliminating the analysis. There may be design problems which compromise or destroy the redundant design.

3. SA can and has been successfully implemented on complete, unspared turbomachinery applications, as well as limited subsystem or functional applications. The analysis is best performed on configurations involving numerous system interfaces and large-size systems. The high number of interfaces, as well as the complex designs, are primary causes of embedded sneak conditions.

4. The applicable systems should be completely specified by component or instruction-level documentation in the form of schematics, drawings, wire lists, and source computer program code so that the analysis can be conducted at the "as-built" and "as-coded" levels, respectively.

5. Detailed analysis of critical systems can be performed by blending various analysis techniques which bring to bear the best features of each analysis in identifying design and fault-related problems. Favorable project results and costs are obtained in blended analyses. Highly critical functions can be identified by other high-level system analyses such as a preliminary hazard analysis or system hazard analysis.

Calculating Project Cost and Allocation of Budget

1. The cost of SA can be computed on the basis of the number and type of hardware components and the number and type of computer program language instructions. Cost figures relating to NASA-sponsored SA projects do not appear to match those experienced in recent petrochemical industry projects. The user should, therefore, base his cost estimate on formal discussions with, and proposals from, qualified contractors.

2. Limited budgets may force scope reductions and restrict a broad program application of SA. The analysis can and has been scoped to individual systems, subsystems, and functions. Excessive scoping, however, could limit the analysis effectiveness by eliminating the detailed function tracking which is typically developed across system boundaries. Acceptable project costs are possible by selection of limited program systems.

3. If project funding and/or documentation are major factors restricting performance of SA, then an incremental contracting approach can be undertaken. Perform SA on one or more of the highly critical systems for which documentation is readily available. In a following fiscal period, contract for SA on the remaining systems, with the stipulation that the analysis includes the new systems and interfaces into the previously analyzed systems. This approach is especially desirable when detailed drawing or code instructions are missing for particular equipment or program modules, respectively. Functional diagrams should be made available to the SA contractor for these missing areas.

4. The purchaser can expect annual costs for SA *and* problem resolution to be lowest in the early development phases. There are significant cost and risk penalties associated with late identification and resolution of system problems.

5. Appropriations for the analysis should be allocated in the formulation of the reliability assurance plan and maintained throughout the development cycle for the desired schedule start time. If the project dollars have not been allocated early, they may not be available when required.
6. Since SA can be effectively blended with other analyses, reduced project costs for the combined analyses can be achieved.

Scheduling Requirements

1. SA should be scheduled so that final project results are obtained and can be adequately evaluated by the purchaser and equipment manufacturers prior to the end of the full-scale testing phase. Project costs to implement system changes increase dramatically after this phase.
2. The preferred start time is prior to the full-scale engineering development phase. This is an ideal time to provide a formal input into the design review process.
3. Timely results can be obtained for all scheduled SA projects and also for those projects which are intended to identify a single test or operational problem. For single-problem-oriented SA, limited system scoping and available documentation can provide project results as soon as 1-2 months into the project schedule.
4. Orderly scheduling of SA can be based on an assumed average 4-6-month project duration. Targeting the analysis to a specific project milestone can be performed by moving the start date back the specified number of calendar months. The two most important items affecting successful performance by the designated program milestone date are data availability and contractor performance.

Monitoring Guidelines for the Purchaser

1. Data acquisition has customarily been handled by the ultimate equipment owner. If data acquisition is assigned to the SA contractor, extra cost is incurred. Proprietary data from vendors and contractors typically requires proprietary data agreements which may require significant time to acquire.
2. SA report evaluation and coordination are important responsibilities of the purchaser. Tracking all reports and their eventual dispositions is an important element in assuring effective program benefits for the project expenditure. The resolution of the identified problems provides a measure of reliability improvement and sneak-finding capability of the performing contractor. Document error reports are typically found early in the project schedule, and once the network tree drawings or diagrams are generated, the primary reports are design and sneak-condition reports.
3. Liaison, contract monitoring, contract modification, and project closeout are the remaining tasks to be handled by the purchaser, his design contractor, or the ultimate equipment owner.

Conclusion

SA is an especially effective potential problem identification tool which is now available to petrochemical plants intending to improve systems and equipment reliability. The analysis tool adds a new dimension to assessing and evaluating reliability of new or mature systems. Most of the reliability tools are fault or failure related. The fault-related analyses are based either on determining eventual system effects produced by specified equipment failures, or alternatively identifying undesirable top-level system events and then determining those functions which can produce the undesired events. Sneak Analysis, however, provides a critical review of systems based on intended modes of operation and assumes no equipment or code failures. The identification of unintended modes of operation and their resultant system effects is the end product of an SA task. As such, SA is complementary to the fault-related techniques. The analysis is best performed by a contractor, agency, or group independent of the equipment or project design group. SA should be based on actual system design produced from "as-built" drawings, schematics, and wire lists for hardware systems, and from "as-coded" computer program source code for software systems.

SA can identify problems before they occur in test or operation so that the cost to modify or redesign should be decreased and the reliability and safety of the system should increase. The analysis method can be specified, funds allocated, and the entire analysis performed early enough in the project development cycle to allow cost-effective system changes to be implemented. Experience shows that regardless of the project development phase, application environment, equipment/software type, criticality ranking, or project cost, SA identifies a significant number of system problems. These taper to lower, but significant, levels in late-development phases. Critical machinery protection systems represent areas of high problem report levels.

References

1. "Sneak Analysis Application Guidelines," Contract No. F30602-81-0053, CDRL No. A002, Initiated by Rome Air Development Center, New York, 1982.
2. Rankin, John P., "Sneak Circuits—A Class of Random, Unrepeatable Glitch," AIChE Manuscript No. 7852, Denver, Colorado, August 1983.
3. "Contracting and Management Guide for Sneak Circuit Analysis," NAVSEA Document TE001-AA-GYD-010/SCA, Washington, D.C., 1980.

Chapter 9

Formalized Failure Reporting as a Teaching Tool

Experienced process plant managers are fully aware of the need to investigate costly equipment failures to prevent, or at least reduce, recurrence of the problem. Failure analysis reports are, of course, a means of presenting the findings in written form for the purpose of informing management. However, the hidden value of formalized reporting should not be overlooked.

A formal failure analysis report is a teaching tool for the investigator. It compels him to think logically, capture all relevant data, consider pertinent background events, and clearly state his conclusions and recommendations. Experienced engineers can testify that the bottom line, or final analysis results contained in their formal written reports, are sometimes quite different from the mentally summarized and verbally enunciated "early" results.

To be useful and effective, formal failure reports do not have to conform to a rigid format. Nevertheless, they must be structured to present the story in a relevant and logical sequence. This sequence typically consists of the following:

1. *Overview.* This is the first paragraph or subsection in a formal failure report. The overview could also be called a "management summary." As this term implies, it contains a synopsis of the entire investigation and answers in a few well-chosen words or sentences the basic questions of what happened, when-where-how, and possibly why.

   Note how these questions are answered in both of the sample failure reports which follow later. In Report No. 1, "Propylene Storage Unit P-2952 HSLF Pump Failure Report," the writers chose to call the lead paragraph an "overview." In Report No. 2, "Primary Fractionator MP-17A Tar Product Pump Failure and Fire Investigation," the first subsection is called "summary" and is comprised of three short paragraphs. However, the lead portions of both reports are about equal in length—roughly 100 words—and manage to convey essential overview information.

2. *Background information and/or summary of events.* A formal failure analysis report will probably be read by management and staff personnel whose familiarity with the equipment, process unit, operation, etc., will vary greatly. It is thus appropriate to include pertinent background information in the failure analysis report so as to make it a "stand-alone" document.

Although many failure analysis reports use the word "background" as the second heading, neither of the two sample failure analysis reports contain this term. Instead, Report No. 1 describes all pertinent background data under "summary of events," while Report No. 2 distributes appropriate background data over two headings, "equipment and process description," and "sequence of events." Recall, again, that even a formal report need not adhere to a rigid, repetitive format. A given topic or writing style lends itself to one or more ways of effective documentation, and both report formats follow a logical thought pattern. Illustrations or graphic data may be included in this section of the report. Very often, these representations are effectively presented directly in the text. Contrary to what we may have learned in college, bunching the representations at the end of a formal report makes very little sense and will, in *all* cases, disrupt the smooth flow of the reader's thoughts and perceptions. Only appendix material or similar supplements should be given last.

3. *Failure analysis.* This is normally the next major heading in a formal report. Detailed data collected and observations made are usually documented under this heading. This is where a competent researcher will put the solid evidence and will prove the soundness of his reasoning. Both of our demonstration reports contain this heading and lay the evidence on the table. Report No. 2 uses such subheadings as "Observations," "Probable Failure Progression," and "Probable Failure Causes" simply because they fit the subject matter and, frankly, the writing style of the committee charged with its issuance.

The bulk of the illustrations or graphic data will probably appear in this section of the failure analysis report.

4. *Conclusions.* The word "conclusion" has several dictionary definitions: the close or termination of something, the last part of a discourse or report, a judgment or decision reached after deliberation. The conclusion of a good failure analysis report should be none of the above. Instead, it should contain the evident and logical outcome or result of the observations, and thought processes documented up to this point. Both of the demonstration reports contain partial restatements of what probably caused the failure, as well as what was judged *not* to have caused the failure.

The investigator should consider it appropriate to list not only the most probable cause but give thought to the perceptions of the uninitiated, or less-experienced readers of his report. Take Report No. 1, for instance. When its conclusions were formulated, the investigating team anticipated questions such as "how do they know the pump didn't explode?" To lay to rest questions of this type, it was decided to let the reader know that this possibility had been looked into and was ruled out. Similarly, the conclusions portion of Report No. 2 lists first the principal conclusion, i.e., statement of failure cause, and follows up by explaining why other scenarios or possibilities must be discounted.

5. *Recommendations.* These are made last. Their ultimate aim is, of course, to prevent recurrence of the problem or failure. However, recommendations can sometimes be separated into near-term and long-term requirements. Also, they can be supplemented with an action plan, or simply a notation of action assignments and action status. Report No. 2 represents this latter format.

Examining the Sample Reports

Much thought was given to the best presentation of practical examples of formal failure analysis reports. The intention is to show actual reports which illustrate a proven approach and have general applicability to the majority of machinery failure situations.

We settled on a format which would first reprint a section of the actual report and then explain the thought processes and rationale which had been employed by the report writer in formalizing his findings.

Writing the report is, of course, the culmination of the investigation and analysis. The writing effort will have been preceded by intuitive processes which chiefly include data collection, examination of the chain of events, gathering and judging of relevant input, fitting the pieces, and defining the root causes of failure.

With this background, we can go directly to the reports. Both reports start with a cover letter, or letter of transmittal. This letter states in a few brief, crisp sentences the principal facts of the failure event, comments on safety-related matters, and reports the status of the investigation (see Figure 9-1). The cover letter is signed by the members of the investigating body and may or may not give their titles or job functions. Both cover letter examples give clues as to the preferred composition of many failure investigating committees in the petrochemical industry: a senior member of the technical staff, line supervisors from operating and support departments, and one or more lower-level operating personnel.

The Case of the High-Speed, Low-Flow Pump Failure

The first of our two example reports deals with a high-speed, low-flow (HSLF) pump which suffered a strange failure in July 1996. The seriousness of the failure in terms of potential injury to personnel and monetary loss to the plant was immediately recognized. As is customary in such cases, plant management requested a formal investigation of the failure event with a view toward establishing the root cause and avoiding a repetition of the failure.

Team Composition. Formal failure investigations start with the formation of the investigative team. The team members are designated by the plant manager or his appointed representative. For the investigation of the HSLF pump failure, management asked for representation by personnel from the operating, mainte-nance, and technical departments. A senior engineer from one of the mechanical technical services (MTS) groups was assigned to chair the effort. The maintenance department was represented by a mechanical supervisor, and the operations department delegated a supervisor and a technician to participate.

August 15, 1996

Propylene Storage Unit P-2952
High-Speed Low-Flow (HSLF) Pump Failure Report
File: 97-4-G

On July 30, 1996, one of the propylene storage pumps (P-2952) blew apart at the casing cover split line. The top half of the pump case, the gearbox, diffuser and motor lifted off and came to rest on the concrete about two feet away. The propane in the suction and discharge lines dispersed rapidly. The incident did not cause any injuries and no fire resulted.

An investigation of this event has been completed. The report of the investigating committee is attached.

John P. Sloper
MTS Engineer, Chairman

George W. Windbreaker
Operating Supervisor

M. E. Person
Mechanical Supervisor

Danny Wright
Operating Technician

HPB:jj
Attachment

cc: ABC DEF GHI JKL MNO
18B18

Figure 9-1. Letter of transmittal (cover letter) for a formal pump failure report. Note composition of the investigating committee.

Selecting participants or team members from the various segments of the plant organization is considered advantageous for two reasons: (1) it brings together a variety of backgrounds and experiences and leads to a supplementing and complementing of ideas, thoughts, and approaches, (2) it ensures consensus among departments or work groups which are known to have an occasional tendency to pin the blame on the other party.

At the conclusion of their formal failure investigation, the team members summarize their findings in a brief cover letter and affix their signatures. The cover letter, or letter of transmittal, is generally quite brief. In the HSLF pump failure example, the letter states "when-what-where" in the first two sentences. In just two more sentences the reader is given the principal results of the event. The four signatories opted not to state the root cause of the failure in the cover letter, although a different team of investigators may well have decided to highlight the most probable failure cause in one or two additional sentences.

Overview. The body of the investigative report starts with an overview giving details of the when-where-what mentioned earlier in Figure 9-2. Time, location, service details, and equipment particulars are described. The overview paragraph gives the failure cause as well as estimated cost of repair and restoration. It is in the overview paragraph of the report that the reader should be shown how and where the failed machinery fits into the process and interacts with other elements, units, piping circuits, or components. A few well-placed illustrations or schematics will generally help the reader to understand the topic better. These representations belong in the body of the report and should be interspersed in the narrative as appropriate. However, the body of the report is rarely the right place for highly detailed illustrations, calculations, or reproductions of pertinent correspondence. All of these belong in the appendix or attachments to the report, as we will see later.

Summary of events. It is of interest to note that "summary of events" does not refer to failure progression or failure mechanism suspected or identified. Instead, the summary of events section of a failure or incident report relates operating conditions, unit status, process parameters, personnel, and similar matters preceding or accompanying the event. More than any other part of the report, this section documents statements, depositions, conversations, interviews, log-book entries, and chart-readings relating to equipment status before and during the failure event. Furthermore, this section may state the rationale for operating in a certain mode, give observations made immediately after the event, and highlight actions initiated by responsible parties and personnel on the scene (see Figure 9-3).

Accordingly, the example report gives times and identifies shift numbers, personnel, process lineup (fluid routing), switching sequence, and fluid pressures. A restatement of relevant equipment physical details and appropriate pictorial representation follows. The report states in what condition the equipment was found after the event and notes observations made by various people.

Failure analysis. The team member representing the plant technical organization normally takes the lead in pursuing the actual failure analysis. He will contribute to the failure analysis to the maximum extent possible. However, the major

Overview

At 2:15 a.m., Tuesday, July 30, 1996, one of the propylene storage pumps (P-2952 on simplified schematic, Exhibit 1) blew apart at the bolt line. The pump is a Webster Model HSLF, 150-horsepower, 18,300-rpm unit. The top half of the pump case, the gear box, diffuser, and motor were lifted several inches and came to rest several feet away. The bottom half of the case remained on the base. There was no fire and no injuries were caused by this incident. The failure event was attributed to sulfide stress corrosion of the bolts. Estimated cost of repair and restoration: $10,000.

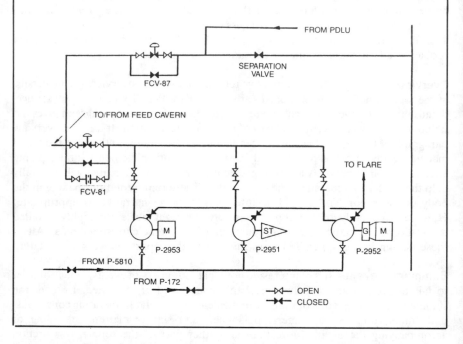

Figure 9-2. Overview section of a formal pump failure report.

contributions of this person are probably best described as perception and resourcefulness. He will have to think about fitting the pieces of the puzzle (sometimes acquiring a thorough understanding of unit operations and processes), separating pertinent observations from extraneous ones, structuring the various findings into a plausible and highly logical sequence, and explaining these matters to the other team members. The person representing the technical organization must show resourcefulness by requesting appropriate input from others in the plant organization, from persons reporting to affiliated plants, colleagues in non-affiliated plants, consultants, outside research organizations, testing laboratories, or perhaps institutions of higher learning.

Summary of Events

The propylene storage unit had been down for a short turnaround. Startup procedures had begun on 11x7 shift Tuesday, July 29. From 11:30 a.m. until 7:30 p.m. on Tuesday. P-2953 had been used to pump the off-test propylene and propane into storage. At 7:30 p.m., the propylene was on-test and routed to our customer.

At that time, P-2953 was shut down and P-2952 was put on line to continue pumping the offtest propane to the feed cavern. The P-2952 was utilized in this service because it is a lower capacity pump. While in this service, P-2952 suction was blocked from the offsite feed line. The depropanizer overhead pump (P-5810) was supplying feed to P-2952. The suction and discharge pressures for P-2952 were 230 psig and 780 psig, respectively. At 1:00 a.m., Wednesday, the propane was on test and routed to the Refinery. P-2952 was shut down at that time.

P-2952 is a high-speed, vertical, Webster pump. The high speed is developed through a gearbox. The pump case is divided into a top half and bottom half which are bolted together with twelve 3/4-inch bolts as indicated on Exhibit 2.

At 2:15 a.m., P-2952 blew apart at the bolt line. The top half of the pump case, the gearbox, diffuser, and motor were lifted up several inches. Evidently, the power cables restrained it from going further. It traveled east approximately two feet. The motor struck and bent a 3/4-inch steam line. The force of the blowout tore the small piping (seal oil, transmission oil, and cooling water) from the case.

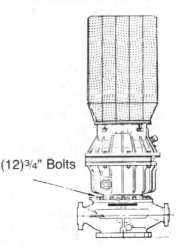

(12)³⁄₄″ Bolts

Exhibit 2. Webster HSLF pump.

The propane in the suction and discharge piping formed a vapor cloud which dispersed rapidly. Immediately after the blowout, Messrs. L.E. Hawker, R.D. Smith, R.W. Runner, and J.D. Scranton (Process Technicians) rushed to the unit to investigate. Mr. W.H. Teckel, Mechanical Technical Service, and Mr. L.N. Navasoti, Night Superintendent, were at the West Gate at the time of the blowout. They also proceeded to the unit. The vapor cloud dispersed prior to any personnel reaching the area.

The technicians blocked the suction and discharge valves at the pump and locked the motor out at the Motor Control Building. Mr. Navasoti requested the area be left as it was so pictures could be taken the next morning.

Figure 9-3. Summary section of a formal pump failure report.

In the example report, the subheading "Failure Analysis" (Figure 9-4), introduces a narrative explaining the testing to which the failed components—in this case nuts and bolts—were subjected. The basis for testing is described as visual inspection after showing pre-existing cracks in several bolts. It is further stated that this observation led the reviewer to suspect stress corrosion cracking. The narrative goes on to list the results of laboratory testing and contrasts as-specified physical characteristics with as-found physical characteristics.

Failure Analysis

Failed bolts and stripped nuts were subjected to extensive physical and metallurgical testing. Visual inspection had shown pre-existing cracks in several bolts, and stress corrosion cracking was suspected. Laboratory examination resulted in the following findings:

1. Nuts proved to be ordinary mild steel (AISI 1010) with inferior strength properties. ASTM 194 Grade 2H nuts are required.
2. Bolting was AISI 4140, but lacked ductility because of incorrect heat treatment. Tested versus required properties for B-7 bolting are as follows:

	Test	Required
Yield point at 0.2% offset:	191,200 psi	105,000
Ult. strength　　　　　:	205,900 psi	125,000 psi
Elongation in 1 inch　　:	8.2%	16%
Reduction of area　　　:	39.2%	50%

3. Fracture surface analysis with scanning electron microscope showed stress corrosion cracking had initiated bolt failures.

   Chemical analysis of residue in contact with the bolt thread showed sulfur present. Also, the hardness of failed B-7 bolts was quite high, Rockwell C-43 average. The combination of high hardness and sulfur points to hydrogen sulfide stress corrosion cracking.

A brief review of sulfide stress corrosion is deemed appropriate. Numerous investigations have shown that the optimum microstructure for resistance to sulfide cracking is tempered martensite resulting from heat treatment by quenching and tempering. These studies have shown that alloy steels having a maximum yield strength of 90,000 psi and a maximum hardness of Rockwell C-26 are not susceptible to sulfide cracking, even in the most aggressive environments. Our plant standards add an extra margin of safety and list RC-22 as an "out of danger" value. As the strength level increases above 100,000-110,000 psi, the threshold stress required to produce sulfide cracking may actually decrease in very severe environments. Exhibit 3, below, shows how AISI 4140 bolts hardened to Rockwell C-43 lie in the 100% failure zone if applied stresses exceed 25,000 psi. Most bolts in pump applications are torqued to stresses in the 30,000-40,000-psi region to which a portion of the internal pressurization stresses may have to be added.

(figure continued on next page)

Figure 9-4. Body of a formal pump failure report.

The following conditions must be fulfilled for sulfide cracking to occur in susceptible materials:

1. Hydrogen sulfide must be present.
2. Water must be present in the liquid state.
3. The pH number must indicate acidity.
4. A tensile stress must be present.
5. Material must be in a susceptible metallurgical condition.

When all of the above conditions are present, sulfide cracking may occur with the passage of time. All of these conditions can be assumed to have been present at P-2952:

1. The automatic transmission fluid in the gearbox of this Webster pump contains 2,600 wppm of sulfur. ATF leakage is a common occurrence on this pump.
2. The bolts are exposed to rainwater.
3. The ATF is slightly acidic.
4. Tensile stresses are present.
5. As shown in Exhibit 3, the material is in a highly susceptible condition.*

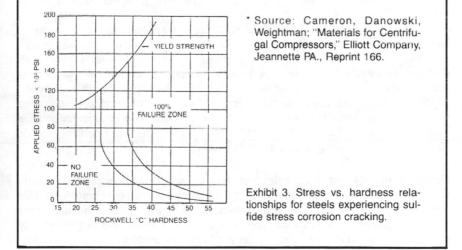

* Source: Cameron, Danowski, Weightman; "Materials for Centrifugal Compressors," Elliott Company, Jeannette PA., Reprint 166.

Exhibit 3. Stress vs. hardness relationships for steels experiencing sulfide stress corrosion cracking.

Figure 9-4. Continued.

At this juncture, there would generally be a temptation to present exact copies of laboratory reports or laboratory correspondence. However, these are generally inappropriate because they lack the narrative ingredient which makes the typical failure report the ''readable'' document it is meant to be. A failure analysis report is written for an audience not normally familiar with laboratory reports. Report writing should always be done with "audience awareness," and failure reports dealing with machinery problem events in the process industries are certainly not exempt from this rule. Thus, laboratory reports and similar supportive documentation should appropriately be placed at the end of the report.

In the HSLF Pump Failure Report, the results of bolt-strength analysis and material verification for nuts are followed by the findings of fracture surface analysis with a scanning electron microscope. The statement relating to these findings further alludes to the fact that a combination of high hardness, as found in the strength-of-materials investigation, and sulfur, as found in the chemical analysis of residue in contact with the bolt threads, could point to hydrogen sulfide stress corrosion cracking. The writers, for the time being, avoid blaming the failure on this mechanism. Instead, they opt for brief review of sulfide stress corrosion cracking because their "audience awareness" told them that many of the readers would benefit from the explanations.

The explanatory paragraph on sulfide stress corrosion shows that, for the purposes of a failure report, even a relatively complex topic can be adequately explained in less than 10 sentences. Note also that the writers were able to allude to "plant standards." This lets the reader know that responsible personnel are well aware of this failure mechanism. A graphical representation of susceptibility ranges is inserted in the text. It is evident that the particular illustration chosen in this instance facilitates the reader's understanding of explanations made in narrative form. It belongs in the text, and the writer's decision not to relegate the illustration to an appendix location was highly appropriate.

With the statement "the combination of high hardness and sulfur points to hydrogen stress corrosion cracking," the failure analysis section of the report set up a hypothesis. This hypothesis must now be tested for validity by probing whether all contributing factors are, in fact, present. The report therefore lists the various conditions which must co-exist for sulfide cracking to occur in susceptible materials. In this particular case, the researchers identified a total of five such conditions and went on to show that all of these contributory conditions must be assumed to have been present at P-2952. The evidence thus strongly supports the belief that sulfide stress corrosion cracking took place and caused the failure.

Conclusions. In formulating their conclusion, the team members decided not only to say what the failure *was,* but also what the failure was *not* (see Figure 9-5). This was again prompted by "audience awareness," and the anticipation that some readers would be inclined to ask questions which the report did not specifically address. Thus the statements about there not being any evidence of a detonation or explosion, and liquid propane and automatic transmission fluid not causing a chemical reaction violent enough to cause excessive pressure rise in the pump casing.

The remaining statements are firm and definitive: bolt failure was initiated by sulfide stress corrosion of excessively hard B-7 material; leakage of sulfur-containing automatic transmission fluid provided the environment necessary to cause cracking of steel in tension. Also, pump vibration observed prior to failure must have accelerated the propagation of pre-existing cracks in the bolts.

The conclusions also address the question of why the pump came apart at standstill, although one would intuitively expect its component parts to be more

---

Conclusions

There was no evidence of a detonation or explosion. Available data shows that liquid propane and automatic transmission fluid would not have caused a chemical reaction violent enough to cause excessive pressure rise in the pump casing.

Bolt failure was initiated by sulfide stress corrosion of B-7 material which had been heat treated to excessive hardness values. Occasional leakage of automatic transmission fluid used in the Webster gearbox provided the sulfide environment conducive to this failure mode.

Pump vibration observed prior to failure is thought to have accelerated the propagation of pre-existing cracks in the bolts.

Minor leakage through the discharge check valve caused the pump casing to be pressurized to cavern pressure. While this pressure is, of course, well within the design pressure of the pump, the separating forces (and tensile stresses) on the casing bolts are higher when the pump is so pressurized than when the pump is running. This explains why failure did not occur while the pump was operating. (See Attachment I).

Similarly, the diffuser is exposed to a net upward force with the pump uniformly pressurized at standstill. This caused the diffuser to be ejected from the pump case.

Figure 9-5. Conclusions section of a formal pump failure report.

highly stressed while the pump is running. In the course of the failure investigation the MTS engineer took a close look at the forces acting on pump components of interest. The calculations—appropriately given as Attachment I (Figure 9-7) to the report—established that structurally sound bolts would safely contain the pressure at either standstill or pump operation. The calculations further established that the net separating forces on the cover at standstill of the pump were higher than the forces exerted while the pump was running. This explains why bolt failure occurred at standstill and not during operation of the pump. As a final check, the investigators verified that the bolts would not have been overstressed even if the casing had been exposed to the full design pressure of 1440 psi.

Recommendations. The main purpose of a failure investigation is to identify steps which could be taken to prevent recurrence of the event. These steps are outlined in a subsection given the heading "Recommendations" (see Figure 9-6). Each recommended step or item must be the logical followup to findings, statements, observations, or conclusions given earlier in the report. It would not be appropriate to make recommendations whose rationale or background would be known to the investigator but would not have been disclosed to the reader of the failure report.

For the sake of illustration and emphasis, test each recommendation to verify that it meets the criteria stated previously:

1. *Immediately check all Webster pumps for excessive vibration.* The report had earlier concluded that pump vibration observed prior to failure could have accelerated the propagation of pre-existing cracks in the bolts.

Recommendations

- Immediately check all Webster pumps for excessive vibration.
- Replace all casing cover bolts in Webster pumps equipped with gearboxes. New bolting shall not exceed a hardness of RC-22 or 250 Brinell. Replacement should be initiated immediately and should be completed no later than September 15, 1996.
- Replace all nuts on Webster pumps. ASTM 194 Grade 2H nuts are required.
- Pump repair shops handling Webster repairs for our plant should be notified, in writing, of the above requirements.
- Specifications for bolts on Webster pumps should be amended so as to *disallow* hardness values in excess of 250 Brinell or RC-22.

Figure 9-6. Recommendations section of a formal pump failure report.

2. *Replace all casing cover bolts in Webster pumps equipped with gear boxes.* The report established sulfur stress corrosion to be the prime initiator of the failure event and further submitted evidence that sulfur was contained in the automatic transmission fluid (ATF) used in Webster gear boxes. Replacement of casing cover bolts is clearly a prudent step for pumps equipped with gear speed increasers. This particular recommendation step also restates the bolt hardness limit given in the body of the report.

3. *Replace all nuts on Webster pumps.* Although the hexagon nuts used in conjunction with the cover bolts did not contribute to the overall failure, they were found not to meet the specified strength properties. They should be replaced in an effort to bring all parts of Webster pumps into compliance with purchase specifications.

4. *Repair shops handling Webster pump repairs for the user's plant should be given formal notification of the new requirements.* Obviously, there is a risk of repair facilities replacing parts in kind. Formal notification of new requirements is appropriate for proper emphasis of urgency and to establish the repair shop's legal obligation to provide proper materials. No mention is made of the possibility that incorrect materials were perhaps substituted for proper ones during a recent repair.

5. *Future specifications should clearly disallow bolts with excessive hardness.* The body of the report contained narrative and graphic-illustrative evidence showing highly stressed bolting with high hardness very susceptible to failure. Accordingly, specifications for future procurement of Webster pumps should clearly indicate allowable hardness limits. No clues are given as to whether bolts with excessive hardness were originally furnished by the Webster Company. The reader is left to conclude that the failure investigation effort was either unable to determine the origin of failed bolts or considered it of academic value only to make this determinaton. It can, however, be inferred that the purchase specification for Webster pumps did not contain any specific requirements regarding bolt hardness.

Appendix material. Except for appendix material, the original failure report was issued in exactly the format shown in this text. For the sake of illustration and completeness, we are providing a listing of the original appendixes without actually reprinting all of them here:

1. *Documentation of separating forces acting on pumpover and diffuser.* Shown above as Attachment I (Figure 9-7), these calculations were given in the appendix of the original report.
2. *Tensile test and chemical analysis of bolts and nuts.* This attachment consisted of certification documents from the metallurgical laboratory.
3. *Tabulation of significant prior failure events on LPG pumps in the United States and Europe.* Significant event reports, fire incidents, etc., are often shared by the petrochemical industry. Major industrial insurance carriers or the central engineering offices of large companies are also in a position to assemble and catalog pertinent event reports for future review and statistical reference.

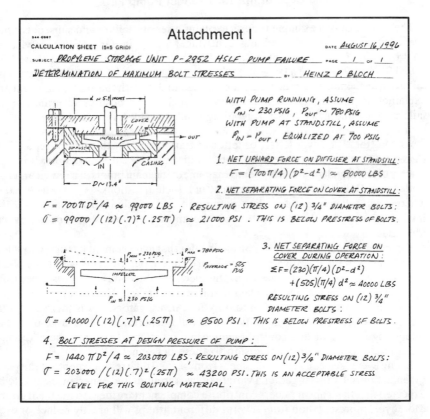

Figure 9-7. Appendix portion of a formal pump failure report would typically include calculations, tabulations, cross-sectional views, etc.

These tabulations were made available in this example case and were appended to the report.

4. *Cross section of Webster HSLF process pump.* It was felt that at least some readers of the original failure analysis report would be interested in a review of the component arrangement found in Webster HSLF pumps. This prompted the investigative team to include a dimensionally accurate cross sectional drawing in the report appendix.

5. *Results of fracture surface analysis (scanning electron microscope examination) made on broken bolt surface.* The findings reported by the research laboratory were appended for future reference and for the purpose of authenticating statements made by the investigating team in this regard.

6. *Specifications for P-2952.* Primarily comprised of the API data sheet, specifications for P-2952 were appended to complete the reference documentation package.

The Case of the Tar Product Pump Fire

The second of two example reports explains a serious failure and subsequent fire event on a single-stage overhung centrifugal pump in tar product service at "Ydrocarbon," a major manufacturing facility. Again, this example is based on actual field experience of a large petrochemical plant in the US Gulf Coast area. The failure investigation and reporting process used to deal with this particular event are explained to show the logic and rationale which led to positive identification of failure mode and guidelines for avoiding similar problems in the future.

Cover Letter. The cover letter, or letter of transmittal, for this failure report (see Figure 9-8) is not very different from that utilized in our earlier example. It is very brief, purely intended to give higher management the main points in a few clear, crisp sentences. Here, the cover letter answers such questions as "when—what—where—why—how" in a four-line paragraph and concludes by stating that the investigation is hereby complete and the most probable sequence of events identified.

From the list of signatories, we can again infer that management has opted to select the members of the investigative team from different departments within the plant. We can identify representatives from the mechanical technical services area, from operating departments, and the plant maintenance organization. There need not be a firm rule to limit representation to only four investigators, nor is there any great merit in restricting team composition to certain departments. If a large plant has safety engineering or loss control engineering functions, there may be additional incentive to assign team membership to these personnel.

In any event, selecting team members from several different segments of the plant organization will bring together a variety of backgrounds and experiences. It is to be expected that the team members contribute complementary ideas to the team effort. Moreover, representation from several different groups will greatly reduce or even eliminate the possibility that others do not accept the findings and recommendations or that one group will blame the other for the failure event.

ydrocarbon

November 26, 1996

MP-17A Tar Product Pump
Failure and Fire Investigation
File: 7-1-F

On November 15, 1980, one of the primary fractionator bottoms pumps was involved in a fire incident. The fire was the result of thrust bearing distress and a mechanical seal deficiency. The incident did not cause any injuries to personnel.

An investigation of this event has been completed and the most probable sequence of events identified. The report of the investigating committee is attached.

Kevin McDermott
MTS Engineer, Chairman

Chris Smith
Operating Supervisor

Pat Jones
Mechanical Supervisor

Clovis Deen
Operating Technician

HPB:jj
Attachment

cc: ABC DEF GHI JKL MNO STV WZF
1UB13

Figure 9-0. Letter of transmittal (cover letter) for a formal pump failure report. Note composition of the investigating committee.

Summary. The reader will recall our earlier statement indicating that failure reports are not necessarily locked into a rigid format. In keeping with this observation, the second example report starts out with the subheading "Summary" (see Figure 9-9) whereas the first sample report used the term "Overview." The summary lists the time at which the subject machinery—MP-17A—experienced catastrophic failure. The lead sentence states that failure of mechanical components resulted in a fire and goes on to report how and when the fire was brought under control. With personnel safety of paramount importance in the petrochemical industry, the investigative team reported "no injuries" in the first paragraph of the summary. Next, the report summary states why the mechanical components failed and concludes with a brief description of extent of damage and cost of restoration.

Equipment and process description. This subsection of the failure report (Figure 9-10) is usually quite straightforward and easy to compile. One or more pictorial representations or schematics such as Figure 9-11 might help the reader to visualize

Summary

At 19:33 hours, Friday, November 15, 1996, one of the primary fractionator tar bottoms pumps (MP-17A) experienced a thrust bearing and seal failure which resulted in a fire. The fire was extinguished by Team 3 technicians and the area reported secure at 19:45 hours. No injuries were caused by this incident.

The failure event is attributed to inadequate prelubrication of the duplex thrust bearing. When the bearing seized, the mechanical seal opened up.

Physical damage includes a fractured bearing housing, severely bent pump shaft, and burned insulation. Restoration cost is estimated at $7000.

Figure 9-9. Summary section of a formal pump failure report.

Equipment and Process Description

MP-17A is an Atlantic Model BTSG 3 x 13, 250-hp motor-driven, 3560-rpm centrifugal pump. At rated conditions, the pump feeds 320 gpm of tar at temperatures ranging from 520°F to 410°F from the primary fractionator tar boot and coke filter MF-2 through tar coolers ME-13 and ME-15 to tar product tank MTK-31. Pump design inlet and discharge pressure conditions are 10 psig and 300 psig, respectively. Refer to Exhibit 1 for a simplified schematic of the tar system.

Differences between MP-17A and MP-17B/C: The A-pump is made by Atlantic. It did not, at time of failure, incorporate reverse flush. The B and C pumps are made by Unicorn. They incorporate reverse flush (stuffing box to suction).

Figure 9-10. Equipment description portion of a formal failure report.

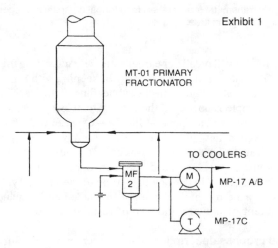

Exhibit 1

MT-01 PRIMARY
FRACTIONATOR

TO COOLERS

MF
2

M   MP-17 A/B

T   MP-17C

Figure 9-11. Schematics will always enhance the equipment description portion of a formal failure report.

specifics of process flow or peculiarities of the equipment whose failure is being analyzed. Process conditions such as flows, pressures, temperatures, and equipment details such as manufacturer, model, size, speed, and flow should be listed.

Significant differences between the failed and any existing satisfactory equipment should be explained if both were used under identical process conditions or if the investigative team has reason to expect that the readers would raise this question. In the case of MP-17A, it was noted that seal failure preceded the fire event. It was thus rightly deemed appropriate to state the differences in seal flush arrangements between the failed MP-17A and the well-functioning pumps MP-17B and C. Whether or not these differences had something to do with the failure of MP-17A is, for the time being, unimportant and may not merit further mention until perhaps later.

Sequence of events. The sequence of events leading to equipment failure must be investigated and documented (see Figure 9-12). In this example case it was

---

Sequence of Events

Process Operations Debriefing

Process operations during the week of November 3, 1996, included tar pump test operations with filters in the circuit. No extreme plugging was experienced. The filters were first bypassed on November 10, 1996, and a seal failure is thought to have occurred during the second shift (late evening) on November 12. At that time, the filters were put back into service.

On November 15, 1996, seal repairs were complete, and the MP-17A motor-driven pump was put in service at 16:30 hours. Team #5 day shift reported the pump in service and operating properly at that time. Team #3 night shift took over duty between 17:30 and 18:00 hours. Satisfactory operation is graphically represented in Exhibit 2, the strip chart obtained from the Console 2 trend recorder for fractionator boot and main vessel level. At 17:00, the chart shows the boot level to be 100%. About 10 minutes later, the boot level is reduced to 35% for approximately 5-10 minutes before returning to 100%. The chart verifies that the boot level never dropped below 35% on the day of the failure incident.

M & C Debriefing

M & C personnel report commencing MP-17A motor solo run on Monday, November 11, 1996. The pump was coupled up on Wednesday, November 13, and run with the tar filter bypassed. It is thought that bypassing the filter caused coke fines to enter the seal cavity. The pump was removed from its field location during Thursday afternoon, November 14, to effect replacement of bearings and seals.

On Friday, November 15, 1996, MP-17A was restarted at 17:00 hours and shut down around 17:10 to replace the discharge pressure gauge. Upon restart, the pump ran for about 10 minutes when it had to be shut down because of seal leakage. The pump was again removed and taken to the shop.

On Saturday, a new seal arrived at 12:00 hours. This new seal and sleeve combination was installed and measurements taken during assembly. At the impeller end, the seal sleeve was found to be free to displace 0.006" (see Exhibit 3). Dial indicator readings taken on the rotating hard face showed an initial out-of-perpendicularity of 0.027" which, through a series of adjustments, was improved to 0.006" upon final installation. At 14:30 hours, the pump was ready for warm-up and hot-alignment checked at 16:00 hours. MP-17A was started at 16:30 and its operation observed until 17:20. During this 50-minute time span, the discharge pressure was observed to go from initially 280 psi to a low of 230 psi.

The pump fire was detected around 19:33 hours and extinguished by operations personnel.

Figure 9-12. Failure reports may include a sequence of events section as shown here.

recognized that MP-17A had recently been repaired and that it was thus necessary to describe two such sequences. Process events were obtained in a debriefing of operators who reported that levels, flows, and pressures had been quite normal until the actual failure event. With adequate NPSH critically important to the safe, cavitation-free operation of centrifugal pumps, the availability of a strip-chart recorder tape showing sufficient level in the suction vessel was considered a particular advantage to the investigators. It is included in the report as Figure 9-13.

A debriefing of Mechanical and Construction (M & C) personnel shed some light on maintenance techniques and assembly quality-control procedures employed during repairs which preceded the final failure event. The report states that bypassing the filters probably caused coke particles to enter the seal cavity. It is implied that this could cause seal malfunctions unless a reverse flushing arrangement, i.e., a line from stuffing box to pump suction, was employed. Recall that under the heading "Equipment and Process Description" only MP-17B and C pumps were listed as incorporating reverse flush.

At this stage, the report does not pass judgment on the adequacy of mechanical work such as seal and bearing replacements but continues to list the sequence of repair events: ran for only 10 minutes on November 14; new seal installed on November 15, but dimensional difficulty was encountered when the installation tolerances were checked. Figure 9-14 is inserted in the report to facilitate visualizing the narrative description.

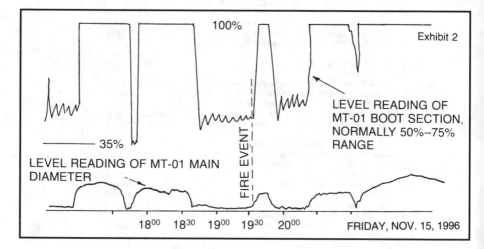

Figure 9-13. Supporting evidence may be in the form of strip charts and should sometimes become part of the formal failure report.

Exhibit 3

Obtained 0.006" Reading
On Dial Indicator

Obtained Out-Of-Perpendicular
Reading Of 0.027" Initially.
Improved it to 0.006" At
Final Installation

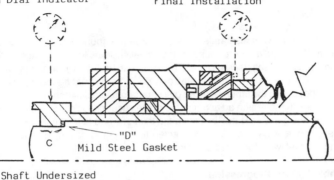

"D"
Mild Steel Gasket

C

Shaft Undersized
at "C"

Figure 9-14. Component details are best shown in sketches.

Failure analysis. The observations, conjectures, and analysis results of the investigative team are finally documented under this subheading (see Figure 9-15). The analyst observed the condition of every mechanical component and included Fig-ure 9-16 to again facilitate the reader's understanding of technical jargon.

It can be seen that the massiveness of the bearing failure lent strength to the investigator's belief that failure of the duplex thrust bearing set in motion the chain of events leading to the fire. Explanatory statements follow: excessive vibration resulted and caused the less-than-optimally installed seal to release pumpage. This pumpage could have ignited on the red-hot bearing.

After establishing a probable failure progression, the investigation now turns to probably failure *causes*. Why would a bearing fail so soon after installation? If it were designed for inadequate life, should it not have failed repeatedly in the past? Could it have been incorrectly installed? The report states that the failure was too massive to determine the exact failure cause from an observation of the failed parts and goes on to list a number of possible deviations which could have played a role in the failure event. Certainly, one of these would be inadequate lubrication.

Plant maintenance personnel were asked to describe how the pump bearings were being lubricated. They explained that new bearings were usually taken out of their oil-paper wraps and given a generous bath of "Western Handy Oil." Next, the bearings are pressed on the pump shaft and the entire assembly installed in the pump casing. After the pump is returned to the field, it is connected to an oil-mist lubrication header which supplies "Oilmist-100" in aerosol form to antifriction bearings throughout the plant.

The investigation now shifted to the properties of the "Western Handy Oil." The manufacturer gives its viscosity as 75 SUS at 100°F (~14 cSt at 38°C), which would

Failure Analysis

Observations

When the pump was dismantled, the seal area was found in clean and undamaged condition. Some solids were found in the impeller. Impeller wear rings and inboard bearing appeared satisfactory. The duplex thrust (outboard) bearings were totally destroyed. Severe metal loss was noted on virtually every bearing ball. Many balls were deeply imbedded in the inner race; the ball separators had disintegrated. The shaft was severely bent in the region adjacent to the duplex bearing lock nut (see Exhibit 4). Two of the four seal gland nuts had backed off, a third one had fallen off. A pedestal support bracket at point A had not been connected to the pump casing. The malleable iron bearing bracket was fractured at point B.

Probable Failure Progression

It must be assumed that failure of the duplex thrust (outboard) bearing set in motion the chain of events leading to the fire. Excessive friction resulting from severe and near-instantaneous bearing failure caused the shaft to bend. Extreme vibration was generated and caused the less-than-optimally installed seal to release pumpage. By this time, the outboard bearing area is thought to have been red-hot, causing spilled pumpage to ignite.

Probable Failure Causes

Near-instantaneous massive failure of rolling contact bearings is most frequently caused by deficient lubrication and overheating. The failure was too massive and too far progressed to allow determination of the origin of overheating. Installation method, housing bore dimensions, shaft dimensions, driver to pump alignment accuracy, class of bearing (i.e., rolling-element tolerance), and sparking action between auxiliary gland and shaft sleeve could have played a role in the event. However, the primary cause of bearing overheating upon initial operation of new bearings at our plant is probably not related to any of the above. Instead, the most probable cause is our shop's practice of prelubricating with "Western Handy Oil." Western does not market this oil for industrial use. Its extreme low viscosity (75 SUS @ 100°F) makes it suitable only for bicycle and door lock-type of lubrication duties. This oil is very volatile and will evaporate at temperatures well below those anticipated for new antifriction bearings operating at relatively high speeds.

Figure 9-15. Body of formal failure analysis report.

put it into the low-viscosity, high-volatility region. Figure 9-17 was used to determine the minimum lubricant viscosity required for adequate bearing protection of MP-17 and a similar pump, MP-03. The bearing pitch diameter d_m is calculated, and the chart entered at the abscissa. The vertical line intersects a diagonal line representing the shaft speed (3570 rpm), and from there moves to the left where it intersects the ordinate at the required viscosity. For MP-17 pumps with d_m = 87.5 mm, the

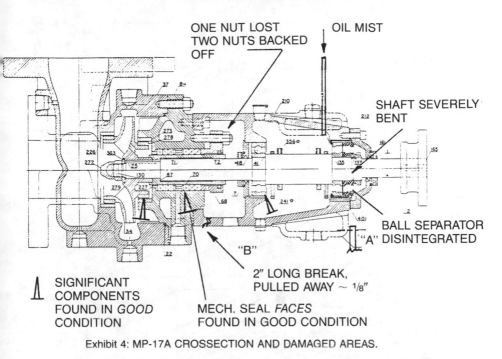

ONE NUT LOST
TWO NUTS BACKED
OFF

OIL MIST

SHAFT SEVERELY
BENT

BALL SEPARATOR
"A" DISINTEGRATED

"B"

2″ LONG BREAK,
PULLED AWAY ~ 1/8″

SIGNIFICANT
COMPONENTS
FOUND IN *GOOD*
CONDITION

MECH. SEAL *FACES*
FOUND IN GOOD CONDITION

Exhibit 4: MP-17A CROSSECTION AND DAMAGED AREAS.

Figure 9-16. Assembly sketches show the respective locations of damaged parts.

lubricant viscosity should always be 8.3 cSt or higher. Next the investigators had to employ Figure 9-18 to find the maximum allowable bearing rotating-element surface temperature at which "Western Handy Oil" would still meet the 8.3 cSt viscosity requirement.

Figure 9-18 represents an ASTM Viscosity-Temperature Chart on which the investigators plotted the viscosity-temperature relationship of "Western Handy Oil" parallel to that of the more typical industrial lubricants. Entering the chart at 8.3 cSt, they found the maximum allowable temperature for MP-17 to be only about 130°F. At bearing operating temperatures higher than 130°F, this oil would have an even lower viscosity than 8.3 cSt. This would certainly *not* give adequate protection to the bearings.

The report brings out these observations and stresses further that newly installed rolling-element bearings will inevitably experience temperatures in excess of 130°F at the loads and operating speeds encountered in centrifugal pumps. Moreover, connecting the pumps to oil-mist headers supplying the superior "Oilmist-100" could not be expected to result in the immediate establishment of an adequate oil film. By the time a dry bearing has been exposed to oil mist long enough for an adequate film to form, the damage would be done.

Continuing under the heading "probable failure causes," the failure analysis report comments on installation or quality control deficiencies. The seal face was

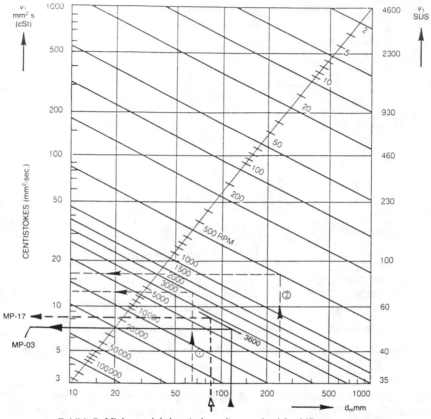

Exhibit 5. Minimum lubricant viscosity required for MP-17 and MP-03.

Minimum Required Lubricant Viscosity
d_m = (bearing bore + bearing OD) ÷ 2
v_1 = required lubricant viscosity for adequate
       lubrication at the operating temperature

Figure 9-17. Lubricant-related failure events often can be better explained if viscosity charts are included in the report.

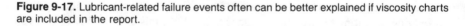

probably not sufficiently square with the shaft centerline. This made the seal inherently more vulnerable, or more likely to leak. Deviations such as an undersized shaft region and an incorrect seal sleeve gasket are pointed out for the record. These deviations were not, however, thought to have caused the seal to fail. Similarly, non-uniform plugging of the impeller had to be ruled out because the impeller wear rings did not show the inevitable contact pattern which would have resulted.

The omission of a bearing housing support bracket at point A of Figure 9-16 is noted, but not considered very significant in terms of the overall failure. But, leaving

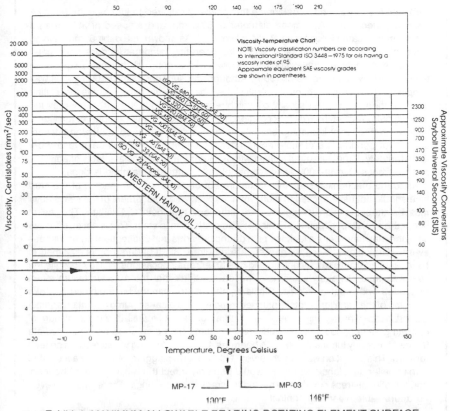

Exhibit 6: MAXIMUM ALLOWABLE BEARING ROTATING ELEMENT SURFACE TEMPERATURE

Figure 9-18. ASTM viscosity-temperature chart and supporting text containing clues to a lubricant-related pump failure event.

out the bracket mounting bolts is an indication of lack of diligent installation followup, to say the least.

Conclusions. There are some noteworthy parallels between the conclusions in the previous failure report and the conclusions written up in this second report (see Figure 9-19). In each of the two reports, the team members decided that the subsection on "conclusions" should not only say what the failure *was*, but also what the failure *was not*. Recall that the former report writers demonstrated "audience awareness" by anticipating that some readers would perhaps ask whether the team had considered this or that possibility before reaching its conclusions. The same thinking went into the conclusions of this second failure analysis report.

Exhibit 5 can be used to determine the viscosity required to lubricate the MP-17A bearings adequately. At a mean diameter of 87.5 mm and a speed of 3570 rpm, a minimum lubricant viscosity of 8.3 cSt is required. As shown in Exhibit 6, maintaining this minimum viscosity is possible only if the operating temperature does not exceed 130°F. Newly installed duplex and double row thrust bearing will, however, experience temperatures well in excess of 130°F. Athough the superior Oilmist-100 lube oil was supplied to the pump bearings via oil mist lube tubing, it could not overcome the diluting effect of the inferior low-viscosity oil which was present in a "trough" formed by the bearing outer race at the 6 o'clock position. The existence of this "trough," or minisump explains why oil mist lubricated bearings in horizontally arranged pumps and drivers survive for periods of 8 or more hours after the oil mist supply has been turned off. Unfortunately, if the minisump is filled with a dilutant, the beneficial effects of applying highly viscous Oilmist-100 cannot come into play until the damage is done.

As to the mechanical seal, the maximum out-of-perpendicular reading of the hard face should not have exceeded 0.001 inch. Excessive displacement of the seal sleeve must be attributed to an undersized shaft dimension near point C, Exhibit 3. The gasket under the seal sleeve (point D, Exhibit 3) was made of mild steel. This is not an effective seal and may have been a leakage path during previous and subsequent pump distress events. Grafoil or soft copper gaskets are required. Shaft undersizing is assumed to have resulted from shaft reconditioning on earlier repair occasions. However, while the seal was thus obviously vulnerable, we do not think that the massive failure event originated with seal problems. Similarly, the failure cannot be attributed to coke plugging of the impeller. A totally plugged impeller would cause pump operation against shutoff head. The resulting thrust and radial loads, while well in excess of normal operating values, could not possibly reach values high enough to cause immediate failure of properly lubricated MP-17A bearings. Uniform plugging would result in no feed forward. This condition was not observed. Non-uniform plugging—the latter associated with impeller unbalance vibration—would primarily affect the radial (inboard) bearing, and impeller distress could be expected to wreck the wear rings. There was, however, no appreciable wear ring contact.

During normal operation, the omission of support-bracket fastening bolts (point A of Exhibit 4) is not considered significant for this style of centrifugal pump. Nevertheless, it is possible that the bearing housing would not have broken had the two bolts been in place.

Figure 9-18. Continued.

The conclusions identify, in the first paragraph, prelubrication with an unsuitable lubricant as the failure initiator. The second paragraph shows why process-initiated impeller unbalance could not have caused this particular problem. Paragraphs three and four anticipate such questions as whether or not a spark could have started the fire, and show why the failure analysis had to rule out this possibility.

In paragraphs five through seven, the writers address component or systems weaknesses which, while in no way responsible for the failure event, could have amplified its severity. Correcting these weaknesses might reduce the possibility of other, not necessarily related, failures in the future.

Conclusions

- It was concluded that failures after a two-hour run length of MP-17A on brand new bearings, and two and eight hour runs of other pumps at our plant on brand new bearings in the same general span of four or five work days follow a pattern pointing to possible commonality of failure causes. The common link in all failures is bearing prelubrication with a lubricant approaching the characteristics of penetrating oil.
- Extreme unbalance vibration originating at the impeller would have been expected to cause wear ring and inboard bearing defects. Seal failure preceding bearing failure is inconsistent with the surprisingly clean appearance of the seal after the fire.
- Sparking action between the non-standard auxiliary gland (made of 316 SS) and seal sleeve (made of 410 SS) is possible but should have resulted in severe galling of the softer of the two materials. No such galling was observed.
- Tar leakage between shaft and sleeve has probably occurred, but is not thought to have started the fire. Experience shows that pump fires brought on by seal distress must reach a very high intensity before outboard thrust bearings disintegrate catastrophically. A low-level fire lasting for 5-10 minutes simply does not fit this scenario.
- Thrust bearings using formed-steel ball separators will work fine as long as the pumps are operated well within intended flow, pressure, NPSH, and lubrication recommendations. However, their margin of safety is inherently lower than that of rolling-element bearings with machined bonze cages.
- Throttle bushings and auxiliary gland plates must be made from non-sparking materials. Both API 610 and Plant Standard P-3002 prohibit the use of ferrous materials.
- Bypassing of filters could have sent higher-than-usual quantities of coke fines into pump and seal cavity. It can be postulated that dead-ending the seal (i.e., the seal is surrounded by pumpage which cannot escape from the seal cavity) is satisfactory for operation with the tar filters in the circuit. However, operation without filters may have, at times, caused excessive amounts of coke fines to accumulate in the seal cavity. It would thus seem appropriate to implement reverse flush on MP-17A to bring it into compliance with the flush method originally specified and actually used on MP-17B and MP-17C.

Figure 9-19. Conclusions portion of a formal pump failure report.

Recommendations. Many of the recommendations made here have already been stated earlier (see Figure 9-20). Others are the logical result of observations made in the report or have, perhaps, been implied in the preceding report text. For instance, recommendations 1, 7, 8, and 9 had been stated either as findings or deviations from good practice. Recommendation 2 represents guidance for the procurement of better components, while recommendations 3 and 10 could be considered steps to improve shop practices. Recommendations 4 and 6 relate to better, more-effective training for maintenance personnel. Finally, recommendation 5 asks for the correction of an installation oversight which, although lubrication-related, did not contribute to the failure being discussed in the report.

Other than the figures described in the main text, the report does not contain any appendix material.

Recommendations

1. Discontinue use of "Western Handy Oil" for prelubrication of pump antifriction bearings. Use Oilmist-100 instead.
   Action: Had discussions with all millwrights on November 25, 1996 (PQR). Follow up by ABC/DEF.
2. To strengthen general quality improvement, implement bearing procurement and incoming inspection procedures aimed at verifying that bearings comply with plant requirements as outlined in PS/P-3002:

   - AFBMA-1 Class C-3 clearance fit for radial bearings.
   - Use machined bronze or phenolic cages for all bearings in hydrocarbon services.
   - Disallow double row antifriction bearings in thrust locations.
   - Disallow thrust bearings with filler notches.
   - Disallow snap rings as a means of axially locating thrust bearings.

   Action: ABC/DEF
3. Develop mechanical seal face and sleeve installation and tolerance criteria (ask Whooper Company to provide, then post in prominent shop location).
   Action: GHI has initiated and will follow up.
4. Arrange for additional training opportunities in bearing installation and maintenance.
   Action: PQR has contacted SKF Company to obtain information on February 1997 training course. SKF reply will be forwarded to JKL.
5. Verify use of directed oil-mist fittings on pumps with inner-race velocities more than 2000 fpm (MP-17 operates at 2600 fpm but did not have directed mist fittings).
   Action: ABC/DEF
6. Conduct a 30-minute information interchange session once every month. This discourse should be between MTS machinery engineers and millwrights.
   Action: First session held on 11/24/96, RSQ. Subsequent sessions with MNO and GHI to be arranged by ABC/DEF.
7. Steel auxiliary gland to be replaced with identical bronze part.
   Action: Initiated by ABC.
8. Sleeve gaskets to be made of soft copper.
   Action: Initiated by ABC.
9. Implement API Flush Plan 13 (reverse flush from stuffing box to suction line) on MP-17A.
   Action: Initiated by DEF.
10. Develop a "pump repair check board" for prominent display in our shop. Include, in large letters, such items as:

    - Auxiliary glands made of non-sparking material?
    - Bearings prelubricated with Oilmist-100?
    - Seal sleeves installed with shaft gasket?
    - Seal faces installed perpendicular?
    - Viton O-rings disallowed for MP-10, MP-18 and MP-38.
    - No riveted steel cages as ball separators?
    - Directed oil-mist fitting if shaft diameter >2.2 inches and speed >3000 rpm?

    Action: ABC/DEF to implement. GHI to assist in completing the above list.

Figure 9-20. Recommendations may be appropriate for inclusion in a formal pump failure report.

Failure Investigation May Require Calculations

Incomplete, inconclusive, or even erroneous identification of failure cause can result if the analysis neglects to perform calculations or, as a minimum, takes into consideration pertinent numerical data from equipment manufacturers. The analysis of thrust bearing failure events on large compressor speed-increaser gears will serve to illustrate the point.

A large North American petrochemical complex, "Petronta," operates a motor-driven refrigeration compressor train as shown in Figure 9-21. After a few years' operation, the train was brought down for general inspection and major preventive maintenance such as cleaning of motor windings. Approximately six months after reassembly and startup, the gear unit suffered a serious axial locating bearing (ALB) failure.

To find out what happened and why, we must resort to both observation and calculation.

The observations can be in the form of a service report written by vendor representatives or maintenance work forces. Here is a pertinent excerpt from a failure analysis report written by the resident machinery engineer and the machinery maintenance section head:

1. . . . the compressor train developed again high axial motor vibrations that started to exceed levels experienced prior to the last shutdown by a factor of (2). Also, serious gearbox vibrations were noted, whereas the compressor itself ran exceptionally quiet without any signs of distress.

Exhibit 1

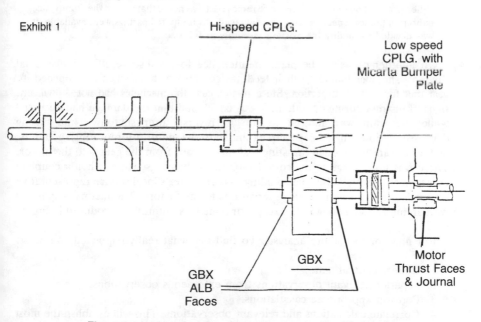

Figure 9-21. Motor-gear-driven compressor train at PETRONTA.

2. A task group looking at the train's past history and the vibration patterns recommended that it be shut-down for an inspection, including the motor, and the correction of possible misalignment between the motor and gearbox.
3. The train was subsequently shut-down. Ensuing inspection work revealed a severely damaged bull gear bump shoulder at the non-drive end, marginal but tolerable alignment between motor and gear, and severely locked up high-speed and low-speed couplings.
4. No signs of distress were observed when compressor thrust and journal bearings were examined.
5. Thorough inspection of the motor internals under direction of OEM personnel revealed no problems.
6. Both couplings were replaced. LS coupling was expedited from OEM, HS coupling was on hand. It had been ordered but not installed during the last turnaround due to good condition of the original coupling.
7. Gearbox bump shoulder arrangement was redesigned and repaired.
8. The unit was started up after resetting and realigning the gearbox. Motor and gear were operating at satisfactory vibration levels.

The gear manufacturer's service representative gave the following account of the incident:

> The gear unit cover was next removed. It was noted that the low-speed gear was fully against the blind end bearing 'bump' face and spalling of babbitt was visible. It was further noted that operating heavy contact was visible on both driving, driven, *and* following flanks of gear and pinion teeth.

> When gear and pinion assemblies were removed it was noted that indeed the 'bump' face babbitt of the low-speed bearing was damaged. As no spare bearings were available, it was decided to modify the existing assembly as follows: . . .

On another occasion the plant maintenance forces discovered a similar axial "bump" shoulder failure on their feed-gas compressor train when they opened its gear unit for routine inspection. Since in this case the machines had not shown any signs of distress during operation, it was concluded that the damage had occurred while the train was allowed to run in reverse after shutdown and during depressurization. The damage also extended to the motor rotor locating shoulders, and it appeared as though thrusting had taken place from the gear into the motor. There was no damage to the compressor thrust bearings. Here, when a coupling defect was suspected also, the coupling manufacturer's local service representative was called in and reported his observations to the factory. The factory responded with an interoffice note which, except for names, is faithfully reproduced in Figure 9-22.

But now for the failure analysis. To find out what really happened, we must:

1. Collect general data.
2. Separate relevant observations from extraneous observations.
3. Perform appropriate calculations.
4. Correlate calculations and relevant observations. This will establish the most probable cause and sequence of failure events.

Superior Coupling Manufacturing Company
Interoffice Correspondence

TO: J. Localrep
FROM: J. F. Manager

Subject: Petronta, Ontario
Date: July 7, 1997

   The thrust bearing failure in a herringbone gear box was originally attributed to the high-speed gear coupling. Petronta is using our coupling as shown in Exhibit 1 and quoted a technical paper entitled "Why Properly Rated Gears Still Fail" to substantiate their conclusion. After visiting the plant and discussing the failure with plant personnel, we conclude that the failure was caused by improper axial installation of the driving motor. The bearing failure evidenced that the gear was pulled toward the motor. Two possible forces could do that:

1. The compressor shaft expands toward the gear box. If the high-speed coupling has a high friction coefficient, it will transmit to the pinion an axial force. This force will unload one side of the gear mesh and overload the other side. It will be almost totally transmitted to the thrust bearing. Inspection of the unit should reveal some differential wear between the two sides of the herringbone teeth. As we were told, no such uneven wear was visible, and the high-speed coupling teeth were found to be in excellent condition. Hence, we can safely eliminate this possibility.
2. The low-speed coupling between the gear box and the motor is a "limited end float" coupling. The coupling is designed to allow the electric motor rotor to seek its magnetic center, but to prevent it from bumping into the end housings when the power is cut-off. If we assume that the motor was installed too far from the gear box, the coupling will prevent the rotor from going to the magnetic center, and the large forces pulling the rotor will load the thrust bearing. As the low-speed coupling was not available for inspection, we could not confirm this possibility, but as we cannot see a third possibility, we have to assume that improper axial installation caused the bearing to fail.

J.F. Manager
J.F. Manager

Figure 9-22. A coupling manufacturer's "quickie" analysis of a serious component failure.

Collecting Data

   For the train configuration shown in Figure 9-21, the user organization listed the following data:

Motor output:	8400 hp.
Bull gear speed:	1190 rpm.
Pinion speed:	6198 rpm.
High-speed coupling pitch diameter:	7.38 inches.
Possible axial float of bull gear:	0.005 inches.

Motor bearings thrust face area: 19.4 sq. in., babbitt on steel.
Bull gear bearing thrust face area: 19.6 sq. in., babbitt on steel.
Motor rotor magnetic centering force: 4200 pounds.

Calculating Thrust Forces

Gear couplings are capable of transmitting axial thrusts equal to *(T/r)* μ, where torque *T* is (63025) (hp)/rpm, *r* is the pitch radius of the gear coupling, and μ is the coefficient of sliding friction between the coupling teeth. For a given compressor train, all of these values, except for μ, are assumed to be constant. The coefficient of friction is known to vary from perhaps 0.05 in a well-lubricated, new coupling to as high as 0.25 or more after an appreciable operating time with sometimes marginal lubrication. We conclude from this that any kind or degree of gear-coupling deterioration initially manifests itself in an increase of coefficient of sliding friction. The worst case is usually referred to as friction lock-up, where μ will momentarily become unity, i.e., equals one.

As long as the motor shaft journals are not contacting the motor bearing thrust faces, the bull gear axial bump faces may be exposed to a combined thrust load composed of motor rotor magnetic centering force and axial thrust transmitted by the high-speed gear coupling. The magnetic centering force can be calculated from

$$F_M = 0.0117A \ (60/f)E_oI_{MO}(l_o/l)^2[dl/dx]$$

where:

A = number of phases.
f = frequency, cycles per second.
E_o = phase voltage, with motor in reference position.
I_{mo} = phase current, effective mutual value.
ℓ_o = effective stator (rotor) length, inches (reference position).
ℓ = effective rotor length, inches (displaced position).
dl/dx = rate of change of effective machine lengths when stator (rotor) overhangs rotor (stator) at each end.

Alternatively, we can use a rule of thumb for determining the approximate axial thrust generated by the magnetic centering force in induction motors with two, four or six poles.

---

two-pole motors: 0.1 pounds per hp.
four-pole motors: 0.35 pounds per hp.
six-pole motors: 0.5 pounds per hp.

We suspect this is how the owner obtained the reported 4200 lbs in the data summary for this motor.

If we assume the high-speed coupling to have a coefficient of sliding friction of 0.15, the transmitted axial thrust will be

$$F_{HS} = (63025)(8400) \ (0.15)\big]\big/\big[(3.69)(6198)\big]$$
$$= 3472 \text{ pounds}$$

The total combined thrust is, therefore,

$$F_t = 4200 + 3472 = 7672 \text{ pounds}$$

and the load on the axial bump faces of the gear $7672/19.6 = 391$ pounds per sq. in. This pressure load is substantially in excess of the 250-psi load normally permitted for simple babbitted thrust bumpers.

Does this mean that the thrust bearing failed because of improper axial installation location of the motor rotor? Not necessarily. Let us assume the high-speed coupling did, in fact, have a coefficient of friction of 0.15. This value is not at all unusual in our experience. Then

$$(3472 + F_M) / 19.6 \text{ in}^2 = 250 \text{ pounds per in}^2$$

and

$$F_M = 1428 \text{ pounds}$$

This value may already be reached when the motor rotor is displaced a fraction of an inch from its magnetic center and coupled, via limited end float low-speed coupling, to the bull gear.

What, then, is responsible for causing the thrust bearing failure? Why did the installation in the case of the refrigeration train given reliable service for several years but failed relatively soon after turnaround work was performed on the compressor train? The previous calculations contain all the clues to the answer.

We can see that the gear thrust bump faces are small, 19.6 in². Their maximum allowable pressure loading of 250 pounds per in². is easily exceeded; all it takes is for the total axial thrust to exceed $(19.6)(250) = 4900$ pounds. The total axial thrust is made up of coupling-generated and motor rotor-generated components. Therefore, the gear thrust bumper faces failed because either the coupling coefficient of friction or the motor rotor magnetic center-seeking force, or perhaps both, were too high. The coupling undoubtedly had a wear pattern which gave a reasonably low coefficient of friction during its earlier, long-term operation. At the time of the turnaround, a new wear pattern had to be established, or the teeth were perhaps driven against a ridge representing the boundary of the original wear pattern. This could have caused the coefficient of friction to go up.

One question in this context will always be difficult to answer, namely: "Were both high- and low-speed coupling assembly match marks observed to avoid random reassembly?" Often this is well understood by mechanics as far as high-speed couplings are concerned because of the impact on coupling balance. In the case of low-speed gear couplings ($n \leq 1800$) less emphasis is encountered. Yet, the previous question should be pertinent in connection with all gear couplings, because as they get older their teeth will no doubt have deformed in plastic flow. This deformation may be minute, but it will be sufficient to cause the coefficient of sliding friction to increase if, after turnaround, the original mesh is not reestablished. In the case of the refrigeration train, the investigators were unable to find any match marks on the failed low-speed coupling. Also, the motor was removed from its pedestal for cleaning work during the turnaround. When it was reinstalled, the rotor was probably no longer in exactly the position where it had been previously. It probably exerted a different axial thrust force on the coupled system.

For installations of this type, it is prudent to insist on larger gearbox thrust bearings. Also, turbomachinery trains require gear couplings whose coefficient of friction is guaranteed to be very low—especially upon reuse after a turnaround. A properly designed non-lubricated disc or diaphragm coupling will probably give more reliable service than a gear-type coupling. As for the motor installation procedure, it should ensure that after coupling up, the rotor will be located directly at the magnetic center of the stator. In the case of the refrigeration train, OEM service representatives were very unclear about the location of the motor's magnetic center. Old plant records showed the magnetic center location of the motor to be at a ⅝-inch nondrive-end shaft protrusion. This was the shaft location during the distress sequence. After repairs, the motor was running without problems with a ⅜-inch protrusion.

In conclusion, we have shown that calculation can be of tremendous help to the engineer investigating equipment failures. Even an educated guess may not be sufficient or may, as the coupling manufacturer's report shows, be only partly correct.

The Short-Form Failure Analysis Report

Every troubleshooter should strive to make himself easily understood. This requires that he be successful in clearly organizing his more or less complex ideas. In the preceding section we offered a few good examples in connection with critical machinery incident reports.

Writing a good technical transmittal is largely a matter of organization. Most people find that when they can arrange their thoughts properly, putting the words on paper is a relatively simple matter. One of the best ways to put your ideas together is by using the typical problem solution process discussed in Chapter 6. In this context steps to organize one's thoughts can be summarized in five points:

1. *Situation analysis.* This first step sets the background for what you want to communicate.
2. *Problem definition.* This arises out of the situation analysis. It may be perfectly clear to you that there was or is a problem and that it has been solved. Make sure that your readers understand exactly what the problem is.
3. *Possible solutions.* This is the possible answer to the problem, and it is stated as short as possible. It introduces Step 4.
4. *Preferred solution.* This details the preferred solution. It also contains all the necessary information to tell the reader why this particular solution was chosen.
5. *Implementation.* This is actually still part of Step 4 but stands on its own. It will show what was done, and what still needs to be done, i.e., it lays out the action plan.

We might want to call this system "SPSI" (Situation-Problem-Solution-Implementation). No doubt, it could well be too cumbersome to be applied in all cases.

There is however, an area where SPSI could be routinely applied—on the smaller, non-critical machinery failure events. Here, a one-page failure analysis report often refreshes memories and keeps the record straight. Figures 9-24 and 9-25 show how so-called "short-form" analysis reports can be easily structured around SPSI.

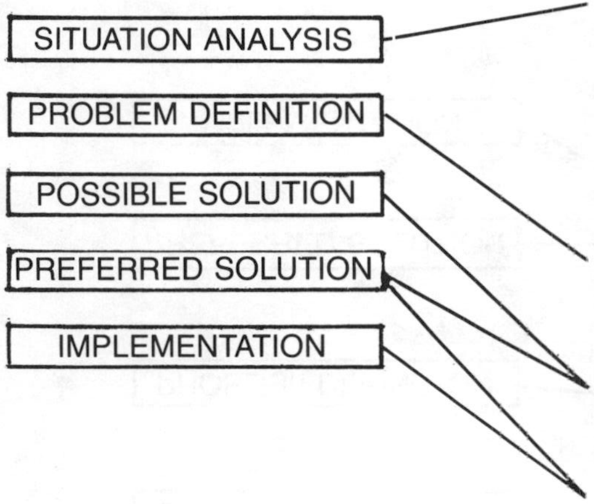

June 4, 1996

CP-11 FAILURE ANALYSIS

To: Superintendent, Process

File: 4040

From: H. Y.

P-11 was transported to Area 3 for repair on Monday, May 20, 1996. It ran satisfactorily on Monday, June 3, 1996. Following is a summary of the failure analysis and work done to repair the pump.

Observation. Pump in as found condition.

- Pump shaft was broken at the thrust bearing.
- Thrust bearing was totally destroyed.
- Radial bearing was severely damaged.
- Very little oil was found in the bearing housing.
- Thrust bearing housing was bored .002" out-of-round.
- Previous bushing for radial bearing was incorrectly machined; No oil return cutouts were made.
- Brazing on OB bearing housing was cracked.

Most Probable Failure Cause. Radial bearing was first damaged by inadequate oil supply. Metal particles were then transmitted to the thrust bearing through oil circulation. Subsequently, thrust bearing was damaged. A lot of heat was generated which also weakened the pump shaft. Eventually the shaft broke off.

Recommendation. The bearing housing has been cracked for some time (exact date not known). The cracks were repaired by brazing. This can only be considered a short-term repair. For long-term reliability, we strongly recommend ordering a new bearing housing for replacement. Replacement cost for a new bearing housing is $1882.

Highlights of Repair Work

- Bearing housing was bored and bushed. Proper cutouts were made for the radial bearing oil return.
- New pump shaft was made out of 4140 material.
- New radial bearing was installed to increase its load-carrying capacity. A SKF 22214C spherical roller bearing was selected for its superior load capacity and also for its ability to adjust for internal misalignment.

/cac

SITUATION ANALYSIS

PROBLEM DEFINITION

POSSIBLE SOLUTION

PREFERRED SOLUTION

IMPLEMENTATION

Figure 9-23. Example of short-form failure analysis report using SPSI concept.

MEMORANDUM

To: Supervisor, Process

April 8, 1997
CP-117 FAILURE ANALYSIS
File: 1750
From: J. V.

SITUATION ANALYSIS
{ SYMPTOMS:
  OBSERVATIONS:

Motor tripped out and would not restart.
Motor and conduit run contained water. Motor bearings showed signs of rust.

PROBLEM DEFINITION
{ DIAGNOSIS:

Motor burned out due to insulation breakdown caused by excessive water in motor and electrical conduit.

TYPE OF PROBLEM:

Operating conditions.

PREFERRED SOLUTION
{ RECTIFICATION:
  RECOMMENDATIONS:

Motor rewound at Delta, bearings changed.
Water spilling over motor and near it must be stopped. This failure is caused by the same problem referred to in C-117 "Failure Analysis" dated 2 March 1997. Reactor cleaning drains spill into lower level of recovery room as they are not piped to any drain system. Drains should be installed or the affected equipment made "waterproof" by covers similiar to those used on LB-110.

IMPLEMENTATION

/cac

c.c. J. J.
     N. G.
     H. W.

Figure 9-24. Short-form failure analysis report.

Chapter 10

The "Seven Cause Category Approach" to Root-Cause Failure Analysis

In Chapter 1, we listed seven meaningful classifications of failure causes. We also made the statement that the basic agents of machinery component and part failure are *always* force, time, temperature, and a reactive environment. One or more of these mechanisms may combine and hasten component degradation. These observations are often applied by failure analysts intending to quickly get to the bottom of a given failure event.

Like virtually any other well-thought-out failure analysis method, the "Seven Root-Cause Category Approach" is aimed at uncovering the sometimes elusive failure sequence—and thus root cause—of the events leading up to the equipment failure. It assumes that *all* failures, without exception, belong to one or more of the seven cause categories, or failure-cause classifications, found on page 3. They are:

- Faulty design
- Material defects
- Fabrication or processing errors
- Assembly or installation defects
- Off-design or unintended service conditions
- Maintenance deficiencies (neglect procedures), and, finally,
- Improper operation

Contributing or interacting factors are all part of a system; consequently, the entire system must be subjected to review and scrutiny.

Using these premises, we can introduce a straightforward approach that has assisted the authors in identifying the root causes of many costly failures involving centrifugal pumps or other machines in process and utility services. Six such failures are examined in greater detail:

- Repeat bearing failures that were attributed to vendor *design* error.
- Extreme vibration and deterioration of grease-lubricated sleeve bearings in large sea-water intake pumps traceable to *operations* error.
- Repeated and costly thrust bearing failures in a mining slurry pump caused by mistakes in *parts documentation, fabrication,* and *processing.*
- Loss of life in a U.S. Gulf Coast plant, possibly caused by simple *maintenance* oversight.
- Reduced life and catastrophic failure of electric motor bearings due to fabrication and assembly errors.

Checklists and Failure Statistics Can be Helpful

It would be difficult to think of machinery troubleshooting tasks that would not benefit from a structured approach. Time is saved, accuracy improved, and the risk of encountering repeat failures is reduced whenever the troubleshooter makes use of a comprehensive checklist such as the one compiled by Igor J. Karassik, Tables 4-6 and 4-7. (Ref. 1) Another approach was highlighted in Chapter 4 (pages 275 through 307), and given the description "Matrix Approach."

However, while the use of checklists and/or probability rankings is strongly recommended, the person engaged in machinery failure analysis may do well to remember that all problems can be assigned to one or more of the seven cause categories mentioned earlier. In addition, the troubleshooter should keep in mind the basic agents of machinery component and part failure mechanisms, i.e., force, time, temperature, and a reactive environment.

It is doubtful whether statistics have been compiled to show the overall distribution of failures as they relate to the seven cause categories given in the introduction of this chapter. At best, the reviewer might expect to find failure cause and failure mode distributions for critical components (Chapter 2, p. 14) or entire machine categories such as gears (Chapter 3, p 152).

Systematic Approaches Always Valuable

Considerable involvement with pump maintenance and repair would lead us to estimate a failure cause distribution for centrifugal pumps in U.S. process plants, as shown in Table 10-1. Our failure analysis and troubleshooting approach attempts to focus on the estimated cause distribution for a given machine type or equipment family. In other words, we might generally endorse an approach that seeks to first find the root causes of failures in the categories with the highest probability ranking. Our approach does not, however, overlook the need to:

Table 10-1
Failure Cause Distribution Estimate for
Centrifugal Pumps in U.S. Process Plants

Cause	% Incidence	Probability Ranking
Maintenance Deficiencies (Neglect, Procedures)	30%	1
Assembly or Installation Defects	25%	2
Off-Design or Unintended Service Conditions	15%	3
Improper Operation	12%	4
Fabrication or Processing Errors	8%	5
Faulty Design	6%	6
Material Defects	4%	7

1. Start at the beginning by
   - Reviewing the pump cross-section drawing
   - "Thinking through" how the individual parts function or *mal*function.
   - Understanding the process loop and process operations.
2. Take a systems approach. Never lose sight of the fact that
   - The pump is only part of the overall loop.
   - The part that failed is very often not the root cause of the problem and unless we find the root cause, repeat failures are likely to occur.
3. Collect all the pieces. The missing part may contain clues that must be examined and that may have had an influence on failure cause or failure progression.
4. Use a calculation approach while not, of course, neglecting the intuitive or prior experience-utilization approach.

With this in mind, we can now examine the first of several problems.

Faulty Design Causes Premature Bearing Failures

Not too long ago, a 125 hp, 3,560 rpm, 310 gpm, 670-ft-head, single-stage overhung impeller centrifugal pump in hydrocarbon service experienced frequent bearing failures. With "Faulty Design" ranking next to last in the Failure Cause Distribution listing of Table 10-1, it was certainly not logical to immediately suspect a fundamental design error to vendor-related engineering problem. Because of the probability ranking, maintenance-type causes were pursued first. Table 4-7 was consulted (items 77 through 85) and the supplementary 54-item bearing problem checklist (Chapter 3, pages 115-127) used to ascertain that faulty assembly or maintenance could also be ruled out. Next the failure analysis review focused on "Off-Design Conditions" and "Improper Operation." When no problems were found in any of these areas and it was further established that there were no material defects in the rolling element bearings, the investigation began to concentrate heavily on the possibility of a vendor error, i.e., "Faulty Design."

Pump owner and pump manufacturer agreed to perform a field test on this failure-prone pump. A special test rig, Figure 10-1, was designed and fabricated by the pump manufacturer. It consisted of means to allow the pump rotor-bearing assembly to move in the axial (impeller thrust) direction. The total axial movement was limited so as not to exceed permissible impeller travel. Also, the axial thrust value was measured by three load cells ("LC," Fig. 10-1), whose connecting cables are visible in the field test setup shown in Figure 10-2.

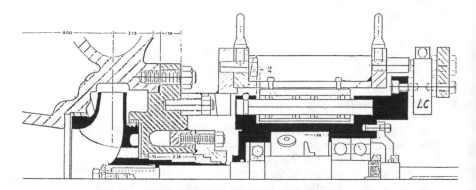

Figure 10-1. Cross-sectional view of test rig allowing impeller thrust to be absorbed by load cell (LC).

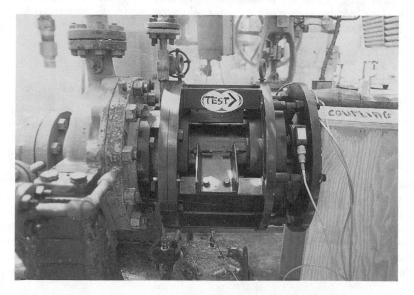

Figure 10-2. Test setup for determining thrust load on pump.

Test results were plotted and compared to the manufacturer's calculated and originally anticipated thrust values for this pump. As indicated in Figure 10-3, the experimentally verified thrust at shutoff was 2.6 times greater than anticipated. Since ball-bearing life varies as the cube of the load change, the life of this pump bearing would thus be reduced by a factor of 17.

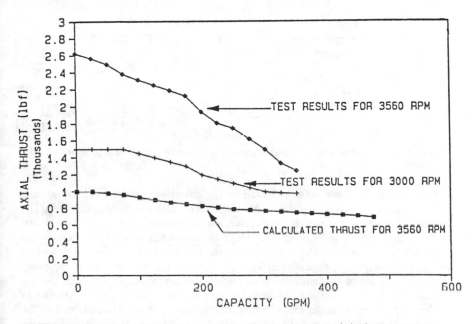

Figure 10-3. Calculated and experimentally verified axial loading of single-stage pump rotor.

The test results convinced both the operating company and the pump manufacturer that the pump internals had to be redesigned to limit hydraulically induced thrust values to more reasonable limits. Obviously, the basic agent of the bearing failure mechanism was excessive force.

Fabrication and Processing Errors Can Prove Costly

There is an interesting story behind a long series of randomly occurring thrust-bearing failures in one particular type of slurry pump in service at a South American bauxite mine. Apparently the thrust bearings would sometimes fail after a few days or, at other times, after a few weeks of operation. Before the mechanics produced a crossectional view similar to the simplified version depicted in Figure 10-4, the visiting troubleshooter had been told that it was often necessary to rebush and line-bore the bearing housing. The relevance, accuracy, or importance of this verbal description becomes evident only when the drawing is examined in detail.

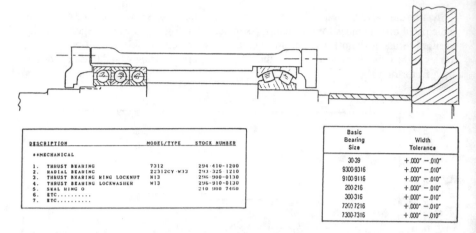

Basic Bearing Size	Width Tolerance
30-39	+.000″ −.010″
9300-9316	+.000″ −.010″
9100-9116	+.000″ −.010″
200-216	+.000″ −.010″
300-316	+.000″ −.010″
7200-7216	+.000″ −.010″
7300-7316	+.000″ −.010″

Figure 10-4. Cross-sectional view of slurry pump with failure-prone thrust bearing configuration.

With the impeller inverted so as to reduce the differential pressure across the shaft packing area, it is immediately shown that the primary thrust is from right to left. The two angular contact bearings on the extreme left are correctly oriented to take up the predominant load. However, their outer rings are completely unsupported because the fabricator had somehow decided to overbore the housing in the vicinity of these two bearings. Consequently, the entire radial load acting on the coupling end of the pump had to be absorbed by the remaining third angular contact thrust bearing. This bearing was thus overloaded to the point of rapid failure and was, of course, prone to rotate in the housing. Using a double row spherical roller bearing at the hydraulic end of the pump would normally make for a sturdy, well-designed pump. In this case, however, the spherical rotation or compliance feature tended to further increase the radial load transferred to the one remaining outboard bearing. The basic agent of the component failure mechanism was, of course, force.

An equally serious burden was imposed on this pump by the well-intentioned person who, in an effort to link the spare parts requirements of the North and South American plants of this major aluminum producer, added to the drawing the parts list partially reproduced in the lower left-hand corner of Figure 10-4. Having left off the appropriate alphanumeric coding behind the bearing identification number 7312, this plant and its sister facilities would receive thrust bearings in other than matched sets. A quick look at the bearing manufacturer's dimension tables (see insert, Figure 10-4) shows simple type 7312 bearings to have a width that may differ from the next bearing by as much as 0.010 inches. Mounting two such bearings in tandem may cause one to carry zero to 100% of the load, while the other one would simultaneously carry 100 to zero % of the load. On the other hand, matched sets intended for tandem mounting would be precision-ground for equal load sharing (50/50%) and would be furnished with code letter suffixes to indicate this design intent.

Did we go through our seven cause categories to identify the above root causes? Frankly, no. When both the fabrication sketch and the procurement documentation—"information *processing*"—show two very obvious errors, it is reasonable that rectification of these deviations should be a prerequisite to further fine-tuning. Which is just another way of saying that if it looks like a duck, walks like a duck, and quacks like a duck, we ought to call it a duck and dispense with further research into the ancestry of the bird.

Operations Errors Can Cause Pumps To Malfunction

The next problem involved a 2000 kw (2660 hp) vertical pump in seawater service. The equipment owner reported flow delivery problems occurring somewhat randomly, but more typically at startup or commissioning of downstream fixed equipment. Associated failures often involved pump bearings.

Our analysis strategy was representative of the "Seven Root Cause Analysis" approach: Collect all available information, eliminate the less likely causes, concentrate on the one or two most probable ones. We reviewed the plant's vibration records, repair history, and spare parts consumption. Next, we examined the wear patterns on bearings and impellers. Since the pumps had been designed and manufactured by an experienced North American manufacturer and were known to perform quite well at three competitor's plants, the cause category "Faulty Design" was ruled out first. Since the operating facility had neither fabricated nor reprocessed any of the pump components, "Fabrication or Processing Errors" and "Assembly or Installation Defects" were also considered unlikely contributors.

The average maintenance requirements for this vertical pump were straightforward and component materials of construction were in harmony with satisfactory experience elsewhere. This took care of cause categories "Maintenance Deficiencies" and "Material Defects," leaving us to pursue only "Off-Design or Unintended Service Conditions" and "Improper Operation." Figure 10-5, plotting pump torque and delivery head against the rate of flow, rapidly led to the answer, as did an operator's comment that "these pumps are being started just like all other pumps."

While centrifugal pumps arranged for manual control are usually started against a closed delivery valve, this is not the case with propeller pumps. The starting torque of these pumps, T_s, may be considerably greater than the full-load torque at operating point B. Propeller pumps may thus have to be started against open delivery valves. The customary check valve in the discharge piping opens at a comparatively early stage, ensuring that flow commences before the full operating speed has been reached. An example is given in Figure 10-5, with all values plotted as a percentage of the conditions

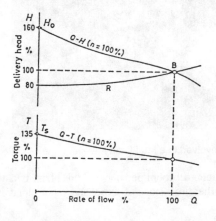

Figure 10-5. Plot of torque and delivery head vs. flow rate in vertical seawater intake pumps.

prevailing at point B. Assuming the characteristic resistance curve R starts at 80% of the anticipated total head at B, and with the pump characteristic curve H rising to 160% at zero discharge, the pump delivery will be delayed. With the check valve still closed, pump speed must be increased to the point where H_o has reached 80% of the total head at point B.

The principal cause category in this failure example was improper operation. Tackling the problem from a parts failure point-of-view, we could have examined the slightly worn bearings first. Noting an absence of discoloration, we would surely conclude that temperature excursions were not at fault. Comparing the relatively short bearing life to demonstrably superior run lengths achieved under identical conditions elsewhere would allow us to rule out the time factor, and verification of the suitability of the bronze bearing composition for sea water would have eliminated reactive environment from our "FRETT" list. Occasional bearing distress should thus be attributed to force. High forces on pump bearings could originate with unbalance, low flow, cavitation, and inadequate available NPSH. The randomness of vibration is the clue to flow-related operational problems.

Maintenance Omissions Can Cause Loss of Life

One of the more tragic pump failure incidents occurred at a hydrocarbon processing plant in the U.S. Gulf Coast area in 1982. It involved the pump shown in the foreground of Figure 10-6. Note also the pump baseplate, Figure 10-7.

Figure 10-6. The pump in the foreground was involved in a fatal flash fire event.

When a pump malfunction was detected by control room personnel at this plant, two operators went to the area and realized from the dimensions of a propane vapor cloud

Figure 10-7. Pump baseplate after removal of defective pump. Note coupling element; note piping without expansion loop.

that the pump had to be shut down. As they approached the equipment, the vapor ignited, causing both men to suffer extreme burns. One of the two operators later died.

As is customary in such cases, an effort was later made by a local expert to reconstruct the event and determine the cause of the fire. His report noted that the mechanical seals had received flush liquid via API Plan 31, i.e., recirculation from the pump case through a cyclone separator delivering clean propane to the seal and, in his own words, certain amounts of entrained water to the pump suction. After examining the pump internals (Figure 10-8), the local experts determined

> that a failure of the pump occurred and that the dynamic forces distorted the pump in order to throw the rotating section between the two bearings out and away from the center line of the shaft, causing it to wear on exactly the same side throughout the length of the pump inside the pump casing.
>
> It is the opinion of the [local expert] that a pump failure of this type with these results could not be anticipated by the operating personnel of [the owner's plant] and that there was no failure of adequacy of instruction to operating personnel at the plant. It is the further opinion of the [local expert] that the two leaks that were present and that ignited after the pump fire occurred were the type of leaks that are normal to the operation of a plant of this nature and that they, in and of themselves, are not indicative of any failure of either good engineering practices or proper maintenance.

Figure 10-8. Multistage pump rotor after failure incident. Note portion of ball-bearing cage missing (foreground).

Authorized by the owner, the local expert supervised the removal of seals, bearings and impellers (Figures 10-9 through 10-13), in efforts to find a crack or cracks in the shaft material, which, he reasoned, might have initiated the catastrophic failure of the pump.

Together with 5 or 6 other equipment and component manufacturers, a repair shop that had worked on the pump 4 years prior to this incident had to defend itself in court. The attorney representing this pump-repair shop engaged one of the authors and requested reviews of depositions and photographs to prepare defense arguments.

Although the plaintiff's expert had already gone on record with the statement that pump failure originated with a crack somewhere in the pump shaft, our review effort was again aimed at eliminating at least some of the seven principal failure cause categories by assembling as much pertinent data as was possible at this late stage in the investigation. The category "Off-Design or Unintended Service Conditions" was ruled out on the basis that similar pumps had been installed elsewhere and had operated well under similar conditions. "Improper Operation" did not appear likely since the unit was running normally and the pump in question was not in a startup or shutdown phase at the time of the incident. "Faulty Design" was not judged likely in view of the age and experience record of this pump model. "Fabrication or Processing Errors" and "Assembly or Installation Defects" were ranked somewhat more likely, and "Maintenance Deficiencies" and "Material Defects" tentatively and somewhat arbitrarily put at the top of our list. *(text continued on page 585)*

Figure 10-9. Mechanical seal surface indicating contact with stationary pump internals.

Figure 10-10. Double row (outboard) bearing and seal gland.

Figure 10-11. Impeller after careless disassembly.

Figure 10-12. Careless disassembly of defective bearing.

(text continued from page 582)

Next, a site visit was arranged. As is appropriate when using a conscientious systems approach that includes a review of all relevant component parts, a box of broken parts was examined at the plant site (Figure 10-13), but no coupling components were found.

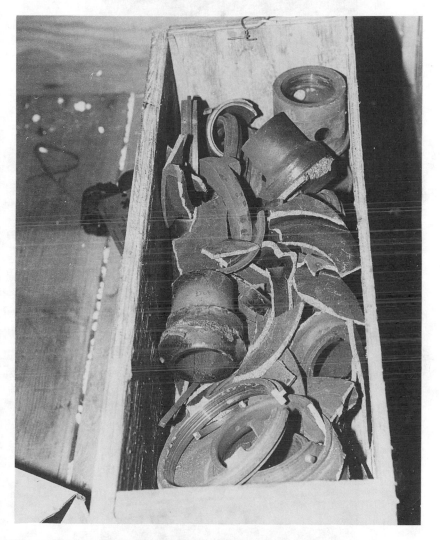

Figure 10-13. Impellers after careless removal from shaft.

Since a motor-to-pump coupling is, of course, part of the system, we judged it important to review its condition. Fortunately, the coupling was found on the ground in close proximity to the pump base (Figure 10-14) and serious wear was immediately evident. Figure 10-15 shows a wear pattern similar to the one found on this coupling. Other

Figure 10-14. Coupling part found in close proximity to pump base.

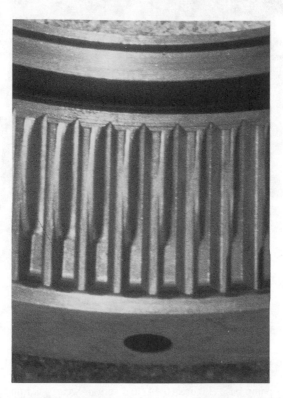

Figure 10-15. Coupling tooth wear patter of the type observed in failure incident.

pertinent observations rapidly followed and led to a rather concise summary report of the most likely sequence of events at this facility. Note our "points of evidence":

1. A combination of misalignment and lack of lubrication in the gear coupling very probably led to excessive vibration.
   Points of evidence: Severe ridges were visible in the softer of the two mating gears; no traces of lubricant were found in the drive-side coupling on the failed pump set; also we found no traces of lubricant in the couplings of both adjacent identical pump sets.
2. High levels of vibration and severe misalignment probably caused crack propagation in each of the four pump support legs. It was noted that a similar crack had been repair-welded on one support leg of an adjacent identical pump.
3. At this time a combination of shaft misalignment and coupling inflexibility is thought to have caused amplified vibration, which led to shaft bow and internal rubbing.
4. Severe internal misalignment now caused the ball separator on the radial bearing (near drive end) to disintegrate (Figure 10-8).
5. Vibration next caused fatigue failure of a pipe nipple connecting the seal gland to the cyclone separator (Figure 10-10). This caused a massive spill of pumpage and also deprived the mechanical seal of flush liquid.
6. The bearing now reached a temperature of approximately 800°F; at about the same time, the mechanical seal faces began to heat-check and secondary leakage started to develop.
7. Hydrocarbon vapors and/or liquids with an auto-ignition temperature ranging between 450°F and 650°F ignited and a flash fire resulted.

Additionally, we noted that the entire pump installation had to be considered vulnerable due to the lack of thermal expansion capability of pump suction piping, lack of seal welding on pipe nipples filled with flammable liquid, and lack of coupling lubrication. Even visual observation allowed the observer to conclude that the shaft was bent. This bend was located near the hot bearing and followed the classic pattern observed by machinery engineers on the vast majority of pumps involved in this type of failure progression. Had the shaft been bent to begin with, this 3,600-rpm pump set would have exhibited abnormally high vibration from the time of commissioning. Evidence of rubbing of internal parts and also the cracking pattern on shaft sleeves were completely as anticipated in this particular event and were judged the consequence of the sequence indicated above.

A number of valuable lessons are contained in this story. First, very few failure events are the result of a single error or omission. What if the pump suction piping had been designed more flexibly and had not frequently have pushed the equipment out of alignment? What if someone had seen to it that the relatively heavy cyclone separator had been braced or supported differently or, better yet, had challenged its highly questionable usefulness in the first place? What if someone had decided that highly flexible non-lubricated or elastomeric couplings should be used on these pumps? Or what if someone had simply greased the gear couplings twice a year? This certainly would have vastly reduced the probability of *time* emerging as the basic agent of the failure mechanism causing serious coupling distress.

And, finally, from a failure analysis and troubleshooting point of view, how much more quickly would the most probable failure sequence and its root causes have been uncovered if someone had used a more reasonable and well-structured failure analysis approach?

Awareness of Off-Design and Unintended Service Conditions Needed to Prevent Failures*

In a failure incident in early 1990, the twin screws of a large rotary positive-displacement compressor (Figure 10-16) experienced serious damage during restart after an

Figure 10-16. Rotary positive-displacement machine (screw compressor) being opened to inspect damage.

unrelated downtime event. On this particular machine, the male screw is driven by a steam turbine through a reduction gear box, and has four lobes wrapped axially in a helical manner around a central core/shaft. The male rotor drives the female rotor through timing gears located at the back of the rotors. These gears set the timing as well as the clearance of the rotors. The operating clearance between the male and female rotor is approximately 0.010″ (0.250 mm).

The speed range of the male rotor is 1,362 rpm to 3,710 rpm. This is equivalent to a speed range of 2,533 rpm to 6,900 rpm for the steam turbine.

On the day of the incident, the compressor was lined up and put on slow roll at 900 rpm. It slow rolled for about 90 minutes when the speed was gradually increased with

---

*Source: Hurlel Elliott, Saudi Petrochemical Company. Adapted by permission.

the trip/throttle valve (TTV) up to 2,500 rpm, at which point the TTV was wide open. The deck operator then gave control to the console operator to continue increasing the compressor speed. About 13 minutes later, during the speed increase, radial-vibration alarms were experienced, followed by high-discharge-temperature alarms. The machine subsequently tripped on high discharge temperature, and locked up following the trip.

Efforts to free the rotors were partially successful; however, a tight spot was evident while turning the rotors manually. The discharge silencer was removed for further inspection, at which time metal shavings were found and the decision made to remove the upper half casing. On removal of the upper half casing, both rotors were found damaged and had apparently contacted each other. Damage was confined to the rotors; all other components were in excellent condition (Figure 10-17).

Figure 10-17. Damaged male rotor (top) and female rotor (bottom) after extended operation in the torsional critical speed range.

Since this machine and similar compressors had been giving very satisfactory service before, a basic *design flaw* was considered unlikely. *Material defects* were ruled out because rotor metallurgy and surface conditions were almost certainly within acceptable limits.

Unless new parts are installed in an existing machine, or a new machine has not undergone a suitable precommissioning test, *fabrication or processing errors* are not a priority concern. Similarly, it was reasoned that *assembly* or *maintenance deficiencies* were not a factor in the failure event of a straightforward rotary screw compressor that had always been properly maintained by an experienced workforce.

Using the "seven cause category approach," the focus now shifted to the possibilities of *improper operation* and *off-design or unintended service conditions*. Operating parameters in the time frame leading up to, or at the time of, failure were reviewed with process technicians. It was established that compressor speed variation was accomplished through a mechanical speed governor by inputs from either the process computer or a mechanical knob located on the steam turbine governor. The compressor had *lateral critical speeds* outside the normal speed range; however, there were two *torsional critical speeds* within the operating speed range. Lateral critical speeds are speeds where the rotor's natural frequencies are excited, causing the rotor to vibrate in a side-to-side/up-and-down manner. Torsional critical speeds are similar to lateral critical speeds, except that the rotor vibrates in an oscillatory/twisting fashion about its polar axis. The speed ranges that are considered critical during start up are 600 rpm to 820 rpm and 1,700 rpm to 3,000 rpm.

Figure 10-18 represents data captured by the process computer. These one-minute snapshots indicate that the compressor train was operating in the torsional critical speed range for approximately 19 minutes. Vibration excursions experienced during this time period are shown in Figure 10-19; these would indicate severe rotor contact in the

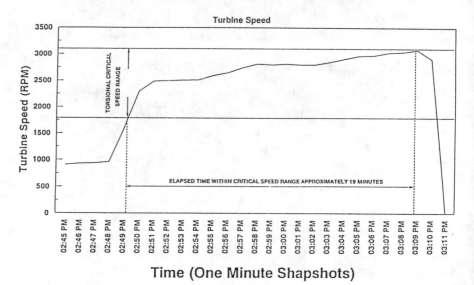

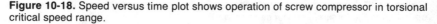

Figure 10-18. Speed versus time plot shows operation of screw compressor in torsional critical speed range.

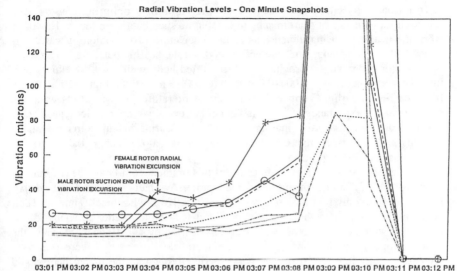

Figure 10-19. Vibration excursions during operation of screw compressor in torsional critical speed range.

3-minute time span preceding shutdown of the machine. The extent of torsional-vibration-induced damage can be seen in Figure 10-17.

The equipment owners knew that the elapsed time in going through critical speeds should be held to less than one minute. Staying within the torsional critical range for a considerably longer time caused the rotors to contact. Contact of the rotors results from one rotor (male) oscillating in an angular/twisting fashion, changing the rotor clearance and causing both rotors to touch.

The root cause of the problem was conclusively traced to an erroneous setting of the minimum governor speed, e.g., 2,500 rpm instead of 3,000 rpm.

Reduced Life and Catastrophic Failure of Electric Motor Bearings

A 200 hp induction motor suffered from bearing failures every 10 to 18 months. The seven possible root causes of reduced life and, ultimately, catastrophic failure of the coupling-side rolling element bearing were examined; the intent was to eliminate cause categories that were judged unlikely to apply. The first causes to be eliminated were "Operator Error" and "Unintended Service," because a running motor does not usually require operator attention, and it was easily established that the service conditions did not deviate from the accepted and prevailing norm. "Maintenance Deficiencies" were ruled out because grease type and regreasing intervals met expectations. A fundamental "Design Flaw" would have made little sense; numerous motors of the same make and characteristics had accrued decades of successful service at this plant site. And "Material Defects" on a succession of 65 mm bore Conrad bearings from the world's leading manufacturer was judged too unlikely to pursue. This left "Assembly and Installation" and "Fabrication and Processing."

To retroactively check on "Installation," the vibration monitoring records were reviewed for evidence of shaft misalignment, but no such data were found. However, when the bearing housing dimensions came under close investigation, it was determined that "persons unknown" had accommodated the bearing outer ring to an over-bored motor end cover by providing a shrink-fitted hoop with a wall thickness of ⅛ inch. This steel sleeve was 0.004–0.005 inch (0.1–0.13 mm) tight. Figure 10-20, taken from the SKF Bearing Company's commercial literature (Ref. 3) alerts us to the exponential life expectancy effect of having either excessive internal looseness or, as was the case here, excessive internal preload. The bearing manufacturer calculated that this preload, or negative clearance, amounted to 0.0023 inches (0.05 mm) and was responsible for the problem.

We might add that the immediate cause of this problem could have been found by applying "FRETT" in the manner explained earlier. We had established that *components* fail due to only four possibilities: *F*orce, *R*eactive *E*nvironment, *T*ime, or *T*emperature. Of these, we might have quickly eliminated "Time"—after all, many identical motors with identical bearings outlived this bearing by 4 to 10 years. Because the bearings operate under the same environmental conditions and were regularly regreased with a lubricant that gave good service throughout the plant, there would have been no reason to suspect the "RE"-portion of "FRETT," and this would have left "Force" and "Temperature." Either of these actions is associated with bearing load and, ruling out coupling-induced axial or vibration-related force, we would have pursued all possible measurements of the various bearing fits. We can be certain to have identified the problem in this manner, as well.

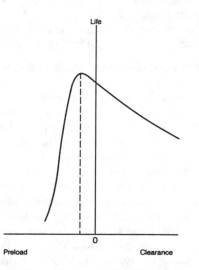

Figure 10-20. Excessive preloads and/or clearances will adversely affect bearing life.

The next step is for the failure analyst or machinery troubleshooter to remember the seven principal cause categories and to rank them in logical order. Using a process of elimination, the most probable cause categories or perhaps the ones that are most easily and rapidly screened are investigated first. The four basic agents of machinery component and part-failure mechanisms also have to be kept in mind. The final and most important review will then almost naturally focus on the one area that contains the root cause of a failure event.

References

1. Karassik, I. J., "Centrifugal Pump Clinic," *Chemical Processing*, September 1988, pp. 122–128.
2. Bloch, H. P., *Improving Machinery Reliability, Second Edition*, Gulf Publishing Company, Houston, Texas, 1988.
3. SKF America, Inc., General Catalog, Form 4000US, King of Prussia, Pennsylvania, 1991.

Chapter 11

Cause Analysis by Pursuing The Cause-and-Effect Relationship*

In recent years the reliability engineering movement has come into its own. The most successful companies have aggressively embraced the wisdom of prevention. They have learned that operating and maintaining reliable equipment requires new skills, new tools, and new knowledge—knowledge that includes predictive tools like Weibull analysis, histograms, Pareto charts, and effective failure analysis. Companies have learned that by understanding the mechanisms of failure, such as force, reactive environment, temperature, and time (FRETT), they not only prevent failures from *recurring,* they prevent failures from *occurring.* With this wisdom of prevention also comes insight into problem solving.

Successful companies continuously seek new approaches to problem solving. Effective problem-solving methods focus on creative solutions that go beyond our current paradigms and favorite solutions. To paraphrase Albert Einstein, we cannot use yesterday's thinking to solve today's problems.

Creative solutions add value because they meet our goals and objectives better than our favorite solutions. Many world leaders in process industries have been implementing a new problem-solving approach developed by Apollo Associated Services of Richland, Washington. They have broken out of the old-fashioned methods of brainstorming and other tired root-cause analysis schemes. They no longer use the guessing and voting of fishbone diagrams; instead, they are using a simple structured approach based on fundamental principles that can be understood and used by everyone in the organization.

This approach involves better problem definition, a simple charting process, and a new set of criteria to help us define the best solution. It applies structure to the creative process without restricting it.

With this approach, cause analysis by pursuing the cause-and-effect relationship, effective problem-solving skills now belong to everyone, although specific knowledge

*Courtesy of Dean Gano, Apollo Associated Services, Richland, Washington. Adapted from *Proceedings of the 1995 Process Plant Reliability Conference,* Houston: Gulf Publishing Company, 1995.

is still provided by various experts from the many craft workers to the plant manager. Less time is spent telling stories and analyzing the problem, and the time spent solving the problem becomes more productive. Companies using this problem-solving approach report non-recurring problems, decreased variability in production, and a 50% reduction in the number of corrective actions required. Furthermore, the corrective actions are more specific, with fewer time-consuming reviews and costly analyses.

In the following discussion, two types of problems will be examined that will lead us to a new and unique understanding of the cause-and-effect principle. With this new understanding, we will see how this problem-solving strategy not only provides effective solutions but more creative ones. Several examples will enhance our understanding of this sound approach.

Two Types of Problems

It is important to understand that there are two types of problems: specific-knowledge problems and event-based problems. Specific-knowledge problems, sometimes called rule-based problems, are founded on a pre-defined set of rules and, therefore, can have single correct answers. Event-based problem solving has no predefined set of rules and, therefore, may have many "right answers."

Rule-based problems are born of science and our need to control things. We establish conventions such as language, mathematics, and equations defining the laws of science. We agree that $2 + 2 = 4$, and that the words you are reading have a certain meaning, but this is only true because we have agreed on it. In rule-based problems, we seek the correct answer and the notion of truth is understood to be quantifiable.

Event-based problems, on the other hand, do not follow any convention or predefined set of rules. These are the day-to-day problems we all face, and range from finding a way to feed ourselves to how to fix a leaky valve. In event-based problems, we seek answers that work for us, not the "right answer." As an example, if our problem is hunger, the answer is not predefined nor is it singular. There is no true answer to the problem of hunger. While the solution appears to be to eat, there is much more to it, such as how to get food, what will it be, will it sustain us, and so forth.

The failure to understand these differences in problem types seems to keep us using the same old problem-solving strategies. In industry most problem solvers tend to believe all problems can be reduced to a rule-based problem type. Hence, rule-based problem-solving techniques are inappropriately used to solve event-based problems. We sometimes create forms and checklists based on predefined categories with the belief we will find the "right answer." By looking for the "right answer" to event-based problems, we subsequently limit our ability to find creative solutions. This rule-based thinking ignores the most fundamental of all principles: the cause-and-effect principle. This simple principle dictates a continuum of effects, an infinite set of causes, and many opportunities for creative solutions.

The Cause-and-Effect Principle

The cause-and-effect principle is not new. Experience in teaching it to many different people around the world indicates a large majority of adults, perhaps 80% or more, know about the law of cause and effect but not how to apply it. Consequently, effective solutions are only found about 30% of the time.

The characteristics of the cause-and-effect principle are:

1. Cause and effect are the same thing.
2. Causes exist as part of a cause-and-effect chain.
3. For most effects, there are at least two causes.
4. There is an infinite set of solutions to event-based problems.

Cause and Effect Are the Same Thing

When we look closely at causes and effects, we see that a "cause" and an "effect" are the same thing; like the sides of a coin. They are only different by the way in which we look at them.

For example, if we use an injury as the first, or primary, effect, and ask "why?," the answer may come back as: "because he fell." If we continue to ask "why," we may find that the cause of falling was a slick surface, and the cause of the slick surface was water, which was caused by a leaky valve, which was caused by an inadequate maintenance activity. As we look at the cause-and-effect process a little closer, we see that we cannot ask "why" of a cause, we must ask "why" of an effect. That is, the fall has to become an effect before we can ask, why did he fall? The cause, "slick surface," then becomes an effect, and the cause-and-effect chain continues. This can best be seen if we put the causes and effects in columns as shown below:

Effects		Causes
Injury	Caused by	Fall
Fall	Caused by	Slick surface
Slick surface	Caused by	Water
Water	Caused by	Leaky valve
Leaky valve	Caused by	Inadequate maintenance

While this characteristic may seem trivial, it is a cornerstone of the cause-and-effect principle. This characteristic explains the continuum of causes and effects on either side of our starting point. As a result, causes always exist as part of a continuous cause-and-effect chain.

Causes Exist as Part of a Cause-and-Effect Chain

Since cause and effect are the same thing, each time we ask "why," a chain of causes is created. Regardless of where we begin, we are always between causes. To understand the values of this characteristic, let's establish an arbitrary starting point and call it the primary effect. The primary effect is the effect we do not want to occur and is completely dependent on our perspective.

In the example above, for the safety engineer, the primary effect might be the injury. For the maintenance engineer, it may be the leaky valve. For the injured person, it may be pain. The primary effect is just an arbitrary starting point, and is your choice based on your own perspective of the problem. By recognizing that all causes exist as part of a continuum, we are no longer restricted by the rule-based approach that requires us to "recognize the right problem," before solving it. We can start with any reasonable effect (the pain or the leaky valve) with the full knowledge that by the time we are finished identifying the important cause-and-effect relationships, all causes will fit into our cause-and-effect chart.

As we continue to ask "why," we eventually come to one or more causes that, if acted upon, will prevent the primary effect from occurring. We call these causes root causes, and have now defined the key elements of the cause-and-effect process.

Primary Effect: The defined problem, or effect, we want to prevent from occurring.

Root Causes: The causes that if removed, changed, or controlled will prevent the primary effect from occurring.

There can be more than one primary effect in any event. Likewise, there can be more than one root cause for any primary effect. The primary effect is your choice, and is the beginning of asking "why?" A root cause is not necessarily at the end of a chain of causes, but is where you find solutions that act upon this cause to prevent problem occurrence.

It might be worth mentioning that the concept of one root cause at the end of a chain was developed by the U.S. Navy in the early 1960s as a means of categorizing the most probable cause. It was only used for tracking and trending causes, not analyzing them. In the search for better problem-solving methods, another government agency inappropriately adopted this simple categorization scheme. By the 1980s, without knowing the purpose of this scheme, the American Quality movement had adopted it as a root-cause analysis methodology.[1]

For Most Effects There Are at Least Two Causes

The next characteristic of the cause-and-effect principle is equally simple and profound: every effect can have at least two causes: an action or a condition. While the tendency is to associate causes with actions, finding causes associated with conditions provides a more complete picture.

For every effect we choose to focus on, we will normally find at least two answers to "why." Figure 11-1 illustrates this principle for the causes of a fire. A fire requires three conditions and one action as shown on the following page. The action brings the conditions together to cause the fire.

For the injury example discussed in the preceding section, we could also say the fall was caused by gravity and by being in an awkward position. Furthermore, we could say that being in an awkward position was caused by the person's not watching where he was going and not wearing shoes. For each cause/effect we list, we may find two or more subsequent causes, unless one of the causes is a fundamental law, like gravity; then we are forced to stop asking "why," because we do not know what causes gravity. Admitting we lack knowledge is very important. If we don't

[1]Management Overview and Risk Tree; INEL & EG&G Services, Inc., 1981.

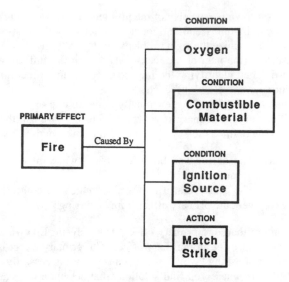

Figure 11-1. Cause-and-effect principle.

know why, we must stop! We need to learn that it is acceptable to not know. If we do not know why, then we should develop a plan to find the answers.

At this point, the cause-and-effect relationship for the injury example may look something like Figure 11-2.

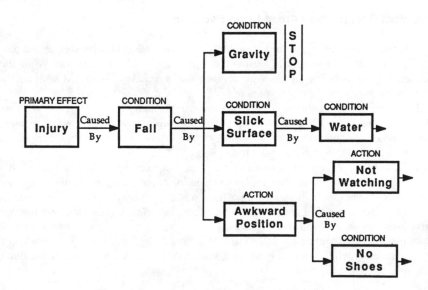

Figure 11-2. Expanded cause-and-effect relationships.

There is an Infinite Set of Solutions to Event-Based Problems

Since, in any given event, each "why" question can have two or more answers, the cause-and-effect relationship is not just a simple chain of causes and effects as first presented, but rather it is theoretically an infinite set of causes that grows exponentially. Figure 11-3 illustrates this concept.

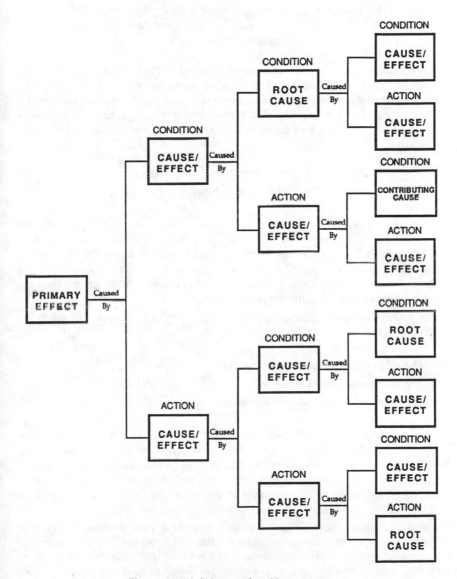

Figure 11-3. Infinite set of problem causes.

With this set of infinite causes comes an infinite set of solutions. This concept is important because it helps us overcome the ingrained belief that problem solving must lead to a single "right answer."

Figure 11-3 shows three root causes along with one contributing cause. Contributing causes are important, and are distinguished from root causes in that they only contribute to the problem.

Contributing Cause: A cause that if removed, changed, or controlled would have an important effect on a problem but would not prevent the problem from occurring.

Some caution is necessary at this point. Situations will arise where we may find several causes that must be removed or acted upon. In this situation, labeling a cause a root cause or contributing cause may create conflict. In this situation, do not waste time categorizing. Identify all effective solutions and proceed to implement them. After all, the purpose of problem solving is to implement solutions, not categorize causes.

Effective Solutions

"With an infinite set of solutions, it must be very time-consuming and tedious trying to find the right answer." Such a statement is the common response to the infinite-set concept, and reflects the inherent belief there is a single "right answer," when in fact, there is not. Again, we are looking for effective solutions. In rule-based problems, there is often a single right answer, but in event-based problems, there are often many right answers; some are just better than others.

After years of trying to define what we mean by the "right answer," the following criteria seem to best describe it:

1. The solutions should prevent recurrence.
2. The solutions should be within our control.
3. The solutions should meet our goals and objectives.

Since effective solutions exist where we find them, we need to be mindful of these three criteria at all times. It may not be fruitful to start down a path of causes when we know we have no control over them. It also makes sense to avoid a chain of causes that does not meet our goals and objectives, such as a solution that will cause other significant problems. Using these criteria is an important aspect of effective problem solving, and helps us know when to stop asking "why."

During a visit to a manufacturing plant where these new methods had been in place for almost a year, the plant manager confided, "It is a little scary letting go, but if I can successfully apply these criteria to the corrective actions I get from my employees, then I let it go—regardless of my gut feeling or opinions. I have not been disappointed yet."

Creative Solutions

At its lowest level, event-based problem solving only requires control of one cause that meets our objectives—this is often our favorite solution. It is the creative solutions, however, that provide the best return on investment while exceeding the customer's expectations. By using the cause-and-effect principle along with a few other tools, we can break out of the favorite solution box that restrains our thinking. After clearly defining each problem by identifying the What, When, Where, and Significance, a Cause-and-Effect Chart can be developed similar to the one shown in Figure 11-4.

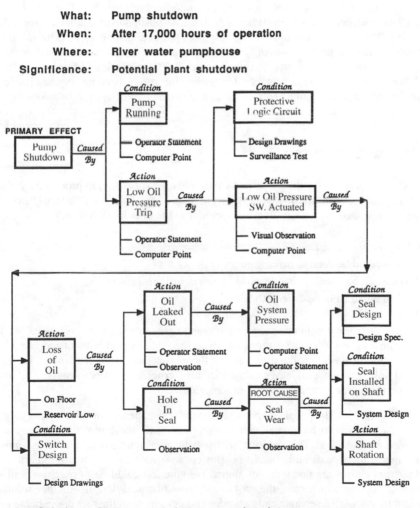

What: **Pump shutdown**
When: **After 17,000 hours of operation**
Where: **River water pumphouse**
Significance: **Potential plant shutdown**

Solution: Replace seal during annual maintenance outage.

Figure 11-4. Cause-and-Effect Chart format.

Developing the Cause-and-Effect Chart is a simple three-step process as follows:

1. Identify a primary effect and begin asking "why."
2. Connect all causes with the words "caused by."
3. Provide evidence for each cause.

For a given event, one or more problem statements are defined and include the primary effect. For each primary effect, we simply start asking "why." We continue to ask "why" until causes are found to which we can attach a solution and prevent the primary effect from occurring (root causes). The connecting words "caused by" are used to establish and confirm the cause-and-effect relationship. The requirement to show evidence levels the playing field because unsupported claims are soon discarded. This method precludes using opinions, political power, or favorite solutions when engaged in problem solving. Providing evidence demands a clear statement of how we know a cause exists. Rationalization like "engineering judgment" or "it is obvious" are no longer acceptable. Guessing and voting become meaningless and fall by the wayside. Blame and punishment are no longer offered as solutions unless they can be shown to prevent recurrence.

Challenging the Causes

After preparing a Cause-and-Effect Chart, creative solutions are proposed by challenging each cause. Starting from the right side of the Cause-and-Effect Chart, we spend a few moments asking questions similar to the ones listed below:

- Why is this cause here?
- What could be done to make it go away or not act?
- What can be done to remove, change, or control this cause?
- What could be done differently?

Without judging right or wrong or assigning value to the various solutions, we let the team's synergy flow. Solutions are expressed openly and without judgment. At some point in time, usually after most of the causes have been challenged, the proposed solutions are evaluated against the three criteria for effectiveness. The ones that best meet the criteria are then implemented.

In Figure 11-4, the people responsible for the consequences of failure have judged the best solution to be annual seal replacement, and, therefore, the root cause is defined as seal wear. This solution will prevent recurrence, is within their control, and meets their goals and objectives. But because root causes are solution-driven, other root causes are possible. An alternative solution could be to purchase a more durable seal, in which case, the root cause would be seal design. Another solution might be to install a separate and redundant oil system, with the root cause being loss of oil. Creative solutions are found by first knowing the causal relationships, challenging these relationships, and offering solutions that best meet the three criteria.

Success Story

The example below illustrates the outcome of putting this effective problem-solving approach to use. When the problem first occurred, one group of engineers followed conventional wisdom right to their favorite solution.

The problem was a high thrust-bearing temperature on an 800 horsepower river water pump motor. This is a vertical pump and motor, at a remote location, and the original investigation found river debris restricting the orifice in the cooling water line to the oil reservoir. The cooling water came from the pump discharge. To "solve" this problem, the engineering staff implemented their favorite solution for plugged water lines. They installed a Y-strainer. With this solution, each time a high temperature alarm occurred, operations would send an operator several miles to the pumphouse to clean the strainer.

A few years later, an Apollo-trained reliability engineer formed a new team to investigate this problem. Using out-of-the-box thinking and the Cause-and-Effect Chart, they found a much better solution. This is shown in Figure 11-5:

After evaluating the various proposed solutions, the one that best met the three criteria (prevents recurrence, is within their control, and meets their goals and objectives) was to use a different source of cooling water, which they found about 3 feet away. The new source of cooling water was not only clean, it was at low pressure, which allowed them to remove both the restricting orifice and the Y-strainer. Contrary to the favorite solution of adding a Y-strainer, this one did not add to the maintenance and operational workload. It also improved the reliability of the motor by removing equipment and providing a better cooling water source.

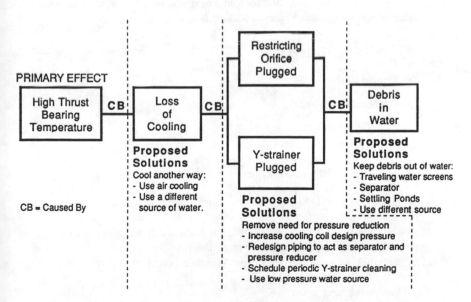

Figure 11-5. Creative solutions may solve problems more completely.

Try to get out of your box and find more creative solutions. Challenge the causes in the completed Cause-and-Effect Chart, offer solutions to each cause, and evaluate the solutions against the three criteria.

Integration of Different Failure Analysis Methodologies

There are many instances in which the application of the cause-and-effect principle has resulted in finding optimum solutions to problems. The authors found it especially helpful to integrate the cause-and-effect approach in the Generalized Failure Analysis and Seven Root Cause Analysis methodologies. By way of summary, then, we offer the following:

1. Define the problem (Situation Analysis)
   What's the problem? Identify areas requiring your action.
2. Select 1 or 2 of only seven groupings of possible cause categories: (Preliminary Machine Failure Cause Analysis through Process of Elimination)
   Why did it happen? Discover the critical cause-effect relationship. *All* failures are traceable to one or two of these seven:
   - Faulty design
   - Material defects
   - Fabrication or processing errors
   - Assembly or installation defects
   - Off-design or unintended service conditions
   - Maintenance deficiencies, including neglect, procedures
   - Improper operation

   Components fail *only* due to four possibilities: (Preliminary Component Failure Analysis through Process of Elimination)
   - Force
   - Reactive environment
   - Time
   - Temperature

   Ask repeatedly "why?"

   > Cause → Effect
   >> Cause → Effect
   >>> Cause → Effect
   >>>> Cause → Effect

   Analyze the data. Do some "what if" questioning. Is it plausible or extremely far-fetched?
3. What are some innovative, corrective, opportunistic, adaptive, selective, preventive, or contingent solutions and remedial steps?
4. Decision making, reaching consensus: Which action steps best meet your objectives?
   When do we have to act?
   Who is doing what?

5. Planning for change: Develop an action plan that goes beyond the obvious. Ask what could go wrong and how do we protect against this eventuality.
   When do we have to act?
   Who is doing what?

As more personnel within an organization learn this simple structured, and yet comprehensive, approach to problem solving, a questioning attitude begins to permeate the work culture. With this new attitude comes the understanding that things do not just happen and that if we look at causes and effects closely, we can prevent problems from occurring in the first place, which is the goal of any reliability program. At the same time, the time and money saved not having to resolve emerging issues can be used for positive business activities such as increasing productivity or upgrading equipment.

Chapter 12

Knowledge-Based Systems for Machinery Failure Diagnosis

In Chapter 4, page 267, we alluded to the possibilities of computerized machinery troubleshooting and fault diagnosis. Because of the relationship that exists, for instance, between the vibration pattern and machinery distress, it is possible to correlate results of a vibration spectrum analysis directly with the causes of machinery problems. We have shown that it may be relatively simple to arrive at accurate and detailed spectrum analysis results, but it can be difficult to interpret them quickly and successfully within the context of a specific machine and its particular operating environment. We know that there is a great deal of knowledge and experience required, which usually belongs to the domain of senior specialists and consultants. This requirement can be both time consuming and costly, and if we are confronted with a problem that does not occur all that frequently, even an expert has to go through another learning period.

to the domain of senior specialists and consultants. This requirement can be both time consuming and costly, and if we are confronted with a problem that does not occur all that frequently, even an expert has to go through another learning period.

Recent developments in an area called *artificial intelligence* have made it possible to collect domain expertise, in our example, vibration diagnostics, and apply it whenever or wherever it is required in the form of a computer software program. A computerized collection of machinery troubleshooting "rules," for instance, can thus provide expertise and judgment to inexperienced personnel in order to raise their level of efficiency.

What then is artificial intelligence?

"Artificial intelligence is an attempt to program computers to perform tasks, which, when they are performed by people, are considered to be intelligent behavior—playing checkers, chess, understanding sentences, finding mineral deposits, diagnosing diseases, and so on."[1]

Figure 12-1 illustrates how artificial intelligence covers a wide range of areas, from language processing to robotics. In machinery troubleshooting and distress diagnostics, expert systems or knowledge-based systems appear to have a most promising future.

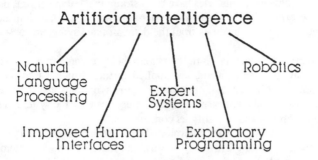

Figure 12-1. The commercial offspring of artificial intelligence[2].

A knowledge-based system has two primary components: an inference engine and a knowledge base. The inference engine is the simple reasoning mechanism that the computer uses: it looks for rules in the knowledge base to apply and then performs the appropriate actions. The knowledge base will contain a series of "if . . . then" rules. For example, one might have the following rules:

- If the vibration on all bearings is excessive, then replace the shaft.
- If the vibration on only one bearing is excessive, then replace that bearing.

Other rules might well be the rules—based on likelihoods—offered in the troubleshooting tables of this text.

Using rules of the style mentioned above, knowledge-based systems have been written for medical diagnosis, computer configurations, electronics troubleshooting, predictive control systems for plant operations, maintenance planning, and many other areas.

To date, most knowledge-based systems for machinery fault diagnosis or condition monitoring require the user to read values off various test instruments and manually type them into the knowledge base. When the system requires more information, it requests the user to measure the required values; these measurements must be made manually and the value entered. With current computer and interface technology, the requirement to enter data manually will soon be eliminated. Most recent developments indicate that on-line expert systems are about to become reality.[3,4]

Examples of Knowledge-Based Systems

The *Rotating Equipment Vibration Advisor* (*REVA*)[5] is an off-line diagnostic expert system. REVA will typically first ask the user for some general information about the particular piece of equipment (such as running speed and mechanical configuration). It follows this basic data input stage by asking for certain vibration data. Either the user enters this data manually, or the system accepts them directly from a vibration analyzer in the form of a vibration frequency spectrum.

Next, the system interprets the data to determine if there is a vibration problem; if there is, it makes a detailed analysis to uncover the cause. If the system needs additional information, it asks for it, just as a consultant would. It asks such things as whether or not the machine was difficult to balance and when it was last overhauled.

If equipment maintenance histories have been stored in a computer, this database can be linked to the expert system. Thus, the system has access to measurements from the data collector, maintenance histories from the database, and symptoms observed by the user.

Similar to our troubleshooting charts the system then presents a ranked list of probable causes. Statistical confidence factors are applied to this listing, based on the human expert's assessment of the relative importance of the symptoms and their relationships with the observations and the measurements. The user can ask the system to explain the reasons for its questions and to justify its conclusions.

ROMEX[6] is an expert system being developed by the Rotating Machinery and Controls (ROMAC) Laboratory of the University of Virginia. It is a rule-based system developed for vibration-oriented diagnosis of turbomachinery, for fault identification and for predictive maintenance. The system is implemented in a PC-based PROLOG environment, with belief functions utilized for evidential support of hypotheses. PROLOG is used for knowledge representation, rule interpretation, control strategy, and user interaction.

The vibration-fault diagnosis system is considered to be one component of a comprehensive system for turbomachinery. The framework of this comprehensive system comprises hierarchical levels of generic rules or surface knowledge, and generic analytical simulation models, also referred to as "deep knowledge." The root level includes the surface and deep knowledge for vibration, bearings, lubricants, seals, gears and couplings, and mechanical/metallurgical aspects of fault detection. Another level comprises the generic but specific knowledge base for various categories of turbomachinery, i.e., pumps, compressors, turbines, and engines. The third level includes the installation-specific rules, maintenance, repair, and troubleshooting logs, and other specific use experiences. Each component of the comprehensive system can be considered a distinct expert system that can be developed and utilized independent of the other subsystems.

From the foregoing it becomes clear that knowledge-based systems will gain increasing importance in many areas of problem solving. Rather than dwelling on the advantages and disadvantages of a particular present system, it would be well to consider some basic rules for the application and selection of knowledge-based systems.

Identification and Selection of Knowledge-Based System Applications*

The first application of a new computer technology to industry has a high probability of failure, especially within an industry as conservative as the electric utility industry. Artificial intelligence technology is especially risky since its glamorous reviews by the media may lead to inflated expectations by laypersons. An example of a similar situation is the first application of mainframe computers to control of power plants during the 1960s. This application of computers did not meet the industry's expectations and was considered a failure. That failure caused, one decade later, the delayed application of distributed control technology to the power industry.

It is important for managers and technologists to carefully select first applications of expert systems and to nurture the fledgling applications. In the following, we are showing a semirigorous process for the identification and selection of candidate expert systems to improve the success rate of applications of expert systems. The process quickly focuses

---

*See Ref. 7. Adapted by permission.

on the applications that have the highest probability of success. It also fosters a good understanding of the candidate applications and allows the team to rank expert systems. From the information gathered during the process, estimates can be made of the resources and costs required to develop the systems.

Front-End Planning

The most important tasks in the selection of expert-system applications are the planning of the project and definition of goals. These goals are used to establish the expert-system selection criteria and criteria weighting factors. The planning facilitates cost-effective and proper communication with the sponsors, candidate champions, and end users. All of these factors are critical to successful application of a system that changes the normal work process or the flow of information, which typically includes expert system applications.

Expert system selection criteria are:

1. To select expert systems applications that would demonstrate the applicability of the technology to the organization.
2. To select applications with very high probabilities of success.
3. To not necessarily select the "best" applications for the company.
4. To use a methodology that would be defendable.

Candidate Identification Process

The process used to identify candidate expert systems will vary from organization to organization depending upon its structure, experience in using expert system technology, preferences of management, attitudes of end users, and many other factors. In general, a method for identifying the "champions" or sponsors of the expert systems candidates must be identified; candidate selection criteria must be established; and, based upon the selection criteria, an expert system candidate application form with questions must be developed.

A broad cross-section of a user's organization might be canvassed for applications by holding educational seminars to introduce the concept of expert systems. The seminars could be followed by brainstorming sessions, during which the attendees and the selection project team openly exchange ideas. All ideas should be encouraged in an informal and positive setting. One application form may be filled out for each application.

Selection Criteria

The development of the selection criteria draws heavily upon a set of attributes indicative of a good application of expert systems. Those attributes led to the development of a questionnaire that was used by the expert system sponsor and knowledge engineer to define and evaluate the expert system candidate. The evaluation team used the selection criteria to assign a weight factor to each question. The weight factors were used to determine scores for each application.

Each organization using this methodology should assign weight factors to each question. The weight factors should support the selection criteria. So, for example, if a criterion for selection was to pick an application that demonstrated the technical adequacy of a certain expert-system shell (a "shell" is a commercially available computer program to accept the knowledge base of a given domain expert), then the questions relating to "payback" might have a small weight factor. The essence of the attribute set and the selection criteria are embodied in the "Candidate Description Application Form," the six-question "Go/No-Go Filter," and the 44-question "Expert-System Questionnaire," which follow.

The process for selecting the candidate ideas consisted of applying the rough Go/No-Go Filter to quickly eliminate the less viable applications. The remaining ideas were evaluated through two iterations of interviews and ranking using the Expert-System Questionnaire.

The highest ranking ideas passed one more level of interviewing and scoring prior to recommendation for development and funding.

Candidate Description Application Forms. This one-page form was designed to help focus the thoughts of those submitting the candidate application, and to provide information to be used later to perform the first rough-screening process.

The following questions were posed for each application:

1. Describe the problem and how and by whom it is currently solved.
2. Describe the benefits of the expert system.
3. What is the length of analysis time required to solve the problem?
4. What group of employees would use the completed expert system?
5. Will they be receptive to using it?
6. Who are the current experts?

Seminars were developed to introduce employees, including management, to expert-system technology and to explain its limits and applicability. The seminars were designed to stimulate employees to generate ideas for consideration as candidate applications.

Next, expert systems that were then presently being developed and applied elsewhere within the industry were described. The content and purpose of the Candidate Description Application Form and the questionnaires also were explained.

Brainstorming sessions were then held following each seminar to permit employees to discuss potential applications with the project team, a group of recognized authorities in the development of expert systems. These sessions served to apply the initial concepts presented in the seminar to specific employee ideas, helping them to more accurately complete both the Candidate Description Application Form and the Expert-System Questionnaire.

Six-Question Go/No-Go Filter. A "Go" or "Yes" response was required for all six questions for any candidate to be considered for further and more detailed evaluation. The filtering process allowed the evaluation team to quickly focus on only those candidate applications with the highest probability of success as an expert system.

A discussion of each of the six questions follows:

1. **Does a human expert currently solve the problem?** In general, expert systems are not designed to solve problems that are intractable to human experts. Thus, it is inappropriate to select such an application as a demonstration of the first generation of the technology in an organization.

   On the other hand, a good application for expert systems is one in which a few recognized experts solve a problem very well. The intent is then for their knowledge to be encoded into the expert system and to be made available to the users of the expert system.

2. **Is a recognized expert available?** If the person solving the problem is not recognized as an expert, then the advice or knowledge encoded in the expert system will be suspect and will not be accepted by either the end users or the users' management. If more than one expert is recognized, then only one must be selected as arbitrator of what knowledge is recorded in the expert system. This arbitration will preclude conflicts of knowledge.

3. **Are conventional algorithms inadequate?** If conventional algorithms are adequate, they should be used instead of expert systems because they will be more readily accepted and they will have a greater likelihood of producing a successful solution.

4. **Is the application of appropriate complexity?** A trivial problem is not a good demonstration of the value of the technology and would not foster its acceptance. An unduly complex problem solution generally is expensive and has a higher probability of failure. A problem that takes an expert several minutes to actually solve once all of the information and data are available is considered optimal.

5. **Is there end-user support?** Since expert systems are new and considered by some to be a threat, special attention must be given to providing friendly user interfaces. Presupposing that there is a friendly user interface and that the end user is incapable or unwilling to use the system, the system will fail. If the end user does not think he or she has a problem or does not think the expert system proposed solution will benefit him or her, the expert system will not be used.

6. **Is there obvious payback to the sponsor?** Payback resulting from the application of each expert system is considered essential.

Expert-System Questionnaire. A 44-question questionnaire was developed to facilitate the ranking of the candidates who passed the Go/No-Go filter. The questionnaire was designed to be used by a knowledge engineer while interviewing the sponsor(s) of an application. The questions facilitated discussions by generating additional information needed by the evaluation team to more fully understand the problem and to more accurately determine the candidate's suitability for expert-system development. The 44 questions were grouped into six categories. Each question and a seventh category, "Subjective Assessment," were assigned scoring points as summarized in Table 12-1.

Table 12-1
Scoring of the Expert System Questionnaire

Candidate Subject Category	Number of Questions	Point Score
Overall Project	4	20
Problem Solved	15	50
Source of Expertise	10	30
Data Availability	3	16
End User	9	30
Pay Back to Corporation	3	24
Subjective Assessment	0	30
Totals	44	200

Following is a brief discussion of each of the candidate subject categories:

- *Overall Project.* Is the overall project well defined and will it result in an expert system? If the expert-system development fails to be completed on schedule, will it have a serious negative effect on the corporation?
- *Problem Solved.* How well suited is the problem for solution by expert systems? Is the application of appropriate complexity? What are the consequences of the expert system generating bad advice or errors? Is the domain stable? Is the problem focused?
- *Source of Expertise.* Is the expertise available and how valuable is it? Is the expert cooperative and capable of communicating knowledge? Will there be a conflict among multiple experts?
- *Data Availability.* Are data available to test and operate the expert system? How expensive is the data to acquire?
- *End User.* Is there management support for the development and use of the system? Will there be end user acceptance of the system? Does the system involve politically sensitive issues or data?
- *Payback to Corporation.* Will there be measurable short- and long-term payback?
- *Subjective Assessment.* Are there serious flaws in the candidate? Is there positive support from key players including sponsor, user, and manager? Does the manager have funds to develop the candidate? This category was used to adjust scoring for exceptional cases such as where error could cause loss of life.

As can be seen from the relative values of the categories in the questionnaire, the most important criteria used for selection of a candidate for development must be the appropriateness of the problem for solution by expert-systems technology, with the "problem solved" category accounting for 25% of the score.

The questions were posed to determine not only the ranking in each category but also if there was a fatal flaw in the candidate. The questions can be answered with a simple "yes" or "no," with room provided for recording additional information presented during the interview. Each question also has an explanatory sentence or paragraph to further define the nature of the question and the implication of the response. An example is:

Question:
"Is the project on the critical path for any other development effort or goal and is there any absolute milestone for completion?"
Explanatory note:
"The use of expert systems technology for real corporate applications is still relatively new, so any development has some risk. Thus, the less dependent the application is upon other activities the better."

All 44 questions are listed on pages 615-621.

Project Implementation

In a large public utility company, Houston Lighting & Power Company, dissemination of announcements and flyers about the project, along with two days of training and brainstorming sessions, proved to be sufficient to identify a set of superior candidates.

Seventy-two candidate ideas were generated during the two days of seminars and brainstorming sessions.

Twenty-eight ideas passed the initial filtering. The first set of interviews was conducted for the 28 candidate ideas. Several ideas were found to have serious shortcomings that precluded further analysis. The Expert-System Questionnaire was used to conduct another more detailed set of interviews. Based upon those interviews, the candidates were ranked, and the top 15 candidates were selected.

Re-ranking of the top 15 based upon additional information gathered during the interviews resulted in the top six candidates. A more detailed analysis including a budgetary cost estimate was performed for the top six. All six candidates were recommended for implementation and funding.

Review

Previous approaches to determining the feasibility of using expert-systems technology for a given problem have typically been informal. Usually the managers responsible for the decision assemble their most knowledgeable expert-systems resource (possibly an outside consultant) and some of the relevant personnel from the problem domain area. Following an informal discussion of the problem, the decision is then based upon the advice of the expert-systems developer. Based on what can be learned about the problem during the informal discussions and on prior development experience, the expert-system specialist can often make a sound judgment regarding feasibility. In some cases, sets of criteria have been developed for guiding the discussions and the decision process. However, in most cases, the decisions are still based largely on a gut feeling response.

On the other hand, a related area of research, "choosing alternatives," has developed a very rigorous methodology that could be applied to expert-systems feasibility decisions. Formal methods and algorithms for quantifying the decisions have been used extensively. Normalization across alternatives is one of the techniques utilized. These highly formalized methods have been reported enthusiastically by their developers. However, they require a high level of confidence in the assigning of weights and scoring of alternatives. (It may be that this high level of confidence is not attainable in the expert-systems feasibility decision-making process at this time.)

Rigorous methods were applied to this project, where appropriate, but the methodology was not completely formalized. Scores were assigned for each criterion and were normalized across applications, but were not done with the rigor described above. Since the methodology was semiformal, interpretation of results was also semiformal. The final numeric scores derived for each application were not viewed as highly accurate. Rather, applications were categorized as either recommended for funding, held for later consideration, or rejected due to a fatal flaw.

Benefits of Semiformal Evaluation

Until the initial expert-system development projects have been completed and evaluated, it will not be possible to evaluate the selection process, but it is the evaluation team's impression that applications were recommended with confidence, indicating successful meeting of goals. The effort invested in the more rigorous methods resulted in the following benefits:

Completeness. Every criterion is considered for each application. Informal discussions naturally consider only those criteria that come to mind. A formal checklist used to guide the discussion brings up each criterion for thorough evaluation.

Careful Consideration. Assigning a score for each criterion (quantification) forces the evaluation team to consider the criterion in greater depth. Not only does quantification force one to seriously think about how the individual criterion applies to the application under consideration, but it also motivates deeper thinking about the criteria themselves than was originally intended.

Normalization. Cross application consideration (normalization) forces the evaluation team to compare the scores and justifications for each application to the other applications. Normalization aids in the detection and correction of drift in scoring perceptions. It is not uncommon for evaluators to experience slow changes in the score they might give for a criterion during the course of evaluating multiple applications.

Independence. Normalization also promotes independence of criteria. Intense concentration on one application alone can introduce biases from one criterion to the next. By studying a criterion by itself, across multiple applications, the biases can be spotted and compensation can be made.

Understanding. The additional concentration and attention to details of each criterion leads to an improved understanding of the application.

Lessons Learned

In this company, the conclusion of this project was that expert-system technology is an appropriate solution to certain problems encountered in the company's operations and management. As mentioned, the top six candidates were recommended for implementation and funding.

This project concentrated on lower management and end users. More emphasis should be placed upon education of upper management from whom funding and support must be secured. This education process should be done prior to the training and brainstorming of the sponsors. Upper- and middle-management support for implementation would be minimal without a better understanding of the technologies' benefits and limitations.

Question 3 of the Go/No-Go Filter proved difficult to answer. Most responses included the elapsed time required to set up the problem and to submit the solution. In order to estimate the size of the problem, a more useful measure, yet difficult to ascertain, is how long the actual mental process takes to solve the problem.

For those systems candidates that are to be seriously considered, equal emphasis should be placed on the interviews with the sponsor, the end user, and the manager. Lack of enthusiastic support or misunderstanding of the scope of the application by any one of the principals will most likely cause problems during development and use of the application.

It is recommended that an interview be conducted with two interviewers concurrently. It is also helpful if one of the interviewers has a good understanding of the domain. The interviewers can then gather information and knowledge from the interviewed expert more quickly and can jointly assess the end user acceptance of the proposed expert system. This environment proves to be more synergistic than the one interviewer–one interviewee format. The interviews for each of the final six candidates took 16 to 24 hours to perform over a several week period. Liberal schedules are required to coordinate the availability of all principals.

Laypersons sometimes want to apply expert systems to very difficult problems such as real-time control or transient situations, resolution of political problems, automation of large complex designs, and solution of problems that previously have not been solved. Education of the sponsor and review of the application by a knowledge engineer will alleviate this predilection.

The introduction of a new technology, like expert-systems development, needs nurturing and protection to be used to its fullest and most cost-effective advantage. The semirigorous selection process just described may prove successful. The process provides an education to selected individuals and dollar savings that might otherwise have been spent on false starts.

Expert-System Questionnaire

Overall Project

Proj-1. Does the system development have the goal to deliver a knowledge-based expert system for actual use?   **Yes/No**

If the proposed system will not actually be used, its value as a demonstration will be limited. If the proposed system is the first stage of a more ambitious system, it is desirable for the first stage itself to be usable in order to properly demonstrate its value.

Proj-2. Is the task clearly defined?   **Yes/No**

At the project outset, there should be a precise definition of the inputs, outputs, and reasoning of the system to be developed. This is a good attribute of any task. However, it is not necessary that the task definition be fixed for all time. As the system evolves and as situations change, it should be possible to change the task definition accordingly.

Proj-3. Does an alternative solution to the problem exist, is one being pursued, or is one expected to be pursued?   **Yes/No**

If yes, explain.
If a solution is available off the shelf, it will probably be cost-effective to purchase it. If you are already pursuing this problem through conventional means, the need for a knowledge-based solution isn't as great. However, if a project goal is to compare expert-system technology to other technologies, this may be just what is desired.

Proj-4. Is the project on the critical path for any other development, and does the project have any absolute milestones for completion?   **Yes/No**

The use of expert-systems technology for real corporate applications is still relatively new, and so any development has some risk. Thus, the less dependent other activities are, the better.

Problem Solved

Prob-1. **Are conventional programming (algorithmic) approaches to the task satisfactory?**   **Yes/No**

Describe why or why not.

If a conventional approach will work well, there is usually less technical risk to using it rather than an expert-system approach. If most of the problem can be handled using calculation, database retrievals, and so forth, then it may not warrant the use of expert-systems technology. Note, however, that expert-system methodology may offer some additional advantages over conventional techniques, such as the expected ease of updating and maintaining a knowledge base and the ability to explain results. Symbolic processing of numerically computed values can be advantageous. If the effort is properly defined and partitioned, the best of both technologies may be exploited on the problem.

Prob-2. **Is the domain characterized by the use of expert knowledge, judgment, and experience?**   **Yes/No**

If the task requires the expert to reason about the facts available to him using heuristics, e.g., rules of thumb, strategies, etc., then it can probably benefit from the use of expert-systems technology. Cite some examples of the expert reasoning required.

Prob-3. **Does the expert typically make decisions based upon incomplete information?**   **Yes/No**

If all of the data is not available at a given time and a decision is still required, an expert system can be useful by reasoning as best it can despite the missing data.

Prob-4. **Does the expert typically make decisions based upon data that has degrees of certainty or confidence associated with it?**   **Yes/No**

Expert-systems techniques are available to reason about uncertain data when confidences can be assigned to the data values. If this is the case, an expert system might be appropriate.

Prob-5. **Is the task decomposable?**   **Yes/No**

If the task is modular, allowing relatively rapid prototyping for a closed small subset of the complete task, and then slow expansion to the complete task, this makes development much easier.

Prob-6. **Will errors made by the system be catastrophic for this problem?** **Yes/No**

Some percentage of incorrect or nonoptimal results should be tolerated by the system. The more tolerance for incorrect results, the faster the system can be deployed and the easier it will be to win system acceptance. For example, in a domain where even the best experts are often wrong, system users will not be as upset by an incorrect result from the system.

Prob-7. Is the problem domain stable?   Yes/No

Describe any changes to the domain knowledge that can be foreseen.

Expected changes may be such that they utilize the strength of expert systems (e.g., ease of updating or revising specific rules in the knowledge base), but will not require major changes in the reasoning process. An unstable domain may yield a situation where a large number of previously developed knowledge structures (e.g., rules) are no longer valid but cannot easily be changed without redoing the entire development process. Are unforeseen changes likely?

Prob-8. How long does it take the human expert to solve the problem (in minutes)?

If it takes the expert less than a couple of minutes to solve the problem, it is probably too easy. If it takes longer than two hours of actual work on the problem, it may be too difficult.

Prob-9. Does the problem require the expert to consider a large number of possibilities?   Yes/No

The knowledge base must be large enough to be interesting, or a decision tree might be a more appropriate approach.

Prob-10. What is a rough estimate of the number of important concepts related to the problem?

Even though expert systems have been built that deal with thousands of rules, it is better if they are limited to several hundred to ensure success.

Prob-11. Is the task sufficiently narrow and self-contained?   Yes/No

The aim is not for a system that is an expert in an entire domain, but for a system that is an expert in a limited task within the domain.

Prob-12. Is the skill required by the task taught to novices?   Yes/No

If the task is "teachable," and there is some experience with teaching the domain knowledge (and, ultimately, the system) to neophytes, such as the project team, then this usually means that there is an organization to the knowledge that can prove useful (at least initially) in building the system.

Prob-13. Has the problem solution been documented previously?   Yes/No

If there are books or other written materials discussing the domain, then an expert has already extracted and organized some of the domain expertise. As in the previous point, this organized knowledge might prove useful (at least initially) in building the system. Note, however, that one benefit of capturing an expert's domain knowledge might be to make a step toward formalizing a domain that has not been treated in a formal manner before.

Prob-14. Is the task similar to that of a successful existing expert system? Yes/No

Prob-15. Is there a primary requirement for a real-time response by the system?
Yes/No

Though it is certainly possible to develop a system for a problem's real-time requirement, bear in mind that this will divert effort from the primary task of knowledge acquisition.

Source of Expertise Category

Exper-1.   Is the expertise scarce?   Yes/No

If there is a dependence on overworked experts, then the system may help to relieve a serious corporate problem.

Exper-2.   Is the expertise expensive?   Yes/No

Exper-3.   Is the expertise available today, but will not be available, or be less available, in the future?   Yes/No

If the expertise is not available on a reliable or continuing basis, then there is a need to capture the expertise.

Exper-4.   Who is the expert that will work with the project? Will the expert be able to commit a substantial amount of time to the development of the system?   Yes/No

The best experts, in the most important corporate areas, are usually the ones that can be least spared from their usual position.

Exper-5.   Is the expert cooperative?   Yes/No

The expert should be eager to work on the project, or, at worst, should not be antagonistic.

Exper-6.   Is the expert capable of communicating his or her knowledge, judgment, experience, and the methods used to apply them to the particular task?   Yes/No

It is important to find an expert who not only has the expertise, but also the ability to impart it to the project team, whose members may know little or nothing about the subject area. The expert should be able to be introspective, to analyze his or her reasoning process, and then should be able to describe the reasoning process clearly to the project team, and to discuss it with them.

Exper-7.   Is the expert well respected?   Yes/No

The expert's knowledge and reputation must be such that if the expert system is able to capture a portion of the expert's expertise, the system's output will have credibility and authority. Otherwise, the system may not be used. (This may not be necessary in a domain where an accepted test for "goodness" of result exists.)

Exper-8. **Has the expert built up expertise over a long period of task performance?** Yes/No How long?

If so, the expert has had the amount of experience necessary to be able to develop the insights into the area that result in heuristics.

Exper-9. **Will it be easy to work with the expert?** Yes/No

The project team and the expert will be spending a lot of time together.

Exper-10. **Will the expertise for the system, at least that pertaining to one particular subdomain, be obtained primarily from one expert?** Yes/No

This avoids the problem of dealing with multiple experts whose conclusions or problem-solving techniques do not agree. However, there may be some advantages to using multiple experts, e.g., strength of authority and breadth of expertise in subdomains. If multiple experts contribute in a particular subdomain, one of them should be the primary expert with final authority. This allows all the expertise to be filtered through a single person's reasoning process.

Data Availability Category

Data-1. **Will the data required by the proposed system be available during the use of the system?** Yes/No

If the development of a new sensor system is necessary to provide crucial data, it could delay or impair the effort.

Data-2. **Are test cases available to be used during development of the system?** Yes/No

The development process depends upon heavy use of a develop and test cycle that requires test cases. If they are not available, the quality and reliability of the knowledge is compromised.

Data-3. **Will input data be available on-line?** Yes/No

If the majority of the data used by the system are already on electronic media, then the system will require less user intervention.

End-User Category

User-1. **Is there a strong managerial commitment of the time of the expert?** Yes/No

User-2. **Do the managers in the domain area agree upon the need to solve the problem that the system attacks?** Yes/No

If the specific task is agreed upon jointly and supported by management, then success is more likely.

User-3. **Will potential users welcome the completed system?** Yes/No

If users desire the system and have solicited its development, it is more likely that it will actually be used, and thus be considered a success.

User-4. Can the system be developed with a minimum of disturbance to the user community?   Yes/No

User-5. Do domain-area personnel understand that even a successful system will likely be limited in scope, and, like a human expert, may not produce optimal or correct results 100% of the time?   Yes/No

Personnel in the domain area must be realistic, understanding the potential of an expert system for their domain, but also realizing that thus far few expert systems have resulted in actual production programs with major industrial payoff. The system recipients should not be overly optimistic nor overly pessimistic. The project team may have to educate them to understand what are reasonable expectations.

User-6. Can the system be smoothly introduced and integrated with the existing practices?   Yes/No

User-7. Will the introduction of the system be politically sensitive or controversial?

Yes/No

If so, the potential resulting problems should be considered in advance. One typical problem is when the control or use of the system goes across existing organizational boundaries.

User-8. Will the knowledge contained by the system be politically sensitive or controversial?   Yes/No

For example, there may be certain practices, embodied in heuristics, which may prove embarrassing if written down, such as how certain customers are treated relative to other customers.

User-9. Will the system's results be politically sensitive or controversial?   Yes/No

If there will be corporate parties who challenge the system if its results do not favor them politically (e.g., on appropriation of funds), then it will be much harder to gain system acceptance.

Payback to the Corporation Category

Pay-1. What will be the short-term payback for the company if the proposed system is completed and used?

Applications selected for expert-systems development should be the ones that best meet the overall project goals for payoff versus risk of failure. For example, a project could benefit Houston Lighting & Power Company tangibly in the form of increased megawatt production, less fuel, more efficient engineering, fewer equipment failures, and so forth.

Pay-2. What will be the long-term payback to the company?

If a task has a projected need that extends far enough into the future, it may be possible to make long-term financial gains, as well as intangible benefits associated with capturing expertise, positioning for future needs, etc.

Pay-3. Will the payback to the company be measurable?   **Yes/No**

If yes, how will it be measured?

References

1. Chapnick, P., "What is Artificial Intelligence," *AI Expert,* Vol. 2, No. 10, p. 5, 1987.
2. Harmon, P., et al., "Expert Systems Tools and Applications," John Wiley & Sons, New York, 1988.
3. Musgrove, J. G., and P. N. Di Domenico, "Expert System–Operator Advisor for Particulate Control Equipment," ASME, New York, 1988.
4. Eisen, D. M., and R. von Konigslow, "Predictive Control Systems for Plant Operations," *Engrg. Dimensions,* Toronto, 1990.
5. Finn, G. A., and K. F. Reinschmidt, "Experts Systems in Operation and Maintenance," *Chemical Engineering,* February 1990.
6. Lewis, D. W., and P. N. Sheth, "ROMEX—An Expert System Testbed for Turbomachinery Diagnostics," Proceedings of the 19th Turbomachinery Symposium, Dallas, 1990.
7. Antinoja, R., P. Di Domenico, M. Joyce, and J. Musgrove, "Identification and Selection of Expert System Application," ASME, New York, 1988.

Chapter 13

Training and Organizing for Successful Failure Analysis and Troubleshooting

The reader probably need not be reminded that most machinery systems and components in the process industries will sooner or later experience defects, failures, and malfunctions. Unless the root causes of these failures or abnormalities are identified and rapidly remedied, a given plant may risk costly outages or production curtailments. Consequently, the process industries tend to employ specialist personnel to recognize potential failures, determine the failure mode, and perform troubleshooting tasks in efforts to define the many factors causing an unexpected or undesirable event.

Specialist Training Should Be Considered

The beginning of training should be a well-focused, written role statement that explains to both manager and managed their respective perceptions of the technical employee's role. Is he or she a parts changer or innovator? A fixer or improver? A person who is expected to react to problems or anticipate problems? The role statement must, at least, allude to a training plan. The technical person and his or her supervisor should discuss both role statement and training plan initially and, of course, during scheduled future performance reviews.

The detailed training plan may well be a separate document. A good one will give firm guidance and yet leave a lot of room for individual initiative. Its aim will be the achievement of proficiency in a technical skill or craft.

Technical training for a new engineer. Let's say your plant employs four maintenance/machinery engineers or technicians. You could get them to engage in worthwhile self-training by obtaining subscriptions—many of them free of charge—to such periodicals as *Hydrocarbon Processing, Power, Oil and Gas Journal, Plant Services,* and *Compressed Air.* (For a more complete listing of available periodicals, see References at the end of this chapter.) Each technical employee's name would be at the top of the distribution list of two or three of these periodicals, and he or she would be required to screen the content of the periodical(s) for relevant material. The employee would not have to read the articles, but would be expected to recognize from headings or abstracts the present or possible future usefulness of the writeup. Copies would have to be made of these writeups and sent to the other "professionals-in-training" on the "PIT" distribution form. One copy would be filed in the plant's central library under appropriate headings, which might follow a simple numbering system to enable easy retrieval via a straightforward, well-cross-referenced PC-based software program.

"Dig-up-the-facts." The second phase of training might be called the "dig-up-the-facts" phase. Each "PIT" would be asked to present periodically scheduled briefings or information sharing sessions to mechanical workforce personnel assigned to shop or field (e.g., millwright) tasks. Tacked on to the ubiquitous safety meetings, these 10- to 20-minute briefings or information-sharing sessions would deal with such topics as "How to Install Rolling Element Bearings," "Proper Lubrication Procedures for Pumps and Motors," "Why Four Different Types of Couplings Are Used at Our Plant," "When to Use Bellows Seals Instead of Pusher Seals," etc. There are literally hundreds of worthwhile topics to research and disseminate. The "digging-up-the-facts" process would compel the presenter to do some homework instead of guesswork, communicating with vendors and manufacturers instead of reinventing the wheel, and perhaps even rediscovering one or more of the many good technical textbooks that are generally available at a fraction of the cost of making a single mistake. The researcher would educate himself or herself in the process while at the same time contributing significantly to the development of team spirit, and enhancing mutual respect and cooperation among the many job functions in the plant.

From here, the phased approach to training could move to in-plant courses by competent presenters with both analytical and practical knowledge in machinery maintenance and reliability improvement procedures, and then progress to well-defined, known-to-be-relevant outside seminars or symposia.

Professional Growth: The Next Step

Ensuring professional growth of employees should really be a priority for process plant managers. In the early 1980s, a VP of a major multinational chemical company expressed the belief that through judicious use of outside contacts, participation in technical societies and publication of pertinent material, we can be sure that our technical productivity will continue to improve.

At roughly the same time, the president of a Houston-based engineering firm voiced a similar sentiment when he pleaded with other managers to support their employees' active participation in professional societies. He defined active participation as:

- Attendance at 75% of the meetings that a society has in an area.
- Attentive participation in the technical fellowship.
- Absorbing the technical data presented.
- Constructive criticism of the activity.

By actively participating in a professional society, engineers maintain a built-in resistance to technical obsolescence. Societies provide a place where professionals may share technical problems and discuss technical advancements in an atmosphere of fellowship that can be beneficial, both personally and professionally.

Personal benefits. What are the personal benefits that the employee will gain from participating actively in a professional society?

- Technical knowledge (by associating with his or her peers and absorbing the technical matter presented).
- The motivation of knowing that colleagues are interested in the same subject matter.
- The personal satisfaction of realizing that his or her ideas are on par with those of others.
- New ideas that are applicable to the profession and job.
- Personal acquaintances with serious-minded professionals.
- A source of reference from fellow professionals on materials, processes, methods, vendors, etc.
- State-of-the-art solutions to problems where such solutions are not yet published or available to the general public.
- Inspiration to enroll in and attend special courses, thus accumulating continuing education units.

Employer benefits. But what about benefits the company can expect if it supports its employees' participation in professional societies? There are several:

- Ideas such as improvements in systems operations that could substantially reduce production costs or result in new uses of material.
- Locating personnel for a particular job or vendors for difficult materials or equipment.
- Ideas for better application of equipment or fewer errors in new installations through the technology learned from the technical programs.
- Employee professionalism through pursuit of continuing education requirements.
- Exposure of employee and company capabilities to peer groups.
- Employee pride in being able to represent the company in a professional society.
- Favorable public relations exposure.

Most companies recognize that they are dependent on the well-being of their employees. Therefore, it is appropriate for management to recognize that people value these professional society rewards. The company should value them too.

In 1978, the company executive mentioned earlier examined a hypothetical 10-year program. He assumed that active participation requires one three-hour meeting per month, nine months of the year, plus two hours per month of reading and/or committee activities. Using 1996 data, the meetings are assumed to cost $40/month, plus annual dues of $90. In 10 years, this represents an investment of approximately 450 hours by the employee and roughly $4,500 by the company. In this period, the employee would have attended 90 meetings and, assuming attendance was an average of 30 per meeting, would have been exposed to peers 2,700 times.

From this, it appears that any company should support the cost of participation in a society of the engineer's choice and expect to profit from the investment. It is also apparent that the employee should invest his or her time to take advantage of the opportunity, because the personal rewards would be significant.

Organizing for Failure Analysis and Troubleshooting

Although very costly, dangerous, or environmentally unacceptable failure events are often analyzed by composite teams of investigators whose backgrounds and organizational assignments cross interdepartmental or intra-affiliate boundaries, the more routine analyses are generally handled by one or more members of a so-called mechanical technical support (MTS) or reliability engineering group. The typical functions of an MTS group were described in Volume 1 of *Practical Machinery Management for Process Plants* and include two tasks which relate to failure analysis and troubleshooting. These tasks are the investigation of special or recurring maintenance problems, and the review of maintenance costs and service factors. Accordingly, the systematic definition and reduction of costly repeat failures through in-depth failure analysis and troubleshooting is one of their priority tasks.

How the skills of qualified MTS machinery engineers can be utilized is best demonstrated by example. With centrifugal pump failures making up the bulk of machinery problems in the process industries, we will consider the steps necessary to reduce systematically the number of pump failure incidents in a typical petrochemical plant.

Setting Up a Centrifugal Pump Failure Reduction Program

Centrifugal pump repairs account for high maintenance expenditures in most petrochemical companies and other process plants. Recent statistics show that large refineries with 3,000 installed pumps experience approximately 600–800 failure events per year. A highly conservative estimate assumes average repair costs of $5,000 per event, although in 1996 many petrochemical plants reported a more believable average cost of $9,000 after burden, shop space, field labor, engineering analysis, and miscellaneous charges have been added.

The incentive to reduce pump failures is quite evident and hardly merits further discussion. An effective program of reducing pump failures is always desirable and should have the following objectives:

1. To improve centrifugal pump reliability through more accurate determination and subsequent elimination of failure cause.
2. To reduce process debits resulting from pump outages.
3. To reduce centrifugal pump maintenance costs by effective analysis of component condition, replacing only those parts which actually require change-out.

However, the success of a pump failure reduction program is influenced by a number of factors. These range from administrative-managerial to technical-clerical and could include such items as management commitment, willingness of technical staff to follow-up on tasks previously handled by maintenance work forces, and clerical support for better documentation and record keeping.

Definition of Approach and Goals

There are two basic implementation methods for a program of this type:

1. *The team approach.* This method requires the formation of a team of troubleshooting specialists who, over a short period of time, will make a concentrated effort to define and eliminate sources of frequent pump failures.

2. *The engineering support approach.* Here the effort is centered around the field maintenance organization, supplemented by support and guidance from a staff engineer trained in pump selection and failure analysis. The staff engineer belongs to the MTS group.

Although still popular with some petrochemical companies, the team approach was found to have a number of significant shortcomings:

1. It lacks continuity. Team members are often unavailable if follow-up needs should arise.
2. The team is sometimes considered as "outsiders" effort. Operating personnel and field maintenance forces fail to develop the necessary rapport with the team.
3. A lack of commitment and reluctance to communicate have sometimes been observed. The field forces may tend to stand back and let the "brains" struggle with the problem. Field personnel see the team as a vote of non-confidence in their own capabilities or past efforts.
4. Extra manpower is generally required with this approach.

In view of these drawbacks, petrochemical and other process plants would do well to consider the "engineering support" approach for a pump failure reduction program. Let us examine how this works.

Based on an analysis of frequency and cost and severity of pump failures, the plant mechanical support service group should designate a certain number of pumps as "problem pumps." These could be pumps that failed more than twice in a 12-month time period, or pumps that have a history of one failure per year but cost more than $10,000 or some other significant amount to repair. Of course, these guidelines are for illustrative purposes only and can be modified to suit any given need or situation.

A plant or operating unit should set itself an improvement goal. One typical goal is to strive for reducing the number of "problem pumps" by 50% in one year's time, or to reduce pump maintenance expenditures by a certain amount in a designated time period. In the spirit of participative management, the initiative for goal setting and program execution should originate with field personnel rather than office or headquarter staff personnel. While being guided by staff MTS personnel, field forces would make the program work and be credited with its successful implementation.

Action Steps Outlined

A number of action steps are required to get the pump failure reduction program going, and management must designate responsibilities for the execution of these steps. Responsible MTS personnel will have to:

1. Designate "problem pumps" after reviewing past failure history, cost of repairs, cost of product losses, etc.
2. Identify these pumps for record purposes and by actually tagging them in the field. Tagging will alert shop and field foremen to the need to arrange for a designated technician or engineer to be present whenever work is performed.

3. Update pump records to include failure history for past 18 months. Develop and include step-by-step instructions for assembly, disassembly, tolerance checks, etc., as necessary.
4. Follow the disassembly and reassembly of all problem pumps, record measurements, and revise detailed instructions, as required.
5. Identify where, within the plant or elsewhere, pumps identical or similar to the "problem pumps" are operating. Compare their operating histories and examine significant deviations for clues to the source of the problem.
6. Provide checklists and tabulations of critical mechanical and hydraulic parameters, e.g., bearing fits, piping strains, excessive suction specific speeds, operation away from BEP (best efficiency point), inadequate clearance between impeller periphery and cutwater, etc.
7. Provide checklists and guidance on critical process parameters (solids or polymers in pumpage, non-optimized seal flush arrangements, equipment start-up and shutdown, preheating, cooldown, venting, etc.).
8. Assist mechanical work force in finding the true failure causes and train mechanical personnel in identifying ways of eliminating these causes.
9. Monitor progress of program and communicate results to management. To fulfill this responsibility, the MTS engineer should:

- Issue monthly activity summaries.
- Tabulate interim results after six months.
- Report accomplishments after one year.
- Develop plans for next phase of program.

Documentation and Reference Data Requirements

It is hoped that pump maintenance data folders already exist at the plant which is interested in implementing a pump failure reduction program. If data folders are not available, this would be a good time to at first put them together for problem pumps and then expand the data bank to cover all pumps.[1]

Ideally, the data folders should contain the following documents:

1. API data sheets, or data sheets originally provided by the owner.
2. Performance curves.
3. Supplemental specifications invoked at time of purchase.
4. Dimensionally accurate cross-section drawing (actual dimensions need not be shown).
5. Assembly instructions and drawings.
6. Seal drawings with installation dimensions and specific instructions.
7. Manufacturing drawings of shaft and seal gland.
8. Small-bore piping isometric sketch.
9. Spare-parts list and storehouse retrieval data.
10. Maintenance instructions, including critical dimensions and tolerances, running clearances, shop test procedures for seals and bearings, and impeller balancing specifications.

11. Repair history, including parts replaced, analysis of parts condition, and cost of repair.
12. Design change documentation.
13. Initial startup and check-out data.
14. Lubrication-related data.
15. Special installation procedures.
16. Alignment data.
17. Vibration history.

Needless to say, the data files should be as complete as possible, kept current, and readily available to every person involved in the implementation of the pump failure reduction program.

Development of Checklists and Procedures

Checklists, procedures, or similar guidelines must be used if the objectives outlined earlier are to be reached in an expeditious and well-defined fashion. These checklists or procedures should cover topics which fall into roughly three classification categories:

1. Fieldwork-related reviews.
2. Shopwork-related reviews.
3. Technical support reviews (failure prevention, procurement guidelines, and pre-purchase reliability assurance topics) for MTS use.

In the first category, fieldwork-related checklists must be developed and applied to accomplish a number of necessary steps which will aid in problem identification and thus will lead to a reduction of problem incidents in the future. These checklists must address problem pump:

1. Lubrication instructions.
2. Routine surveillance instructions.
3. Data-taking requirements preparatory to shutdown for subsequent shop repair.
4. Installation precautions and data requirements prior to restart after shop repair.
5. Special commissioning instructions such as cool-down of cold service pumps, air-freeing of seal cavities, etc.

Checklist development can be a joint effort involving plant operating, technical service, and pump vendor personnel. Table 13-1 shows one such joint-effort procedure; it gives guidance on the commissioning of a vulnerable cold-service pump[2] and serves as a typical example.

In the second category, shopwork instructions and procedures must be developed and conscientiously used to ascertain both quality workmanship and uniformity of

Table 13-1
Field-Posted Commissioning Instructions*

Barrier fluid system dry out procedure (mechanical technicians)
1. Open flanges on barrier fluid return line.
2. Verify reservoir isolation valves in barrier fluid system piping are open.
3. Pump methanol into barrier fluid reservoir drain until clean methanol flows from open flanges.
4. Block reservoir drain valve and remove barrel pump hose.
5. Follow Pump casing/process piping dry out procedure below.
6. Prime barrel pump with fresh barrier fluid until clean fluid flows from open flanges.
7. Reconnect hose to reservoir drain valve and unblock valve.
8. Pump barrier fluid into system until clean fluid flows from open flanges.
9. Tighten open flanges using new gasket.
10. Continue pumping barrier fluid into system until reservoir is filled.
11. Block reservoir drain valve and remove pump hose.
12. Repeat procedure for remaining barrier fluid system.

Pump casing/process piping dry out procedure (process technicians)
1. Block all valves into pump casing, suction, and discharge process piping.
2. Open atmospheric vent valves A1 on casing drain, A2 on seal flush piping, A3 on suction piping, and A5 on discharge check valve bypass. Verify vent valve A4 on check valve bypass closed.
3. Open casing drain valve D1, verify drain valve D2 closed.
4. Open seal flush piping vent valve V1, verify vent valve V2 to reflux drum firmly closed.
5. Open methanol injection valve M1 into pump casing and begin filling with methanol. CAUTION: DO NOT PRESSURIZE PUMP CASING WITH METHANOL!
6. Block individual vent valves as methanol appears. Continue filling.
7. Block methanol injection valve M1 when methanol appears at vent valve A5 on discharge check valve bypass.
8. Block vent valve A5 on discharge check valve bypass.
9. Slow roll turbine at 50-60 rpm for about 5 minutes.
10. Open atmospheric vent A2 on seal flush piping briefly to clear collected air.

Pressure out procedure (process technician)
1. Open seal flush piping vent V2 to reflux drum to pressurize pump casing.
2. Open pump casing drain D2 to cold blowdown.
3. Allow pump casing to blowdown until casing drain piping begins to frost.
4. Verify methanol completely purged by cracking vent valve A1 on casing drain.
5. Block pump casing drain D1, D2 out of cold blowdown when light frost begins to form on pump casing.

Cold soak procedure (process technician)
1. Crack suction block valve B1 2-3 turns.
2. Open suction block valve B1 completely when casing is well frosted.
3. Open discharge block valve B2 completely.
4. Allow pump to cold soak for 2 hours.
5. Verify pump shaft turns freely by hand.

Startup/shutdown (process technicians—follow normal operating procedures)

*For vulnerable, cold-service pump. Condition: pump is completely blocked in, cleared, and depressurized.

product inprovement efforts. Typical instructions and checklists should make maximum use of technical information which is routinely available from pump, bearing, and mechanical seal manufacturers. Included in this second category are such guidelines as:

1. Wear ring and throat bushing clearances.
2. Antifriction bearing housing fit tolerance to be used for pumps, motors and turbines.[3]
3. Antifriction bearing shaft fit tolerance to be used for pumps, motors and turbines.[3]
4. Dimensional checking of mechanical seal components.
5. Maximum allowable rotor unbalance for pumps and small steam turbines.
6. Dimensions for steel bushings converting standard bearing housings to accept angular contact bearings.
7. Others (see Tables 13-2 and 13-3).

Material in the third category is intended for failure analysis use by plant engineering or technical service groups. It is a "memory jogger" and should incorporate helpful hints from the numerous publications dealing with pump reliability improvements.[4-18] The plant MTS engineer is encouraged to structure his review in the following manner.

Table 13-2
Typical Shop Repair Instructions*

Brg. bore, mm	Shaft sizes		Hsg. ID, mm	Housing fits		
	Min	Max		Min	Mean	Max
35	1.3781	1.3785	40	1.5744	1.5749	1.5754
40	1.5749	1.5753	42	1.6531	1.6536	1.6541
45	1.7718	1.7722	47	1.8500	1.8505	1.8510
50	1.9686	1.9690	52	2.0467	2.0474	2.0480
55	2.1655	2.1660	55	2.1649	2.1656	2.1662
60	2.3623	2.3628	62	2.4404	2.4411	2.4417
65	2.5592	2.5597	68	2.6767	2.6774	2.6780
70	2.7560	2.7565	72	2.8341	2.8348	2.8354
75	2.9529	2.9534	75	2.9523	2.9529	2.9535
80	3.1497	3.1502	80	3.1491	3.1498	3.1504
85	3.3466	3.3472	85	3.3460	3.3467	3.3474
90	3.5434	3.5440	90	3.5428	3.5435	3.5442
95	3.7403	3.7409	95	3.7397	3.7404	3.7411
100	3.9371	3.9377	100	3.9365	3.9372	3.9379
105	4.1340	4.1346	110	4.3302	4.3309	4.3316
110	4.3308	4.3314	115	4.5271	4.5278	4.5285
115	4.5277	4.5283	120	4.7239	4.7246	4.7253
120	4.7245	4.7251	125	4.9207	4.9215	4.9223
125	4.9214	4.9221	130	5.1175	5.1183	5.1191
130	5.1182	5.1189	140	5.5112	5.5120	5.5128
140	5.5119	5.5126	145	5.7081	5.7089	5.7097
150	5.9056	5.9063	150	5.9049	5.9057	5.9065
160	6.2993	6.3000	160	6.2986	6.2994	6.3002
170	6.6930	6.6937	170	6.6923	6.6931	6.6939
180	7.0867	7.0874	180	7.0860	7.0868	7.0876

*For pumps with rolling element bearings.

Table 13-3
Detailed Pump Assembly and Disassembly Instructions
for a Specific Pump

A. Field disassembly
1. Remove coupling spacer.
2. Drain pump completely.
3. Disconnect all external piping (flush, steam tracing, etc.) and interfering insulation.
4. Remove casing and foot support bolts.
5. Separate casing from distance piece by means of pusher screws. Leave bearing housing connected to distance piece unless bearing replacement is contemplated.

B. Shop disassembly
1. Unscrew impeller nut and remove impeller.
2. Remove stuffing box housing by carefully applying pressure from two crowbars contacting seal plate.
3. Remove seal plate from stuffing box housing and also remove stationary and rotating seal elements.
4. Remove shaft sleeve without undue force. Apply heat, if necessary.
5. Examine seal parts for warpage, fractures, and solids build-up on faces.

C. Shop reassembly
1. Verify flatness of seal faces to be within two light bands.
2. Measure inside diameter of throat bushing and outside diameter of shaft sleeve.
3. Machine inside diameter of throat bushing to achieve 0.0200-0.030-in. clearance with shaft sleeve O.D.
4. Assemble stationary seal ring and seal plate, taking care to insert required gaskets.
5. Carefully tighten plate hold-down nuts. Use depth micrometer to verify that stationary seal ring is installed parallel to product side of stuffing box cover within .0005 in.
6. Place stuffing box cover on distance piece.
7. Place rotating seal assembly on shaft sleeve and slide over shaft. Make sure all dimensions are per CHEMEX Dwg. B-541599.
8. Insert keys.
9. Assemble impeller, taking care to place gaskets between hub and shaft sleeve, and between shaft and impeller nut.

D. Measurements required for record
Before and after replacing worn parts with new parts, please measure and record the following dimensions:

	Manufacturer's recommended value, mils	With old parts, mils	With new parts, mils
Shaft runout (impeller region)	1 mil or less		
Shaft runout (seal region)	Not to exceed 2 mils		
Seal face compression	Per CHEMEX DWG. B-541599		
Depth micrometer Check to ensure stationary seal is parallel to seal gland.	0.0005 in. or less		
Clearance between impeller disc vanes and stuffing box housing	0.032-0.047 in.		
Clearance between shaft sleeve and throat bushing	0.020 in. diametral. (max 0.032")		

E. Field reassembly
1. Back out pusher screws and bolt distance piece to casing. Make sure proper gasket is placed in between.
2. Replace foot support bolts.
3. Reconnect all auxiliary piping.
4. Refill bearing housing with Synesstic-32 lubricant.
5. Hot-align coupling to zero-zero setting (max. allowable deviation: place motor 0.002 in. higher than pump shaft) while observing dial indicators on casing.
6. Replace coupling spacer. Ascertain coupling disc stretch or compression does not exceed 4 mil.

Date _OCTOBER 17, 1996_

Anytown Chemical
Winddrift, Texas

FAILURE ANALYSIS DATA FORM
(BREAKDOWNS OTHER THAN PLANNED PREVENTIVE MAINTENANCE)

° IDENTIFICATION: UNIT _HEPA-2_   COST CENTER _48217_

Equip. Type _CENT. PUMP_   Ident. No. _P-2257_   Yard No. _225-17101_

Serial No. _GOOFER-DURR 832121_   System (If Not Number Equip.) _ACID PREWASH_

° MECHANICAL AVAILABILITY DATA: Total Hours Unavailable _48_

	Date	Time
Equipment Down	10/12/96	9:00 AM
Ready for Service	10/14/96	8:20 AM

° BRIEF DESCRIPTION OF FAILURE: _SEAL FLUSH PIPING DEVELOPED CRACK, BROKE OFF._
SEAL FAILURE CAUSED BEARING WASHOUT . BEARING SEIZED, CAUSING
SHAFT TO BREAK

° COMPONENT FAILURE ANALYSIS ATTACHED

° SERIOUSNESS OF FAILURE   YES   NO
Safety   [X]
Production Loss   [X]
Air or Water Quality   [X]
  Problem
Conservation Loss   [X]

° REPAIR ANALYSIS
[X] Routine
[ ] Delay For Parts
[ ] Shop Quality Control Problem
[ ] Other Delay

° CORRECTIVE ACTION   Complete
Design Change   [X]
Revise Operating Instr.   [ ]
Additional Training   [ ]
Revise P.M. Schedule   [ ]

Follow Up Action Required: _NONE . HAVE_
CHANGED TO SCH. 160 PIPING AND HAVE
SEAL WELDED

° DIRECT COST DATA:
Estimate [X]
Actual [ ]

	Field	Central Shop	Contract	Subtotals
Labor	2000	4000		6000
Material	200	2000		2200

Product Loss _____ x _____ = _____
                 Hours $/Hour
                              Total = _8200_

° BASIC CAUSE OF FAILURE:
1. °MAINTENANCE
   A. _____ Improper Repair
   B. _____ Inadequate P.M.
   C. _____ Faulty analysis of trouble
   D. _____ Other
   E. _____ Design and/or selection
   F. _____ Improper field installation

2. °OPERATING
   A. _____ Overload
   B. _____ Improper start-up, shutdown,
              or oper. tech.
   C. _____ Lubrication lacking, wrong,
              contamination
   D. _____ Failure to release for
              scheduled P.M.
   E. _____ Other _____

3. °NOT SUITED FOR SERVICE DUE TO
   A. ✓ Faulty design assumptions
   B. _____ Incorrect material
   C. _____ Change in oper. cond. since
              design
   D. _____ Unexpected corrosion
   E. _____ Other

4. °MANUFACTURING DEFECTS
   A. _____ Design
   B. _____ Workmanship
   C. _____ Material
   D. _____ Other _____

5. °END OF DESIGN LIFE
   A. _____ Normal Wear
   B. _____ Reached retiring
              thickness
   C. _____ Other _____

Distribution List:
D. P. SONDERBERG
S. A. YONDREWICZ
F. T. ARBENZ

Analysis By _Elmer Breckhorn Jones_
Operating Supervisor _George Thublexnor III_

Figure 13-1. Failure analysis form used in pump failure reduction program.

General Review

1. Use the machinery problem-solving approaches that were described in Chapters 6, 10, and 11 to assist in defining the most probable failure cause.
2. Identify identical pumps in same or similar service within affiliated plants. Analyze maintenance history for these pumps.
3. If no identical pumps exist, or if identical pumps are applied differently, compare construction and design details. Do not neglect installation details such as piping, etc.
4. Verify that rotating elements are dynamically balanced per applicable checklist or procedure.
5. Review applicable field and shop checklists for accuracy and supervisor's signature at conclusion of work.

The results of the failure analysis should be documented on a short form similar to the one shown in Figure 13-1.[4] Far more detailed forms have been devised and are appropriate for plants with computerized failure records.

Review of Hydraulic Parameters*

1. Examine hydraulic unbalance potential of single volute pumps at low flow.
2. Verify that low flow operation is not a problem (e.g., excessive heat build-up, internal recirculation, thermal distortion, vaporization of fluid, etc., have all been ruled out).

Metallurgy Review

1. Examine component condition. Initiate comparison study, as required.
2. Verify that wear ring materials and clearances reflect best applicable experience. High-temperature pump clearances may have to be as large as [(API clearance) + (0.02 x °F)] mils, diametral.

Installation Review

1. Evidence of pipe stresses (due to routing, non-sliding supports, incorrect hangers, etc.)
2. Coupling misalignment and lockup problems.
3. Baseplate resonance, out-of-parallelism, grout defects.
4. Preferential flow, if suction elbows are too close to double-suction pump.

Bearing and Lubrication Review

1. Ascertain bearing fits comply with applicable checklist.
2. Verify filler notch bearings have been replaced by angular contact bearings per applicable checklist. No spacers between duplex rows!
3. Check to see that thermal expansion of bearing outer race is not restricted by coolant surrounding it.

---

* This review, as well as the ones following, represents only a partial listing. The failure analysis staff should expand this checklist to include topics pertinent to their own situations.

4. Calculate bearing L-10 life under normal operating conditions.
5. Verify that oil-mist lubrication reaches every row of bearing balls. Directed-mist fittings may be required instead of plain-mist application fittings. Consult Reference 19 for specific guidance.
6. Verify that motor windage does not impair effective venting of oil mist.
7. Ensure that approved grade oils only are used in oil-mist lubrication systems. Recall that paraffinic mineral oils and oils with high pour point can cause plugging of applicator fittings. Consider approved synthetic lube at temperatures below 35°F.

Seal and Quench Gland Design and Troubleshooting Review

1. Verify API flush plan conforms to applicable experience. Reverse flush may not be possible if balance holes are used (see earlier comment).
2. Ascertain flush fluid has adequate lubricity, vapor pressure at seal face temperature, supply quantity and pressure, and is free from solid particles. How about during abnormal operation? At startup?
3. Tandem seal lubrication and installation details should conform to latest experience. Consult pages 346–353 of Reference 20.
4. Seal installation tolerances should be verified.
5. Verify that only balanced seals are used at pressures above 75 psi.
6. Ascertain that quench steam or water is applied only if required to keep solid particles from forming external to seal components.

Tackle your mechanical seal troubleshooting tasks by:

1. Examining the entire seal.
2. Examining the wear track.
3. Examining the faces.
4. Inspecting the seal drive.
5. Checking spring condition.
6. Checking the elastomer.
7. Checking for rubbing.

Vertical Pump Considerations

1. Verify that critical speed is not a problem.
2. Check for proper lubrication of column bearings.
3. Some vertical pumps may need vortex baffles in suction bell.

Safety Considerations Must Not Be Overlooked

1. If minimum flow bypasses are used, verify each pump has its own bypass arrangement.
2. Do not allow excessively oversized pumps.
3. Verify isolation valves are accessible in case of fire. Are weight-closing valves possible?

4. Piping must be braced and gusseted. Flexible instrument connections may be advisable—but only if well engineered.
5. Is onstream condition monitoring instrumentation feasible? Should high- or low-frequency vibration monitoring be used? Consult Reference 20, pages 382–415.
6. Verify that appropriate coupling guards are used.

Armed with this checklist and after expanding it and then reading the reference material cited, plant engineers and mechanical technical support personnel should feel confident in their ability to diagnose recurring pump problems.

Program Results and Conclusions

The results of centrifugal pump failure reduction programs are generally very significant. One large plant reported a 29% reduction in failures after the first year. Although the failure reduction rates and percentages leveled off at 37% after three years, the ensuing savings exceeded this percentage because the reductions were concentrated on many maintenance-intensive problem pumps.

Failure rate comparisons are given in Figure 13-2 for a medium-size chemical plant. They are again quite representative of typical first-year comparisons.

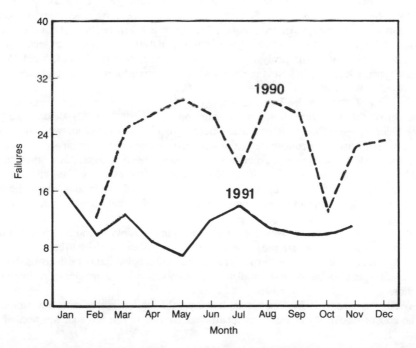

Figure 13-2. Centrifugal pump failure rates before and after implementing a failure reduction program in a medium-size chemical plant.

Experience shows that significant reductions in centrifugal pump failure events are possible without, in most cases, adding to the work force of typical petrochemical and other process plants. The key to the successful implementation of a failure analysis and reduction program can be found in management commitment, thorough involvement of plant technical personnel, and good documentation.

Management commitment manifests itself in many ways. Briefing sessions and reporting requirements can be arranged so as to lend high visibility to the program. The thoroughness of plant engineering and technical personnel is exemplified by checklist and procedure development and by their ability to convince mechanical or field work forces of the need to follow these guidelines conscientiously. Finally, the importance of good documentation cannot be overemphasized. Computerized failure records are ideally suited to keep track of a pump failure reduction program, but the computer output can only be as good as the input. Manual data keeping is perfectly acceptable. Close cooperation between technical and clerical staff have gone a long way toward ensuring the overall success which, happily, can be reported for centrifugal pump failure reduction programs using the approach outlined here.

Postscript: How to Find a Reliability Professional

Every two or three months we receive a call from professional employment agents—headhunters, for short. Typically, they are looking for a machinery engineer with about 10 years of experience in the HPI. His or her qualifications should include, but not be limited to, a knowledge of stress analysis, measurement techniques, instrumentation, vibration analysis, materials, machine shop procedures, fluid flow, rotor dynamics, machinery field erection and startup procedures, and an understanding of effective maintenance management.

This list of qualifications keeps reminding us of an ad that appeared in the classified section of the *New York Times* many years ago (January 2, 1972). It described a machinery engineer's job as one that would: "require a personable, well-educated, literate individual with college degree in any form of engineering or physics. . . . Job requires wide knowledge and experience in physical sciences, materials, construction techniques, mathematics and drafting. Competence in the use of spoken and written English is required. Must be willing to suffer personal indignities from clients, professional derision from peers in more conventional jobs, and slanderous insults from colleagues.

"Job involves frequent physical danger, trips to inaccessible locations throughout the world, manual labor and extreme frustration from lack of data on which to base decisions.

"Applicant must be willing to risk personal and professional future on decisions based on inadequate information and complete lack of control over acceptance of recommendations. . . ."

Well, that was in 1972. Since then, however, the job has not become any simpler. The cost of machinery outages and repairs has escalated. The prerequisites needed to perform as a machinery reliability engineer may well exceed those of the machinery engineer of the 1970s and 1980s.

We sense from these discussions with various "talent scouts" that highly trained, competent machinery engineers are not easily found. It also appears that some employers are beginning to notice the loss of qualified old-timers due to early retirement and similar attrition. Others seem to understand that machinery engineers can be effective and productive when their analytical skills are consistently developed and used, and can be ineffective and nonproductive if the application of sound engineering practices is discouraged.

Chances are the search for reliability professionals willing to change jobs will often prove difficult, time consuming, or both. Knowing the value of a skilled machinery reliability engineer makes one appreciate the potential and *real* loss suffered by a company that has to do without having this critically important job function filled.

The idea that contractors can always fill the gap surely lacks merit and forethought. Where did the contractor's young engineers receive their training? Look again at the perceived qualifications.

Here's our advice to the process plant manager who understands the value of a thoroughly well-trained equipment reliability engineer: develop one. Start by compiling a role statement, then map out a training plan. Interview a number of interested candidates, select the right one. Give him or her periodic performance feedback; defend his or her goals and contributions as necessary and appropriate. Don't ever allow the *trained* reliability professional to become just a pair of hands, or a person whose entire time is spent fighting repair deadlines rather than being immersed in proactive failure prevention. Groom this reliability professional's abilities, judgment, and motivation by encouraging access to his or her peers. Ask this person to use analytical skills to the utmost, to read, to write, to communicate. Rest assured that all parties will benefit from the experience if you carefully and consistently implement this "grow-your-own" formula.

References

1. Bloch, H. P., "Machinery Documentation Requirements For Process Plants," ASME Paper 81-WA/Mgt-2, presented in Washington D.C., Nov. 1981.

2. Sangerhausen, C. R., "Mechanical Seals In Light Hydrocarbon Service: Design and Commissioning of Installations For Reliability," paper presented at ASME Petroleum Mechanical Engineering Conference, Denver, Colorado, September 14-16, 1980.

3. SKF Bearing Company, *Engineering Data Manual,* King of Prussia, Pennsylvania.

4. James, R., "Pump Maintenance," *Chemical Engineering Progress,* Feb. 1976, pp. 35-40.

5. Taylor, I., "The Most Persistent Pump—Application Problem For Petroleum and Power Engineers," ASME Paper No. 77-Pet 5, presented in Houston, September 1977.

6. Nelson, W. E., "Pump Curves Can Be Deceptive," *Proceedings of NPRA Refinery and Petrochemical Plant Engineering Conference,* 1980, Pages 141-150.

7. Eeds, J. M., Ingram, J. H., and Moses, S. T., "Mechanical Seal Applications—A User's Viewpoint," *Proceedings of the Sixth Turbomachinery Symposium,* Texas A&M University, Pages 171-185, 1977.

8. Bush, Fraser, Karassik, "Coping With Pump Progress: The Sources and Solutions of Centrifugal Pump Pulsations, Surges and Vibrations." *Pump World,* 1976, Volume 2, Number 1, pp. 13-19 (A publication of Worthington Group, McGraw-Edison Company).

9. McQueen, R., "Minimum Flow Requirements for Centrifugal Pumps." *Pump World,* 1980, Volume 6, Number 2, pp. 10-15 (Ibid).

10. Bloch, H. P., "Improve Safety and Reliability of Pumps and Drivers," *Hydrocarbon Processing,* January through May, 1977.

11. Bloch, H. P., "Optimized Lubrication of Antifriction Bearings for Centrifugal Pumps," ASLE Paper No. 78-AM-1D-1, presented in Dearborn, Michigan, April 17, 1978.

12. Sprinker, E. K. and Patterson, F. M., "Experimental Investigation of Critical Submergence For Vortexing in a Vertical Cylindrical Tank," ASME Paper 69-FE-49, June, 1969.

13. Bloch, H. P., "Mechanical Reliability Review of Centrifugal Pumps For Petrochemical Services," Presented at the 1981 ASME Failure Prevention and Reliability Conference, Hartford, Connecticut, September 20-23, 1981.

14. Bussemaker, E. J., "Design Aspects of Baseplates For Oil and Petrochemical Industry Pumps," IMechE (UK) Paper C45/81, Pages 135-141.

15. Karassik, Igor J., "So, You Are Short on NPSH?" Presented at Pacific Energy Association Pump Workshop, Ventura, California, March, 1979.

16. Makay, Elemer and Szamody, Olaf, "Survey of Feed Pump Outages," FP-754 Research Project 641 for Electric Power Research Institute, Palo Alto, Ca. 1978.

17. Fraser, W. H., "Flow Recirculation in Centrifugal Pumps," *Proceedings of the Tenth Turbomachinery Symposium,* Texas A&M University, College Station, Texas, Dec. 1981.

18. Ingram, J. H., "Pump Reliability—Where Do You Start?" Presented at ASME Petroleum Mechanical Engineering Workshop and Conference, Dallas, Texas, September 13-15, 1981.

19. Bloch, H. P., "Large Scale Application of Pure Oil Mist Lubrication In Petrochemical Plants," ASME Paper 80-C2/Lub 25, Presented at Century 2 ASME/ASLE International Lubrication Conference, San Francisco, California, August 18-21, 1980.

20. Bloch, H. P., *Improving Machinery Reliability, Second Edition,* Gulf Publishing Company, Houston, Texas, 1988.

Periodicals

The following periodicals are among the many we found useful for technical training for new engineers:

Hydrocarbon Processing	*Hydraulics & Pneumatics*
Power	*Cogeneration*
Oil and Gas Journal	*Maintenance Technology*
Plant Services	*P/PM Technology*
Compressed Air	*Gas Turbine World*
Noise and Vibration	*Lubrication World*
Machine Design	*Water Engineering and Management*
Chemical Engineering	*Pumps and Systems*
Chemical Processing	*Compressor Tech*
Mechanical Engineering	*World Pumps*

Appendix A

Machinery Equipment Life Data[1,2]

Table A-1
Life Spans of Selected Machinery Components and Equipment

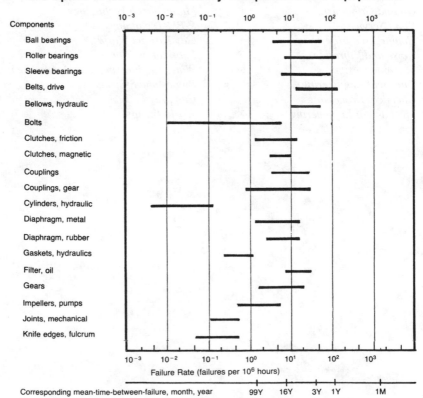

Failure Rate (failures per 10^6 hours)

Corresponding mean-time-between-failure, month, year

Table A-1 (cont.)

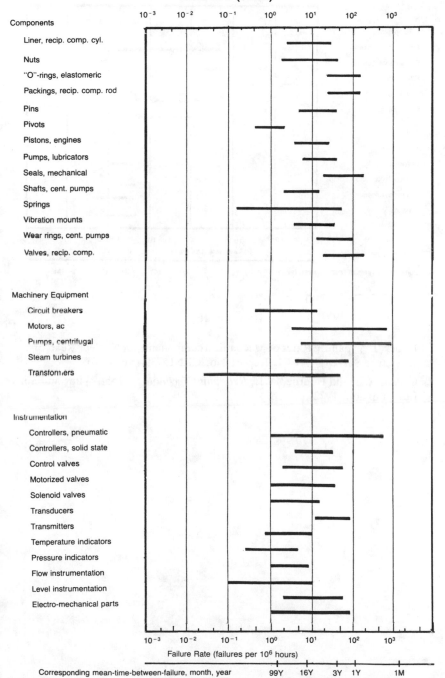

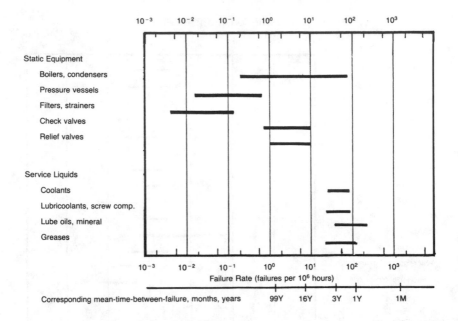

References

1. Hauck, D., "Failure Rates of Mechanical components for Nuclear Reactors: A Literature Survey," AECL Report No. CRNL-739 (SP-R-10).

2. Green, A.E. and Bourne, A.J., *Reliability Technology,* John Wiley and Sons Ltd., London, 1972, p. 538.

Appendix B

Theory of Hazard Plotting*

This appendix contains a presentation of theory underlying hazard plotting. Also, theoretical properties of the exponential, Weibull, normal, log normal, and extreme-value distributions are presented briefly to aid us in deciding which hazard paper to use for a set of data. In the formulas given in the appendix, ln x denotes the natural logarithm (base e) of the number x, and log x denotes the common logarithm (base 10).

General Theory and Application

General theory underlying hazard plotting is presented here and then applied to the exponential, Weibull, normal, log normal and extreme-value distributions.

A probability model for the distribution of time to failure is given by a cumulative distribution function $F(x)$, which is the probability of failure by time x. $F(x)$ is defined for $x \geq 0$, where time zero is the starting time, and increases from zero to one as x goes from zero to ∞. The derivative of a cumulative distribution function is called a probability density function and denoted by $f(x)$. That is,

$$f(x) = \frac{d}{dx} F(x) \qquad (B1)$$

or, equivalently,

$$F(x) = \int_0^x f(x) \, dx \qquad (B2)$$

---

*See Chapter 7, Reference 10. By permission.

643

The hazard function $h(x)$ of a distribution is defined by $x \geq 0$ by

$$h(x) = \frac{f(x)}{1 - F(x)} \qquad \text{(B3)}$$

and is the conditional probability of failure at time x given that failure has not occurred before then. In other words, $h(x)**$ is the instantaneous failure rate at time x; that is, in a short time Δ from x to $x + \Delta$, a proportion $\Delta \times h(x)$ of the units at age x can be expected to fail. The cumulative hazard function of a distribution is defined for $x \geq 0$ by

$$H(x) = \int_0^x h(x) \, dx \qquad \text{(B4)}$$

From (B3) and (B4), one gets

$$H(x) = -\ln [1 - F(x)] \qquad \text{(B5)}$$

and, equivalently,

$$F(x) = 1 - e^{-H(x)} \qquad \text{(B6)}$$

This relationship allows one to convert probability paper for a distribution into hazard paper. A cumulative hazard value H is located on probability paper at the cumulative probability value F given by (B6). For example, the cumulative hazard value 100% is located on probability paper at the cumulative probability value $F = 1 - e^{-1.00} = 63.2\%$. The relationship (B6) between the probability and hazard scales can be seen on all hazard papers. In actuality, the hazard and probability scales on the hazard papers were obtained from numerical calculations, rather than probability paper, to ensure accuracy.

For small values of the cumulative hazard function $H(x)$, (B6) simplifies to the approximation

$$F(x) \approx H(x) \qquad \text{(B7)}$$

That is, the cumulative probability value and the cumulative hazard value coincide for small probabilities, say, less than a few percent. This can be seen on the various hazard papers.

The principle underlying plotting on hazard paper is essentially that underlying plotting on probability paper. For a complete sample of n failures plotted on probability paper, the increase in the sample cumulative probability distribution function at a failure time is equal to the probability $1/n$ for the failure. Then the sample cumulative distribution function, based on the sum of the probabilities of failure, approximates the theoretical cumulative distribution function, which is the

---

**$h(x)$ is also referred to as $r(t)$; see Figure 7-1.

integral of the probability density. Similarly, for a sample plotted on hazard paper, the increase in the sample cumulative hazard function at a failure time is equal to the conditional probability $1/K$ for the failure, which is equal to one divided by the number K of units in operation at the time of the failure, including the failed unit. Then the sample cumulative hazard function, based on the sum of the conditional probabilities of failure, approximates the theoretical cumulative hazard function, which is the integral of the conditional probability of failure.

It must be emphasized that the hazard plotting method is based on the assumption that the censoring time associated with a censored unit is statistically independent of the time when the unit would fail. In other words, the distribution of potential failure times of censored units is the same as the distribution of failure times of uncensored units.

The theoretical distributions and their hazard plotting papers are presented next. These results are obtained from this general theory.

Exponential Distribution

The exponential cumulative distribution function is

$$F(x) = 1 - e^{-x/\theta} \quad x \geq 0 \tag{B8}$$

where $\theta > 0$ is the value of the mean time to failure. The probability density function is

$$f(x) = \frac{1}{\theta} e^{-x/\theta} \qquad x \geq 0 \tag{B9}$$

The hazard function is

$$h(x) = 1/\theta \qquad x \geq 0 \tag{B10}$$

which is constant over time. $\lambda = 1/\theta$ is called the failure rate. Because the failure rate (B10) for the exponential distribution is constant over time, the distribution is suitable for "random failures" rather than for wearout or infant mortality failures. The cumulative hazard function is

$$H(x) = x/\theta \qquad x \geq 0 \tag{B11}$$

Thus, from (B11), time to failure x as a function of the cumulative hazard H is

$$x(H) = \theta H \tag{B12}$$

Since time to failure is a linear function of the cumulative hazard, x plots as a straight-line function of H on square-grid graph paper. Thus, exponential hazard paper is merely square-grid graph paper. The slope of the line is the mean time to failure θ of the distribution. More simply, θ is the value of the time for which $H = 1$; this fact is used to estimate θ from data plotted on exponential hazard paper.

For more detailed discussions of the theoretical properties of the exponential distribution, refer to Reference 1. For sketches of the previous functions for the exponential distribution, see Figure B1 for the functions labeled $\beta = 1$, substituting θ for α.

Weibull Distribution

The Weibull cumulative distribution function is

$$F(x) = 1 - e^{-(x/\alpha)^\beta} \qquad x > 0 \qquad (B13)$$

where $\beta > 0$ is the value of the shape parameter of the distribution and $\alpha > 0$ is the value of the scale parameter. α is a characteristic time to failure, since it is the 100 x (1 $- e^{-1}$) = 63.2 percentile of the distribution regardless what the value of the shape parameter is. As a special case with $\beta = 1$, the Weibull reduces to the exponential distribution. The probability density function is

$$f(x) = \frac{\beta}{\alpha^\beta} x^{\beta-1} e^{-(x/\alpha)^\beta} \qquad x > 0 \qquad (B14)$$

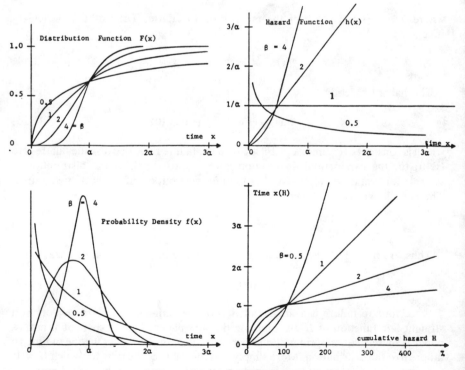

Figure B1. Weibull distributions.

The hazard function is

$$h(x) = \frac{\beta}{\alpha^\beta} x^{\beta-1} \qquad x \geq 0 \qquad \text{(B15)}$$

which is just a power of the time x. The failure rate $h(x)$ increases with time for values of the shape parameter β greater than one and decreases with time for values less than one. Thus, the Weibull distribution is flexible enough to describe either an increasing or a decreasing failure rate according to the value of the shape parameter. For $\beta = 1$, the exponential distribution results and the failure rate is constant. The Weibull distribution is an extreme-value distribution for the smallest time to failure from a large sample of times from some failure distribution. In other words, if a unit can be regarded as a system of many parts (each with a failure time from the same distribution) and if the unit fails with the first part failure, then the Weibull distribution may be appropriate for time to failure of such units. The cumulative hazard function is

$$H(x) = (x/\alpha)^\beta \qquad x \geq 0 \qquad \text{(B16)}$$

which is a power function of time x. Then, from (B16), time to failure x as a function of the cumulative hazard H is

$$x(H) = \alpha H^{1/\beta} \qquad \text{(B17)}$$

or, taking logarithms (base 10),

$$\log x = (1/\beta) \log H + \log \alpha \qquad \text{(B18)}$$

By (B18), x plots as a straight-line function of H on log-log graph paper, since $\log x$ is a linear function of $\log H$. Thus, Weibull hazard paper is merely log-log graph paper. The slope of the straight line equals $1/\beta$; this fact is used to estimate β from data plotted on Weibull hazard paper. Also, for $H = 1$, the corresponding time x equals α; this fact is used to estimate α from data plotted on Weibull hazard paper.

For sketches of the functions for Weibull distributions, see Figure B1.

Normal Distribution

The normal cumulative distribution function is

$$F(x) = \frac{1}{\sigma \sqrt{2\pi}} \int_{-\infty}^{x} e^{-(x-\mu)^2/2\sigma^2} \, dx \quad -\infty < x < \infty \qquad \text{(B19)}$$

where μ is the value of the mean time to failure and is the center of the distribution, and $\sigma > 0$ is the value of the standard deviation of the distribution. This distribution is defined for all values of time to failure x both positive and negative. If μ is at least two or three times as great as σ, then the distribution gives a probability of failure at a negative time that is small enough to be neglected, and the distribution may serve as a satisfactory model for time to failure. The probability density function is

$$f(x) = \frac{1}{\sigma \sqrt{2\pi}} e^{-(x-\mu)^2/2\sigma^2} \tag{B20}$$

The hazard function is

$$h(x) = \frac{f(x)^*}{1 - F(x)} \tag{B21}$$

where $F(x)$ and $f(x)$ are given by (B19) and (B20). The failure rate is an increasing function of time. Because the failure rate for the normal distribution is strictly increasing, this distribution may be appropriate to wearout types of failure. The normal distribution is symmetric about the mean value μ; this symmetry should be considered in deciding on the appropriateness of the distribution for a set of data.

The functions

$$\Phi(u) = \frac{1}{\sqrt{2\pi}} \int_{-\infty}^{u} e^{-u^2/2} \, du \tag{B22}$$

and

$$\varphi(u) = \frac{1}{\sqrt{2\pi}} e^{-u^2/2} \tag{B23}$$

are called the standard normal distribution function and probability density function and are widely tabulated. They are used to give

$$F(x) = \Phi\left(\frac{x - \mu}{\sigma}\right) \tag{B24}$$

$$f(x) = \frac{1}{\sigma} \varphi\left(\frac{x - \mu}{\sigma}\right) \tag{B25}$$

and

$$h(x) = \frac{\frac{1}{\sigma}\varphi\left(\frac{x - \mu}{\sigma}\right)}{1 - \Phi\left(\frac{x - \mu}{\sigma}\right)} \tag{B26}$$

---

*Also referred to as r(t).

The cumulative hazard function is

$$H(x) = -\ln\left(1 - \Phi\left(\frac{x - \mu}{\sigma}\right)\right)$$

(B27)

From (B27), time to failure x as a function of the cumulative hazard H is

$$x(H) = \mu + \sigma\Phi^{-1}(1 - e^{-H})$$

(B28)

Φ^{-1} is the inverse of the standard normal distribution function Φ defined in (B22) and gives the values of the percentage points of the standard normal distribution. By (B28), time x plots as a straight-line function of $\Phi^{-1}(1 - e^{-H})$. Thus, on normal hazard paper, a cumulative hazard value H is located on the horizontal axis at the position $\Phi^{-1}(1 - e^{-H})$, and the vertical time scale is linear. The slope of the straight line equals the value of the standard deviation; this can be used to estimate σ from data plotted on normal hazard paper. Also, for $H = \ln 2$ or, equivalently, for the 50% point on the probability scale, the corresponding time x equals μ; this fact is used to estimate μ from data plotted on normal hazard paper.

For more detailed discussions of the theoretical properties of the normal distribution, refer to Reference 1. For sketches of the functions for the normal distributions, see Figure B2.

Log Normal Distribution

The log normal cumulative distribution function is

$$F(x) = \Phi\left(\frac{\log x - \mu}{\sigma}\right) \qquad x > 0$$

(B29)

where the standard normal distribution function Φ is defined in (B22), μ is the value of the mean log time to failure, and $\sigma > 0$ is the value of the standard deviation of log time to failure. Note that $\log x$ is a base 10 logarithm. The probability density function is

$$f(x) = \frac{0.4343}{x\sigma} \varphi\left(\frac{\log x - \mu}{\sigma}\right) \qquad x > 0$$

(B30)

where the standard normal density φ is defined in (B23). The hazard function is

$$h(x) = \frac{\dfrac{0.4343}{x\sigma} \varphi\left(\dfrac{\log x - \mu}{\sigma}\right)}{1 - \Phi\left(\dfrac{\log x - \mu}{\sigma}\right)} \qquad x > 0$$

(B31)

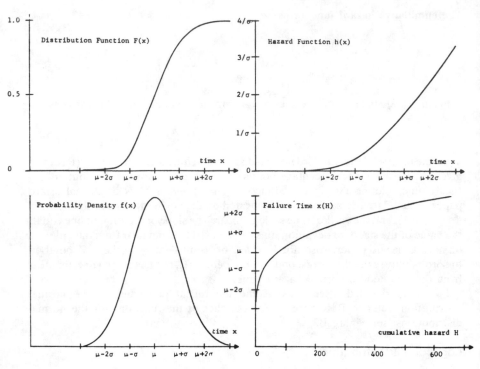

Figure B2. The normal distribution.

The failure rate (B31) is zero at time zero, increases with time to a maximum, and then decreases back down to zero with increasing time. The failure rate decreasing to zero after some time is a property of the log normal distribution that may make it an unsuitable model for wearout failure. Weibull cumulative distribution functions with a shape parameter value less than two are close to certain log normal distribution functions in the lower tail. Thus, it may be difficult to choose between a Weibull plot and a log normal plot for a set of data, since both may fit the data equally well in the lower tail. This may be a problem, since the two distributions can lead to different conclusions about the failure rate in the upper tail of the distribution of time to failure. The cumulative hazard function is

$$H(x) = -\ln\left(1 - \Phi\left(\frac{\log x - \mu}{\sigma}\right)\right) \qquad x > 0 \qquad \text{(B32)}$$

Note that base e and base 10 logarithms are used in (B32). From (B32), the log of the failure time x, as a function of the cumulative hazard H, is

$$\log x(H) = \mu + \sigma\Phi^{-1}(1 - e^{-H}) \tag{B33}$$

where Φ^{-1} is the inverse of the standard normal distribution function Φ defined by (B22). By (B33), $\log x(H)$ is a straight-line function of $\Phi^{-1}(1 - e^{-H})$. Thus, normal hazard paper also serves as log normal hazard paper, if the vertical time scale is made logarithmic rather than linear. The slope of the straight line is the value of the parameter σ; this fact can be used to estimate σ from data plotted on log normal hazard paper. Also, for $H = ln\ 2$ or, equivalently, for the 50% point on the probability scale, the corresponding $\log x$ equals the value μ which can be read directly from the logarithm scale; this fact is used to estimate μ from data plotted on log normal hazard paper.

For more detailed discussions of the theoretical properties of the log normal distribution, refer to Reference 1. For sketches of the functions for log normal distributions, see Figure B3.

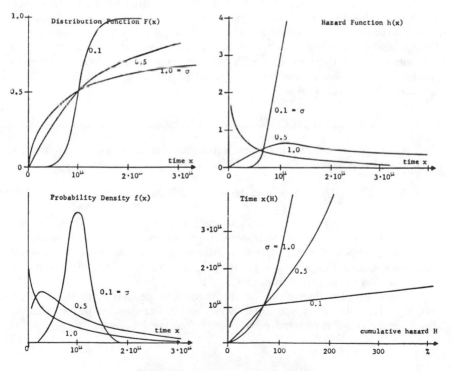

Figure B3. Log normal distributions.

Smallest Extreme-Value Distribution

The cumulative distribution function of the smallest extreme-value distribution is

$$F(x) = 1 - \exp\left[-e^{(x-b)/a}\right] \quad -\infty < x < \infty \tag{B34}$$

where b is the value of location parameter of the distribution, and $a > 0$ is the value of the scale parameter. b is a characteristic time to failure, since it is the $100 \times (1 - e^{-1})$ = 63.2 percentile of the distribution regardless what the value of the scale parameter is. The probability density function is

$$f(x) = \frac{1}{a} e^{(x-b)/a} \exp\left[-e^{(x-b)/a}\right] \quad -\infty < x < \infty \tag{B35}$$

The hazard function is

$$h(x) = \frac{1}{a} e^{(x-b)/a} \quad -\infty < x < \infty \tag{B36}$$

By (B36), the failure rate for the smallest extreme-value distribution increases exponentially with time. Like the Weibull, the smallest extreme-value distribution is an extreme-value distribution for the smallest time to failure for a large sample of times from some failure distribution which is far from the time origin. In other words, if a unit can be regarded as a system of identical parts with times to failure from a distribution and if the unit fails with the first part failure, then the smallest extreme-value distribution may be appropriate for time to failure of the units. The cumulative hazard function is

$$H(x) = e^{(x-b)/a} \quad -\infty < x < \infty \tag{B37}$$

which is an exponential function of time. From (B37), time to failure x as a function of the cumulative hazard H is

$$x(H) = b + a \ln H = b + 2.3026 \, a \log H \tag{B38}$$

By (B38), time x plots as a straight-line function of log H. Thus, smallest extreme-value hazard paper is merely semilog graph paper. The slope of the straight line equals $2.3026 \, a$; this fact is used to estimate a from data plotted on smallest extreme-value hazard paper. Also, for $H = 1$, the corresponding time x equals b; this fact is used to estimate b from plotted data.

For more detailed discussions of the theoretical properties of the smallest extreme-value distribution, see References 1 and 2. For sketches of the functions above for the smallest extreme-value distribution, see Figure B4.

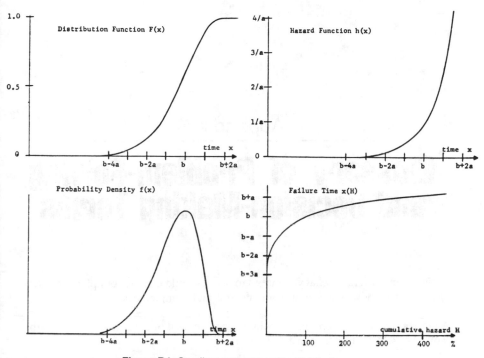

Figure B4. Smallest extreme-value distribution.

References

1. Hahn, G. J. and Shapiro, S. S., *Statistical Models in Engineering,* Wiley, New York, 1967, Chapter 3.
2. Gumbel, E. J., *Statistics of Extremes,* Columbia University Press, New York, 1958.

Appendix C

Glossary of Problem-Solving and Decision-Making Terms

Action Plan   This is what you intend to do. It should include your planned action, against what, how much, and when started and completed.

Adaptive Action   Steps taken to allow achievement of goals in spite of a trouble. You live with it, rather than fix it.

Ex.: A man with a bad leg may decide to buy a cane rather than get the leg fixed.

Cause   Usually refers to a past event or events which resulted in a problem. It is often a compound series of things which created a specific non-standard effect.

Cause Analysis   The systematic search for the cause of a problem. It is a questioning technique to gather the relevant facts and analyze them.

Contingent Action   Your fall-back plan. A plan set up ahead of time to be implemented if trouble occurs.

Ex.: A sprinkler system is set up ahead of time. It does not operate unless there is a fire.

Decision Analysis   A systematic technique to select the action which best fits your criteria or your objectives.

Fix   What you do to really take care of a trouble. It ought to be *safe,* be *legal,* and *remove the cause.* A good fix is not a patch and does not cause additional problems.

Interim Action   Temporary action to meet goals while the problem is worked on. A patch.

Ex.: Hiring temporary workers while cleaning up a spill.

Planning for Change   A systematic approach to identifying future troubles or opportunities and making an Action Plan to avoid them, reduce their effects, capitalize on them, or to reinforce their benefits.

Preventive Action   A plan to remove causes of future problems. (This is different from contingent action—here we want to avoid fire by removing its cause—for instance, piles of oily rags, etc.)

Situation Analysis   The breaking apart of a general situation into more specific problems, picking the most important ones and working on them. It generally includes selecting appropriate analysis techniques, if any.

Appendix D

Gear Nomenclature

Symbol Definition

C	Center distance (inches).
C_p	Influence coefficient, depending upon the elastic properties of the materials.
C_H	Hardness ratio factor.
C_f	Surface condition factor.
C_L, K_L	Life factor.
C_m, K_m	Load distribution factor.
C_o, K_o	Overload factor.
C_R, K_R	Safety factor.
C_S, K_S	Size factor.
C_T, K_T	Temperature factor.
C_V, K_V	Dynamic factor.
d	Diameter, pitch, operating pinion (inches).
E	Young's modulus.
F	Face width, effective (inches).
h	Film thickness, minimum (inches).
J	Geometry factor (bending fatigue).
m_N	Ratio, loading sharing.
N_G	Number of teeth, gear
N_P	Number of teeth, pinion
n_p	Revolutions per minute, pinion number.
P_d	Diametral pitch transverse, operating.
S	Surface finish, RMS, (after running in).
S_{ac}	Contact stress number.
S_{at}	Bending stress number.
S_c	Contact stress number, calculated.

S_t	Tensile stress number, calculated.
T_b	Temperature, gear blank.
T_f	Flash temperature index (°F).
T_i	Temperature, initial (°F).
w	Load per inch of face.
W	Specific loading.
W_t	Transmitted tangential load, (at operating pitch diameter).
W_{tc}	Tangential load, effective (lbs).
z_t	Factor, scoring geometry.
α	Coefficient, pressure Viscosity.
η_o	Viscosity.
θ	Roll angle at point of contact.
λ	Specific film thickness.
ρ_p	Radius of curvature pinion.
ρ_G	Radius of curvature, gear.
ϕ_n	Pressure angle, normal (degrees).
ϕ_t	Pressure angle, transverse (degrees).
ψ	Helix angle.

$$C_c = \frac{\cos\phi_t \, \sin\phi_t}{2} \left(\frac{m_G}{m_G + 1} \right)$$

$$I = C_c/m_N$$

$$m_G = N_G/N_P$$

$$Z = \frac{.0175 \, [(\rho_p^{.5} - (N_P/N_G^{.5} \, \rho_G)] \, P_d^{.25}}{\text{COS} \, \phi_t^{.75} \, [(\rho_P \, \rho_G)/(\rho_P + \rho_G)]^{.25}}$$

Index

A

B

failure statistics for antifriction
bearings, 480
fan, 367
fluting, 110, 114–115
fragment denting, 88, 92
fretting, 96
fretting corrosion, 88–90
frosting, 100–102
glazing, 101
housing, 367
improper fit damage, 96–98
inboard, 366
interference fit, 86
life dispersion curves, 79–81
load distribution, 84–88
misalignment, 86, 91–92
mounting defects, 94–95
out-of-roundness, 87–80
outboard, 366, 367
parasitic load, 94–96
pitting, 112–113
rolling element defects, 387–394
    ball pass frequency inner (BPFI), 387, 388
    ball pass frequency outer (BPFO),
        387, 388, 392, 394, 403, 405, 406,
        407, 408
    ball spin frequency (BSF), 387, 388, 406,
        407, 408
    fundamental train frequency (FTF),
        387–388
    scoring, 97
    seat defects, 88
    smearing, 91–92, 104
    spalling, 83–84, 88, 93
    vibration spectra. *See* Vibration spectra.
Belt drives, 319–320
Bending fatigue breakage, 179–180
Bending stress of gear teeth, 177–178
Benefit-to-cost ratio, in lube oil reclama-
    tion, 233
Beta analyzer, 315, 319
Beta Machinery Analysis, Ltd, 412, 414, 415,
    416, 418
Bin, definition, 425
Black box, 422
Black scab damage, 139–140
Blade pass frequency, 425
Blistering, as cause of seal failure, 208–209
Blowers. *See* Centrifugal compressors.
Bolted joints
    analysis of, 19–24
    measures to increase effective length, 22
    preload requirements, 23
Boundary lubrication, 106

Brinelling
    false, 110
    true, 111
Brittle failures, in metals, 13–15
Bump test, 401

C

Calculation(s)
    efficiency, 411
    failure analysis, 565–570
    power cost, 370–371
Calibration, definition, 426
Cap screws, 21
Case crushing fatigue, 173–175
Cause analysis, 440–448, 594–605, 654
    cause-and-effect
        diagrams, 442, 443
        chain, 595–596
        principle, 594–595
    communication, 446–448
    deviations, 445–446
    diagrams, 442, 443
    histograms, 440–441
    process steps in, 440
    run charts, 440, 441
    steps, 442–445
Cause categories, of equipment failure, 573–574
Cause, definition, 654
Causes of failure
    machinery failures, 2–9
    in metals, 13
Cavitation
    corrosive effects of, 48–50
    erosion, 137–138
    material loss due to, 48, 50
    pitting, 48
Censoring times, 495
Centrifugal compressor
    train failure, 565
Centrifugal pump(s)
    failure distribution, 575
    failure reduction program, 625, 635–636
    problems, check chart for, 272
    repair statistics, 271
    troubleshooting, 271–278
    vertical turbine type, 279–283
Centrifuges, for water removal from lube oil,
    217–220
Change, planning for, 470–476, 655
Channel Industries, 421
Check sheets, used in problem solving, 435,
    436, 437
Checklist approaches to failure analysis, 574

Fragment denting, of bearings, 88
Free water, 215, 216
Frequency
   analysis module (FAM), 406, 407, 408
   definition, 426
   domain, 426
   fundamental, 394
   interaction, 388
   natural, 399, 428
   response curve, 400
   span, 379
   slip. *See* Slip frequency.
FRETT, 593
Fretting, of bearings, 142–143, 147–148
Fretting corrosion, 45–48, 199–200
Fretting wear, 45, 48
Friction oxidation, 45
Frosting
   of bearings, 102
   of gears, 163–164
Fundamental train frequency (FTF), 387–388

G

G, definition, 428
Gas turbine(s)
   troubleshooting, 270, 336–338
Gear(s)
   allowable operating cycles, 168
   bending fatigue, 179–180
   failure analysis, 149–193
   failure distribution, 152
   failure modes, 151–152
   frosting, 163–164
   load capability, 153
   lubrication, 154
   lubrication flash point, 163
   manufacturing-induced problems, 171–175
   mesh frequency, 427
   metallurgical analysis, 153–154
   nomenclature, 656
   oil pump, 385
   overload breakage, 180
   pitting, 153, 168
   random failures, 175–176
   rim and web failure, 177
   scoring, 162–167
   spalling, 171
   tip and root interference, 166, 167
   tooth bending stresses, 176–177
   vibration
      defects, 383–387
      spectra. *See* Vibration spectra.
      velocities, 384

Gear couplings. *See* Couplings.
Generator fan failure statistics, 497
Glazing, of bearings, 100–101
Glitch, 524
Glossary of vibration terms, 425–430
Governors, troubleshooting, 333–336
Grease
   lithium-based, 237
   synthetic polyurea, 237
Grease depletion, effects in gear
   couplings, 193
Grease failure analysis, 237
   contaminant sources, 237
Grinding cracks, in gears, 172
Ground dome (double-ground)
   topograph, 532
Grouped data, statistical, 513

H

"H" node ("H" pattern) topograph, 533
Hanning window, 427
Hardness measurements, 53, 174, 191
Harmonics, 371, 374, 385, 394, 405, 409,
   424, 427
Hazard analysis, in SCA, 536
Hazard plotting, 496–505, 643–653
   general theory, 643–645
Heat checking, 193, 207
Heat treatment effect on fastener steels, 23
Heating, uneven, 379
Hertz (Hz), definition, 427
Hertzian stresses, excessive, 173
Highlight(s), 524
High-speed, low-flow pump failure,
   541–552
   analysis, 543–548
   conclusions, 548–549
   overview, 543
   recommendations, 549–550
   summary, 543
Histograms, used in cause analysis, 440–441
Horizontal measurement, 427

I

Imbalance, 363–368
   characteristics of, 363
   definition, 427
Impeller
   balance, 366
   spare, 364
   wear, 365
Inches per second (IPS), 428

R

Radial measurement, definition, 429
Random censoring, 507–511
  failure of gears, 175–176
Ratcheting, thermal, 148
Reciprocating compressor
  failure cause statistics, 308
  troubleshooting, 308–320
  valves, 308–314
Reciprocating pumps, 285–287
Recording of surface damage, 54
Refrigeration systems, 342, 346–348
Reliability professional, finding a, 636–637
Repairs, distinguishing bad repairs from bad
  designs, 514–521
Reports, failure, formalized, 539–572
Resistance, internal high, 373
Resonance, 397–401, 429. *See also* Frequency,
  natural
  frequency spectrum, 398
  time domain response plot, 398
Rim failure of gears, 177
RMS (vibration), 429
Rolling-element bearings, 81–127
  conditions resulting in failures, 116–117
  failure modes, 81
  hard turning on shaft, 126–127
  impact damage, 94
  looseness on shaft, 126
  noisy bearings, 119–121
  overheating, 117–119
  performance deficiencies, 124–125
  replacements too frequent, 121–123
  vibration, 123–124
Root-cause failure analysis, 573–593
Root mean square, 429
Rotating Equipment Vibration Advisor (REVA),
  607–608
Rotor, rotor bars, 373, 374, 375, 429
  out of balance, 379, 382
  vibration spectrum. *See* Vibration spectra.
Rotary pumps, 284–285
Royal Purple Company, 420, 421
Rubber components, distress in mechanical
  seals, 201–202
Runout, 429
Run charts used in cause analysis, 440, 441

S

Scoring, 165–167
  of gears, 162–167
  risk, 162

rolling-element bearings, 97
  tilt-pad and journal bearings, 128, 146
Scrubbers, compressed gas, 313
Scuffing, 4, 6
Seal face deflection, as cause of seal failure, 203
Seal face distortion, as cause of seal failure,
  201–203
Seal failures. *See* Mechanical seal failures.
Sealing, ineffective, 106–109
SEE (Spectral Emitted Energy) technology,
  402–409, 429
Selection, elements of in F/A and T/S, 262
Selection-related problems, 274
Selective processes
  combination, 262
  comparison, 262
  encoding, 262
Sensor (vibration), 429
Seven cause category approach to root-cause
  failure analysis, 573–593
Shaft
  alignment, 38
  bent, 367
  bowed, 379
  orientation of stresses, 24
  stress systems, 24–26
Shaft failures, 25–42
  brittle, 25, 26
  causes, 25
  creep, 25
  ductile, 25, 26
  examination of, 40
  fatigue, 25
  hydrogen-induced, 25, 26
  stress rupture, 26–27
Shaft fatigue, 29–36
  bending-type, 31
  corrosion-type, 34–36
  instantaneous zone, 30–33
  rotational bending, 33
  torsional, 36, 38
Shaft stress systems, 26–36
Short-form failure analysis form, 570–572
Shrink fit
  stresses due to, 37
Signal analyzer, 422
Sine wave, 429
Single-line topograph, 532
Single phase failure, 339
Singly-censored data, 495
Situation analysis, 655
Slip frequency, 374, 377
  calculating, 375
Smallest extreme-value distribution, 652–653

Vibration severity charts, 389, 390
Vibration severity, definition, 430
Vibration spectra
    after cutting rotor bar, 376, 377
    after misalignment, 370
    after repair, 369
    amplitude
        changes in, 363
        excursion, 382
    baseline, 395
    bearing, 390, 391, 393, 404, 405, 406,
        408, 409
    bell, 398
    bump test, 401
    compressor, 410, 411, 418
    cylinder, 414, 415
    gas engine, 412, 413, 414, 415,
        416, 417
    gears, 384, 385, 386
    electric motor, 378, 381, 407, 408
        in good condition, 376, 380
        with degraded insulation, 379
        with high-pitched noise, 406
        with one unsupported foot, 372
        time domain, 380
    fan, 381, 383, 392
    log scale, induction motor, 374, 375
    loose parts, 373, 383
    spray pump, 364
    steel mill, 399
    valve, 395, 396, 397
Viscosity test, for lube oil, 227–228
Visibility techniques
    pictorial, 454
    stair-step, 454
Voltage measurements, converting to db, 375

W

Water contamination of lube oil
    causing corrosion of journal bearings,
        135–136
    general, 212, 215–219
    tests for, 225–226
Water leak, in internal room cooler, 380
Water levels, ideal in lube oil, 217
Water separation in machinery, 226
Wave form analysis, 411
Wear
    abrasive, 44, 106–108, 159
    adhesive, 4, 44
    analysis procedures, 52–53
    corrosive, 42, 44, 160–162
    debris, 51, 57
    environments, 50–51
    failure analysis, 50–57
    failure modes, 55–56
    of gears, 155–162
    laboratory analysis, 52
    problems and solutions, 52
    resistance, of pump materials, 50
Wear-out mode, for bearings, 79
Wear particle analysis, 233–237
Weibull
    distribution, 646–647
    functions, 479–493, 499–511
    parameter estimates, 507–511
"Western Handy Oil," 557, 559, 564
Wheel of quality, 3
Wiped bearing surfaces, 130–132
Wire wool damage, 139–140
Worm tracking of gear couplings, 183